DATE DUE

MARINE GEOLOGY
AND OCEANOGRAPHY OF
ARABIAN SEA
AND COASTAL PAKISTAN

CONTRIBUTORS

- S. I. AHMAD, University of Washington, Seattle, Washington
- F. AZAM, Scripps Institution of Oceanography, La Jolla, California
- K. BANSE, University of Washington, Seattle, Washington
- M. A. A. BEG, PCSIR Laboratories, Karachi, Pakistan
- D. BLASCO, Bigelow Laboratory, West Bootbay Harbor, Maine
- H. N. CAPPEL, Marathon Oil Company, Damascus, Syria
- J. M. COLEMAN, Louisiana State University, Baton Rouge, Louisiana
- F. COUMES, Elf Aquitaine, Pau, France
- J. A. CRAME, British Antarctic Survey, Cambridge, U.K.
- K. A. DE JONG, University of Cincinnati, Cincinnati, Ohio
- R. C. DUGDALE, University of Southern California, Los Angeles, California
- A. FARAH, Geological Survey of Pakistan, Quetta, Pakistan
- R. J. FLEMING, National Oceanic and Atmospheric Administration, Rockville, Maryland
- D. C. FRANCIS, Marathon Oil Company, Singapore
- B. U. HAQ, Exxon Production Research Company, Houston, Texas
- J. C. HARMS, Harms and Brady, Inc., Littleton, Colorado
- O. HOLM-HANSEN, Scripps Institution of Oceanography, La Jolla, California
- D. C. HURD, Shell Development Company, Houston, Texas
- V. JOHNSON, University of Washington, Seattle, Washington
- A. H. KAZMI, Gemstone Corporation of Pakistan, Peshawar, Pakistan
- S. L. KING, University of Washington, Seattle, Washington
- V. KOLLA, Superior Oil Company, Houston, Texas
- P. KROOPNICK, Exxon Production Research Company, Houston, Texas
- R. D. LAWRENCE, Oregon State University, Corvallis, Oregon
- J. K. LEGGETT, Imperial College of Science and Technology, London, U.K.
- J. D. MILLIMAN, Woods Hole Oceanographic Institution, Woods Hole, Massachusetts
- J. B. MOODY, Battelle, Columbus, Ohio
- D. NANCE, Ohio University, Athens, Ohio
- N. A. OSTENSO, National Oceanic and Atmospheric Administration, Rockville, Maryland
- T. T. PACKARD, Bigelow Laboratory, West Boothbay Harbor, Maine
- J. PLATT, Oxford University, Oxford, U.K.
- V. N. QUADRI, Oil and Gas Development Corporation, Karachi, Pakistan
- G. S. QURAISHEE, National Institute of Oceanography, Karachi, Pakistan
- J. R. SCHUBEL, State University of New York, Stoney Brook, New York
- R. D. SLATER, University of Chicago, Chicago, Illinois
- S.C. SNEDAKER, University of Miami, Miami, Florida
- R. W. STERNBERG, University of Washington, Seattle, Washington
- J. T. WELLS, Louisiana State University, Baton Rouge, Louisiana
- B. L. WILLIAMS, University of Washington, Seattle, Washington
- F. WILLIAMS, University of Miami, Miami, Florida
- T. R. WORSLEY, Ohio University, Athens, Ohio
- R. S. YEATS, Oregon State University, Corvallis, Oregon

MARINE GEOLOGY AND OCEANOGRAPHY OF ARABIAN SEA AND COASTAL PAKISTAN

EDITED BY

BILAL U. HAQ

Exxon Production Research Company
Houston, Texas

JOHN D. MILLIMAN

Woods Hole Oceanographic Institution
Woods Hole, Massachusetts

Van Nostrand Reinhold Company
Scientific and Academic Editions

New York   Cincinnati   Stroudsburg
Toronto   London   Melbourne

This volume is an edited version of the proceedings of a Symposium and Workshop supported by the National Science Foundation under grant no. INT 82-10182. Any opinions, findings, conclusions, or recommendations expressed in this publication are those of the authors and do not necessarily reflect the views of the National Science Foundation.

Published by Van Nostrand Reinhold Company Inc.
135 West 50th Street
New York, New York 10020

Van Nostrand Reinhold Company Limited
Molly Millars Lane
Wokingham, Berkshire RG11 2PY, England

Van Nostrand Reinhold
480 Latrobe Street
Melbourne, Victoria 3000, Australia

Macmillan of Canada
Division of Gage Publishing Limted
164 Commander Boulevard
Agincourt, Ontario MIS 3C7, Canada

15   14   13   12   11   10   9   8   7   6   5   4   3   2   1

LIBRARY OF CONGRESS CATALOGING IN PUBLICATION DATA
Main entry under title:
Marine Geology and Oceanography of Arabian Sea and Coastal Pakistan.
   A collection of papers, most of which were presented at the first
   U.S.—Pakistan workshop on marine science held in Karachi in November
   1982.
   Includes index.
   1. Geology—Arabian Sea—Congresses. 2. Oceanography—Arabian Sea—Congresses. 3. Geology—Pakistan—Congresses.
I. Haq, Bilal U.   II. Milliman, John D.
QE350.52.A73M37   1984   551.46'08'09537   83-25983
ISBN 0-442-23216—0

CONTENTS

E. PALEOCEANS

F. ARABIAN SEA AND COASTAL PAKISTAN BIOLOGICAL OCEANOGRAPHY

G. MARINE NUTRIENT CYCLE

H. PHYSICAL AND CHEMICAL OCEANOGRAPHY

PREFACE

This volume is the culmination of a workshop of U.S. and Pakistan scientists on marine sciences in the Arabian Sea and coastal Pakistan held in Karachi in November 1982 under the joint sponsorship of the U.S. National Science Foundation and Pakistan Ministry of Science and Technology. The worskshop was convened to discuss marine science on both local and regional scales and to summarize the status of various subdisciplines of ocean sciences.

An additional objective was to identify research areas requiring immediate attention of the international scientific community and governmental agencies. Several high-priority research topics were identified that need urgent attention from the national and international agencies. These include: the study of Indus Delta (particularly its reactions to upstream damming of the River); regional and geological reconnaissance of Pakistan's coastal shelf regime; studies of pollution in the harbors and coastal waters; and the assessment of the damage to the mangrove environment in view of the damming of Indus and increased pollution.

The present volume includes the scientific papers presented at the workshop. In most cases, they include details of recent research and up-to-date bibliographies that will aid the readers in further perusal of the subject. The volume also includes five solicited articles that were not presented at the workshop.

The contents of the volume are organized into geological, biological, and physical/chemical oceanographic topics, arranged in eight sections. The first four chapters comprise a section that deals with the geology (and paleontology) of the Makran Coast of Pakistan and its hydrocarbon potential, both on- and offshore. These are followed by four chapters on various aspects of Indus River flow and sediment discharge, including the effects of upstream damming, geology of the Indus Delta, a comparison of this delta to other major deltas of the world, and the seismic structure of the Upper Indus Fan. Two other articles cover more general topics of estuarine and shelf circulation and sedimentation. The section on the tectonics of Pakistan gives important overviews that aid in the understanding of the development of the Makran margin and the paleoceanography of the region. The final geological section contains two general interest articles, one on the global paleoceanography of the past 200 million years, and another on some thought-provoking ideas about the influence of periodic convergence and breakup of the supercontinents on litho-, hydro- and biosphere.

The biological oceanography section also covers a very mixed menu. Articles on the plight of the mangrove (especially the influence of upstream damming of the Indus), and the need for environmental studies to fisheries development and management are followed by a detailed account of the Arabian Sea hydrography and the associated biological phenomena. Another paper outlines the distribution of dissolved oxygen and organic matter in the Arabian Sea. The last two biological papers deal with the marine nutrient cycle, both the role of bacterioplankton and phytoplankton in the food-chain.

All the physical/chemical oceanography articles are of more general interest. The charateristics of coastal upwelling, the implications of El Niño for global climates and marine food chain, and the inorganic and organic interactions in the open ocean and in coastal inlets presented in the last four chapters have relevance to Pakistan and elsewhere.

We hope that this volume, with its diverse contents, will prove to be a useful starting point for students of marine science of the Indian Ocean and Arabian Sea region. The up-to-date summaries of important present-day marine science issues also should be valuable to researchers elsewhere. We hope that the contents of this volume will also be helpful to marine policy-makers, and national and international agencies in recognising the importance of marine science research (and the perils of ignoring it!) in achieving the national and regional aspirations for resource self-sufficiency.

As the workshop coordinators from the U.S. side, we express our appreciation to Dr. M.A. Kazi, Science Advisor to the President, Mr. M. Masihuddin, Secretary, Ministry of Science and Technology, and Dr. M.D. Shami, Chairman of Pakistan Science Foundation for their encouragement and enthusiasm for the workshop. Their personal interest in the activities of the workshop and the follow-through actions are the most important factors in directing the future course of marine sciences in Pakistan and the region.

As editors, we acknowledge the untiring efforts of the staff of National Institute of Oceanography, both during the workshop and in the production of this volume. The director of NIO, Dr. G.S. Quraishee, and senior research officers, Farrukh Mirza and Shahid Amjad, in particular, oversaw the numerous details necessary for the successful running of our workshop, and also saw to the relatively smooth and productive collaboration with Elite Publishers Ltd. during the production of this book. One of us (BUH) would like to thank EPRCO for permitting him to coordinate the workshop and help edit this volume. The color-enhanced landsat image of lower Indus Valley and Delta reproduced on the dustcover of this volume was also obtained through the courtesy of EPRCO.

In closing, we would also like to express our deep appreciation for Dr. Osman Shinaishin of the National Science Foundation for his continuous support and committment to the cause of development of marine and other sciences in this region. This project was supported by funding from the National Science Foundation grants INT82-10182 and INT82-17426.

Bilal U. Haq, Houston, Texas

John D. Milliman, Woods Hole, Massachusetts

[Readers interested in another recently published up-to-date volume on Oceanography of the northwestern Indian Ocean (based on the *Murray* Symposium held in Alexandria, Egypt in 1983) are referred to: M. Angel (Editor): *Marine Science of the North-West Indian Ocean and adjacent waters.* Deep Sea Research, Special Volume, 31 (6-8): 571-1035 pp. 1984.]

A. GEOLOGY OF THE MAKRAN COAST AND OFFSHORE PETROLEUM POTENTIAL

1

The Makran Coast of Pakistan: It's Stratigraphy and Hydrocarbon Potential

J. C. HARMS
Harms and Brady, Inc., Littleton, Colorado,
H. N. CAPPEL
Marathon Oil Company, Damascus, Syria
and
D. C. FRANCIS
Marathon Oil Company,
Singapore

ABSTRACT

Coastal Makran is a subduction zone where the Indian Ocean plate moves northward under continental crust. The structural and stratigraphic history of the area is complex and can be subdivided into three major episodes: The middle Miocene and older phase was dominated by deposition of turbidites on a vast deep sea fan which originated from the east, supplied from uplands raised by the collision of India and Asia. Sediment transport was towards the west and sandstones found along an axial belt of nearly 400 km show little change in texture or bedding characteristics, suggesting a fan rivaling the dimensions of the Bengal Fan.

A second, late Miocene to mid Pleistocene, phase was characterized by slope, shelf and coastal plain progradation. Seaward advance of the sedimentary prism was accomplished incrementally by several generations of shelf-slope lobes, measured 10s of km in dimensions, and supplied by local drainage from the northern coastal ranges. Tectonically, these sediment lobes were soon folded and uplifted on their landward margin, later forming a source of sediments for the next younger lobes. The third, mid Pleistocene to Recent, phase was accompanied by dramatic uplift and local normal faulting. Sub-Recent shoreline sediments in some areas have been raised 500 m, and normal faults with substantial throw have been mapped seismically onshore and offshore.

Spectacular mud volcanoes have been built in several areas along the Makran coast by gas charged water escaping to the surface. The gases are mainly methane, but traces of heavier hydrocarbons and isotopic compositions indicate generation from thermally mature source rocks. Possible reservoir include middle Miocene coarse grained turbidites or late Miocene—Pliocene shelf sand stones, though the latter are mostly very fine grained, burrowed, and low in permeability. Attractive structures can be identified, particularly offshore, from prominent seismic reflectors of possibly middle Miocene age. However, these potential objectives are overlain by overpressured young sediments that are serious obstacles to drilling.

INTRODUCTION

Makran comprises the southern part of Pakistan and Iran between Sonmiani Bay near Karachi and the Straits of Hormuz. Viewed on a regional scale, it is a great festoon of folded and faulted Tertiary sediments extending 800 km from the Las Bela axial fold belt on the east to the Oman Line on the west (Fig. 1). It is well known that these eastern and western boundaries separate Makran from older terranes with deformational styles and histories distinct from that of Makran. India moved northward in its collisional course with Asia along a transform fault expressed as the Owen fracture zone, the Murray Ridge, and the axial fold belt of Pakistan. The Arabian and Iranian plates converged to form the Zagros ranges. In both cases, blocks of continental crust

Figure 1.   Satellite photo mosaic of coastal Makran (Courtesy R.D. Lawrence)

have moved northward against other continental plates, closing and crushing former deep oceanic seaways. Along Makran, it appears that oceanic crust has moved northward, subducted beneath a continental margin composed of small plates such as the Lut and Afgan blocks and complex ophiolite zones. A recent review of regional tectonics is included within a volume on the geodynamics of Pakistan (Farah and DeJong, 1979) and the tectonic map of Pakistan (Kazmi and Rana, 1982).

Coastal Makran and the area to the north is an accretionary wedge of deformed sediments ranging in age from Late Cretaceous to Recent, piled up at an oceanic subduction margin. The structure and depositional setting has been compared to "a typical arc model" composed of upper-slope deposits followed by lower-slope and trench deposits, progressively deformed by continuing subduction (Farhoudi and Karig, 1977). However, as an arc-trench system, Makran is hardly typical; indeed it is perhaps largely anomalous in its characteristics, as pointed out by Jacob and Quittmeyer (1979). The arc-trench gap is on the order of 500 km, far wider than most systems. A possible Benioff zone is weakly developed and shallow, and focal mechanism solutions indicate instances of tension in the oceanic slab. Volcanic centers are widely spaced along the arc feature. Additionally, a very large part of the accretionary prism is exposed, and a significant volume of post-middle Miocene sediments are shallow shelf deposits, not trench and slope deposits as alleged by Farhoudi and Karig (1977).

This paper will report the results of intensive field mapping, reflection seismic studies, and drilling conducted along the Pakistan Makran coast between 1973 and 1977. Detailed studies were concentrated within 60 km landward of the modern coastline and across the shelf to water depths of 200 m. It is within this belt that sedimentary strata and structural development seemed most appropriate for petroleum exploration. Although no commercial discoveries of hydrocarbons were made, some potential remains. Whether or not discoveries are ever made, coastal Makran provides a valuable and perhaps unique opportunity to study a well exposed and young accretionary margin. The depositional and deformational history is clear in at least the coastal area, as compared to other older or more intensely deformed examples. It can be used as one type of accretionary prism and compared in future studies to the many others that are known or must surely exist.

OVERVIEW OF GEOLOGIC HISTORY

The well dated stratigraphic record in coastal Makran begins in early Middle Miocene and extends to Late Pleistocene, based upon calcareous nannofossils. A geologic map, structural cross-sections, and stratigraphic column are shown in Figures 2, 3, and 4. North of well dated beds lies a sequence of much deformed shale and sandstone strata of unknown thickness and age. Near Mand, some 100 km north of the coast, these units contain exotic blocks of Eocene nummulitic limestone which appear to be tectonically emplaced. These relationships suggest that the broad belt of deformed clastic rocks in Makran consists mainly or totally of Cenozoic strata. About 500 to 1000 km north of the coastline, deformed sedimentary and volcanic rocks range in age from Senonian to Oligocene within the Sistan suture zone between the Lut and Afgan blocks (Tirrul et al., 1983). All relationships suggest that northward subduction of the Indian Ocean plate has continued from at least Senonian to present, with India moving relatively faster along the east edge of Makran and Arabia relatively faster along the west edge. The convergence and jostling of these larger continental plates has caused great internal complexity, as along the Sistan suture where the Lut and Afgan blocks have been forced together. Subduction of oceanic crust beneath Makran causes the general northward vergence which would be expected in the accretionary wedge, but crushing between India and Arabia also causes a broad arcuate form and many zones of strike slip motion.

Depositional environments of the Neogene section exposed in coastal Makran changed significantly through time. Thick-bedded sandstone and conglomerate with small proportions of thin shale beds of Middle Miocene age were deposited by turbidity currents. Forams typical of bathyal environments and elaborate grazing spoor suggest deep water sedimentation. The turbidity current-emplaced unit is at least 400 m thick; the sedimentary base is nowhere clearly exposed, and the section commonly terminates at reverse faults or the axial planes of recumbent isoclinal folds. This thick sandy section is commonly stacks of massive graded or graded-to-laminated amalgamated beds 0.3 to 1 m thick, suggesting large, powerful, and frequent turbidity currents. The sand is composed of quartz and carbonate grains and includes large mollusc shells and shelf foraminifers which have obviously been transported. The unit is exposed in the cores of anticlines scattered near the coastline over a distance of 300 km (Fig. 2). Flow directions based on flutes, grooves, and current ripples are consistently westward. All of these attributes suggest that a very large deep-sea fan existed along the coastal Makran area in early to late Middle Miocene time. The thick, coarse, proximal nature of these deposits over such a large area point to a fan of exceptional dimensions. We interpret the Middle Miocene as a deep-sea fan supplied from a quartz-rich terrane to the east, probably part of the collisional belt raised between India and Asia, a supply so large that it dominated the northern Arabian Sea. The undated strata that lie to the north of Middle Miocene exposures are also turbidity current deposits, presumably of Paleogene age. These strata are complexly deformed, steeply dipping, and become phyllitic and develop cleavage north of the Kech Valley. Although not studied by us in any detail, the sequence is more shaly than the Middle Miocene but contains numerous packages of graded sandstone beds with directional features indicating general westward transport. Sparse sampling of shaly units recovered no datable calcareous nannofossils, planktic foraminifers, or palynomorphs. We presume that much of this sediment was also supplied from the India-Asia collisional zone and formed a deep-sea fan to abyssal plain or trench complex on a subducting Indian Ocean plate.

Depositional environments changed significantly beginning in Late Miocene time and onward (Figs. 3 and 4). A sequence as thick in aggregate as 7000 m represents an over-all progradation and shoaling. Progradational episodes are marked by beautifully exposed vertical sequences of slope-shelf-shoreline deposits. Coastal advance over older turbidity current deposits was apparently accomplished by depositional systems that resembled the modern Makran setting of slope, narrow wave-dominated shelf, and coastal plain rimmed by rugged hills. The shoaling phase was not simply and uniformly achieved throughout Makran. Rather, areally limited shelf and slope lobes advanced locally as major stream drainages shifted from place to place along the coastline in response to uplift, folding, and faulting. In most areas, two or three shoaling

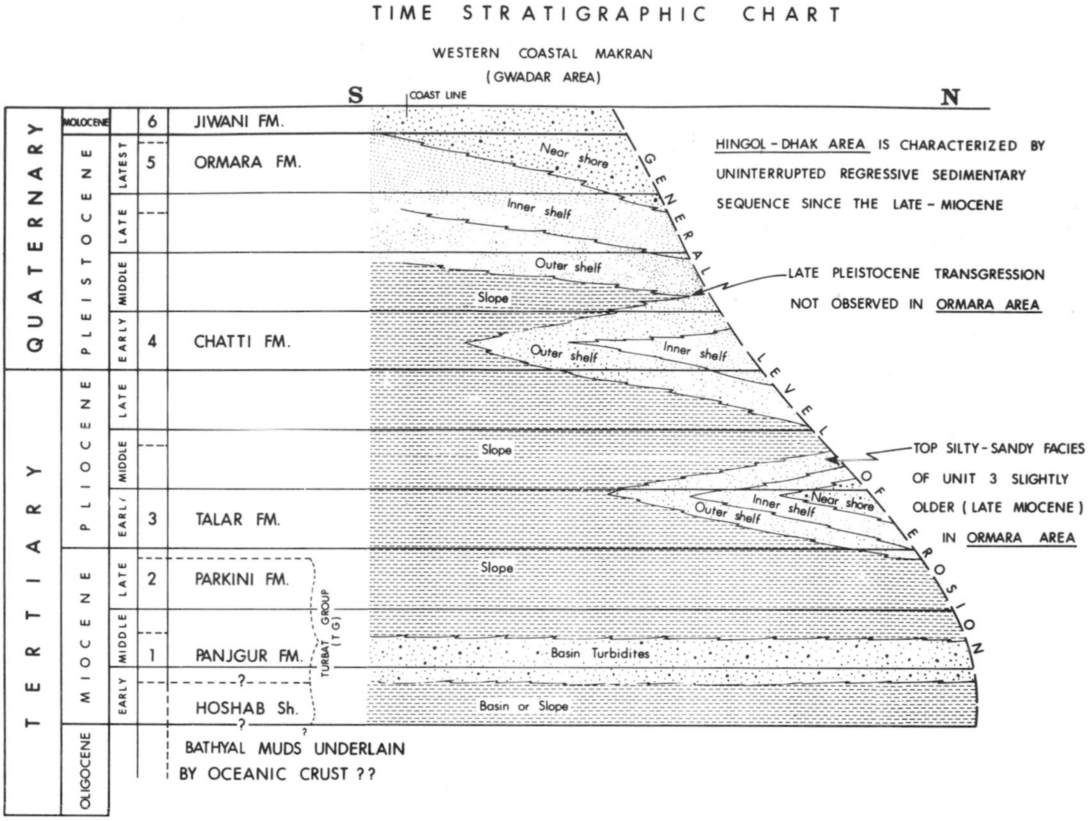

Figure 4. Time-stratigraphic-facies chart for Pakistan coastal Makran. Boundaries between units 1 to 6 are mapped in Figure 2. The time sequence is based on calcareous nannofossils. Formation names in the central column are those employed by the Hunting Survey (1960) for major lithofacies. Ages suggested in 1960 were generally based on foraminifers or stratigraphic position and were generally older than those indicated by calcareous nannofossils.

phases are superimposed or offlapping, each ranging in thickness from 1000 to 3000 m. Each progradational sequence has muddy massively bedded, partly deformed sediments with bathyal benthic foraminifers at the base, interpreted as slope deposits formed at depths beyond reach of storm waves and sand transporting currents. Above these mudstones are sandstone beds, commonly in many individual 20-30 m coarsening cycles capped by coquinas, that overall become coarser upward and contain progressively more robust, shallower water molluscs, interpreted as outer to inner shelf deposits.

The shelf-slope lobes are 90-120 km long in the east-west direction (Fig. 2) and only about 10 km wide. These imbricated or shingled lobes lead to complex time-rock unit relationships that baffle simple approaches to stratigraphic nomenclature. For this reason, the formation names developed during the Hunting Survey (1960), when age dating was based on foraminifers and was rather crude, are largely to be avoided but are indicated

on Figure 4.

Structural deformation was progressive and continuing from at least Middle Miocene time onward, compounding and partly causing the complex stratigraphic relationships. The dominant structural style in coastal Makran is one of east-west trending, very long, narrow, asymmetric to overturned isoclinal anticlines separated by broader asymmetric synclines (Figs. 2 and 3). Figures 5 and 6 are selected examples of multifold seismic data which contributed to interpretations on the cross-sections A-A′ and C-C′ in Figure 3. Overturning and steep fold limbs indicate relative overthrusting from the north. Anticlinal axes are commonly nearly horizontal for large distances, so that the eroded limbs form prominent ridges known as "bands," the local word for wall. Anticlinal folds exposing Middle Miocene beds extend to the coastline in the Pasni area; this zone of relatively greater uplift and deeper exposure plunges southwest and has been called the Pasni anticlinorium by us (Fig. 2).

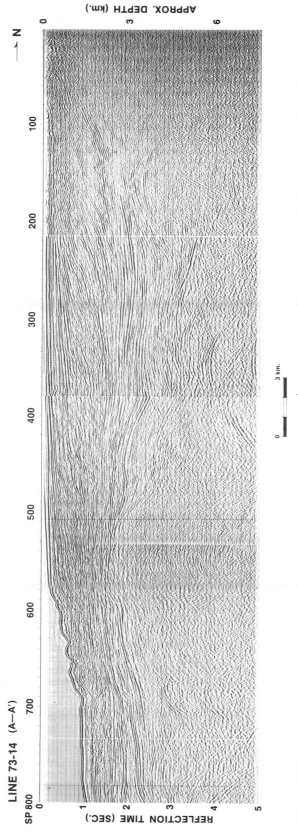

Figure 5. Seismic section in western Pakistan coastal Makran. The section is located on A–A´ of Figure 3. The stratigraphic and facies patterns are extended from onshore control. Reflective units are shelf sandstone beds; the transition to sloping reflectors interpreted as part of the slope is clear in the central part of the section.

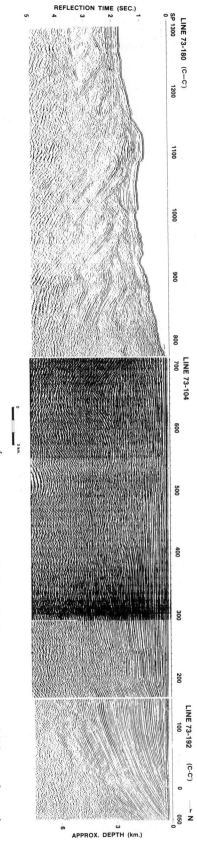

Figure 6. Seismic section in eastern Pakistan coastal Makran. The section is located on C–C´ of Figure 3. The stratigraphic and facies patterns are extended from surface and subsurface control indicated on Figure 3.

Thrust faults along the anticlinal trends are generally absent in surface sections. Even where axial surfaces dip 40 to 60 degrees to the north, the overturned limb is not significantly attenuated or in discordant relationships to adjacent rock units. However, the fold forms and offshore seismic data imply extensive detachment at depth (Fig. 3). Farther inland, the position of steeply dipping, probably older units separated by valleys or linear fracture trends (Fig. 1) suggest that reverse faults may juxtapose older blocks from the north on younger blocks to the south.

The large anticlines have a profound influence on the stratigraphy and the preserved deposits. As reviewed above, the post-Middle Miocene record is composed of offlapping shelf-slope lobes. Each lobe was originally a series of facies belts from north to south of a narrow coastal plain, shoreline, inner-to-outer shelf, and slope. Folding and uplift have destroyed the northern fluvial and shoreline facies in all but the very youngest lobes (Fig. 3). Indeed, it appears that uplift and erosion of the near-shoreline facies supplied a part of the sediment required to build the next younger lobe. Anticlinal folds appear to have partly grown during deposition in some cases; there are gentle convergences of beds on some limbs that appear to be sedimentary rather than attenuation related to slip adjustments between beds. The folds developed progressively from north to south and continued to grow and tighten over significant time spans. This almost peristalsis-like movement fed sediment to the prograding sedimentary prism.

Strike-slip faults are not common in coastal Makran. Only where fold axes swing northeastward near the Hingol River is there displacement of 5 km along such a fault. In this case, slip is left lateral on a N10° E fault, about parallel to the major Ornach-Nal and Chaman faults along the Las Bela axial fold belt. The curve in fold axes trends and left lateral displacement are adjustments to the northward passage of the adjacent Indian continental block.

Although evidence of north-south lateral shortening is abundant in the many asymmetric folds, there has also been recent significant vertical uplift and extension in the coastal area. Several large normal faults with displacement as much as 2500 m have been observed in surface and seismic sections around Gwadar. Late Pleistocene to Holocene shoreline or shallow shelf deposits are in many places along the coastline raised as much as 500 m above modern sea level. Average rate of uplift at Ormara has been estimated as 0.1 to 0.2 cm a year by Page et al. (1979). A large domal feature at Koh Dimak (Fig. 2) suggests local gentle upwarping rather than pronounced lateral shortening.

The normal faults that have been identified all cut quite young sediments onshore and offshore. The abundant seismic activity in Makran is tensional in part (Jacob and Quittmeyer, 1979). It may be that Late Pleistocene and Holocene uplift and extension are a late and unusual phase in the

Figure 7. Middle Miocene thick-bedded turbidity current deposits exposed in an anticlinal fold northeast of Pasni.

Figure 8. An example of a thinner turbidity current bed near the top of the Middle Miocene deeper water sandstone unit (Panjgur). Note the sharp base resting on shale just above the scale and massive to flat laminated structure succeeded by climbing ripples.

long-term subduction history of Makran.

STRATIGRAPHY AND DEPOSITIONAL MODELS

Middle Miocene Turbidity Current Deposits

Middle Miocene strata are exposed near the central coastline in a series of closely spaced folds on the Pasni anticlinorium, extensively inland to the north, and along an anticline near the western border of Pakistan (Fig. 2). In all of these exposures, the Middle Miocene is as much as 400 m thick, although a depositional base is nowhere clearly exposed. It is composed of fine-to medium-grained micaceous quartz sand, which locally is coarse-grained and conglomeratic, in graded beds ranging from 0.3 to 2 m or more in thickness. Most of this thick unit is sandstone, resistant to erosion, and forms prominent ridges along anticlinal cores (Fig. 7). Thin shale interbeds are scattered throughout the sequence but become thicker and more common at the top, as graded beds become thinner and finer grained. This entire unit was called the Panjgur Formation in the Hunting Survey (1960).

All of the features of turbidity current deposits are clearly evident. The beds are consistently texturally graded from coarser at the base to finer upward. Thick beds ranging up to one or more

meters are massive or have thin flat laminated units at their tops and commonly are overlain by other thick graded beds on flat erosional surfaces. These would be the amalgamated a-a or a-b-a sequences in the nomenclature commonly applied to turbidity current deposits. Thinner beds are more common at the top of the sequence, and shaly beds are more abundant and thicker. These thinner sandy units have ripples and ripple lamination at their tops (Fig. 8). Beds commonly have flute or groove casts at the base where they rest on a muddy substrate (Fig. 9). Flutes, ripples, and grooves indicate consistent westward flow. Because these beds crop out on anticlines, in some places with plunging axes, and are partly overturned, we present no detailed directional data. However, in single outcrops, scatter of directions within similarly deformed sequences is slight, so that we believe flow to the west had little variation.

Thin lenses of conglomerate are found at widely scattered localities. Clasts range from pebbles to cobbles of well rounded quartzite, sandstone, and chert. Some clasts are robust snails, clams, or oysters which are only slightly worn and rounded. The conglomerate beds appear to be parts of normally graded sequences.

The turbidity current sequence ranges in age from early to late Middle Miocene, based on recovery from shale interbeds of the clacareous

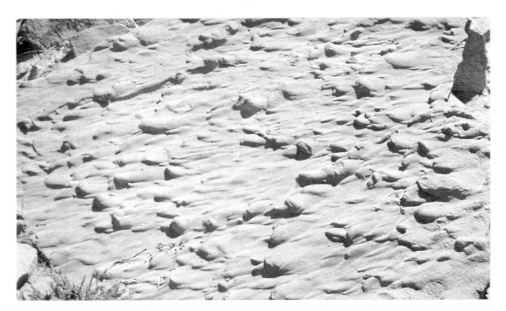

Figure 9. Flute casts on the undersurface of a turbidity current bed. Flute and groove orientations show consistent westward flow of these currents.

nannofossils *S. heteromorphus, C. coalitus* and *D. hamatus.* Benthic foraminifers in these beds suggest bathyal to lower bathyal depths. Bases of thin sandy or silty beds are marked by elaborate spoor such as *Spirorhaphe* and *Paleodictyon* (Fig. 10), commonly associated with deep basinal deposits.

The regional interpretation suggests that during Middle Miocene time a very large submarine fan built westward in a moderately deep oceanic setting, probably across older and

perhaps deeper water turbidity current deposits which contained higher proportions of mud (Fig. 11). The very sandy, thick-bedded nature of these Middle Miocene deposits stretching at least 400 km west of potential source areas at the Indian plate boundary indicate a fan of enormous proportions and energetic character. This sand-dominated event corresponds with the deposition of the Lower Siwalik Group in northern Pakistan (Johnson et al., 1979), and thus may be linked to orogenic events in that collisional zone. Supply of

Figure 10. Grazing spoor on the undersurface of a turbidity current bed. The forms are typical of deeper water deposits.

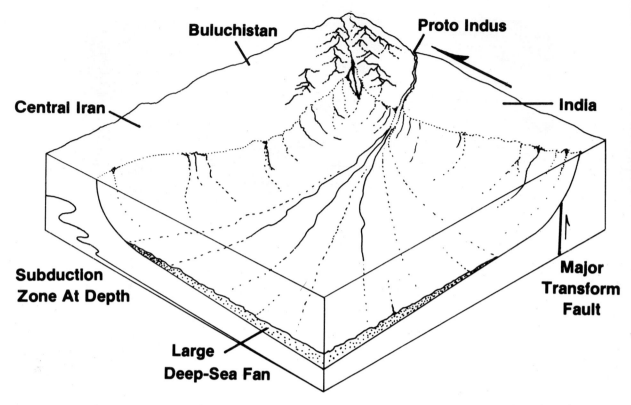

Figure 11. Regional setting of the Makran area during Middle Miocene time. A large deep-sea fan was built by sand and some mud transported westward from a source area east of the major Chaman-Ornach transform zone.

sand waned in late Middle Miocene, perhaps as deformation shifted major distribution toward the south along a route more like the modern Indus River and Cone.

Late Miocene-Pleistocene Shoreline-Shelf-Slope Deposits

The Late Miocene to Pleistocene sediments exposed in coastal Makran are mainly slope and shelf facies, with minor shoreline deposits preserved in only the youngest strata. The composited thickness of this sequence ranges up to 7000 m, measured across sections exposed over 10 to 20 km, so that it is unlikely that a section this thick is represented at any given point. However, seismic data suggest Pliocene to Pleistocene sections as thick as 6000 m, using reasonable velocities to convert transit time to depth.

The post-Middle Miocene is deposited in one or more large regressive cycles composed of slope to outer and inner shelf facies. Figure 4 illustrates in a simple way the facies sequences exposed at the general surface level along north-south transects in three areas; the facies are more complexly exposed across the numerous folds that do exist in these

areas (Figure 3). Formation names are suggested in Figure 4 which combine age and lithofacies; some were defined by the Hunting Survey (1960) but others have been added from particularly good geographic localities to adequately cover the complex time-facies relationships. In the following paragraphs, each major facies is described and illustrated.

Slope Deposits

Slope-deposited sediments are mainly gray, slightly calcareous mudstone, commonly massive or thick-bedded in appearance because of textural homogeneity and eroded in "badlands" forms (Fig. 12). Viewed in ideal outcrops, lamination and bedding is visible in even uniformly fine-grained mudstone, but discordances and probable slip surfaces are common (Fig. 13), as though failure and slumping disturbed bedding soon after deformation. Thin siltstone or very fine-grained sandstone beds are present in some zones. At the Middle Miocene-Late Miocene boundary, such beds are thin graded turbidity current units which form the transition from massive sandstone beds of the older deep-sea fan to the consistently fine-

Figure 12. Regressive slope-shelf sequence north of Ormara. Slope deposits on the left have few resistant beds and weather into a badlands topography. Resistant beds to right are individual coarsening-upward shelf deposits, which vertically become more sandy and resistant, reflecting a progression from outer- to inner-shelf facies. View is toward west.

grained slope deposits. Towards the top of slope units, sandy beds increase in number and thickness, each composed of sharp-based 5-20 cm thick beds with hummocky cross-stratification and wave ripples. These beds are interpreted as storm-wave deposits, based on primary sedimentary structures (Harms et al., 1975)

Slope sediments contain abundant fossils. Age dating is based upon easily recovered calcareous nannoplankton. Benthic and planktic foraminifers are commonly abundant. Micro-fossils are increasingly reworked upward in the section, and stratigraphically displaced and worn forms are common in younger units. Larger fossils are less abundant; thin-shelled articulated clam shells are sparsely distributed and wood, teeth, or bone fragments are seen in varying abundance. Burrows, formed by clams or other organisms

Figure 13. Thin-bedded calcareous siltstone represents slope deposition. Coarse siltstone or fine-grained sandstone beds are rare. Slight discordances and bulges in bedding represent adjustments to mechanical slippage shortly after deposition.

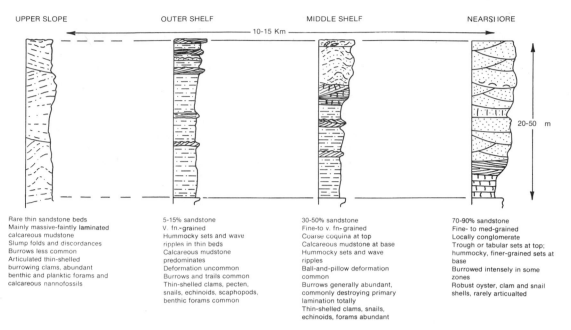

UPPER SLOPE   OUTER SHELF   MIDDLE SHELF   NEARSHORE

|← 10-15 Km →|

20-50 m

| Rare thin sandstone beds
Mainly massive-faintly laminated
calcareous mudstone
Slump folds and discordances
Burrows less common
Articulated thin-shelled
burrowing clams, abundant
benthic and planktic forams and
calcareous nannofossils | 5-15% sandstone
V. fn.-grained
Hummocky sets and wave
ripples in thin beds
Calcareous mudstone
predominates
Deformation uncommon
Burrows and trails common
Thin-shelled clams, pecten,
snails, echinoids, scaphopods,
benthic forams common | 30-50% sandstone
Fine-to v. fn- grained
Coarse coquina at top
Calcareous mudstone at base
Hummocky sets and wave
ripples
Ball-and-pillow deformation
common
Burrows generally abundant,
commonly destroying primary
lamination totally
Thin-shelled clams, snails,
echinoids, forams abundant | 70-90% sandstone
Fine- to med-grained
Locally conglomerate
Trough or tabular sets at top;
hummocky, finer-grained sets at
base
Burrowed intensely in some
zones
Robust oyster, clam and snail
shells, rarely articualted |

Figure 14. Shelf-slope facies transitions from near-shore to upper slope within a single depositional cycle. The mollusc fauna in these facies are discussed by Crame (this volume).

without preserved hard parts, are found but in noting like the abundance of shallower water shelf deposits. Forms are smooth-walled vertical burrows a centimeter or less in diameter or trails best preserved on siltstone bed surfaces. Ochre-weathering iron-carbonate concretions form cylinders around burrows, bones, or wood.

The environment of deposition of these thick mudstone sequences is interpreted as slopes in upper to lower bathyal depths (200-2000 m). Benthic foraminifers, which are generally unworn and appear to be in place, suggest depths of as much as 2000 m but not truly abyssal settings. The thickness of mudstone beds between turbidity current sediments at the base and shelf sediments at the top is also an approximate estimate of water

Figure 15. View of inner-to outer-shelf transition. In the foreground, units are sandier and more resistant, and toward the background, sandy strata are thinner. This represents the changes diagrammatically illustrated in Figure 14. Area is north of Gwadar.

depth. In the least tectonically disturbed sections, mudstone units are as much as 2300 m thick. Of course, this thickness may include the effects of subsidence during deposition, increasing the apparent water depth, or the effects of compaction, diminishing the apparent depth.

Inclined bedding, suggestive of a prograding slope, is not easily established in outcrop. Bedding is commonly disrupted by complex discordances (Fig. 13) interpreted as early slump failures which suggest that a slope did exist but baffle clear observation of clinoform beds on the scale of outcrops. Additionally, the mudstone units are cast into complex tectonic folds and, by virtue of their plastic nature, are more intricately deformed than adjacent sandstone members. However, seismic data in the youngest and least deformed progradational sequences do in some places indicate thick units with inclined bedding (Fig. 5). Conversion of transit time to depth indicates that these clinoforms are 3000 m thick and dip seaward at 15 to 20 degrees.

Slope beds appear to be mainly deposited from suspended material transported from the north into deeper water. Clay minerals are predominantly smectite-illite and kaolinite. The number and thickness of coarse siltstone and very fine-grained sandstone beds increase upward; such beds bear storm-wave generated features such as hummocky cross-strata and wave ripples. Where ripple transport of sand is recorded by cross laminae, sand drift is southward. Heavy mineral assemblages in these sandstones are dominated by chrome spinel, as are the shelf sandstones above. The heavy mineral and clay mineral assemblages are dissimilar to those of the modern Indus Cone and deep Indian Ocean sediments, derived from the Indus drainage (Kazmi, this vol.), but resemble those of Paleogene beds in Baluchistan. Major supply was from uplifted ranges in Baluchistan and northern Makran.

Shelf Deposits

Overlying and in part interfingering with slope deposits are sandstone-mudstone couplets which represent inner- to outer-shelf deposits (Fig. 14). The relationships are in some places elegantly exposed on fold limbs (Figs. 15 and 16). Sandstone units, weathered to dark brown resistant cuestas, can be seen to thin and taper over distances of 8-15 km, replaced by light gray mudstone beds. These interfingering beds represent the transition from sandy inner-shelf to muddy outer-shelf environments.

Individual mudstone-sandstone couplets about 30-50 m thick clearly show the transition from shallow, energetic environments to deeper, somewhat quieter settings (Fig. 14). In the nearshore setting, sandstone is thick-bedded and composes 70-90% of the couplets. Cross-stratification in trough or tabular sets is dominant (Fig. 17), beds

Figure 16. A more distant view of shelf sandstone (dark beds) decreasing in thickness in an offshore direction (lower left). Changes in sand content occur in distances of a few kilometers, as indicated on Figure 14.

Figure 17. Near-shore coarsening-upward sequence about 40 m thick. Upper part is thick bedded sandstone with trough or tabular sets of cross-strata deposited in shallow energetic environments.

are 0.5-3.0 m thick, and commonly occupy broad, shallow scours oriented north-south. Deformed bedding is rare. Grain size is fine to very fine in some units, particularly those deposited in the Pleistocene in the central and eastern part of Figure 2, but in other sequences the grain size is coarse sand with some pebbles. The faunal remains are mainly thick-shelled oysters, ribbed clams, and robust snails. For a more detailed discussion of molluscs, see Crame (1984, this volume). Burrows include smooth-walled and knobby cylindrical forms *(Thallasinoides* and *Ophiomorpha)* and shell-filled excavations (Fig. 18).

In the mid-shelf setting, sandstone-to-mudstone ratios range from 30% to 50%, and the couplets are distinctly asymmetric, sandy and coarser at their tops. The mudstone is commonly poorly exposed but parallel laminated and somewhat burrowed with thin-shelled clams,

Figure 18. Burrowed to cross-stratified sandstone with a robust clam and snail fauna. Photograph is from the base of the most massive cliff-forming unit in Figure 21.

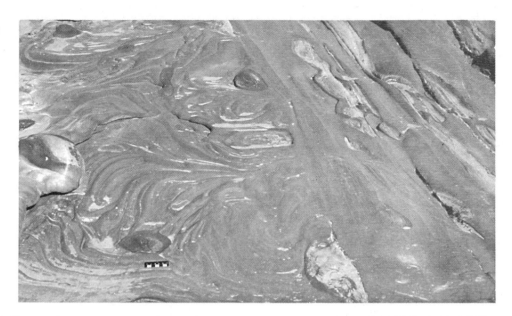

Figure 19. Hummocky cross-strata to right lie above a zone of intense ball-and-pillow convolution in very fine-grained sandstone of mid-shelf facies.

echinoids, sharks' teeth, wood, and bone remains. Sandstone transitionally increases upward in each couplet by sandy beds increasing in number and thickness. Sandstone is fine- to very fine-grained, and horizontally or hummocky cross-stratified. Commonly burrows are so abundant that primary lamination is completely destroyed. Additionally, ball-and-pillow deformational structures may be abundant and intense (Fig. 19). Very commonly the top of the sandy interval is sharp and planar. Ball-and-pillow structures are truncated and capped by a thin coquina, in many places composed of high-spired snails *(Turritella)*, but in other places containing clam and echinoid remains. In some places, these shell lags have small encrusting organisms on their upper shell surfaces, suggesting slow rates of deposition after the formation of the lag.

These coarsening upward couplets capped by coquina suggest shoaling cycles on a storm wave-dominated shelf. Hummocky cross-strata and wave ripples are clearly associated with strong wave motion (Harms, 1979). The coquina lag deposits indicate winnowing events; *Turritella* live with their shells partly imbedded in mud, so their dense concentration in non-life positions indicates erosion and redistribution. The same storm events that form coquinas may also cause the deformed bedding, for it is well known that pressure pulses associated with storm waves cause the ocean bed to become quick and unstable.

Directional features are rather poorly preserved in this facies because of burrowing and deformation. Where wave and current ripples are preserved and internal laminae indicate sediment drift direction, transport to the south is clear.

In the outer shelf setting, the couplets are similar but show increasingly greater proportions of mudstone. Sandstone is very fine-grained, beds are 30 cm or less thick, but concentrated near the top of the couplet. Like the mid-shelf, structures are horizontal or hummocky cross-strata and wave ripples. Deformational structures are uncommon. Sandstone beds are extensive and tabular or are broad lenses.

Fossils are abundant. Clams, especially pecten, are very common, as are benthic foraminifers, snails, echinoids, and scaphopods. Sharks' teeth and vertebrate bones are also abundant enough to be easily found. Spoor are common and include vertical penetrative forms and trails.

The facies couplets described in the preceding paragraphs and summarized in Figure 14 are commonly stacked in thick, repetetive sequences. These sequences in most places show long term regression (Fig. 12), so that inner shelf facies gradually advance over the belts earlier occupied by mid- or outer-shelf facies. These prograding pulses can be 1000 to 3000 m thick, and are composed of many individual couplets. In many places, 20 to 60 couplets are represented, each a somewhat shoaling event dominated and

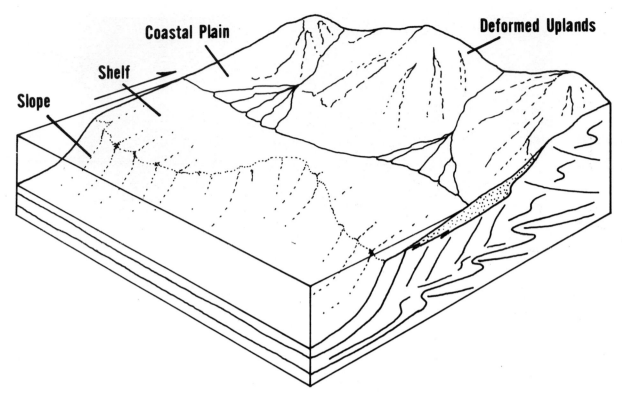

Figure 20. Shelf-slope lobes prograded seaward along the Makran area from Late Miocene to Present. The position of such lobes was perhaps localized by the position of major drainages, and those shifted from time to time giving the complex stratigraphy illustrated in Figure 4.

punctuated by storm events in their upper part. The couplets were deposited in 10,000- to 20,000-year periods, on the average, and indicate average sedimentation rates of 1.0-3.0 m per 1,000 years.

In map view, these facies belts and the major regressive pulses show lobate forms. The inner- to outer-shelf transitions are indicated on Figures 2 and 3, although in only a very general way. Sandy shelf sediments appear to extend 8 or 10 km seaward from inferred shorelines, and form lobes 90 to 120 km long along the shoreline direction. We suggest that the coastline in Late Tertiary time, as now, was supplied sediment by small and fairly widely spaced drainages (Fig. 20). However, because these drainages received large volumes of material from nearby rugged and easily eroded hills, rates of deposition were very high. Each shelf lobe is thought to represent a local focus of supply related to drainage position, where shelf progradation was more rapid and coarser grained near stream mouths and less rapid and finer grained between such lobes.

Shoreline Deposits

Shoreline deposits are poorly represented

except in Late Pleistocene or Holocene sequences because progressive folding, uplift, and erosion has continuously removed the northern edges of the sedimentary prisms. Shoreline deposits are of two kinds. One is very sandy and shows typical coarsening upward cycles of lower shoreface, surf-zone, and beach facies developed on wave-dominated prograding shorelines (Harms et. al., 1975). The other is more calcareous and contains abundant remains of corals, robust clams, and other forms that thrive in shallow wave-agitated water where detrital supply does not smother their growth. These two types presumably represent reaches of shoreline that were either at or near stream mouths or between streams where longshore supply of mud or sand was limited.

A typical sandy shoreline sequence is shown in Figure 21. The example is Late Pleistocene in age but already tilted and eroding. Surf-zone deposits composed of trough cross-stratified medium- to fine-grained sand show orientations indicating strong westward longshore drift, commonly observed on the modern shoreline under the attack of large obliquely approaching Indian Ocean waves. Capping these surf-zone sandstone deposits are well-sorted beach beds, showing

Figure 21. Pleistocene sandy shoreline sequence with burrowed marine near-shore deposits in the lower part and beach deposits at the top of the cliff-forming unit. Area is along the Hingol River.

delicate lamination dipping at low angles seaward in cusp-shaped sets. In some areas, eolian dune fields overlie the beaches, identified by intricately cross-stratified large sets with high local dips and large sweeping sets. Within the shoreline deposits are limpets, cerrithid snails, and large clams that live only at the shoreline today.

Calcareous shoreline deposits contain corals, large clams, oyster colonies, and many other robust wave-resistant forms. Such deposits are relatively thin and represent areas with temporarily low rates of detrital supply. These beds, assuredly young and deposited at sea level, now cap gebels rising some 300 m above modern sea level. They provide excellent evidence of rates of recent uplift.

STRUCTURAL STYLE

Structure in coastal Makran is dominated by narrow overturned anticlines and broader asymmetric synclines with east-west axes (Figs. 2 and 3). On a regional scale, older and more tightly folded Middle Miocene strata are exposed close to the coast along the Pasni anticlinorium (Fig. 2). Younger Neogene strata are somewhat less folded, except to the east where fold axes swing northward into the Las Bela axial fold belt and the Nal-Ornach fault zone. Although intense folding and substantial lateral shortening is evident, significant reverse faults have not been identified

cutting Neogene strata. However, some large normal faults and strike-slip faults can be documented. The following paragraphs describe these structural features.

Folds

The essential character of the folds is illustrated on structural cross-sections (Fig. 3) which are based on a combination of surface and reflection seismic observations. Anticlines are asymmetric and overturned to isoclinal. In the tightly folded massive Middle Miocene sandstone, the axial surfaces dip northward at 15 to 30 degrees. In some places the axial zone is collapsed into smaller chevron kinks, but it is common in the massively bedded sandstone to pass from upright to recumbent limb with little evidence of disruption or slippage. To achieve the form illustrated in Figure 3, a 400 m thick sandy package had to detach cleanly at its base and collapse upon itself in substantial north-to-south relative slip.

The more open but asymmetric folds in younger strata similarly must detach at some deeper level. The cores of these anticlines are commonly formed in slope mudstone facies, commonly in vertical or recumbent attitude (Fig. 22). These cores appear massive on aerial photographs, and might be interpreted as diapiric based on such appearance. However, detailed field observations show bedding is not disrupted or

Figure 22. Recumbant anticlinal limb (left) passing into upright beds of low dip (right) on the south flank of a fold near Pasni. Beds are upper slope-outer shelf facies.

intricately folded on a small scale as would be the case if diapiric flow occurred. Facies changes across the fold axes are common in post-Middle Miocene sequences. The gentler north limb is composed of massive inner-shelf sandstone beds which change to thin bedded outer-shelf or slope facies across the fold width. Some care is required to trace time correlative strata across core areas, as illustrated in Section C of Figure 3.

Anticlinal axes are nearly horizontal for long distances, then plunge abruptly. Erosion of these folds produces long parallel ridges exposing essentially the same stratigraphic interval over substantial distances.

Synclines are more open, although also asymmetric and steep-limbed on the north. Axes plunge more gently, so that erosion produces elliptical basins so evident on aerial views (Fig. 1). Figure 23 shows a synclinal fold in massive inner-shelf sandstone beds underlain by thinner bedded

Figure 23. An open syncline stripped to the general level of a massive near-shore sandstone unit. View is toward the northeast along the shoreline east of Ormara. The northern flank of the syncline is breached where dip steepens on the south limb of the asymmetric anticline to the north.

outer-shelf facies.

The folds evident in onshore exposures continue offshore across the modern shelf and are readily mapped on reflection seismic records. In some areas, folds are slightly disharmonic and separated by detachment surfaces.

Although such detachment surfaces resemble soles of thrust-faulted terranes, reverse faults with significant throw are not recognized in surface exposures or seismic records. As a result, folds do not appear to form over ramps where faults locally cut more steeply through strata, the situation interpreted in many classical thrusted mountain belts. Rather in Makran, extensive sheets appear to have moved over essentially the same horizon, and lateral displacement diminishes progressively southward at that stratal level. Principal zones for detachment are the base of Middle Miocene and lower part of the Late Miocene, both presumably the most extensive shaley horizons in the stratigraphic succession. High fluid pressures encountered in wells indicate that such shales can provide a zone of easily moved detachment.

A large domal uplift, very different from other folds, lies along the coastline between Gwadar and Pasni, which we refer to as Koh Dimak. This large oval uplift, 20 km in diameter, is only half exposed on land (Fig. 2). It appears to be related to broad uplift of the coastal area rather than lateral shortening suggested by the other folds typical of the area. Certainly, there is abundant evidence in the form of raised coastal deposits of vertical uplift along the modern coastline.

Faults

Strike-slip and normal faults with substantial displacement are mapped in several areas, whereas reverse faults are not conspicuous as reviewed above. Only one major strike-slip fault is seen in the area covered by Figure 2, that lying to the east near the Hingol River. This fault trends N 10°E, has left-lateral displacement, and an apparent offset of 5 km. It apparently represents a tear fault in the folded Plio-Pleistocene sheet, required as fold axes swing northward along the Las Bela axial fold belt. In trend and sense of offset, the Hingol River fault is similar to the very large Nal-Ornach and Chaman fault zones, which reflect the more rapid northward drift rate of India relative to Baluchistan.

Several large normal faults are observed in surface exposures or seismic sections in western Pakistan Makran. North of Gwadar, two such faults very clearly crop out. The fault south of the Kulanch syncline has a sinuous map view traced east-west at least 100 km and dips as gently as 30° southward. The low dip may be related to coastal uplift and northward tilting, especially on the north flank of the Koh Dimak Dome, at a time after fault displacement. Displacement may be as much as 2,500 m. Alternatively, it may be that erosion cuts deeply enough into the fault to expose the lower more gently dipping portion of a curved surface.

Other faults in the Gwadar area are certainly more steeply dipping in surface exposures or as interpreted on seismic sections. All of these faults are east-west trending and downthrown to the south. Several can be traced for 30 to 80 km.

The control on normal fault distribution that leads them to be concentrated in the western area is not clear. Uplift of similar amounts has occurred along much of the coast and apparently at approximately similar rates. The largest of the recent earthquakes centered near this area, but any causal relationship between earthquakes and these normal faults is unclear to us. The existence of these large normal faults and suggested tension from earthquake data could be used as evidence that the very recent behavior in Makran is not a simple subduction motion.

MAKRAN AS AN ACCRETIONARY PRISM

The characteristics of coastal Makran are clearly shown in the map and cross sections of Figures 2 and 3. Bathyal depth turbidites, succeeded by bathyal to shoreline slope and shelf deposits, form a sedimentary prism as much as 7000 m thick that has been progressively deformed into southward overturned folds. Truly oceanic trench-depth deposits are not present in the exposed parts of this prism, based upon a good deal of paleontologic and sedimentologic data. Neither are major reverse faults known to cut surface exposures of Middle Miocene or younger rocks, although regional detachment at at least two stratigraphic levels must be invoked as a kinematic model for the folding.

The Neogene and Pleistocene Makran prism differs in these several respects from accretionary prism models proposed for other arc-trench areas by Karig (1974) and by Seely et al. (1974). Figure 3

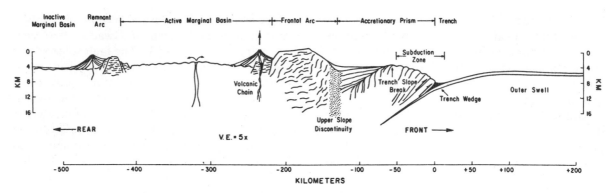

Figure 24. Generalized framework of a western Pacific island arc system from Karig (1974).

can be compared to Figure 24, a diagram illustrating these latter models. The Makran example does not include a bathymetric trench, is very broad and substantially emergent, and is not cut by major thrust faults which would be rotated to steeper inclinations by progressive deformation.

These somewhat different features are perhaps accounted for by the general convergent setting. The Indian-Asian collision zone is relatively near-by and apparently supplied an enormous volume of westward-transported sediment to Makran during the Paleogene. This supply perhaps was sufficient to totally fill any incipient trench, and later uplift of Baluchistan continued the high rates of sedimentation in the coastal area. Whether this sort of enormous supply or the nature of the small plate mosaic to the north was the cause, the

Benioff zone appears to be more gently inclined and more weakly developed than in many arc-trench systems. Whatever the causes, coastal Makran represents a fascinating example of a subduction zone. It and the deeper, more poorly known zones of Baluchistan should be regarded as one of the possible subduction models in what must be a more varied spectrum of types than is commonly acknowledged. As such, the area deserves additional careful study.

PETROLEUM POTENTIAL

The petroleum potential of coastal Makran, although not outstanding, has some promise which has not been fully evaluated. Certainly, there is no dearth of structure; there are many untested anticlines onshore and offshore. Perhaps more

Figure 25. A mud volcano formed by rising hydrocarbon gases, water, and entrained sediment. This example is in the eastern part of Makran close to the shoreline near the Hunt Dak wells (Fig. 2).

critical is an evaluation of source bed quality, degree of thermal maturity of organic material, and reservoir properties.

Source quality has been evaluated using both surface and subsurface samples. Organic carbon content of surface samples is lean, but subsurface samples, although taken from similar facies, are richer, ranging from 0.22% to 0.96% total organic carbon. Although these levels would be considered as only lean to adequate, there is a very large volume of mudstone deposited in slope and outer shelf environments, suggesting a large volume of organic carbon that potentially could be converted to hydrocarbons. Because the pre-Middle Miocene is not exposed in this area, its character is unknown. These sediments may be largely fine grained since they provide a regional detachment zone for folding and were very likely deposited in deep marine environments. Their possible contribution cannot be assessed, but neither should it be totally discounted.

Thermal maturity is more doubtful. Two wells drilled in the area encountered bottom hole temperatures indicating thermal gradients of only 1°F per 100 feet. Maximum burial depths are uncertain because of recent uplift and complex structure. Thermal models commonly used to predict maturation are difficult to apply because of the complex structural history. However, burial is young and temperature gradient low, suggesting limited chances of maturation and hydrocarbon evolution.

Perhaps the single most encouraging feature, suggesting at least some adequate source beds and appropriate temperature regimes for maturation, is the existence of numerous gas seeps. These seeps occur either as spectacular mud volcanoes (Fig. 25) or subdued pools onshore and as bubbling gas and turbid water offshore. Gas sampled from several seeps is flammable, contains mostly methane but traces of heavier hydrocarbons, and isotopically relatively heavy carbon. The characteristics of the hydrocarbons certainly suggest derivation from thermally mature source beds.

Reservoir beds range from potentially good to poor. The sandstone deposited by turbidity currents in the Middle Miocene is the best in terms of fine- to medium-grain size, lack of clay matrix and burrowing, and proximity to the most deeply buried source beds. In most outcrop samples, the sandstone is somewhat cemented by calcite and

porosity ranges from 10% to 25%. Shelf-deposited sandstone of Late Miocene to Pleistocene ages is less attractive as a reservoir. Outer- and middle-shelf sandstone is very fine- to fine-grained, commonly mixed with muddy matrix by burrowing, or cemented partially by calcite. The inner-shelf and shoreline sandstone is coarser and better sorted, but these facies are uplifted and eroded and are commonly not represented at depth on anticlines (Fig. 3).

Four wells have been drilled in coastal Makran (Fig. 2). Three penetrated only to the Pliocene and one to Late Miocene. All encountered very high pressures or drilling problems that would suggest high pressures. Such pressures constitute a serious obstacle to reaching potential Middle Miocene reservoir objectives on many onshore and offshore structures in coastal Makran.

The most attractive area for additional exploration lies offshore near Pasni, the southern flank of the Pasni anticlinorium. There potentially good reservoir rocks lying in proximity to deeply buried source beds form structures which are shallowly buried by Neogene and Pleistocene sediments. In that area, potential reservoirs could be reached without penetrating thick, highly overpressured beds.

ACKNOWLEDGEMENTS

We acknowledge with thanks Marathon Oil Company for permission to publish this paper. Many people within the Marathon organization provided assistance and data which made this publication possible. We also acknowledge Union Oil Company of California which shared in the cost of acquiring the information presented. The interpretation reflects the views of the authors and does not necessarily state the opinion of either Marathon or Union.

REFERENCES

Crame, J.A., 1984. Neogene and Quaternary Mollusca from the Makran Coast, Pakistan. (this volume).

Farah A. and K. A. DeJong, (eds.), 1979. *Geodynamics of Pakistan*, Geol. Surv. Pakistan, Quetta, 361 p.

Farhoudi, G. and D.E. Karig, 1977. Makran of Iran and Pakistan as an active arc system: Geology, 5: 664-668.

Harms, J.C. et al, 1975, Depositional environments as interpreted from primary sedimentary structures and stratification sequences: SEPM Short Course No. 2, SEPM, Tulsa, 161 p.

Harms, J.C., H.N. Cappel, and D.C. Francis, 1979. Geology and petroleum potential of the Makran Coast, Pakistan, Offshore South East Asia Conference, Singapore, 1979, 9 pp.

Hunting Survey Corporation, 1960. Reconnaissance geology of part of West Pakistan: Govt. of Canada, Toronto, 550 p.

Jacob, K.H. and R.L. Quittmeyer, 1979. The Makran region of Pakistan and Iran: trench-arc system with active subduction: *In:* A. Farah and K.A. DeJong. (eds.), *Geodynamics of Pakistan,* Geol. Surv. Pakistan, p. 305-317.

Johnson, G.D., N.M. Johnson, N.D. Opdyke, and R.A.K. Tahirkheli,

1979. Magnetic reversal stratigraphy and sedimentary tectonic history of the Upper Siwalik Group, eastern Salt Range and southwestern Kashmir; *In:* A. Farah and K.A. DeJong, (eds.), *Geodynamics of Pakistan,* Geol. Surv. Pakistan, p. 149-165.

Karig, D.E., 1974. Evolution of arc systems in the western Pacific. Earth and Planet. Sci. Ann. Rev., 2: 51-75.

Kazmi, A.H., and R.A. Rana, 1982. Tectonic map of Pakistan: Geol. Surv. Pakistan.

Page, W.D., J.N. Alt, L.S. Cluff, and G. Plafker, 1979. "Evidence for the recurrence of larger magnitude earthquakes along the Makran coast of Iran and Pakistan": Technophysics, 52: 533-547.

Seely, D.R., P.R. Vail, and G.G. Walton, 1974. Trench-slope model; *In:* C.A. Burke and C.L. Drake, (eds.), *Geology of continental margins,* Springer-Verlag, New York, P. 249-260.

Tirrul, R., I.R. Bell, R.J. Griffis, and V.E. Camp, 1983. The Sistan suture zone of eastern Iran: Geol. Soc. Amer. Bull., 94: 134-150.

2

Status of Petroleum
Exploration Offshore Pakistan

V. N. QUADRI
Oil & Gas Development Corporation,
Karachi, Pakistan

ABSTRACT

Exploration for hydrocarbons along coastal Pakistan has been slow, inspite of favourable conditions and similarities with other offshore oil producing areas of the world. So far only Sun Oil, Wintershall, Husky Oil and Marathon Oil Companies have drilled in the Pakistan offshore areas. However, the government's plans to invite bids for a vast tract of over 306,000 offshore acres holds much promise for the eventual drilling of structures with fair to high petroleum prospects in the coming years.

INTRODUCTION

Energy crisis of the seventies left us with two basic realities: first that conventional hydrocarbons will remain a major source of global energy until at least the first decade of next century, and second that no nation has escaped the energy problem unscathed, nor can hope to resolve it independently. In a developing country such as Pakistan, which pays a colossal bill in foreign exchange for oil import, we must increase petroleum prospecting both on land and at sea. To date, Pakistan offshore cannot boast of important hydrocarbon finds. However, at least two distinct geological provinces, the Indus offshore and the Makran offshore have been defined in this region of a total Exclusive Economic Zone of more than 266, 650 sq km. The former compares favourably with other major oil producing offshore basins, while the latter, albeit with subduction-system-gauche-geology, still offers analogs which are producing oil and gas in other global locations.

MAIN TECTONIC FEATURES AND GENERAL GEOLOGY

Indus offshore basin is an Atlantic-Type Passive Margin basin which straddles the continental crust of extension of Sind platform and Kirthar Foredeep, and oceanic crust of the Arabian Sea, east of Murray Ridge-Owen Fracture Zone plate boundary.

The Indus offshore is dominated by the intricate distributary channels of the Indus, with canyons, channels and Indus Cone extending into the Arabian Sea. Liquid hydrocarbon (HC) genesis in this area is evidenced by Bombay High Oilfields and oil shows in Indian Well GKH-1 to the south and recent discovery of Khaskeli Oilfield on the Sind Platform about 200 km northeast.

There are about 34 seismically delineated structures, with only 7 tested by drilling of these holes; 3 were stopped by technical difficulties, while at least 2 apparently were located on inadequate (no?) traps. Bank/reefal development and roll over traps against growth faults, as well as point bars, barrier islands, pinch outs/facies change beneath unconformities are expected traps. Overpressured zones do occur; but since the pressures are mainly a function of depth and not stratigraphy, they do not pose drilling problems.

The Makran offshore basin, according to Harms et al. (1982), is not typical of an arc-trench system as alleged by some published data. Rather, it is largely anomalous in its characteristics, with a wider arc-trench gap, weak Benioff zone (if present), and a significant volume of post middle Miocene sediments. Oil seepage reported in the north western corner of Baluchistan, and source rock evaluation of Garh-Koh Well-1 drilled just onshore with oil window estimated between 2400-4000 m and gas ejection with oil films from mud

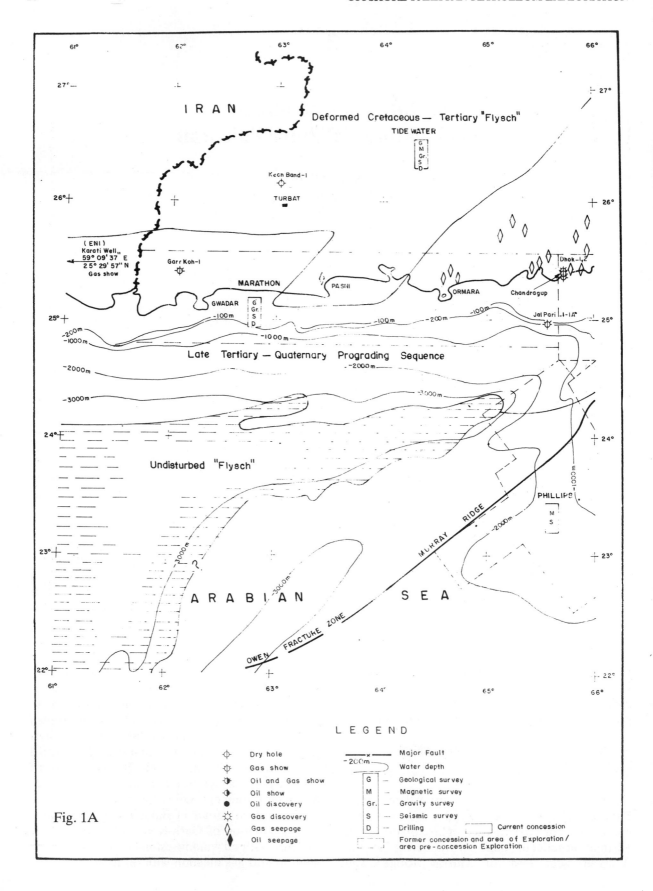

Fig. 1A

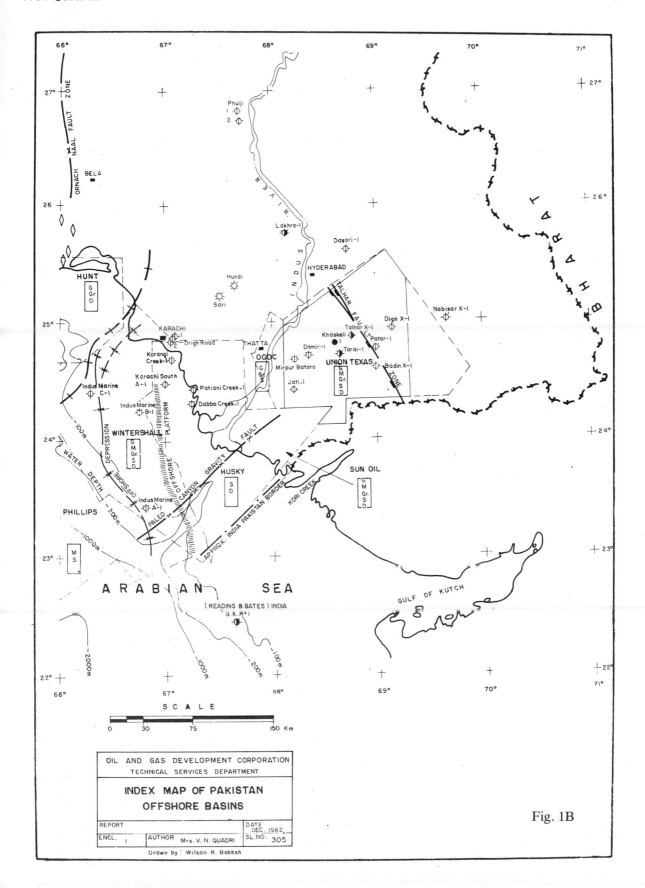

OIL AND GAS DEVELOPMENT CORPORATION
TECHNICAL SERVICES DEPARTMENT

INDEX MAP OF PAKISTAN
OFFSHORE BASINS

| REPORT | DATE DEC. 1982 |
| ENCL. 1 | AUTHOR Mrs. V. N. QUADRI | SL.NO. 305 |

Drawn by : Wilson R. Bakhsh

Fig. 1B

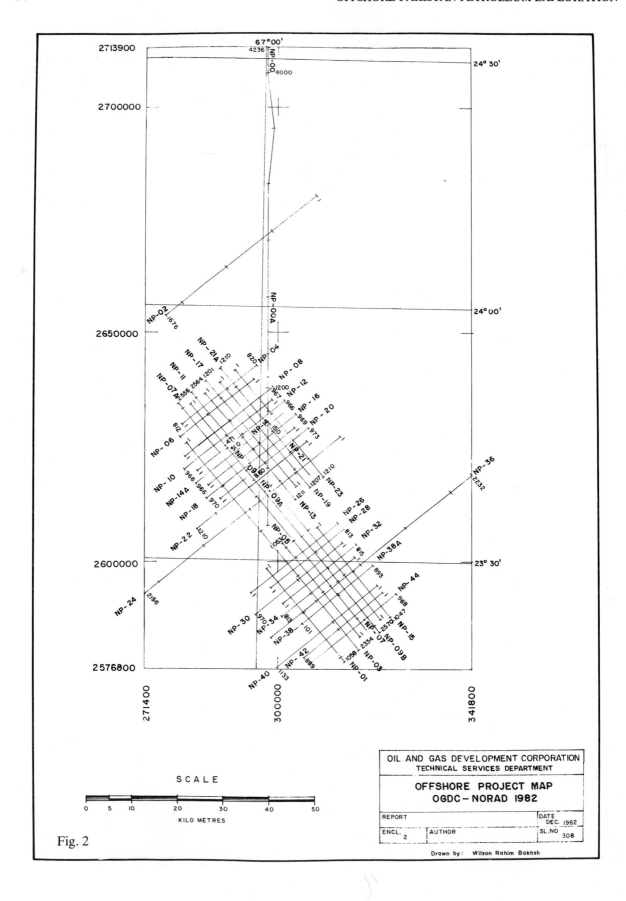

Fig. 2

HISTORY OF PETROLEUM EXPLORATION
PAKISTAN OFFSHORE BASINS

CONCESSIONS:		(All free area at present)			
SEISMIC SURVEY:					
INDUS OFFSHORE BASIN	1 Sun Oil Co.	(Single & double fold only)	(1961-62)	3816 km	
	2. Wintershall	(24 fold)	(1969-72)	9400 km	
	3. Husky	(24 fold)	(1976-78)	2380 km	Total 19026 km.
	4. Phillips	(24 fold)	(1977)	2280 km	
	5. OGDC-NORAD	(48 fold)	(May 1982)	1150 km	
MAKRAN OFFSHORE BASIN	1. Marathon/Union	(24 & 48 fold)	(1973-77)		
			Offshore:	6541 km	
			Coastal:	1541 km	Total 8082 km

WELLS DRILLED:	Name of Well	Company/Year		TD.(m) Age of the Formation Reached	Remarks
INDUS OFFSHORE BASIN	1. Dabbo Creek-1	Sun/1963	4354 (-4335)	L. Cretaceous (?)	D & A. Minor Gas shows. Drilled on down thrown side of fault according to Husky Seismic.
	2. Patiani Creek-1	Sun/1964	2659 (-2643)	U. Cretaceous	D & A. Gas show in U. Cretaceous
	3. Korangi Creek-1	Sun/1964	4140 (-4124)	U. Cretaceous	D & A. 5 gas shows in Paleogene
	4. Indus Marine A-1	Wintershall 1972	2841 (-2831)	M. Miocene	Abandoned due to technical reasons after kicking.
	5. Indus Marine B-1	Wintershall 1972	3804 (-3793)	L. Miocene	Abandoned due to technical reasons after kicking.
	6. Indus Marine C-1	Wintershall 1975	1942 (-1932)	L. Eocene	Abandoned due to high pressure.
	7. Karachi South A-1	Husky/1978	3353 (-3338)	U. Cretaceous	D & A. No structural closure due to absence of fault.
MAKRAN OFFSHORE BASIN	1. Jalpari-A 1	Marathon 1976	2007 (-1987)	L. Pleistocene pressure	Abandoned due to extreme high

volcanoes, are evidence of Liquid HC genesis. The Parkini and Hoshab shales (U. and L. Miocene respectively) are good source rocks, while Miocene Panjgur Turbidites, Pliocene Talar Sandstone and Pleistocene Sandstones may provide adequate reservoirs in structural fault-closure traps, or shoreline deltaic-subsea fan associations. Oil producing analogs are Santa Elena, a trench slope basin, Progeso, a transition between trench slope and forearc, and other broad or narrow ridged forearcs like Cook Inlet and Sacremento Valley, and Talara-Peru. As Hedberg (1974) states 'although area of shale diapirs and mud volcanoes may be unfavourable for HC accumulation because of lack of extensive reservoir, the surrounding facies above, below and laterally adjacent, may be particularly favourable in their relation to both reservoir and source environments'. The primary objective would be the structures like Pasni where Panjgur Turbidites occur at moderate depth. There are almost 22 seismically delineated structures, with atleast 6

lying on the southern flank of the Pasni Anticlinorium and 4 located in the transition zone/plate boundary between the Makran and Indus offshore basin. Only one well (Jalpari 1-A) has been drilled in Makran offshore basin in 1976, on a comparatively shallow highly overpressured shale diapir. This well was abandoned due to uncontrollable overpressure, without reaching objective reservoirs.

STATUS OF EXPLORATION

In the Indus offshore, Sun Oil Co, carried out seismic surveys 3816 km (single and double fold) in 1961-62 and drilled three nearshore wells, Dabbo Creek-1 (1963), Patiani Creek-1 (1964), and Korangi Creek-1 (1964). Wintershall A.G. conducted 9400 km of 24 fold seismic survey from 1969-1972, and drilled three wells; Indus Marine A-1 (1972), Indus Marine B-1 (1972) and Indus Marine C-1 (1975). Phillips Petroleum carried out a prelicense 24 fold seismic survey in deep water

Indus cone area, of 2280 km length, in 1977. Husky Oil digitized and reprocessed Sun Oil's data and conducted 24 fold seismic survey on about 2380 km lines from 1976-1978, and drilled one well (Karachi South A-1) in 1978. The most recent seismic survey has been carried out by OGDC-NORAD in May 1982, 1150 km of 48 fold multiplicity. This survey will have provided by end of first quarter 1983, 3 to 5 drilling sites, as well as a plan for near-future exploration activity in the Pakistan offshore basins.

Only one company Marathon/Union, has worked in the Makran offshore basin. It collected about 8082 km of 24 and 48 fold seismic data out of which 6541 km was offshore, and drilled one well (Jalpari 1-A in 1976-77) en-echelon to the offshore extension of Dhak onshore structure. The well was abandoned due to uncontrollable overpressure at 2007 m, in a Lower Pleistocene formation, without reaching the Talar Sandstone/Panjgur Turbidites, the objective reservoirs.

CONCLUSION

It is envisaged that in 1983-84, the government of Pakistan and OGDC shall be in a position to invite bids for about 70 offshore acreage blocks each 124000 hectares (306,400 acres), accompanied by related seismic and geological/geochemical data package, for joint venture operations. The area covered by the recent NORAD-OGDC detailed seismic survey, and the structures there- in, will be used as gravitational centre of interest, both in the promotion and in the actual licensing. Oil companies or geophysical companies shall be allowed to run non-exclusive or pre-license surveys in different parts of the offshore area. This should ensure drilling of structures of fair to high petroleum prospects in the Pakistan offshore basins, during 1984-85.

REFERENCES

Dickinson W.R. and D.R. Steely, 1979. Structure and Stratigraphy of Fore arc Regions, AAPG Bulletin, 63/1: 2-31.

Forhoudi, G. and D.E. Kraig, 1977. Makran of Iran and Pakistan as an active arc system, Geology 5: 664-668.

Gretener, P.E., 1977. Pore Pressure: Fundamentals, General Ramifications for Structural Geology, AAPG Continuing Education Course Note Series No. 4.

Hardling, T.P. and J.D. Lowell, 1979. Structural Styles their Plate Tectonic Habitats, and Hydrocarbon Traps in Petroleum Provinces, AAPG Bulletin, 63/7: 1016-1085.

Harms, J.C. and P. Tackenberg, 1972. Seismic Signatures of Sedimentation Models, Geophysics, 37. 45-58.

Harms, J.C., H. N. Cappel, D.C. Francis, 1982. 'Geology & Petroleum Potential of the Makran Coast, Pakistan" Offshore South East Asia 82 Conference Exploration III-Geology Session February 1982.

Hedberg, H.D. 1974. Relation of Methane Generation to Under compacted Shales, Shale Diapirs and Mud Volcanoes AAPG, 58: 661-673.

Le Blanc, Sr. R.H. 1977. Distribution and continuity of Sandstone Reservoirs-Parts 1 and 2, Journal of Petroleum Technology: 776-804.

OGDC and HDIP 1980. Abnormal Formation Pressures in Offshore Pakistan.

OGDC 1982. "Preliminary Report on OGDC-NORAD Seismic Survey".

Quadri, V.N. 1982. Role of Public Sector Pakistan Exploration Effort in Meeting the Growing Need for Petroeum. Petroleum Institute of Pakistan Symposium.

Quadri, V.N., Petroleum Prospects of Pakistan and Oil Oriented Exploration Guidelines. Petroleum Institute of Pakistan Symposium.

White, R.S. and K.D. Klitgord, 1976. Sediment deformation and Plate tectonics in the Gulf of Oman Earth and Planetary Sci Letters 32: 199-209.

3

Structural Features of the Makran Fore-arc on Landsat Imagery

J. K. LEGGETT

Imperial College of Science & Technology, London, U.K.

and

J. PLATT

Oxford University, Oxford, U.K.

ABSTRACT

Colour-composite band 5 ERTS satellite images of the southern Central Makran Range and Coastal Makran Range between the longitudes of Gwadar and Ormara reveal the gross pattern of large-scale structural features in the Makran fore-arc with clarity. Using photographs of the best examples, we describe: i) strike-parallel reverse-faults; ii) tight folds (half-wavelength c. 1 to less than 3 km); iii) open synclines (half-wavelength c. 20-30 km); iv) conjugate wrench faults of minor displacement (less than 1 km); v) "terrane-bounding" faults of unknown displacement; vi) anomalous outcrops near the coast in which strata strike northeastward, at a high angle to the east-west regional strike; vii) the transverse (north-south) anticline at Kappar. The degree of deformation decreases progressively upwards in the stratigraphic succession.

INTRODUCTION

In recent years, Deep Sea Drilling Project investigations of fore-arc regions on active plate margins have shown surprisingly diverse responses to subduction. In some fore-arcs such as the Shikoku fore-arc of SW Japan, large volumes of strata are scraped from the sedimentary cover of the subducting oceanic plate, and stacked against the leading edge of the over-riding plate (e.g. Karig et al., 1983). This process, known as subduction accretion, contrasts greatly with behaviour in other fore-arcs such as the Guatemala fore-arc of Central America (von Huene et al., 1982), where the entire sedimentary cover of the subducting oceanic plate appears to be carried below the fore-arc (sediment subduction) and where the underthrusting plate mechanically abrades the leading edge of the fore-arc (subduction erosion).

Northward subduction under the Makran region of eastern Iran and western Pakistan (Fig. 1) has generated between Late Cretaceous and modern times what is apparently the widest-known example of an accretionary fore-arc (Arthurton et al., 1982, Jacob & Quittmeyer 1979). The distance between the trench, buried by sediments several km. thick off the Makran coast (White, 1982), and the Pleistocene volcanoes and older volcanic rock outcrops of the magmatic arc on the Afghanistan border is nearly 500 km. South of the arc, a deformed Late Cretaceous-Paleogene flysch belt in the Ras Koh Range may represent the earliest accreted trench sediments (Arthurton et al., 1982); in the modern fore-arc it acts as a frontal arc (sensu Dickinson & Seely 1979) with a fore-arc basin, the Helmand Basin, to the south (Jacob & Quittmeyer 1979).

South of this, the 170 km-wide Northern Makran (Siahan), Central Makran, and Makran Coast Ranges predominantly comprise deformed flysch (Panjgur Sandstone of Hunting Survey Corporation 1960), of Miocene age in the south and probable Paleogene age in the north (Harms et al., 1982). Overlying the flysch in the Makran Coast Range is a thick mudstone-dominated sequence (Parkini Mudstone of Hunting S.C.), in turn overlain by broad, open, synclinal outliers of varied Neogene sandstones and mudstones (Talar Sandstone and Chatti Mudstone west of Pasni, Hinglaj Formation to the east).

Based on LANDSAT analysis and reconnaissance in the Iranian Makran, Farhoudi & Karig (1977) interpreted the flysch substrate as uplifted trench deposits and the overlying synclinal outliers as uplifted remnants of lower trench slope basins. However, mapping of the coastal Makran

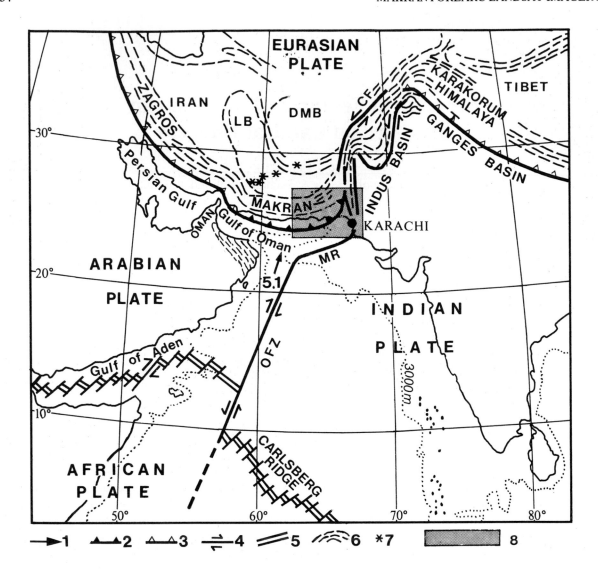

Fig. 1. Plate tectonic setting of the Makran. 1) Estimated vector of relative plate motion and rate in cm./yr. (details in Jacob & Quittmeyer, 1979). 2) Oceanic subduction zone. 3) Continental collision zone. 4) Transform fault with displacement sense. 5) Spreading centre. 6) Illustrative regional strike in Cenozoic fold belts. 7) Volcanic centres relating to Quaternary arc. 8) Area of Figure 2. CF: Chaman Fault. DMB: Dasht-i-Margo (Afghan) Block. LB: Lut Block. MR: Murray Ridge. OFZ: Owen Fracture Zone.

led Harms et al. (1982 and this volume) to interpret the post-Panjgur strata as slope and shelf deposits, recording several late Miocene to Pleistocene transgressive-regressive cycles within an overall prograding shelf regime.

The Makran fore-arc is unique in a number of ways. First, most fore-arcs above active subduction zones are almost entirely submarine (see summary in Dickinson and Seely, 1979), limiting investigation in most cases to geophysics and drilling. Much of the Makran fore-arc, as a consequence of its anomalous width, is emergent. Second, the emergent accretionary complexes

(e.g. Mentawai Islands of Sumatra) and Cenozoic accretionary terranes (e.g. Shimanto Belt of southwest Japan) are commonly shrouded by vegetation. Third, deformation, although intense, has not reduced the exposed strata to the degree of structural incoherence displayed by many of the melange-dominated accretionary terranes of the circum-Pacific. The Makran fore-arc is clearly a key area in studies of tectonic and sedimentary responses to subduction.

In this paper we describe large-scale structural features of the fore-arc as seen on colour-composite band 5 ERTS satellite images. The area

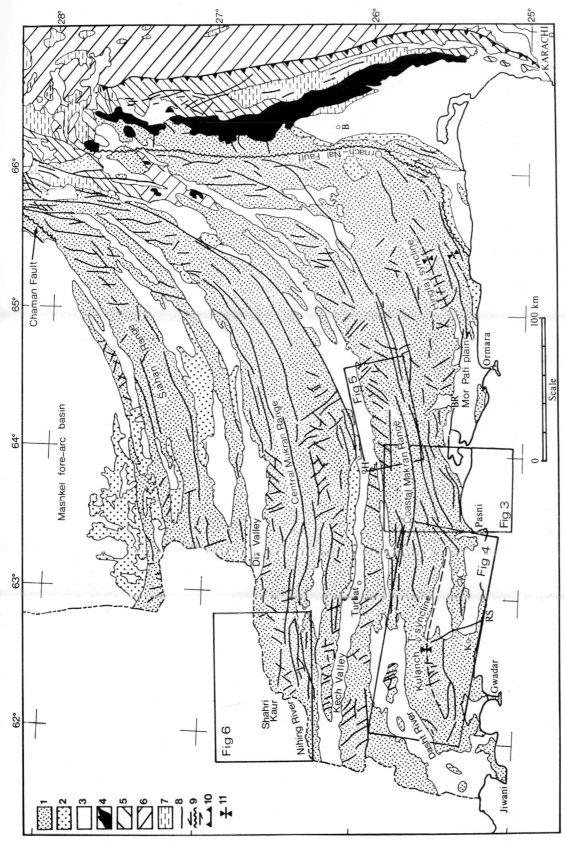

Fig. 2. Tectonic map of southern Baluchistan (after Geological Survey of Pakistan 1: 2000,000 tectonic map of Pakistan). *Tectonostratigraphic units:* MAKRAN UNITS: 1) Eocene (?)—Pliocene deep to shallow-marine strata. 2) M–U Pleistocene, continental in north, shallow marine on coast. 3) U. Pleistocene-Recent (intramontane basins and coastal plain). UNITS E OF ORNACH-NAL FAULT (for further details and definitions see Tectonic Map of Pakistan): 4) Ophiolitic rocks. 5) Rocks affected by "late Himalayan" orogeny. 6) Rocks affected by "middle Himalayan" orogeny. 7) "Pre-orogenic" units. *Structural features.* 8) Faults (undifferentiated). 9) Transcurrent faults. 10) Thrusts. 11) Principal coastal synclines. Place names: B: Bela, BR: Basol River, C: Chakuli, H: Hoshab, K: Kappar, P: Pidarak, RS: Ras Shahid, T: Talar.

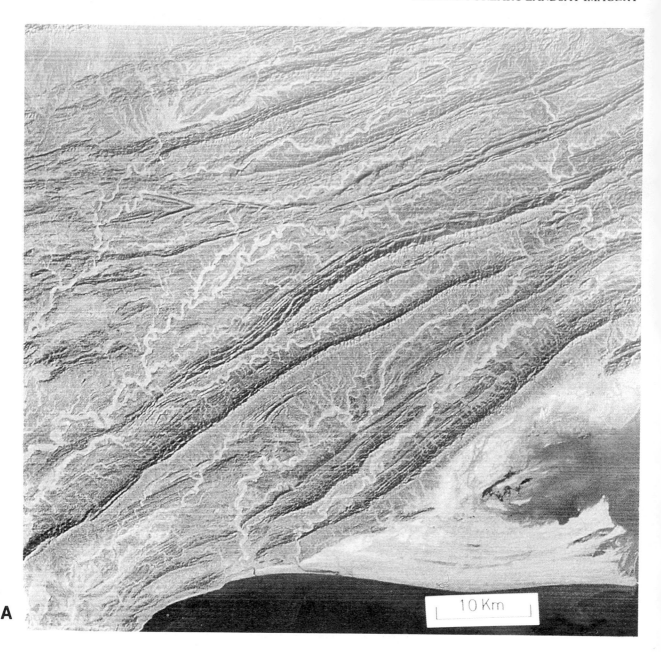

A

surveyed covers ca. 70,000 sq. km. of the Coastal Makran and part of the Central Makran Range between longitude 62° (west of Gwadar) to 65° (east of Ormara), and Latitude 25° 20' (Ras Shahid-Ormara coast) to 26° 40' (through Shahbaz Kalat) (Fig. 2).

STRUCTURAL FEATURES

1. Strike-parallel faults

Faults paralleling regional strike dominate Panjgur and to a lesser extent Parkini outcrops on the Hunting S.C. maps. Most are shown as having reverse displacements. Over much of the study area the general trend is east-west, though ranging from northeast to east-southeast. Figure 3 illustrates the character of the faults in an area northeast of Pasni (the Pasni anticlinorium of Harms et al., 1982. Ridges of Panjgur sandstone are bounded by faults on the south side, overriding subdued ground mainly occupied by mudstone. The pronounced linear fabric in the Panjgur ridges probably reflects grouping of thick-bedded sandstones in packets. The faulted nature of these contacts is shown by the low-angle discordance of the bedding fabric to the faults, and by the

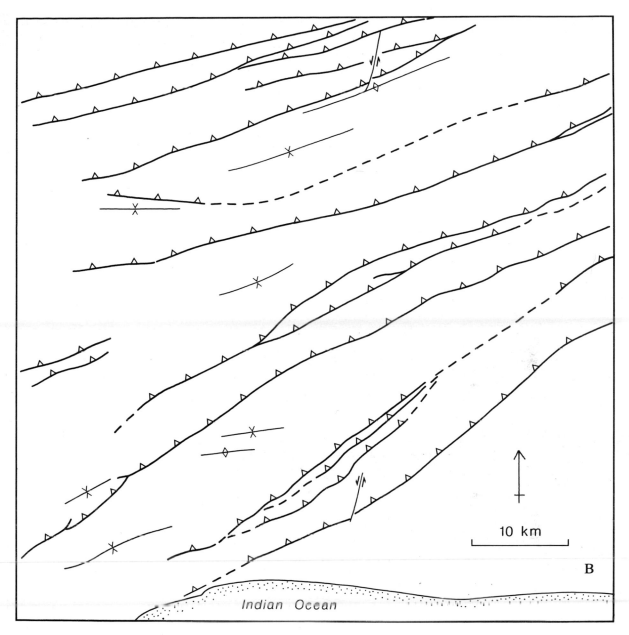

Fig. 3. Landsat image (A) and interpretation (B) of part of the Makran Coast Range northeast of Pasni, showing the pattern of strike-parallel faults and predominantly synclinal folds. For regional location see Figure 2.

presence of lenticular fault slices along them (for example in the south-centre of Figure 3).

From LANDSAT information alone the character of the strike faults is difficult to assess. Regional evidence for compression, such as the tight fold closures evident in the north of Figure 3 suggests that they are reverse faults.

The faults strike parallel to bedding, and their straight outcrop suggests steep inclination. In most cases they are likely to be slightly steeper than bedding since studies of modern accretionary trench slopes show initial decollement of trench turbidites along bedding-plane step-thrusts (e.g., Moore et al., 1982; Karig et al., 1983). Should the turbidites of the Pasni anticlinorium not be trench floor deposits, but ponded lower slope basin turbidites, the ridge-bounding faults are likely to have propagated into the slope sequence from faults in the underlying accretionary substrate deriving from initial decollement. In either case, rotation during subsequent accretion would explain their present attitude.

2. Tight Folds

Folds are relatively common in the Panjgur outcrop, particularly in the Central Makran Range (north of the Kech Valley). All are of less than 3 km half wavelength. Most appear to be synclinal, and the scarcity of anticlinal closures suggest that the latter may be commonly broken through by strike-parallel reverse faults. The folds plunge both east and west, neither direction being obviously predominant, and many folds are periclinal (boat shaped). The angle of plunge is difficult to determine, but in most cases does not appear to be great. The presence of such fold closures implies that many areas of marked bedding fabric on the imagery which would appear to be homoclinal sequences may in fact be tightly folded, with horizontal hinge lines.

Two large synclines with opposite plunges are visible in the massive Panjgur sandstone in the northwest quarter of Figure 3. Note the absence of a clear anticlinal closure between them. Smaller scale folds are visible in the same area, but are not fully resolvable. These folds coincide with a significant divergence of trend which splits the Makran Coast Range in two. The northern part swings from its dominant southwest trend to due west, while the southern part continues on a southwest trend towards Pasni. The Kulanch syncline (section 3) comes in westwards between the two (Fig. 2).

Other large folds, mainly synclines, can be

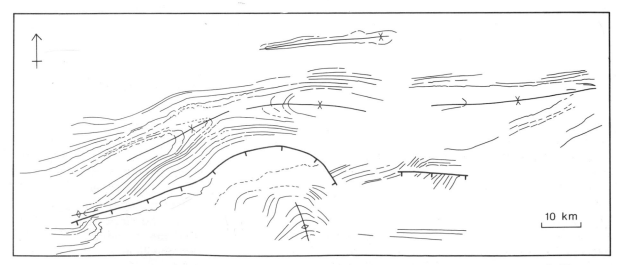

Fig. 4. Landsat image and interpretation of the Kulanch syncline and the north-trending Kappar anticline to the south. Note the major normal faults on the south side of the syncline. For regional location see Figure 2.

picked out in the mudstone-dominated sequences in the centre and south of Figure 3.

3. Open folds

Broad, open synclinal outliers occur along the coastal belt of the Makran Coast Range in Pakistan. Harms et al., (1982, and this volume) report that they involve only post-Panjgur strata of slope and shelf facies. Contacts with the Parkini Mudstone are concordant (Hunting S.C., 1960). In the Iranian Makran, where the synclinal outliers are more numerous and extend further inland, McCall and Kidd (1982) also record that they comprise shallow marine strata.

In the study area synclinal outliers occur north of Kappar (the Kulanch syncline) and north of Ormara (the Hinglaj syncline) where there is also a partly exposed complementary anticline. Figure 4 shows the Kulanch syncline. It closes in the east north of Pasni, plunging west. The axial trace can be traced ca. 150 km. before the structure is obscured by recent alluvium of the Dasht river valley north of Jiwani. The half-wavelength reaches a maximum of 24 km in the west and possibly as much as 30 km if alluvium southeast of Talar is underlain by the uninterrupted south limb of the syncline. The trend of the axial trace shifts from west to west-northwest before reaching a culmination west-southwest of Talar. Beyond that the axial trace swings to west-southwest.

The prominent fabric in the Talar Sandstone probably represents the 20-50 metre thick sand-mud cycles recorded by Harms et al. (1982). This bedding fabric picks out the northward inclination of the syncline clearly. Beds in the north limb dip at moderate to steep angles; beds in the south limb dip at a significantly shallower angle, in part explaining the less pronounced fabric. However, notwithstanding the low sun-angle in this image (elevation 30°, azimuth 145°), this less pronounced fabric probably also reflects a lower proportion of sand in cycles in the south limb. According to the data of Harms et al. (1982 and this volume), across-strike proximal-distal facies changes occur in post-Parkini strata.

Discontinuous outcrops of Chatti Mudstone in the core of the syncline define a clear westward closure on the east side of the culmination south of Talar, but form unhelpful badland topography, reminiscent of the signature of Parkini terrane, in the western part of the structure. We discuss the complex structure south of the Kulanch syncline in section 6.

The Hinglaj structure has a half-wavelength of up to 17 km, but has a faulted contact against Parkini mudstones on its southern limb (Hunting S.C., 1960). The structure is also inclined to the north, though with more gentle dips than the Kulanch syncline. In more extensive outcrops of Hinglaj Formation to the east of the survey area (i.e. the eastern Makran orocline, Figure 2) complementary anticlines with faulted hinges separate broad, open synclines (Hunting S.C., 1960). The south limb of such an anticline is suggested by several outcrops surrounded by recent sands both west and east of Ormara. A complex west to northwest-trending fault must exist between these outcrops and the Hinglaj syncline, since outcrops of Hinglaj Formation in the Mor Pati plain north of Ormara have an anomalous northeast strike. The position of the fault between these outcrops and the south limb of the Hinglaj syncline (itself faulted against Parkini Mudstone) is in all probability picked out by a line of Recent mud volcanoes extruded through the sands of the plain.

4. Wrench Faults

Wrench faults are common throughout the area, affecting all the strata but particularly the Pangjur outcrop. They form a conjugate set trending northeast to north-northeast (sinistral offsets) and northwest to north-northwest (dextral offsets). Displacements are everywhere minor, generally less than 1 km, and the clarity of the wrench faults on the imagery is due in many cases to their having been exploited by drainage.

The wrench faults are clearly visible in the area illustrated in Figure 5, southeast of Hoshab in the Makran Coast Range. Both sets are developed, cutting a strong fabric produced by bedding and strike-parallel faults, but the sinistral set is dominant. The sinistral set is generally predominant in the eastern part of the Makran where the trend swings into the eastern Makran orocline. Further west, where the trend is roughly east-west, the dextral set is more common.

As pointed out by Hunting S.C. (1960), the orientations of the wrench faults are such that they can be explained by the same north-south compression that was responsible for the folds and reverse faults. The wrench faults cut these and so must be late features, perhaps produced during a continuous deformation. Dubey (1978) has used

10 Km

A

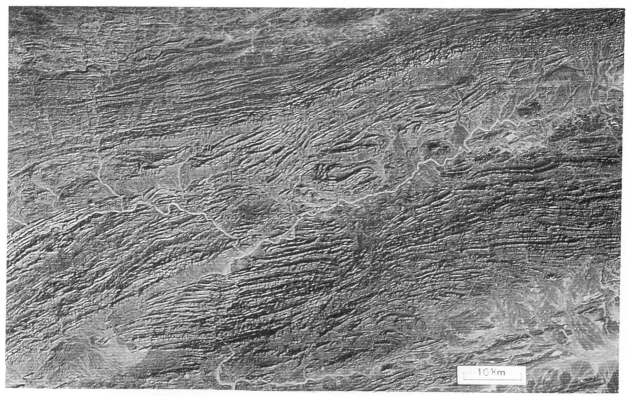

10 Km

A

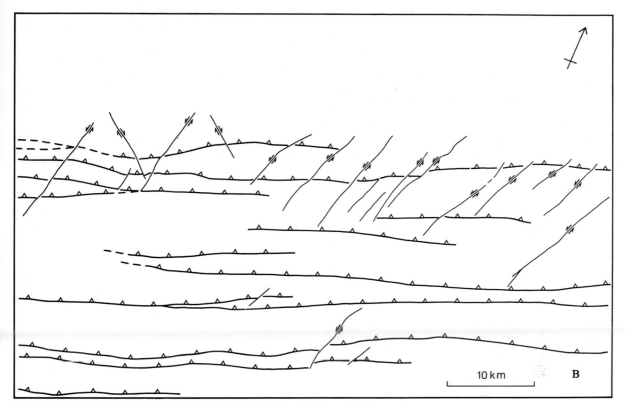

Fig. 5 Landsat image (A) and interpretation (B) of part of the northern Makran Coast Range near Hoshab, showing the NE-trending sinistral wrench faults (predominant), and the the NW-trending dextral set. For regional location see Figure 2.

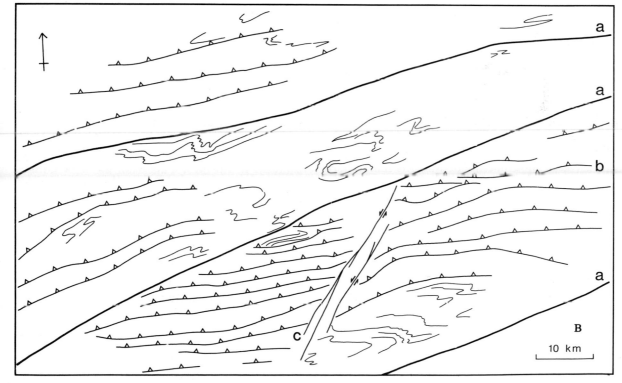

Fig. 6. Landsat image (A) and interpretation (B) of part of the western Central Makran Range, near the Pakistan-Iran border, showing major terrane-bounding faults (a) cutting obliquely across the trend of the strike-parallel faults (b). Note regions of irregular folding, and the zone of sinistral wrench faults (e). For regional location see Figure 2.

experimental models to show how the late development of wrench faults by axial extension during such a deformation can cause isolation and independent evolution of wrench fault-bounded segments during the final stages of the deformation. This mechanism may provide an explanation for alongstrike variations in axial traces and axial plane orientations in the Makran and other coherent accretionary terranes (Leggett and Casey, 1983).

Another feature of the wrench faults is their lack of continuity. Many show clear fabric displacements over limited distances but abut against undisplaced features. A very clear example is visible in the south of Figure 5. Note also the marked concentration of wrench faults in the north of the figure. Only rarely are wrench faults traceable for more than 20 km. This attribute is perhaps to be expected in an accretionary complex, where deformation is progressive and diachronous, so that a package of deforming strata can be bounded on the arc side by an already-deformed, relatively stable package which acts as a buttress.

5. "Terrane-bounding" faults

In several places within the study area prominent fabrics in the Panjgur outcrop, reflecting bedding and/or strike faults, are truncated by lineaments at a low angle. A good example occurs along the Nihing River on the Pakistan-Iran border (Fig. 6). Here a strong east- to east-northeast-trending fabric in Panjgur Sandstone of unknown but probably Paleogene age (Harms et al., 1982) is truncated on the north side by a straight lineament which is remarkably clear on the LANDSAT imagery, but is not marked on Hunting S.C. maps. To the north of this line a northeast-trending fabric in the Iranian Panjgur west of the Shahri Kaur-Nihing River confluence passes eastwards into a patchy wedge-shaped zone which disappears under sub-Recent wadi deposits in the Diz valley. This zone contains remarkable, isolated fold closures giving a swirled texture reminiscent of large-scale broken formation. To the north, a strong east-northeast trending lineament separates the wedge-shaped zone from a terrane with a well-defined fabric parallel to the bounding lineament.

Other, low-angle fabric-truncating lineaments define major Quaternary—Recent intermontane

valleys, such as the Kech Valley, and the valley extending east and west of Pidarak (Fig. 2).

The nature and amount of displacement on these faults are difficult to define. Many could be predominantly dip-slip features (probably thrusts), but from the general appearance of the faults with higher angles of obliquity to regional strike, such as the Nihing River fault, it seems that tracts of Makran strata have been shuffled sideways against each other to a certain extent. One possible explanation for this is that transcurrent movements along the transform boundary which forms the eastern boundary of the Makran geological province (Fig. 1) have propagated into the accretionary complex from the Chaman and Ornach-Nal fault systems. The impressive left-lateral north-south Chaman Fault ends along the eastern boundary of the Siahan Range in the northern apex of the Makran orocline. Several faults splay from near its southern end, including the prominent Panjgur Fault (Kazmi, 1979), which would appear to continue into the Diz Valley directly in line with the lineament shown in Figure 6.

6. Transverse anticlines

In the coastal region of the study area a transverse (north-south) anticline is exposed near Kappar and beds have anomalous transverse strikes in two other places. In the Kappar anticline Parkini Mudstone is arranged in a gentle dome with a north-south axis plunging at a shallow angle to the north (Fig. 4). Further east, a ca. 50 sq. km. outcrop of Talar Formation west-southwest of Chakuli strikes north-northeast (Fig. 4). A similar outcrop in the Mor Pati Plain north of Ormara comprises Hinglaj strata striking northeast, diverging strongly from the east-west strike throughout the rest of that area.

The axis of the Kappar anticline apparently plunges below the Kulanch syncline. The Hunting S.C. map shows a concordant contact with Talar strata, though this is not clear on the imagery. The strike of strata in the main outcrops of Talar Formation in the Kulanch syncline to the north are not notably affected by the Kappar fold. However, the axial trace of the syncline shows a gentle sweep about the Kappar anticline, and the culmination along its axis is opposite the anticline. Hence the overall relationships suggest that the transverse fold, well developed in Parkini outcrops, had a reduced effect at higher stratigraphic levels.

The origin of the Kappar structure is obscure and further complicated by faults along the south side of the Kulanch syncline. On the imagery, these are clear along a discordance where Talar bedding is truncated at the western end of the structure, and by an abrupt east-west termination of the north-northeast-striking outcrop near Chakuli (Fig. 4). The terrane south of the Kulanch southern limb is dominated by a "badland" signature and so the nature and continuity of faulting is difficult to trace on the imagery. Interestingly, Harms et al. (1982) show a normal fault, following an irregular trace because of its low-angle southward dip, linking the two faults detectable on the imagery.

The northeast-striking beds north of Ormara, like those near Chakuli, must be truncated by a fault against the south flank of the Hinglaj syncline. In this case the fault trace is buried below the plain, but marked by a line of mud volcanoes (Hunting S.C., 1960) which suggests that it may be an extensional feature.

These observations, indicate that late stage gravity-collapse features may be important in the outer part of the fore-arc.

DISCUSSION

Structural features on LANDSAT imagery show that the degree of deformation increases progressively from youngest to oldest strata. Reverse strike-parallel faults and tight folds dominate outcrops of the oldest unit (flysch deposits of the Panjgur Formation). These structures are also present, but less well developed, in the overlying Parkini Formation. They do not affect post-Parkini strata, which are arranged in broad, open synclines. Conjugate wrench faults formed at a late stage in response to the same horizontal north-south compression required by the reverse faults, tight folds and open synclines. The wrench faults have smaller offsets and are less extensive at structurally higher levels.

These characteristics equate well with the pattern of deformation which would be expected in a fore arc region. Several studies show that structural features in the accretionary substratum (offscraped trench deposits) below certain modern trench slopes can propagate into slope-mantling cover sequences, with progressively decreasing displacement upward (e.g., Lundberg and Moore, 1982; von Huene, 1983). In the case of Makran, where a thick sequence of neritic sediments (post-

Parkini strata) overlies the supposed trench (Panjgur Formation) and slope-mantling (Parkini Formation) rocks, gravity collapse phenomena are also apparent.

ACKNOWLEDGEMENTS

We wish to thank the U.K. Natural Environment Research Council for financial support of a programme of field research in the Makran, for which this paper is the fore-runner. We are grateful to Dr. John Harms, and colleagues in the Hydrocarbon Development Institute of Pakistan and Geological Survey of Pakistan for fruitful discussions on the geology of the Makran.

REFERENCES

Arthurton, R.S., A. Farah, and W. Ahmed, 1982. The Late Cretaceous-Cenozoic history of western Baluchistan, Pakistan—the northern margin of the Makran subduction complex. In: J.K. Leggett, (ed.), Trench-Forearc Geology, Spec. Publ. Geol. Soc.'Lond. no. 10: 373-385.

Dickinson, W.R. and D.R. Seely, 1979. Structure and stratigraphy of fore-arc regions. Am. Assoc. Petrol. Bull., 63: 2-31.

Dubey, A.K. 1980. Model experiments showing simultaneous development of folds and transcurrent faults. Tectonophysics, 65: 69-84.

Farhoudi, G. and D.E. Karig, 1977. Makran of Iran and Pakistan as an active arc system. Geology. 5: 664-668.

Harms, J.C., H.N. Cappel, and D.C. Francis, 1982. Geology and petroleum potential of the Makran coast, Pakistan. In: Proceedings Offshore SE Asia Conference, 9-12 February, Singapore.

Hunting Survey Corporation, 1960. Reconnaissance geology of part of West Pakistan. Miracle Press, Ontario.

Jacob, K.H. and R.L. Quittmeyer, 1979. The Makran region of Pakistan and Iran: trench-arc system with active plate subduction. In: A. Farah and K.A. DeJong, (eds.) Geodynamics of Pakistan. Geological Survey of Pakistan, Quetta, p. 305-317.

Karig, D.E., H. Kagame, et al., 1983. Leg 87 drills off Honshu and S.W. Japan. Geotimes, Jan. 1983: 15-18.

Kazmi, A.H. 1979. Active fault systems in Pakistan. In: A. Farah and K.A. DeJong, (eds.) Geodynamics of Pakistan, Geological Surv. Pakistan, p. 285-294.

Leggett, J.K. and D.M. Casey, 1983. The Southern Uplands accretionary prism: implications for controls on structural development of subduction complexes. In: J.S. Watkins, and C.L. Drake (eds.) Continental margin processes. Am. Assoc. Petrol. Geol. Memoir, (in press).

Lundberg, N. and J.C. Moore, 1982. Structural features of the Middle America Trench slope off southern Mexico, Deep Sea Drilling Project Leg 66. In: Initial Reports of the Deep Sea Drilling Project, Washington (U.S. Govt. Printing Office), p. 793-822.

McCall, G.J.K. and R.G.W. Kidd, 1982. The Makran, southeastern Iran: the anatomy of a convergent plate margin active from Cretaceous to Present. In: J.K. Leggett, (ed.), Trench-Forearc Geology, Spec. Publ. Geol. Soc. Lond., No. 10:387-397.

Moore, J.C., J.S. Watkins, and T.H. Shipley, 1982. Summary of accretionary processes, Deep Sea Drilling Project Leg. 66: offscraping, underplating, and deformation of the slope apron. In: Initial Reports of DSDP, Washington (U.S. Govt. Printing Office), p. 825-836.

Von Huene, R. 1983. Deformation processes on trench slopes. In: J.S. Watkins and C.L. Drake (eds.) Continental Margin Processes, Am. Assoc. Petrol. Geol. Mem. (in press).

Von Huene, R., J. Aubouin, et al., 1982. Leg 84 of the Deep Sea Drilling Project—Subduction without accretion: Middle America Trench off Guatemala, Nature 269: 458-460.

White, R.S., 1982. Deformation of the Makran accretionary sediment prism in the Gulf of Oman (north-west Indian Ocean). In: J.K. Leggett (ed.), Trench-Forearc Geology, Spec. Publ. Geol. Soc. Lond. no. 10:375-372.

4

Neogene and Quaternary Mollusca from the Makran Coast, Pakistan

J. A. CRAME

British Antarctic Survey, Cambridge

ABSTRACT

As a result of geological mapping by Colombo Plan and Marathon Petroleum geologists, the stratigraphy of the Neogene and Quaternary sediments of the Makran Coast has been placed on a much firmer footing. It is now possible to study the distribution of certain fossil groups, such as the molluscs, in relation to measured sections and assess their value in both biostratigraphic and paleoenvironmental studies.

The lowest lithostratigraphic unit, the Parkini Mudstone, represents a long phase of very deep water deposition within the Miocene. The sparse faunas suggest that there were brief interludes of shallowing but it is likely that the predominant environment was that of the continental slope. There was then a fairly rapid transition into the Talar Sandstone, which is a thick sequence of alternating fossiliferous mudstones and sandstones. The three main types of molluscan assemblages from these lithologies, mudstone-associated, venerid and *Turritella*, indicate regular oscillation between the inner and outer shelf. Species present such as *Turritella angulata*, *Tonna* cf. *zonata*, *T. luteostoma*, *Babylonia spirata*, *Melongena* cf. *gigas* and *Rapana bulbosa* have been taken to indicate a Pliocene age. There is an equally sharp transition at the top of the Talar back into another massive mudstone-siltstone unit, the Chatti Mudstone. Small *Tellin* assemblages from this interval point to a Pliocene age and an outer shelf/slope environment.

It would appear that the Chatti Mudstone grades up into the overlying Ormara Formation. Shell beds and coquinas interbedded within the silty mudstones of this formation have yielded abundant nearshore assemblages. Characteristic elements are; *Turritella bacillum*, *T. cingulifera*, *T.* cf. *fultoni*, *Natica didyma*, *Gyrineum spinosum*, *Tonna* cf. *tessellata*, *Anadara inaequivalvis*, *A. clathrata*, *Larkinia multicostata*, *L. rhombea*, *Trisidos tortuosa*, *Chlamys prototranquebaricus*, *C. townsendi*, *Pecten vasseli*, *P. nearchi*, small oysters, and a variety of mactrids, tellinids and venerids. It is suggested that some of these forms, such as *P. vasseli*, may indicate an Early Pleistocene age. The Ormara Formation in turn grades up into the Jiwani Formation which includes a coral reef of probable Late Pleistocene age.

In the eastern part of the concession area, Talar Sandstone lithologies are succeeded by a thick sequence of nearshore sands, coquinas and littoral sands that are broadly equivalent to the Ormara and Jiwani Formations. There is no direct equivalent of the Chatti Mudstone here and it is thought that its absence may be related to stratigraphic thinning in the proximity of the Las Bela axial zone.

The extremely thick (up to approx. 11 km) Makran coastal sequence promises to become a standard reference section for Neogene and Quaternary molluscan studies in the Indo-Pacific province.

INTRODUCTION

During the course of systematic geological mapping in the Marathon concession area on the Makran Coast, Pakistan (Fig. 1; see also Marathon Sample Location Map for 1974-1976 Field Surveys), rich molluscan faunas were found in a number of areas. As these appeared to range through the greater part of the sedimentary succession and most of the principal lithologies, it was felt that closer investigation of them may expedite both stratigraphic studies and paleoenvironmental reconstructions. Consequently, a short field programme was devised to sample molluscs from a series of sections in the western (Gwadar), central (Pasni) and eastern

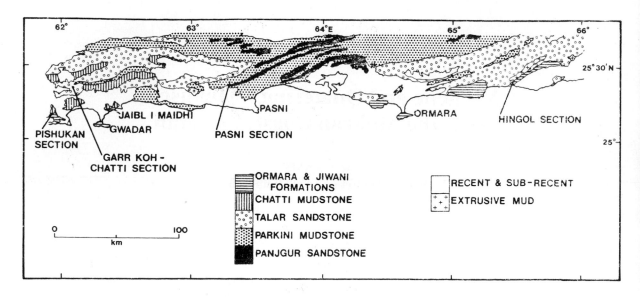

Fig. 1. Sketch geological and locality map of the central Makran Coast. The positions of the principal sections studied are indicated.

(Hingol) parts of the concession area (Fig. 1). Sections were selected that collectively offered an almost complete sequence through the Middle Miocene – Late Pleistocene sediments.

The specimens collected during the field season (January-March 1976) were shipped back to London for identification in the British Museum (Natural History). As the Tertiary molluscan faunas of the Makran Coast are not particularly well known, and as a very large number of specimens (between 4000 and 5000) were collected, it was obviously not possible to identify them all to species level. Some groups could only be subdivided into genera, and one major one, the Tellinacea, could not be taken even this far. There are many small, smooth Tellins that are superficially very similar and extremely difficult to distinguish in the fossil state.

The approach adopted in this study has been to start from the Recent and work backwards through time. Many molluscan species have longer geological ranges than is generally recognized and it is always worthwhile to compare a Neogene fauna with a Recent one from an adjacent area. To this end, the Makran fossils were closely compared with the Recent Mollusca collections from Arabia, Pakistan and India housed in the Dept. of Zoology, B.M. (N.H.) The Townsend collection is particularly helpful in this context (Melvill and Standen, 1901, 1907), as are the specimens from the Trucial Coast described by Biggs (1973). Useful general works for the identification of

Tertiary molluscs are the Bivalvia volumes of the Treatise on Invertebrate Paleontology (Moore, 1969), the standard text-book by Davies (1971, 1975) and the review papers by Eames and Cox (1956) and Shuto (1975). Comparisons were also made with the Neogene faunas of the following regions: East Africa (Cox, 1927, 1930; Nuttall and Sealy, 1961; Kent, et al., 1971), Iran (Cox, 1936), India (Karikal) (Cossmann, 1900, 1903 and 1911), Bangladesh (Vredenburg, 1921), Malaya (Wissima, 1947), Borneo (Cox, 1948; Nuttall, 1961, 1965), Indonesia (Martin 1883-1887; Oostingh, 1935, 1938-40; Pannekoek, 1936), Timor (Tesch, 1915), and Philippines (Shuto, 1969 and 1971).

The most significant advances in the study of the biostratigraphy of the Late Tertiary and Quaternary sediments of this region were made by Blanford (1872) and Vredenburg (1906, 1921, 1925). The latter author assigned the youngest strata of the coastal belt to the Makran Series, which he then subdivided into an older Talar Stage (with Early Pliocene affinities) and a younger Gwadar Stage (with Late Pliocene affinities) (Vredenburg, 1925; Davies, 1975, pp. 373 and 388). The original faunas upon which these stages are based were widely separated by a terrain of complex geology, and thus their true relationship to one another was never fully established. However, subsequent geological mapping, first by Colombo Plan geologists (Hunting Survey Corp., 1960, Companion Geological Maps nos. 3

(Jiwani), 4 (Pasni) and 5 (Ormara)), and then by Marathon Petroleum geologists (1974-1976 Sample Location Map), has done much to elucidate the structure of the region. It is now possible to accurately relate the distribution of Neogene and Quaternary molluscs to the stratigraphy of the Makran Coast for the first time.

Vredenburg's (1925, 1928) monographs on Tertiary molluscs from north-west India are the standard taxonomic reference works for the area. Nevertheless, it should be emphasized that both they, and a number of the papers on Neogene faunas cited above, are in urgent need of revision. Many of the generic names used in the older papers have now been superseded and there are numerous undoubted instances of synonymy. Part of the problem with the study of Neogene molluscan biostratigraphy in the Indo-Pacific province has been the lack of suitable reference sections in which to establish the full ranges of species and their relationships to those of other fossil groups. Thus, the Makran Coast is of particular importance in that its very thick succession with abundant faunas would seem to offer an ideal opportunity for establishing a stratotype for the north-west Indian Ocean Neogene. With this eventual aim in mind, an attempt has been made to present a molluscan biostratigraphy for the region using a series of partially overlapping sections. However, it is important to stress at the outset that the degree of overlap between these sections, and hence the degree of equivalence between the various stratigraphic units, is in some places uncertain. It will become apparent from the descriptions given below that the region is characterized by rapid facies transitions in both lateral and vertical directions; time-stratigraphic boundaries rarely parallel lithologic ones. More comprehensive data on the geology of the Makran Coast are given in a companion paper in this volume by Harms, Cappel and Francis (this vol.). The sections described here can be related to a detailed geological map and section given in this work, and the Colombo Plan lithostratigraphic units to a schematic cross-section of sedimentary facies through time (Harms et al. this vol; figs. 2,3 and 4). Biostratigraphic boundaries in these figures are based on calcareous nannoplankton and differ slightly from those adopted here.

A full species list for each of the sections studied is contained in the Appendix to this paper and some of the key types are illustrated in Plate 1.

Full details of the precise stratigraphic horizons at which specimens were collected are given in Crame (1976) and all the material is now housed in the Department of Palaeontology, British Museum (Natural History).

PALEOENVIRONMENTAL INTERPRETATIONS

As the study progressed, a model of the distribution of molluscan species in the Neogene and Quaternary of the Makran coastal region was gradually built up. This was essentially based on the continental shelf region and the assemblages that were associated with each of the principal sediment types that accumulated across the shelf (Fig. 2). In the construction of the model it was assumed that the shelf was a comparatively broad feature that sloped gently down to depths of around 200 m at its outer edge. Beyond this there was probably a fairly abrupt decline over the continental slope to abyssal depths. For interpreting water depths and relative positions on the shelf or slope, comparisons were made with Recent level bottom communities. Useful comparative data is now available for temperate regions (e.g. Jones, 1950; Thorson, 1957; Sanders, 1958, 1960; Buchanan, 1963; Parker, 1975, 1976), the subtropics and tropics (e.g. Buchanan, 1958; Longhurst, 1958; Parker, 1964a, b; Stephenson et al., 1970; Wade, 1972). Additional data on molluscan bathymetric ranges were obtained from sources such as Babin and Glemarec, 1971; Kira, 1962; Habe, 1964; McAlester and Rhoads, 1967 and Prashad, 1932, as well as specimen labels in the Townsend and Biggs collections.

Comparatively coarse grained sediments predominate along the inner edge of the shelf (Fig. 2). These are for the most part medium to coarse sands, but there are also some patches of pebble and cobble deposition, and even some boulders. Fairly rapidly, however, these coarse sediments are replaced in a seaward direction by much finer grained sands and silts. The greater part of the shelf would appear to have been dominated by fine sands and silts for long periods of time. Some of the sedimentary rocks can also be classified as mudstones and it is likely that towards the outer shelf edge and on the continental slope there were extensive areas of mud deposition (Fig. 2). Two very general rules govern the composition of the molluscan assemblages. The first of these is that on fine grained substrates (i.e. silts and muds)

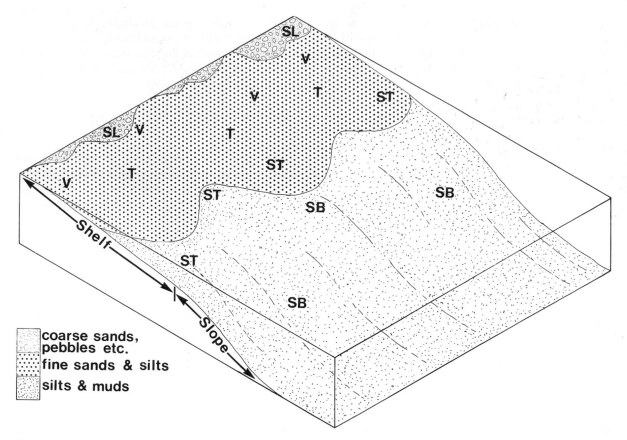

Fig. 2. A model of the distribution of mulluscan assemblages in the Makran coastal region during the Neogene and Quaternary. Key to assemblage types: SL–Shoreline, V–Venerid, T–Turritella, ST–Small Tellin, SB–Sparse Bivalve.

bivalves predominate over gastropods, and the second is that in deep water small, thin-shelled forms take precedence over large ones. In broad terms, the innermost assemblages are characterized by both large bivalves and gastropods, whilst the outermost are mainly composed of small bivalves.

Shoreline assemblages (Fig. 2) typically display patterns of low species diversity but high individual abundances. Large species of *Anadara* and venerids such as *Dosinia* are common, but more diagnostic are certain species of *Chlamys* (such as *C. alexandri* and *C. prototranquibaricus)* and an abundance of oysters. Particularly shallow shoreline assemblages may contain either the surf-living gastropod *Bullia* or the limpet *Cellana*. Venerid assemblages (Fig. 2) are associated with sandy substrates of the inner shelf and water depths that are believed to have been in the 10-50 m range. Large, shallow-burrowing venerids such as *Dosinia, Paphia, Hemitapes, Amiantis* and *Clementia* are prominent, and in certain assemblages families such as the Mactridae,

Tellinidae, Arcidae and Mytilidae are also well represented. With the frequent occurrence of predatory gastropods (especially species of *Natica, Tonna, Purpura, Babylonia* and *Conus),* both species diversity and abundance are usually high in venerid assemblages.

A very distinctive type of assemblage is that dominated by the slender, high-spired gastropod, *Turritella* (Fig. 2 and Pl. 1, Fig. a). There is now considerable evidence to suggest that *Turritella* occurs consistently throughout temperate and tropical regions in water depths between about 10 and 60 m and on muddy sands or silty substrates (e.g. Jones, 1950; Thorson, 1957; Buchanan, 1958, 1963). It seems to preferentially select patches of stiff, sandy silt into which it partly buries the tapering shell before commencing ciliary feeding (Jones, 1950, p. 307; Buchanan, 1958, p. 37). Shell density is always very high in both Recent and fossil *Turritella* communities, even sometimes to the point where almost all other molluscs are excluded. However, the predatory gastropod *Natica* is nearly always present in considerable numbers, and in the Makran assemblages there are also records of *Cymatium* and *Conus*. The small numbers of bivalves present belong mainly to the Veneridae and Cardiidae.

The finer-grained sediments of the more distal parts of the shelf are dominated by Small Tellin assemblages (Fig. 2). Within these, small tellinids and arcids referable to the genus *Anadara* are normally abundant, followed by less frequent occurrences of protobranchiate genera such as *Nucula* and *Nuculana*. The latter, together with some of the tellinids, reflect the more frequent occurrence of deposit-feeding strategies on very fine grained substrates. Shallow-burrowing, suspension-feeding bivalves are represented by a few mactrids (especially *Mactrinula)* and venerids, and there are also some small naticid and turrid gastropods. The Small Tellin assemblages give way to Sparse-Bivalve ones on the outer edge of the shelf and continental slope (Fig. 2). Here, there is a marked reduction in both species diversity and abundance, and those forms that do occur are nearly all very small (with a shell length of less than 1 cm) and thin-shelled. Small Tellins are the commonest types, followed by *Nucula*, *Pitar* (a venerid), *Cuspidaria* (Poromyacea) and some probable representatives of the Verticordiidae. Tiny gastropods include the Trochacean, *Cyclostrema,* adeorbids (Rissoacea) and turrids.

This model is not meant to be a static one, for it is obvious, from the sedimentary evidence, that the shoreline and shelf-slope boundaries will have continually shifted through time. It should also be emphasized that the postulated assemblages may not all have existed together at any one locality or instant in time. The model is more an average of the sedimentary facies and faunas of the Makran coastal region through the Neogene and Quaternary. Within the Recent Arabian Sea molluscan fauna, the Tellinacea is the most widespread and distinctive superfamily, followed by the Veneracea (Melvill and Standen, 1901, 1907).

DESCRIPTION OF THE SECTIONS

Pasni Section

This section, which was the stratigraphically lowest to be studied, is located in the centre of the concession area within the Pasni anticlinorium (Fig. 1). It was run alongside Marathon section No. DRC 3479 (1974-76 Sample Location Map)

from the northern flank of the Zibr anticline northwards onto the southern side of the extensive Kulanch syncline. The base is believed to be only a short distance above the top of the Panjgur Sandstone and the top is close to the base of the Talar Sandstone. Thus, it is likely that the whole section is contained within the Colombo Plan mapping unit known as the Parkini Mudstone. The predominant lithology is a soft, massive, grey-green mudstone with a distinctive pale grey weathering hue. There are occasional thin siltstone intercalations and very occasional sandstones up to 1m thick. Although the total thickness of the Parkini Mudstone in this region was originally thought to be approximately 1200m (Hunting Survey Corp., 1960), measurements by Marathon geologists indicate a figure of at least 3200m, and possibly as much as 3650m.

Only a very meagre molluscan fauna was obtained from the Pasni section. Some of the thirty-five stations sampled yielded no fossils at all and at others there were just isolated occurrences of bivalves such as small Tellins, a venerid close to *Pitar*, or a member of the Nuculacea. The three most distinctive fossil horizons were at the 30, 115 and 2400m levels (Fig. 3). At the first of these a thin band of large venerid bivalves referable to the

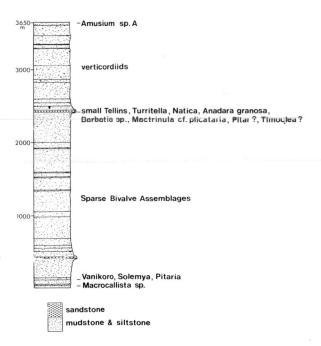

Fig. 3. The Pasni Section. The figures on the left give approximate heights (in metres) in the section; molluscan data is given on the right.

genus *Macrocallista* occurs. The presence of this band in an otherwise barren mudstone is anomalous and may, perhaps, be indicative of a brief phase of shallowing. The same interpretation can be suggested for a fossiliferous assemblage from 115m that includes the limpet-like gastropod *Vanikoro* (Hipponicacea), the protobranch bivalve *Solemya* and a large species of the venerid genus *Pitaria*. At the 2400m level (Fig. 3), a conspicuous fossiliferous horizon comprises a number of small Tellins, *Turritella*, *Natica*, *Anadara granosa*, *Barbatia* sp., *Mactrinula* cf. *plicataria* and probable specimens of *Pitar* and *Timoclea*. Its presence, together with the close proximity of one of the thick sandstone bands, may be attributable to shallowing caused by easterly extension of the sandy Sawar facies. It should be emphasized, however, that the very thick sequences of sparsely fossiliferous mudstone almost certainly indicate the persistence of deep water conditions for very long periods of time. That these conditions still prevailed at the end of the time period represented by this section is indicated by the presence of probable verticordiid bivalves between 2400 and 3650m and *Amusium* sp. A at the 3650m level (Fig. 3). At the present day, verticordiids typically inhabit abyssal muds and *Amusium* sp. A is close to a group of very deep water species of *Amusium* from the Persian Gulf and Gulf of Oman (Melvill and Standen, 1907).

It is concluded that the mudstones of the Pasni section were predominantly deposited under continental slope conditions, punctuated by brief intervals of shallowing. Unfortunately, there are no age diagnostic elements in the fauna. All that can be said is that the genera present, and in particular *Amusium*, *Mactrinula*, *Anadara* and *Timoclea?*, strongly suggest a Neogene age. Micropaleontological dating by both Marathon and Colombo Plan geologists indicates a Middle to Late Miocene age.

Garr Koh-Chatti Section

This long section at the western end of the concession area (Fig. 1) is believed to be an approximate stratigraphic continuation of the Pasni section. Its base is on the southern limb of the Garr Koh anticline at the level where the Parkini Mudstone lithologies are thought to pass conformably up into Talar Sandstone ones. From here, it proceeds roughly due south through Talar lithologies into the overlying Chatti Mudstone

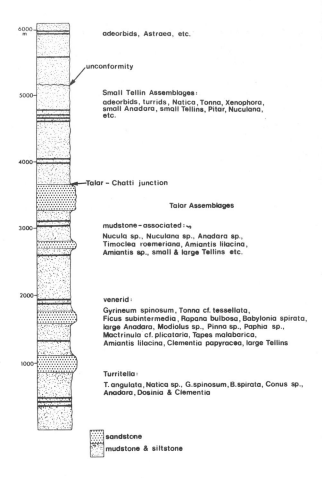

Fig. 4. The Garr Koh-Chatti Section. The figures on the left give approximate heights (in metres) in the section; molluscan data is given on the right.

(Fig. 1 and 1974-76 Sample Location Map). In total, the section comprises approximately 3650m of Talar Sandstone and 2500m of Chatti Mudstone.

The Talar Sandstone is predominantly composed of steeply dipping, grey silty mudstones with thin intercalated grey to brown siltstone and fine sandstone bands. The latter, which are generally 5-20cm in thickness, typically form the caps to small cuesta features and a 'scarp and dipslope' topography predominates throughout the unit. Prominent sandstone beds are first seen at about the 1000m level and then again at 1675m and 2750m; however, they are best developed at the top of the Talar, between 3275m and 3650m (Fig. 4). The sandstones, which are usually medium to coarse grained, massive, and typically weathered to a deep brown colour, often reveal ball and pillow structures and small slump folds. The regular occurrence of these sandstones, especially within the topmost 375m, has been taken as evidence of cyclic sedimentation.

Three main types of molluscan assemblage can be identified in the Talar Sandstone. The first of these is a mudstone-associated one that is typically dominated by bivalves such as *Nucula* sp., *Nuculana* sp., *Anadara* sp., *Timoclea roemeriana*, *Amiantis lilacina*, *Amiantis* sp., and small and large Tellins. Other important genera are; *Mactrinula, Dosinia, Pitar, Modiolus, Hemitapes, Tapes* and *Solecurtus*. Tiny scaphopods are often present and there are gastropods belonging to the genera *Turritella, Natica* and *Conus,* together with some small turrids. Venerid assemblages, which are associated with more sandy substrates, typically have a higher specific diversity. This is principally due to the presence of many more types of gastropod. Common species are; *Gyrineum spinosum, Tonna* cf. *tessellata, Ficus sub-intermedia, Rapana bulbosa* and *Babylonia spirata,* and these are usually joined by several species from each of the genera, *Turritella, Natica, Tibia, Melongena* and *Conus*. The principal bivalves are: *Anadara* (usually several large species), *Modiolus, Pinna, Paphia, Tapes, Amiantis* (especially *A. lilacina), Clementia, Mactrinula* and a variety of large Tellins. Less frequent, but nonetheless characteristic, genera are *Cardium* and *Chlamys*. The third assemblage type, that dominated by *Turritella,* is strongly associated with the tops of thin siltstone bands, where dense accumulations frequently form shell beds several centimetres thick. Where these accumulations are less dense, there are usually occurrences of gastropods such as *Natica* sp., *Gyrineum spinosum, Babylonia spirata* and *Conus* sp., and bivalves belonging to the genera *Anadara, Dosinia* and *Clementia*.

The three types of assemblage in the Talar Sandstone alternate in a regular fashion, and, as might be expected, frequently intergrade. The venerid and *Turritella* assemblages are thought to have developed on the inner shelf in the 10-60m depth range, with the mudstone-associated ones representing slightly deeper, outer shelf environments. The presence of *Cyclostrema* (Trochacean gastropod), tiny turrids and bivalves such as *Cuspidaria* and *Corbula* in some of the sparser assemblages may even indicate the occasional existence of outer slope conditions (Fig. 2).

The junction between the Talar Sandstone and Chatti Mudstone is marked by a pronounced change in lithologies above the 3650m level (Fig. 4). The thick sandstones that characterize the upper levels of the Talar are abruptly eliminated and superseded by thick monotonous sequences of pale grey-green mudstone that are punctuated only by thin siltstone interbeds. Although this is a sharp lithological change, there is faunal evidence to suggest that it may still be a genuine transition. Many of the large molluscs from the top of the Talar persist into the base of the Chatti and are then gradually eliminated up the section. Typical Chatti Mudstone assemblages are low diversity and abundance ones in which small forms predominate. Small species of *Anadara (A. granosa, A. ferruginea* and *A. burnesi?* See Pl. 1, fig. k) are particularly common and there are many small Tellins and venerids (most of the latter being referable to the genus *Pitar).* Other bivalves present are: *Nuculana, Raeta, Paphia, Dosinia* and *Corbula,* whilst gastropods include adeorbids, turrids, *Natica, Tonna* and *Xenophora.* These Small Tellin assemblages strongly indicate outer shelf conditions for the 3650m-5160m interval. At about the latter level there is evidence of a slight angular unconformity in the section (Fig. 4), but, as the lithologies either side of it are essentially similar (they are both uniform grey mudstones), its precise significance is not immediately apparent. Directly beneath the break, a sparse molluscan fauna and a few isolated occurrences of the solitary coral, *Caryophyllia,* suggest a deep water, outer shelf environment. If anything, the very sparse assemblages containing adeorbids and the spinose Trochacean, *Astraea,* from above the unconformity indicate even deeper water. The suggestion has been made that this stratigraphic break may be due to sedimentation over an original topographic feature such as a deep water fan.

Ormara and Jiwani Formations

It is possible to extend the Garr Koh-Chatti section to higher levels by moving its line approximately 15km to the west. Here, the topmost (6100m) level can be correlated with a position approximately midway along Marathon transect no. DRC 3500 (Fig. 1 and 1974-76 Sample Location Map). This transect, which runs roughly parallel to the previous one, extends south to a point approximately 11km west of the village of Pishukan.

The first few hundred metres of this section reveal pale grey mudstone and siltstone lithologies

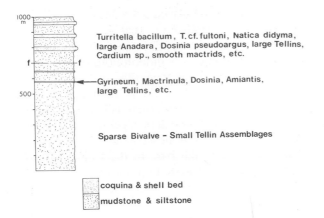

Fig. 5. The Pishukan Section. The figures on the left give approximate heights (in metres) in the section; molluscan data is given on the right. Fault marked by letter f.

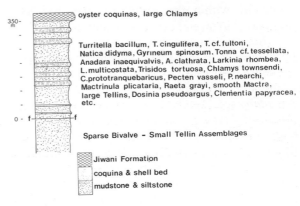

Fig. 6. A composite section for Jaibl-l-Maidhi and Gwadar Headland. The figures on the left give approximate heights (in metres) in the section molluscan data is given on the right. Fault marked by letter f.

identical to those in the top of the Garr Koh-Chatti section, as well as sparse molluscan faunas indicating an outer shelf/slope environment. The first significant change occurs between 575 and 600m (Fig. 5) where many loose, large fossils were found. These belong to types such as *Mactrinula, Dosinia, Amiantis,* large Tellins and *Gyrineum,* and can be traced to several thin (10-20cm) shell seams within the monotonous mudstone units. The seams become progressively more prominent and by the 800m level have reached a thickness of 1m (Fig. 5). They occur somewhat abruptly within the mudstone units and in places are so tightly packed with shells and fragments that they may be termed coquinas. A typical fauna from one of these coquinas would be: *Turritella bacillum, T.* cf. *fultoni, Natica didyma,* large species of *Anadara, Dosinia pseudoargus* (Pl. 1, fig. f), large Tellins, *Cardium* sp. and large, smooth mactrids. These are all typical elements of the Colombo Plan mapping unit known as the Ormara Formation (Hunting Survey Corp., 1960, p. 223).

The foregoing evidence suggests that there may have been a smooth transition from the Chatti Mudstone into the Ormara Formation. The shell beds gradually become thicker and more frequent and most likely indicate the progressive imposition of shallower water conditions. Nevertheless, it must be borne in mind that a major fault probably passes through the section at about the 650-750m level (Fig. 5), and thus the succession may be foreshortened. It would appear though, that the junction between the Chatti and Ormara is either a transition or a fault and not an unconformity as marked on the Colombo Plan map (Map no. 4,

grid ref. K4). This feature is the one described at the 5160m level in the Garr Koh-Chatti section and does not represent the true Chatti-Ormara boundary.

The Ormara Formation is better represented in coastal promontories at Gwadar and Jaibl-I-Maidhi (Fig. 1). At the latter of these localities two short sections were run southwards from points approximately 1km north of the Jaibl-I-Maidhi massif to the coast. Both these commence in typical upper Chatti Mudstone lithologies and have sparse molluscan faunas characterized by adeorbid gastropods, small Tellins, *Anadara, Arcopsis, Amusium, Dosinia, Pitar, Mactrinula* and *Corbula.* At a stratigraphic level corresponding to the northern edge of Jaibl-I-Maidhi, a transition in lithologies and faunas is encountered similar to that just described in the Pishukan section (Fig. 6). Prominent shell beds come into the sequence and these are soon followed by coquinas up to 1m in thickness. Typical exposures at high levels in both sections reveal 5-10m sequences of uniform pale grey-green mudstones and siltstones, separated by distinct white to brown shell beds and coquinas. The faunas, which readily indicate an inner shelf environment, are dominated by the following species: *Turritella bacillum, T. cingulifera, T.* cf. *fultoni, Natica didyma, Gyrineum spinosum, Tonna* cf. *tessellata, Anadara inaequivalvis, A. clathrata, Larkinia multicostata, L. rhombea, Trisidos tortuosa, Chlamys prototranquebaricus, C. townsendi, Pecten vasseli* (Pl. 1, figs. b and d), *P. nearchi* (Pl.1, figs. i and j), *Mactrinula plicataria, Raeta grayi,* large smooth Mactra, large Tellins, large *Cardium, Dosinia pseudoargus, Clementia papyracea, Laternula* sp. and small oysters.

Again, it is not possible to conclude that the Chatti Mudstone grades into the Ormara, for a major fault runs along the northern edge of Jaibl-I-Maidhi (Fig. 6). Indeed, it is probably the same fault as the one running east-west through the line of section DRC 3500 (Pishukan Section) at approximately the 650-750m level (the Jaibl-I-Maidhi fault on 1974-76 Sample Location Map). No accurate measurements of sedimentary thicknesses were made along these sections, but it is estimated that the Ormara Formation at Jaibl-I-Maidhi has a thickness of 350m.

The transition between the Ormara and the overlying Jiwani Formation is best seen in Gwadar headland (Fig. 1). Exposures on the northern flanks of this feature reveal thick (1-2m) shell and coquina beds set in massive mudstone units up to 15-20m thick. As these lithologies are traced upwards they become noticeably more sandy before giving way eventually to a sequence of fine sands, pebbly sandstones, coquinas and conglomerates that constitute the Jiwani Formation (Fig. 6). These sediments, which have an estimated maximum thickness of 30m, form a distinctive and resistant brown cap on top of the paler grey-brown Ormara lithologies. They yield abundant lithological and paleontological evidence of very shallow water deposition and a littoral environment is strongly indicated. Some of the coquinas are almost solid accumulations of oyster shells and there are numerous whole shells and fragments of a large species of *Chlamys*. Right on the summit of the headland, at an estimated height of 145m, a narrow coral reef tract is preserved on top of a sequence of pebbly sandstones and conglomerates (Fig. 6). This reef, which is essentially a shallow-water one, has a width in the 100-150m range and thickness of 1-2m. In places a complete lateral transition can be demonstrated from cross-bedded beach sands, through an inner *Acropora*-dominated coral assemblage to an outer *Porites*-dominated one. Typical shallow-water molluscs from the reef are; *Turbo intercostalis, Cerithium obeliscus, Cypraea arabica, Conus flavidus, Barbatia* sp., *Chlamys townsendi, Tridacna* sp., *Codakia* sp., *Periglypta puerpera* and *Gafrarium dispar*.

A number of species in the Talar Sandstone point to a Pliocene age. *Turritella angulata* (Pl. 1, fig. a) has been recorded from the Upper Miocene and Pliocene of Indonesia and Burma (Noetling,

1901; van der Vlerk, 1931; Shuto, 1974), as well as the Pliocene of Bangladesh (Vredenburg, 1921) and Iran (Cox, 1936). *Tonna* cf. *zonata* occurs in strata of probable Pliocene age in Malaya and Borneo (Wissima, 1947; Cox, 1948) and *T. luteostoma* (Pl. 1, fig. e) is present in the Pliocene of East Africa and Taiwan (e.g., Nuttall and Sealy, 1961). There are records of *Melongena gigas* and *Anadara tambacana* (see Pl. 1, figs. c and l) from the Upper Miocene and Pliocene of Java and Sumatra and *Babylonia spirata* from the Pliocene of the same region and southern India. Other species suggestive of a Pliocene age are: *Conus brevis? C. cosmetulus, Chlamys* cf. *tjaringinensis* (Pl.1, figs. g and h), *Dosinia* cf. *peralta* and *D. subpenicillata* (comparative data from Cossmann, 1900, 1903 and 1911; Vredenburg, 1928; van der Vlerk, 1931; Oostingh, 1938-40; Nuttall, 1965). There are fewer age-diagnostic species from the overlying Chatti Mudstone, but it may yet be shown that one or more of the small species of Arcacean bivalves which characterize this unit (such as *Anadara burnesi?* (Pl. 1, fig. k) and *Arcopsis bataviana*) are restricted to the Pliocene (comparative data from Noetling, 1901; Tesch, 1915; van der Vlerk, 1931; Nuttall, 1965).

It is suggested that a useful criterion for defining the base of the Pleistocene may be the incoming of the Ormara Formation, which obviously reflects a regional shallowing. Although this formation contains a high proportion of extant species, there are also a number of extinct forms within it (such as *Tonna* cf. *zonata, Rapana bulbosa, Anadara* cf. *ferruginea, Chlamys prototranquebaricus, Pecten vasseli, P. nearchi* and *Dosinia pseudoargus* (see Pl. 1, figs. b, d, f, i and j) which could be taken as Late Pliocene indicators. Nevertheless, it is becoming increasingly apparent that some at least of these Late Pliocene forms may also range into the Early Pleistocene; *Pecten vasseli* (Pl. 1, figs. b and d) is a particularly good case in point. It can be traced along much of the East African coast in beds that immediately precede the extensive Pleistocene reefs, and then further north through the Red Sea region to Iran and Pakistan (Cox, 1927, 1930, 1936; Eames and Cox, 1956; Nuttall and Sealy, 1961; Kent et al., 1971). Its age has consistently been given as Upper Pliocene, but, as it is now apparent that the East African reefs are largely Late Pleistocene in age, it could equally well be at least partly Early Pleistocene in that region. The thin reef tract on top of the Jiwani coquinas at Gwadar headland is most likely Late Pleistocene in age. Reefs do not occur at the present day in the northern Arabian Sea, and its presence would seem to be the product of a marked expansion in the range of Indian Ocean coral reefs. Such an expansion last occurred during the last Interglacial period (i.e. at approximately the base of the Upper Pleistocene; e.g. Hopley, 1982, p. 160).

In this study the Miocene-Pliocene boundary has been placed at the base of the Talar Sandstone, where there is an abrupt change in both lithologies and faunas. The Pliocene-Pleistocene boundary is placed at the Chatti-Ormara junction, and it is believed that the bulk of the latter formation is contained within the Lower and Middle Pleistocene. The littoral Jiwani Formation is assigned to the Upper Pleistocene.

Hingol Section

The section studied in the eastern part of the concession area is exposed along the lower reaches of the Hingol River. It runs in a south-easterly direction from the core of the Garr-Koh anticline (which is largely obscured by extrusive mud) across a broad, abandoned river valley and then through the towering Jabal Haro ridge (Fig. 1; Section DRC 3515 on 1974-76 Sample Location Map). Sedimentary and faunal evidence suggest that this section records a gradual transition from very deep water facies to shoreline ones.

Although the line of the continuous section begins close to the south-western end of the anticline, some stratigraphically lower samples were obtained from a reconnaissance section approximately 17 km to the north east (in a tributary of the Sham Kaur River; Marathon sample no. DRC 3518). The lowest of these were from a series of uniform pale grey mudstones that are exposed in vertical beds close to the anticlinal axis. These lower mudstones, which are estimated to be 500m thick, are massive and unfossiliferous, and have their greatest resemblance with the Parkini Mudstone. As they are traced south-eastwards into the Sham Kaur valley they seem to grade up into a unit composed of thin alternating beds of grey silty mudstone and brown fine to coarse sandstones. This is the lithology exposed at the base of the continuous section (Fig. 7).

As the beds at the base of DRC 3515 are still vertical, the thin resistant sandstones form a series of narrow prominent ridges and the thicker mudstones low depressions. This topography

alters to one of a series of low cuestas as the beds are traced across the abandoned valley floor and the dip gradually decreases. The sandstones, which are up to 3m thick, display features such as ripple cross-lamination, hummocky cross-stratification and seams of mudstone pebbles. In addition, some of them are conspicuously burrowed and contain thin layers of shell debris. The higher-energy, shallower-water conditions suggested by these features are also reflected in abundant and diverse molluscan faunas. A typical early mudstone assemblage contains species such as; *Anadara antiquata, A. ferruginea, A. burnesi?* (Pl. 1, fig. k), *Modiolus* sp., *Mactrinula plicataria, Dosinia* cf. *peralta, Paphia* sp., *Pitar* cf. *belcheri* and various small Tellins. A typical early sandstone, on the other hand, yields types such as; *Turritella* cf. *fultoni, Babylonia* sp., *Nassarius persicus, Terebra* sp. A, *Chlamys* cf. *tjaringinensis* (Pl. 1, figs. g and h), *C.* cf. *prototranquebaricus, C.* cf. *singaporinus, Cardium* sp., *Timoclea roemeriana, T.* cf. *cochinensis* and *Martesia* sp. Further species which come in higher up the section are; *Tibia* sp., *Tonna dolium, Gyrineum spinosum, Natica* cf. *didyma, Rapana bulbosa, Conus* sp., *Anadara granosa, Chlamys* cf. *mekranica, Pinna* sp., *Dosinia* cf. *peralta, Paphia* cf. *textile, Amiantis erycina* and *Hemitapes gallus*. Overall, there is a strong resemblance in both lithologies and faunas between this mudstone-sandstone sequence and the Talar Sandstone. In particular, the following molluscs appear to be restricted to these two units: *Turritella* sp. A, *Tibia* sp., *Anadara burnesi?, A.* cf. *tambacana* (Pl. 1,

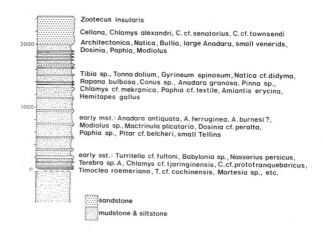

Fig. 7. The Hingol Section. The figures on the left give approximate heights (in metres) in the section; molluscan data is given on the right.

fig. 1), *Chlamys* cf. *tjaringinensis, Dosinia peralta, D. subpenicillata?. Timoclea cochinensis* and *T. roemeriana.*

A pronounced change occurs in the section at the 1980m level (Fig. 7). This horizon corresponds to the base of the Jabal Haro scarp where thick sandy-shelly beds abruptly replace the mudstone-sandstone ridges. This junction was mapped by Colombo Plan geologists as an unconformity (Map No. 5, grid refs. G7 and G11) and there does indeed seem to be a slight angular discordance here. Nevertheless, several lines of evidence suggest that this break is not a major one. Firstly, a steady increase in the density of sandstone beds between the 1220 and 1980m levels (Fig. 7) suggests a gradual transition to sandy facies. Oysters appear in the upper levels of the mudstone-sandstone sequence and at 1960m there is a coquina very similar in composition to those that appear above the unconformity.

The lowest lithologies in the Jabal Haro ridge are thickly bedded (1-2m) sandstones, shelly sandstones and coquinas. These have distinct greeny-brown weathering hues and are usually arranged in regular alternations of plain sandstone and shell-rich layers. Whereas the sandstones are in places strongly cross-bedded and bear seams of small pebbles, the shell layers are often strongly burrowed. Essentially similar features can be traced into higher levels and it is clear that the environment oscillated between very near shore sands and more off-shore shelly beds. The very regular development of trough crossbedded units between horizontally laminated sands around the 2195m level indicates the development of prograding shoreline cycles with a thickness of 10-15m. These continue up to the highest levels of the section where there is good evidence of fluvial channelling (Fig. 7).

The molluscan assemblages from the Jabal Haro sandstones are typically low diversity, high abundance ones indicative of nearshore environments. Common sand-associated gastropod genera are; *Architectonica, Natica* (a small species of which is extremely abundant at some levels) and *Bullia,* which is predominantly a surf-living form. Amongst the bivalves, large species of *Anadara* are especially prominent, and in places form monospecific shell bands. However, even more numerous are certain shallow-burrowing venerids, including an indeterminate small smooth form and two species of *Timoclea.* There are also large venerids such as *Dosinia* and *Paphia* and in

places the mytilid *Modiolus* is common. Most of the *Chlamys* specimens have been fragmented, but it is possible to identify at least three species; *C. alexandri, C.* cf. *senatorius* and *C.* cf. *townsendi.* The limpet *Cellana,* which is present throughout the section, is particualrly indicative of very shallow water conditions, as are the bands of both large and small oysters. Specimens of a terrestrial snail, *Zootecus insularis,* from the highest levels confirm the presence of fluvial channels, although it should be emphasized that they do occur in close proximity to marine fossils.

It is concluded that the upper 457m of nearshore to beach sands in this section (Fig. 7) are broadly equivalent to the Ormara and Jiwani Formations. Species apparently restricted to this unit and the latter two formations include: *Natica didyma, Anadara inaequivalvis, Pecten nearchi* (Pl. 1, figs i and j) *Chlamys* cf. *singaporinus* and *C.* cf. *townsendi.* If this correlation is correct, it is evident that Talar Sandstone lithologies must have passed directly up into the Ormara without any intervening Chatti Mudstone stage. The latter could, of course, have been removed by erosion before the deposition of the Ormara, but it is thought unlikely that the unconformity at the base of the Jabal Haro ridge represents such a long time period. Because of the difficulties of correlating the lithologies in this region with those observed further west, Colombo Plan geologists collectively assigned them to one general group, the Hinglaj Group (Hunting Survey Corp., 1960 and Map no. 5).

The following species from the mudstone-siltstone ridge sequence can be taken as probable Pliocene indicators: *Babylonia spirata, Anadara* cf. *tambacana, A. burnesi?, Chlamys mekranica, C. prototranquebaricus, C.* cf. *tjaringinensis, Dosinia peralta, D. subpenicillata?* and *D. pseudoargus.* There are also four more gastropods whose age-ranges support this conclusion. *Clavilithes* cf. *verbeeki* and *Nassarius* cf. *verbeeki,* by analogy with species ranges in Java, Sumatra and Borneo (e.g. van der Vlerk, 1931; Oostingh, 1938-40; Nuttall, 1965) suggest either an Upper Miocene or Pliocene age, whilst *Tonna losariense* is Pliocene in Java (van der Vlerk, 1931) and *Nassarius persicus* is Pliocene in Iran (Cox, 1936). The base of the Pliocene in the Hingol area is tentatively placed at the level in the core of the Garr-Koh anticline where massive unfossiliferous mudstones pass up into the fossiliferous mudstone-sandstone ridge sequence (Fig. 7). Its top is placed at the junction between the latter unit and the sandy facies of Jabal Haro (Fig. 7), which is again thought to reflect regional shallowing. The majority of the species from the upper sandstone unit are, as would be expected, extant, but it is noticeable that the faunas do contain some extinct species. These include *Chlamys alexandri, Pecten nearchi* and at least three species of *Bullia* (C.P. Nuttall, pers. comm. 1976). Within this unit the base of the Upper Pleistocene is provisionally placed at the 2195m level where there is good evidence of persistant littoral sedimentation (Fig. 7).

STRATIGRAPHIC SYNTHESIS AND SUMMARY

A thick sequence (approximately 3650m) of Parkini Mudstone in the central part of the concession area (Pasni section, Fig. 8) is interpreted as a very deep water deposit. There is some evidence of occasional brief periods of shallowing, but the predominant Sparse Bivalve assemblages indicate long periods of outer slope conditions. Although there are no age-diagnostic molluscs within this interval, the bivalve genera present are suggestive of a Neogene age. There

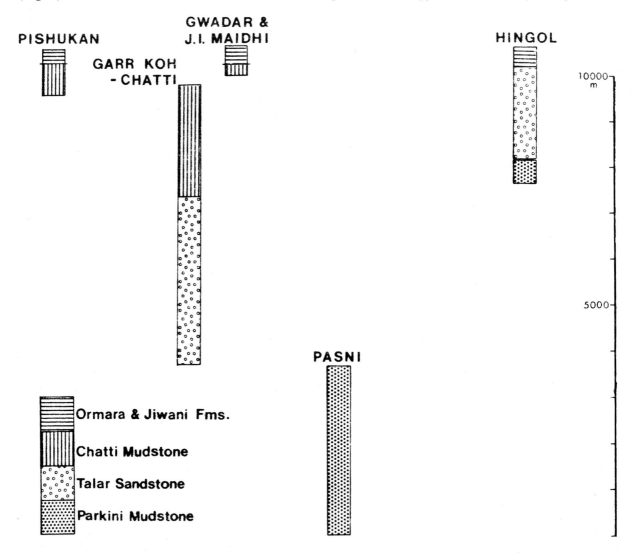

Fig. 8. A tentative stratigraphic correlation of the principal sections. It should be emphasized again here that, due to marked lateral facies changes, the precise degree of overlap of the sections is uncertain. The scheme has been constructed to show the minimum possible overlap and the maximum possible stratigraphic thickness; almost certainly, such a thickness of sediment does not exist at any one locality. This figure should be used in conjunction with the geological map and cross-sections presented by Harms et al. (this volume, figs. 2,3 and 4).

then appears to have been a fairly rapid transition from the Parkini Mudstone up into the Talar Sandstone. In the Garr Koh-Chatti transect an approximately 3650m thickness of the latter unit is present (Fig. 8) and it is apparent, from both the lithologies and faunas, that conditions fluctuated regularly between those of the inner and outer shelves. In addition there is some evidence, from mudstone faunas, of occasional deepening onto the outer slope. The three main types of molluscan assemblage present within the Talar stage are; mudstone-associated, venerid and *Turritella*. Between them, these have yielded probable Pliocene indicator species such as *Turritella angulata, Tonna* cf. *zonata, T. luteostoma, Babylonia spirata* and *Melongena gigas*.

There is some evidence, from the composition of the molluscan faunas, that the junction between the Talar Sandstone and the overlying Chatti Mudstone in the Garr Koh-Chatti section (Fig. 8) is a genuine transition. The latter unit has a thickness in the region of 2500m and is characterized by monotonous grey mudstone-siltstone lithologies and Small Tellin molluscan assemblages. Outer shelf environments are indicated by these molluscs but again there is some possibility of periodic overlap onto the outer slope. The slight angular unconformity present in the upper levels of the Chatti Mudstone is not thought to be of regional significance. Molluscan species from above and below it, and in particular certain types of *Anadara*, most likely indicate a Pliocene age.

By transferring from the Garr Koh-Chatti to Pishukan sections (Fig. 8), it is possible to trace the Chatti Mudstone up into the Ormara Formation. However, it should be re-emphasized that this transition could have been foreshortened by a major fault. It would appear that the Ormara stage is in the region of 350m thick in the Jaibl-I-Maidhi and Gwadar areas and consists of uniformly interbedded silty mudstones and coquinas. High abundance and diversity molluscan assemblages from it typically contain several species of each of the following genera: *Turritella, Natica, Gyrineum, Tonna, Anadara, Larkinia, Chlamys, Pecten, Mactrinula, Mactra, Raeta, Cardium, Dosinia* and *Laternula*. Small oysters are locally abundant and essentially shallow, inner shelf conditions are indicated. The range of extinct species within the Ormara Formation has some distinct Pliocene affinities, but there are also indications from some of the species present, such

as *Pecten vasseli*, that a truer age determination may be Lower Pleistocene. On Gwadar headland, the upper levels of the Ormara become increasingly sandy before grading up into the littoral Jiwani Formation. Estimated to be about 30m thick, this uppermost lithostratigraphic unit has been assigned to an Upper Pleistocene age.

In the eastern part of the concession area (Hingol section, Fig. 8) approximately 500m of Parkini Mudstone-type lithologies probably pass directly up into mudstone-siltstone-sandstone ones that are strongly reminiscent of the Talar Sandstone. Such a resemblance is strengthened by the restriction of certain species of *Turritella, Tibia, Anadara, Chlamys, Dosinia* and *Timoclea* to both these units, and the occurrence of further Pliocene indicator forms such as *Clavilithes* cf. *verbeeki, Nassarius* cf. *verbeeki, Tonna losariense?* and *Nassarius persicus*. Approximately 1950m of these Talar lithologies is succeeded in the Hingol section by 450m of sandstones, shelly sands and coquinas (Fig. 8). The junction between these two lithological divisions appears to be a slight angular unconformity, but it is thought unlikely that this represents an extensive time period. The upper sandy-shelly nearshore and shoreline deposits yield low diversity, high abundance assemblages rich in genera such as *Cellana, Architectonica, Natica, Bullia, Anadara, Modiolus, Chlamys, Dosinia, Paphia, Timoclea* and *Zootecus*. Correlation with the Ormara and Jiwani Formations further west is suggested by the occurrence of the following species: *Natica didyma, Anadara inaequivalvis, Pecten nearchi, Chlamys* cf. *singaporinus* and *C.* cf. *townsendi*. No precise division can be made here into the Ormara and Jiwani. Formations, but it is suggested that the base of the Upper Pleistocene could be placed at the 2195m level where there is good evidence of persistent littoral sedimentation.

The Miocene-Pliocene boundary is tentatively placed at the base of the Talar Sandstone (Fig. 8) where there is an abrupt change in both lithologies and faunas. The top of the Pliocene in the west is placed at the Chatti-Ormara junction, and in the east at the base of the upper sandy-shelly unit in the Hingol section (Fig. 8). It is thought that both these abrupt transitions could reflect the same regional shallowing event. If the positions of these boundaries and the east-west correlations are substantially correct, it is apparent that the sedimentary succession is considerably reduced in the eastern part of the concession area. In the

Hingol section the Talar Sandstone is some 1700m thinner than in the Garr Koh-Chatti section and the Chatti Mudstone is completely missing (Fig. 8). The most likely explanation for this is attenuation of the sedimentary basin the proximity of the Las Bela axial zone.

ACKNOWLEDGEMENTS

I would like to thank J.C. Harms, J. van Dillewijn, D. C. Francis, S. Siddiqui and S. Yusuf for their help in the field and hospitality in Pakistan. H. W. Ball kindly gave permission for me to work in the British Museum (Natural History) and C.P. Nuttall, N. J. Morris and J. D. Taylor generously provided facilities in the Departments of Palaeontology and Zoology; I am particularly grateful to the latter three for passing much useful information on to me. Finally, my thanks go to the British Antarctic Survey for assistance with the publication of this paper.

REFERENCES

Babin, C. and M. Glemarec, 1971. Ecologie et paléoécologie des bivalves marins des sediments meubles. Haliotis, 1: 105-125.

Biggs, H.E.J., 1973. The marine Mollusca of the Trucial Coast, Persian Gulf. Bull. Br. Mus. nat. Hist. (Zool.), 24: 343-421.

Blanford, W.T., 1872. Notes on the geological formations seen along the coasts of Biluchistan and Persia from Karachi to the head of the Persian Gulf, and on some of the Gulf islands. Rec. geol. Surv. India, 5:41-45.

Buchanan, J.B., 1958. The bottom fauna communities across the continental shelf of Accra, Ghana (Gold Coast). Proc. zool. Soc. Lond., 130: 1-56.

Buchanan, J.B., 1963. The bottom fauna communities and their sediment relationships off the coast of Northumberland. Oikos, 14: 154-175.

Cossmann, M., 1900. Faune Pliocene de Karikal. J. Conch. Paris, 48: 14-69.

Cossmann, M., 1903. Faune Pliocene de Karikal. J. Conch. Paris, 50: 105-173.

Cossmann, M., 1911. Faune Pliocene de Karikal. J. Conch. Paris, 58: 34-86.

Cox, L. R., 1927. Neogene and Quaternary Mollusca from the Zanzibar Protectorate. *In: Report on the Palaeontology of the Zanzibar Protectorate*, HMSO, London, pp. 13-102.

Cox, L.R., 1930. Miocene Mollusca; Pliocene Mollusca; post-Pliocene Mollusca. *In: Reports on the geological collections from the coastlands of Kenya colony.* Monogr. geol. Dept. Hunter. Mus., 4: 103-163.

Cox, L. R., 1936. Fossil Mollusca from southern Persia (Iran) and Bahrein Island. Mem. geol. Surv. India Palaeont. indica, 22: 1-69.

Cox, L. R., 1948. Neogene Mollusca from the Dent Peninsula, British North Borneo. Abh. schweiz. paläont. Ges., 66: 1-70.

Crame, J. A., 1976. Late Tertiary and Quaternary molluscan faunas from the Mekran Coast, Pakistan. Marathon Petroleum Pakistan Ltd., unpublished report, 43pp.

Davies, A. M., 1971. *Tertiary Faunas. Volume 1: The Composition of Tertiary Faunas.* Second Edition. (Revised by F. E. Eames). George Allen & Unwin, London, 571 pp.

Davies, A. M., 1975. *Tertiary Faunas. Volume 2: The Sequence of Tertiary Faunas.* Second Edition. (Revised by F. E. Eames with R.J.G. Savage). George Allen & Unwin, London, 447 pp.

Eames, F. E. and L. R. Cox, 1956. Some Tertiary Pectinacea from East Africa, Persia and the Mediterranean region. Proc. malac. Soc. Lond., 22: 1-68.

Habe, T., 1964. Shells of the Western Pacific in Color. Vol. II, Hoikusha,

Osaka, 233 pp.

Hopley, D., 1982. *The Geomorphology of the Great Barrier Reef: Quaternary Development of Coral Reefs.* John Wiley & Sons, New York, 453 pp.

Hunting Survey Corp., Ltd., 1960. Reconnaissance Geology of Part of West Pakistan. A Colombo Plan Co-Operative Project. Maracle Press, Oshawa, Ontario, 550 pp.

Jones, N. S., 1950. Marine bottom communities. Biol. Rev., 25: 283-313.

Kent, P.E., J.A. Hunt and D. W. Johnstone, 1971. The geology and geophysics coastal Tanzania. Inst. geol. Sci. lond., geophys. Pap., 6: 1-101.

Kira, T., 1962. *Shells of the Western Pacific in Color.* Hoikusha, Osaka, 224 pp.

Longhurst, A. R., 1958. An ecological survey of the west African marine benthos. Fishery Publs. colon. Off., London, 11: 1-103.

McAlester, A. L. and D. C. Rhoads, 1967. Bivalves as bathymetric indicators. Mar. Geol., 5: 383-388.

Martin, K., 1883-1887. Palaeontologische ergebnisse von Tiefbohrungen auf Java. Samml. geol. Reichmus. Leiden 1: 1-380.

Melvill, J.C. and R. Standen, 1901. The Mollusca of the Persian Gulf, Gulf of Oman and Arabian Sea, as evidenced mainly through the collections of Mr. F. W. Townsend, 1893-1900. Proc. zool. Soc. Lond., (1901): 327-460.

Melvill, J.C. and R. Standen, 1907. The Mollusca of the Persian Gulf, Gulf of Oman and Arabian Sea, as evidenced mainly through the collections of Mr. F. W. Townsend, 1893-1900. Part II. Pelecypoda. Proc. zool. Soc. Lond., (1906): 783-848.

Moore, R.C. (ed.), 1969. *Treatise on Invertebrate Paleontology Pt. N. Mollusca 6: Bivalvia* (2 Volumes). Lawrence, Kansas, The Geological Society of America Inc. and the University of Kansas, 952 pp.

Noetling, F., 1901. Fauna of the Miocene Beds of Burma. Mem. geol. Surv. India Palaeont. indica, 1: 1-378.

Nuttall, C.P., 1961. Mollusca from the Togopi Formation (Upper Caenozoic) of North Borneo. Ann. Rept. Brit. Borneo Geol. Surv. (1960): 83-96.

Nuttall, C.P., 1965. Report on the Haile collection of fossil Mollusca from the Plio-Pleistocene Togopi Formation, Dent Peninsula, Sabah, Malaysia. Mem. geol. Surv. Dep. Br. Terr. Borneo, 16: 155-192.

Nuttall, C.P. and D.L.F. Sealy, 1961. On the age of a molluscan fauna from the Neogene of the Tanga district, Tanganyika. Rec. geol. Surv. Tanganyika, 9: 61-75.

Oostingh, C.H., 1935. Die mollusken des Pliozäns von Boemiajoe (Java). Wet. Meded. Diest. Mijnb. Ned.-Oost-Indie, 26: 1-247.

Oostingh, C.H., 1938-40. Die mollusken des Pliozäns von Sud-Bantam in Java. De Ingenieur in Ned.-Indie, 5-7.

Pannekoek, A., 1936. *Beiträge zur Kenntnis der Altmiocänen Mollusken-fauna von Rembang (Java).* N.V. Noord-Hollandsche Vitgeversmaatschappij, Amsterdam, 80 pp.

Parker, R.H., 1964a. Zoogeography and ecology of some macro-invertebrates, particularly mollusks, in the Gulf of California and the continental slope off Mexico. Vidensk. Medd. Dansk. Nat. Foren. Bd., 126: 1-178.

Parker, R.H., 1964b. Zoogeography and ecology of macro-invertebrates of Gulf of California and continental slope off western Mexico. Am. Assoc. Pet. Geol., Mem., 3: 331-376.

Parker, R.H., 1975. *The Study of Benthic Communities. A Model and a Review.* Elsevier Oceanography Series, 9, Amsterdam, 279 pp.

Parker, R.H., 1976. Classification of communities based on geomorphology and energy levels in the ecosystem. *In:* R. W. Scott and R. R. West (eds.) *Structure and Classification of Paleocommunities,* Dowden, Hutchinson & Ross, Stroudsburg, Pennsylvania, pp. 67-86.

Prashad, B., 1932. The Lamellibranchia of the Siboga expedition. Systematic Part II, Pelecypoda. Siboga Exped., 53C: 1-353.

Sanders, H.L., 1958. Benthic studies in Buzzards Bay. I. Animal-sediment relationships. Limnol. Oceanogr., 3: 245-258.

Sanders, H.L., 1960. Benthic studies in Buzzards Bay. III. The structure of the soft-bottom community. Limnol. Oceanogr., 5: 138-153.

Shuto, T., 1969. Neogene gastropods from Panay Island, the Philippines. Mem. Fac. Sci. Kyushu Univ. Ser. D, Geol., 19: 1-250.

Shuto, T., 1971. Neogene bivalves from Panay Island, the Philippines. Mem. Fac. Sci. Kyushu Univ. Ser. D, Geol., 21: 1-73.

Shuto, T., 1974. Notes on Indonesian Tertiary and Quaternary gastropods mainly described by the late Professor K. Martin. I. Turritellidae and Mathildidae. Geol. Palaeont. SE Asia, 14: 135-160.

Shuto, T., 1975. Preliminary correlation of the Neogene molluscan faunas in southeast Asia. Geol. Palaeont. SE Asia, 15: 289-301.

Stephenson, W., W. T. Williams and G. N. Lance, 1970. The macrobenthos of Moreton Bay. Ecol. Monogr., 40: 459-494.

Tesch, P., 1915. Jungtertiäre und Quartäre mollusken von Timor. I Teil. Paläontologie von Timor, 5: 1-70.

Thorson, G., 1957. Bottom communities. *In:* J. Hedgpeth (ed.) *Treatise on Marine Ecology and Paleoecology, Vol. 1, Ecology.* Mem. geol. Soc. Am., 67: 461-534.

Vlerk, I.M. van der, 1931. Caenozoic Amphineura, Gastropoda, Lamellibranchiata, Scaphopoda. Leid. Geol. Meded., 5: 206-296.

Vredenburg, E.W., 1906. The classification of the Tertiary system in Sind with reference to the zonal distribution of the Eocene Echinoidea described by Duncan and Sladen. Rec. geol. Surv. India, 34: 172-198.

Vredenburg, E.W., 1921. Notes on marine fossils collected by Mr. Pinfold in the Garo Hills. Rec. geol. Surv. India. 51: 303-337.

Vredenburg, E.W., 1925. Description of Mollusca from the post-Eocene Tertiary formation of north-western India: Cephalopoda, Opisthobranchiata, Siphonostomata. Mem. geol. Surv. India. 50: 1-350.

Vredenburg, E.W., 1928. Descriptions of Mollusca from the post-Eocene Tertiary formation of north-western India: Gastropoda (in part) and Lamellibranchiata. Mem. geol. Surv. India, 50: 351-506.

Wada, B.A., 1972. A description of a highly diverse soft-bottom community in Kingston Harbour, Jamaica. Mar. Biol. Berlin, 13: 57-69.

Wissima, G.G., 1947. *Young Tertiary and Quaternary Gastropoda from the island of Nias (Malay Archipelago).* L.H. Becherer, Leiden, 212 pp.

APPENDIX

Molluscan species from each of the principal lithostratigraphic units.

Parkini Mudstone (Pasni Section):

Adeorbid gastropods, *Turritella* sp., *Vanikoro* cf. *cancellata (Lamarck)*, Natica sp., Rapana? sp., turrid gastropods, *Solemya* sp., *Nucula?* sp., *Nuculana* sp., *Anadara granosa* (Linnaeus), *A.* sp. A, *Barbatia* sp., *Amusium* sp. A, *Mactrinula* cf. *plicataria* (linnaeus), *M.* sp., small Tellins, *Dosinia* sp., *Pitaria* sp. (large), *Pitar* sp. (tiny), *Macrocallista* sp. A, *Timoclea?* sp., *Corbula* sp., verticordiid bivalves.

Talar Sandstone (Garr Koh-Chatti Section):

Cyclostrema sp., *Architectonica* cf. *perspectiva* (Linnaeus), *A.* sp., *Turritella angulata* Sowerby, *T. bacillum* Kiener?, *T. cingulifera* Sowerby, *T. columnaris* Kiener?, *T.* cf. *fultoni* Melvill, *T. illustris* Melvill?, *T.* sp. A, *T.* sp., *Cerithium kochi* Philippi?, *Tibia* sp., *Natica lineata* Lamarck, *N.* cf. *melanostoma* (Gmelin), *N.* cf. *ponsonbyi* Melvill, *N.* sp., *Gyrineum spinosum* (Lamarck), *Tonna* cf. *dolium* (Linnacus), *T. luteostoma* (Kuster), *T. tessellata* (Lamarck), *T.* cf. *zonata* (Green), *T.* sp., *Ficus subintermedia* (d'Orbigny), *F.* sp., *Rapana bulbosa* Solander, *R.* sp., *Phos?* sp., *Babylonia spirata* (Lamarck), *B.* sp., *Melongena* cf. *gigas* (Martin), *M.* sp. A, *Nassarius* sp., mitrid gastropod, *Conus brevis* Sowerby?, *C. cossmetulus* (Cossmann), *C.* cf. *eburneus* Hwass, *C. figulinus* (Linnaeus), *C.* cf. *malaccanus* (Bruguière), *C. radiatus* Gmelin, *C.* sp. B, *C.* sp. C, *C.* cf. turrid gstropods, *Nucula* sp., *Nuculana* sp., *Anadara antiquata* (Linnaeus), *A. burnesi* (d'Archiác & Haime)?, *A. clathrata* (Reeve), *A. ferruginea* (Reeve), *A.* cf. *inaequivalvis*

(Bruguière), *A.* cf. *tambacana* (Martin), *A.* cf. *uropigimelana* (Bory), *A.* sp., *Larkinia rhombea* (Born), *Arcopsis* sp., *Cucullaea labiata* (Solander), *C.* sp., *Barbatia helblingii* (Bruguière)?, *B.* sp., *Modiolus* sp., *Pinna* sp., *Chlamys* cf. *prototranquebaricus* Vredenburg, *C.* cf. *singaporinus* (Sowerby), *C.* cf. *tjaringinensis* Martin, *C.* sp. A, *C.* sp., *Plicatula* sp., *Crassostrea* sp., *Diplodonta?* sp., *Loripes?* sp., *Cardium* sp., *Hemicardium* sp., *Mactrinula* cf. *plicataria* (Linnacus), *M.* sp., *Mactra* sp. (smooth), *Raeta grayi* (H. Adams), *Solen* sp., small Tellins, large Tellins, *Zozia* sp., *Solecurtus* sp., *Gari* sp., *Dosinia peralta* Vredenburg, *D. radiata* (Reeve)?, *D. subpenicillata* Vredenburg?, *D.* sp. A, *D.* sp., B, *D.* sp., *Paphia* sp., *Hemitapes gallus* (Gmelin), *H. cor* (Sowerby)?, *H.* sp., *Tapes malabarica* (Chemnitz), *T.* sp., *Amiantis lilacina* (Lamarck), *A.* sp., *Callista florida* (Lamarck)?, *Meretrix meretrix* (Linnaeus), *Chione?* sp., *Clementia papyracea.* (Gray), *Sunetta* sp., *Sinodia excisa* (Roding)?, *Lioconcha?* sp., *Pitar* cf. *obliquata* (Hanley), *P.* cf. *belcheri* (Sowerby), *P.* sp., *Timoclea roemeriana* (Issel), *T.* sp., *Corbula* sp., *Laternula* sp., *Lyonsia* sp., *Cuspidaria* sp.

Talar Sandstone (Hingol Section):

Cyclostrema sp., *Architectonica perspectiva* (Linnaeus)?, *A.* sp., *Turritella cingulifera* Sowerby, *T.* cf. *fultoni* Melvill, *T.* sp. A, *T.* sp., *Cerithium* sp. A, *Epitonium* sp., *Tibia* sp., *Strombus* sp., *Natica lineata* Lamarck, *N.* cf. *didyma* (Bolten), *N.* cf. *melanostoma* (Gmelin), *N.* cf. *ponsonbyi* Melvill, *N.* cf. *strongyla* Melvill, *N.* sp., *Sigaretus* sp., *Gyrineum spinosum* (Lamarck), cymatiid gastropod, *Tonna dolium* (Linnaeus), *T. losariense* (Martin)?, *T.* sp., *Ficus subintermedia* (d'Orbigny), *F.* sp., *Rapana bulbosa* Solander, *Purpura* sp. A, *P.* sp. B, muricid gastropod, *Babylonia spirata* (Lamarck), *B.* sp., *Nassarius* cf. *mucronatus* (A. Adams), *N. persicus* Cox, *N.* cf. *verbeeki* (Martin), *N.* sp., *Clavilithes* cf. *verbeeki* (Martin), *Bullia* cf. *kurachensis* Angus, *B.* cf. *mauritiana* Gray, *B.* cf. *nitida* Sowerby, *B.* cf. *tahitensis* (Gmelin), *B.* sp. A, *B.* sp. B, *B.* sp. C, *B.* sp., *Bursa?* sp., *Oliva* sp., *Conus* sp. D, *C.* sp., *Terebra* sp. A, turrid gastropods, *Nucula* sp., *Nuculana* sp., *Anadara antiquata* (Linnaeus), *A. burnesi* (d'Archiac & Haime)?, *A. clathrata* (Reeve), *A. ferruginea* (Reeve), *A. granosa* (Linnaeus), *A.* cf. *tambacana* (Martin), *A. inaequivalvis* (Bruguière), *A.* sp., *Scapharca vellicata* (Reeve), *S.* sp., *Arcopsis* sp., *Cucullaea labiata* (Solander), *C.* sp., *Modiolus* sp., *Pinna* sp., *Chlamys mekranica* Eames & Cox, *C. prototranquebaricus* Vredenburg, *C.* cf. *singaporinus* (Sowerby), *C.* cf. *tjaringinensis* Martin, *C.* sp., *Pecten nearchi* Vredenburg, *Anomia* sp., *Crassostrea* sp. (large), small oysters, *Loripes* sp., *Venericardia* sp., *Cardium* sp., *Cardium* sp. (small), *Hemicardium?* sp., *Mactrinula plicataria* (Linnaeus), *M.* sp., *Raeta grayi* (H. Adams), *Solen* sp. (small), small Tellins, large Tellins, *Solecurtus* sp., *Gari* sp., *Zozia* sp., *Dosinia histrio* (Gmelin)?, *D. peralta* Vredenburg, *D. radiata* (Reeve)?, *D. subpenicillata* Vredenburg?, *D. pseudoargus* d'Archiac & Haime, *D.* sp., *Paphia* cf. *textile* (Gmelin), *P.* sp., *Hemitapes gallus* (Gmelin), *H.* sp., *Amiantis erycina* (Linnaeus), *A.* sp., *Pitar* cf. *belcheri* (Sowerby), *P.* sp., *P.* sp. (tiny), *Chione?* sp., *Clementia papyracea* (Gray), *Timoclea cochinensis* (Sowerby), *T. roemeriana* (Issel), *T.* sp., *Tivela* sp., *Corbula* sp., *Clavilithes* cf. *verbeeki* (Martin), *Bullia* cf. *karachensis* Angus, *B.* cf.

Chatti Mudstone (Garr Koh-Chatti, Pishukan and Jaibl-I-Maidhi Section):

Emarginula sp., *Astraea* sp., adeorbid gastropode, *Architectonica* sp., *Turritella angulata* Sowerby *T.* sp., *Epitonium* sp., *Xenophora* sp., *Natica* sp., *Sinum?* sp., *Sigaretus?* sp., *Gyrineum spinosum* (Lamarch), *G.* sp., cymatiid gastropod, *Tonna* cf. *luteostoma* (Küster), *T. tessellata* (Lamarck), *T.* sp., *Ficus* cf. *subintermedia* (d'Orbigny), *F.* sp., *Rapana* sp., muricid gastropod, *Phos* sp., *Nassarius* sp., mitrid gastropod, *Oliva* sp., *Conus* sp. A, *C.* sp., turrid gastropods, *Nucula* sp., *Nuculana* sp., *Anadara antiquata* (Linnaeus), *A. burnesi* (d' Archiac & Haime)?, *A.* cf. *clathrata* (Reeve), *A. ferruginea* (Reeve), *A. granosa* (Linnaeus), *A.* cf. *rectangularis* (Cossmann), *A.* sp., *Arcopsis bataviana* (Martin), *Scapharca* sp., *Trisidos tortuosa* (Linnaeus), *Cucullaea* sp. A, *C.* sp., *Modiolus* sp., *Pinna* sp., *Chlamys* sp. A, *C.* sp., *Amusium* sp. A, *Anomia* sp., *Diplodonta* sp., *Lucina* sp. A, *L.* sp., lucinid bivalve, *Cardium* sp., *Hemicardium?* sp., *Mactrinula* sp., *Raeta grayi* (H. Adams), *Lutraria* sp., *Solen* sp., small Tellins, large Tellins, *Zozia* sp., *Solecurtus* sp., *Dosinia peralta* Vredenburg, *D. pseudoargus* d'Archiac & Haime, *D.* sp., *Paphia* cf. *textile* (Gmelin), *P.* sp., *Hemitapes* sp., *Tapes* sp., *Clementia papyracea* (Gray), *Lioconcha?* sp., *Amiantis* sp., *Pitar* cf. *belcheri* (Sowerby), *Paphia* sp., *Timoclea cochinensis* (Sowerby), T. sp., venerid bivalves, *Corbula* sp., *Cuspidaria* sp.

Ormara and Jiwani Formations (Pishukan, Jaibl-I-Maidhi and Gwadar Sections):

Euchelus asper (Gmelin), *Calliostoma* sp., *Cyclostrema?* sp., *Architectonica perspectiva* (Linnaeus)?, *A.* sp., *Turritella bacillum* Kiener, *T.* cf. *fultoni* Melvill, *T. cingulifera* Sowerby, *T. columnaris* Kiener, *T. terebra* (Linnaeus)?, *T.* sp., *Cerithium kochi* Philippi, *C.* sp., *Cirsostrema* sp., *Cypraea histrio* Gmelin, *Natica didyma* (Bolten), *N.* cf. *antoni* Philippi, *N.* cf. *ponsonbyi* Melvill, *N.* sp., *Sigaretus cuvieranus* Récluz, *Gyrineum spinosum* (Lamarck), cymatiid gastropods, *Tonna dolium* (Linnaeus), *T.* cf. *lutestoma* (Küster), *T.* cf. *tessellata* (Lamarck), *T.* cf. *zonata* (Green), *T.* sp., *Ficus ficus* (Linnaeus), mitrid gastropod, *Oliva* sp., *Conus* cf. *amadis* Martin, *C.* sp., turrid gastropod, *Nucula* sp., *Nuculana* sp., *Anadara antiquata* (Linnaeus), *A. clathrata* (Reeve), *A. granosa* (Linnaeus), *A.* cf. *ferruginea* (Reeve), *A. uropigimelana* (Bory), *A.* sp., *Larkinia* cf. *rhombea* (Born), *L. multicostata* (Sowerby), *L.* sp., *Scapharca vellicata* (Reeve), *S.* cf. *japonica* (Reeve), *Trisidos tortuosa* (Linnaeus), *Chlamys prototranquebaricus* Vredenburg, *C.* cf. *singaporinus* (Sowerby), *C.* cf. *townsendi* (Sowerby), *C.* cf. sp. A, *C.* sp., *Pecten vasseli* Fuchs, *P. nearchi* Vredenburg, *P. dorothea* Melvill, *Plicatula* sp., *Anomia* sp., *Crassostrea* sp. (small), large oysters, *Cardium* sp. (large), *Hemicardium?* sp., *Codakia* sp., *Diplodonta* sp., *Raeta grayi* (H. Adams), *Lutraria* sp., *Mactra* sp. (smooth), *Mactrinula plicataria* (Linnaeus), *M.* sp., *Solen* sp. (large), small Tellins, large Tellins,

Solecurtus sp., *Gari* sp., *Dosinia pseudoargus* d'Archiac & Haime, *D.* sp., *Pitar* cf. *obliquata* (Hanley), *P.* sp., *Pitaria* cf. *umbonella* (Lamarck), *Paphia* cf. *textile* (Gmelin), *P.* sp., *Circe corrugata* (Chemnitz), *Clementia papyracea* (Gray), *Tivela* cf. *ponderosa* (Philippi), *T.* sp., *Timoclea siamensis* (Lynge), *Laternula* sp.

Ormara and Jiwani Formations (Hingol Section):

Diodora? sp., *Cellana karachensis* Winckworth, *C.* sp., *Architectonica* sp., *Xenophora* sp., *Natica didyma* (Bolten), *N* sp. (small), *N.* sp., cymatiid gastropod, *Morula?* sp., *Terebra* sp., *Zootecus insularis* Ehrenburg, *Anadara antiquata* (Linnaeus), *A* cf. *uropigimelana* (Bory), *A. inaequivalvis* (Bruguière), *A.* sp. (large), *A.* sp., *Barbatia obliquata* (Wood), *B.* sp., *Modiolus* cf., *Chlamys alexandri* Vredenburg, *C.* sp., *senatorius* (Gmelin), *C.* cf. *townsendi* (Sowerby), *C.* sp., *Pecten nearchi* Vredenburg, *Plicatula* sp., *Anomia* sp., *Crassostrea* sp. (large), small oysters, *Mactra* sp. (smooth), *Raeta grayi* (H. Adams), small Tellins, large Tellins, *Dosinia radiata* (Reeve)?, *D.* sp., *Paphia* sp., *Pitar* sp., *P.* sp. (tiny), *Lioconcha picta* Lamarck?, *L.* sp., *Timoclea cochinensis* (Sowerby), *T. scabra* (Hanley), T. sp., venerid bivalves, *Martesia* sp.

Coral Reef on Gwadar Headland:

Turbo intercostalis (Menke), *Cerithium obeliscus* (Bruguière), *Cypraea arabica* (Linnaeus), *Natica* cf. *didyma* (Bolten), *Mancinella mancinella* (Lamarck), *M. tuberosa* (Roding), *Purpura rudolphi* (Lamarck), *Conus flavidus* Lamarck, *Barbatia* sp., *Chlamys* cf. *townsendi* (Sowerby), *Tridacna* sp., *Codakia* sp., *Periglypta puerpera* (Linnaeus), *Gafrarium dispar* (Dillwyn).

Plate 1. Some typical molluscs from the Neogene and Quaternary of the Makran Coast: a. *Turritella angulata* Sowerby, Talar Sandstone (approx. 1600m level, Garr Koh-Chatti section), x1; b. *Pecten vasseli* Fuchs, Ormara Formation (Gwadar Headland), right valve, x1; c. *Melongena* cf. *gigas* (Martin), Talar Sandstone (approx. 1000m level, Garr Koh-Chatti section), abapertural view of incomplete specimen, x1; d. *Pecten vasseli* Fuchs, Ormara Formation (Gwadar Headland), left valve, x1; e. *Tonna luteostoma* (Kuster), Talar Sandstone (approx. 3600m level, Garr Koh-Chatti section, abapertural view, x1½; f. *Dosinia pseudoargus* D'Archiac & Haime, Ormara Formation (Jaibl-l-Maidhi), right valve, x1; g. *Chlamys* cf. *tjaringinensis* Martin, Talar Sandstone (approx. 3500m level, Garr Koh-Chatti section) right valve, x 2; h. *Chlamys* cf. *tjaringinensis* Martin, Talar Sandstone (similar level to specimen g.), right valve, x2; i. *Pecten nearchi* Vredenburg, Ormara Formation (Gwadar Headland), left valve, x1; j. *Pecten nearchi* Vredenburg, Ormara Formation (Gwadar Headland), right valve, x1; k. *Anadara burnesi* (d'Archiac & Haime)?, Chatti Mudstone (approx. 4800m level in Garr Koh-Chatti section), right valve, x2; l. *Anadara* cf. *tambacana* (Martin), Talar Sandstone (approx. 1675m level in Garr Koh-Chatti section), left valve, x2. These specimens are part of an unregistered collection housed in the Department of Palaeontology, British Museum (Natural History).

B. INDUS RIVER, DELTA AND FAN

5

Sediment Discharge from the Indus River to the Ocean: Past, Present and Future

J. D. MILLIMAN
Woods Hole Oceanographic Institution, Woods Hole, Massachusetts,
G.S. QURAISHEE
National Institute of Oceanography, Karachi,
and
M.A.A. BEG
Pakistan Council of Scientific and Industrial Research Laboratories, Karachi.

ABSTRACT

Until the late 1940's the Indus River apparently discharged more than 600mt of suspended sediment annually from the Himalaya Mountains. Assuming that most of the coarse fraction settled from suspension as the river flowed over the broad Indus Plain, somewhat less than 250mt reached the Indus delta annually. Since then, various man-made structures have greatly decreased the sediment load of the Indus. Today, less than 50mt reaches the estuary and, by the end of this decade, it may be practically nothing. To date, downstream barrages have accounted for most of the sediment decrease; dams have apparently only affected upper portions of the river basin. Such a dramatic drop in sediment and water flux can have serious impact upon the coastal environment.

INTRODUCTION

The Indus River is one of the world's largest rivers in terms of drainage area, river discharge and sediment load. Yet few people in Pakistan and far fewer in the western world know much detail about the water or sediment discharge from this river. Previous estimates of Indus sediment loads vary from 675 million tons (mt) before 1950 (WAPDA letter to G.S. Quraishee, December, 1982) to 435-440mt (Strahkov, 1961; Holeman, 1968) to 300mt (Kazmi, this volume). How these values relate to the actual amount of Indus sediment reaching the ocean depends upon where, when and how they were obtained (see e.g., Milliman and Meade, 1983). Now that the Indus is increasingly dammed and channeled, the applicability of these divergent values is particularly questionable; yet it is precisely at this time that an accurate understanding of the river is needed to predict possible impact upon the coastal areas from decreased sediment and water discharge.

In this paper we present a synthesis of available data from the Indus and also speculate on probable effects that damming will have upon the river and the adjacent marine environment. Several papers in this volume (e.g., Snedaker; Kazmi; Wells and Coleman; Schubel) deal with other aspects of the Indus, for which cross reference is recommended.

Most of our data are taken from the Pakistan Water and Power Development Authority (WAPDA), particularly data reports issued in 1975 and 1978, and as well as an unpublished report by the Civil Engineering Department of the Peshawar University (1970).

THE INDUS RIVER

The Indus drains the arid to semi-arid western Himalaya Mountains, with headwaters at elevations greater than 4000 m. The Indus and one of its tributaries (Sutlej River) are trans-Himalayan in origin, while the other main tributaries drain Afghanistan to the west (Kabul

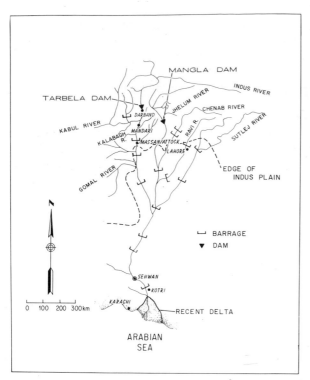

Figure 1. Map of Indus Basin, showing locations of barrages and dams.

River), and the southern side of the Himalayas (Jhelum, Chenab and Ravi rivers) to the east (Fig. 1). Upon leaving the mountains below Attock, the Indus River traverses across the broad Indus Plain for nearly 1000 km before reaching the Indian Ocean. Total drainage basin area is 970,000 km².

Although there is considerable variation in water flow between various stations in the river basin, flow is generally low in November to mid-spring, at which point snow melt (from the mountains) increases discharge. Highest discharge occurs in July, coincident with the peak of the rainy season. The six months, from May through October, account for more than 80 percent of the river discharge (Beg, 1977).

Quoted mean discharge for the Indus ranges from 5550 to 6700 to 7500 m³ sec⁻¹ (Lisitzin, 1972; Meybeck, 1967; UNESCO, 1978). WAPDA data for Sehwan, the most oceanward gauging station on the Indus, however, show that the mean for 1968-75 was 1810 m³ sec⁻¹; upstream at Massan, flow during 1972-75 was 3440 m³ sec⁻¹. During these times maximum discharge at the two stations was 18,200 and 22,300 m³ sec⁻¹, respectively, while minimum discharges were 28 and 2000 m³ sec⁻¹ (WAPDA, 1977).

These numbers point to many of the problems inherent in discharging the Indus. One is that both discharge and sediment load change markedly in a downstream direction. Giving average discharge without quoting the station greatly hinders the applicability of the values. Moreover, as can be seen in Fig. 2, discharge varies greatly from year to year, meaning that short-term averages may or may not be indicative of longer-term discharge at a particular station. Finally, discharge has decreased dramatically since construction of dikes, channels, barrages and dams in the past 30 years. The discharge values first quoted (5550-7550 m³ sec⁻¹) presumably represent upstream discharges for the years prior to wide-scale implementation of man-made structures along the river. In contrast, the smaller estimates are post-dam figures for the early 1970's; only half the water passing Massan reached Sehwan downstream, the rest being used for irrigation or lost to ground water aquifers (Beg, 1977). These average discharge values are even smaller at present with continued construction of dams and barrages.

Sediment Load of the Indus

Considerable amounts of sediment carried by the Indus come from glacially-derived material in the Himalayas as well as reworked fluvial deposits (Beg, 1977; van Aart, 1977). These unconsolidated materials plus the poor soil-holding capacity of the sparse vegetation (arid climate and geologically young terrain) result in large loads. As with water discharge, however, sediment loads vary so much in both time and distance along the river

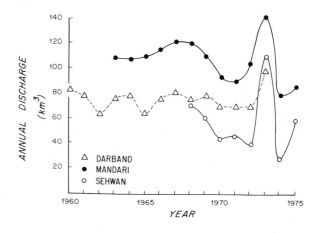

Figure 2. Annual discharge values for Darband, Mandari and Sehwan (see Figure 1 for locations). Data from WAPDA (1979).

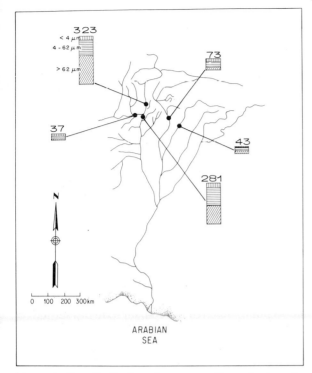

Figure 3. Size distribution of average sediment load on the Indus and tributaries 1960-68. Data from Peshawar University, 1970.

that previously quoted values are more or less meaningless. Even values quoted for the same station and the same year disagree between reports, presumably because of different sediment-rating curves used. For example, Peshawar University estimated sediment discharge at Darband in 1966 to be 347 mt, while WAPDA, using the same raw data, calculated a load of 295 mt.

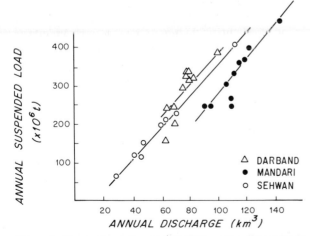

Figure 4. Comparison of annual discharge and sediment loads for the Indus for stations at Darband, Mandari and Sehwan. Data from WAPDA (1979).

Using, for the time being, the Peshawar University figures, between 1960 and 1968 more sediment passed Darband (since covered by the lake created by the Tarbela Dam) than passed Mandari downstream (Fig. 3). Combining the 37 mt entering from the Kabul River with the Indus load, an estimated 80 mt of sediment settled out from the Indus annually between Darband and Mandari. The settled sediment appears to be primarily sand-size material, presumably deposited along this relatively flat valley floor. The decreased competence of Indus River water at Mandari is seen by a lower annual rating curve (i.e., higher discharge required to maintain an equal load to that at Darband—Fig. 4). WAPDA sediment load values show that the load at Mandari can be either greater or lower than that at Darband, depending upon the year (see Fig. 5). Whether these or the Peshawar values are more realistic has not been resolved at this time.

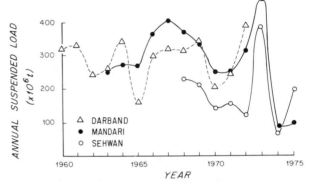

Figure 5. Yearly variation in suspended load passing Darband, Mandari and Sehwan; Darband observations were discontinued in 1973 with the completion of the Tarbela Dam. Data from WAPDA (1979).

To summarize the natural system, much of the Indus sediment load comes from the upper reaches of the drainage basin which, because of the close proximity to the mountains, means a substantial percentage of sand (60 percent at Darband; Fig. 3). Some of this sediment settles downstream of Darband as the river flattens into a relatively low-gradient valley. Below Mandari the gradient increases and a number of large tributary rivers empty into the Indus, increasing the sediment load. As the river begins to flow across the Indus Plain, however, coarse sediment again settles out; by Sehwan, still some 300 km from the sea, 20 to 50 percent of the total load has been lost. Kazmi (this volume) reports much of the present river valley is underlain by more than 100 m of late Quaternary sediment, and McDonald (1966, in

Peshawar Univ., 1970) estimated a vertical accretion in the flood plain of 9 m in the past 5000 years and a seaward growth of the delta of 80 km in 2000 years. Thus, it is not difficult to conclude that a large portion of the suspended load in the Indus is deposited prior to reaching the ocean. If the sand constitutes 60 percent of the load at Darband, then perhaps 60 percent or more of the total load may be deposited over the flood plain.

Man's Influence on Sediment Discharge

Before the 1950's the discussion presented in the previous sections was probably a fairly representative of Indus discharge patterns. Since then, however, human activities have greatly altered discharge patterns of the Indus and, therefore, transport of sediment. Four general types of engineering activities have occurred along the Indus; 1) channels which transfer water to and from various river branches as well as to irrigate farmland; 2) barrages which aid river control by diverting river flow to channels; 3) embankments and dikes which prevent river overflow, thereby restricting flow to the main channels; and 4) dams which are used for hydroelectric power, irrigation and flood control.

Wide-scale alteration of the Indus began in the 1940's with construction of barrages and channels. Prior to this time, annual sediment discharge (at Darband or Mandari ?) may have been the 675 mt quoted above. Assuming that at least 60 percent (sand) plus some of the finer load was deposited along the flood plain, less than 250 mt should have reached the Indus estuary. Neither the source nor the years of the values of 435-440 mt quoted by Lisitzin and Holeman (the latter taken from Fornier, 1960) is clear, but we assume that these may represent sediment discharge at Darband or Mandari in the 1950's. The difference between these and earlier estimates, therefore, would represent the sediment lost to the river system from channels and barrages. The effect of barrages on river flow can be seen at Kotri, 200 km from the sea, where Indus discharge during autumn, winter and early spring months was practically eliminated after barrage construction (used for irrigation during these dry months) and flow during summer months was only high during periodic peak flow (Fig. 6).

Embankments and levees have been utilized since British occupation. While they have helped control flooding of vast areas of the Indus flood ·

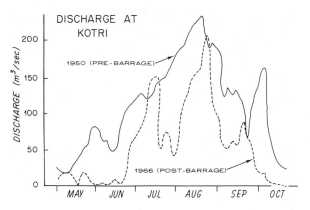

Figure 6. Comparison of discharge values at Kotri before and after barrage construction. Data from WAPDA (1979).

plain, they also have increased salinities in ground waters (Beg, 1977). Moreover, by not allowing suspended sediment to settle over the surrounding flood plain, much of the coarser load has settled on to the river bed, thus progressively elevating the river bed relative to the flood plain. It is not clear how (or if) the embankments have actually altered total sediment discharge from (or along) the Indus.

Two huge dams were constructed on the Indus system and have been operating for the past 10-15 years. Each has had a major impact upon sediment load. The Mangla, damming the Jhelum River, was completed in 1967. Soon after completion, the sediment load of this river fell from an average of 45 mt to less than 0.5 mt (WAPDA, 1978). Down-dam erosion of river banks and the channel presumably restored some of the Jhelum load. The Tarbela Dam, on the Indus near Darband, was completed in 1974. Sediment loads downstream decreased sharply in both 1974 and 75, the last two years for which we have records. At Mandari, loads averaged less than 100 mt compared to the previous three years when loads averaged more than 300 mt (Fig. 5). At Sehwan, sediment loads were somewhat greater than at Mandari, suggesting net erosion of the river channel/banks, and, therefore, little to no net accumulation along the intervention flood plain and river system.

Future of the Indus River System and Estuary

Although the actual sediment load reaching the Indus estuary has not been documented in the past and is not known today, it is safe to assume that as recently as 30 years ago perhaps somewhat less than 250 mt of sediment may have reached the estuary annually. As of 1974-75, this amount was down to less than 100 mt. Presumably it is even less today, and within the next 10 years further construction along the Indus may completely cut off sediment supply reaching the ocean. Effects of this diversion are considerable and should increase in the future.

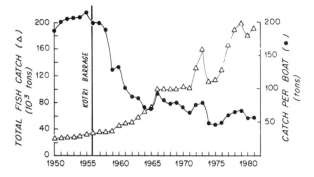

Figure 7. Yearly variations of total fish catch and fish catch per boat along the Sind coast. Construction of the Kotri barrage is shown in 1955. Modified from Quraishee (1975).

Channel and embankment construction have resulted in increased salinities in ground waters, often, negating the positive aspects associated with irrigation. The decreased amount of river flow also has increased salinities in the Indus estuary, thereby killing or hindering mangrove communities, a point dealt with by Snedaker in this volume. Finally, the decreased influx of nutrient-rich river waters, undoubtedly, will affect offshore fisheries. Yearly catches of fish off the Sind Coast increased 7-fold between 1950 and 1981, but becuase of a 23-fold increase in the effective number of fishing boats in this area, the actual catch per boat fell by nearly 3½-fold (Fig. 7). This decrease might be explained by the increased number of boats, catching a more or less constant supply of fish, but the decrease in catch per boat occurred with the construction of the Kotri Barrage in 1955, several years before the number of boats increased substantially. This suggests that decreased river discharge across the barrage may have played a role in the decreased yield (Quraishee, 1975).

Upstream construction of dams on other large rivers also prevented suspended sediment from reaching the ocean, most notably the Colorado and Nile Rivers, both of which had annual loads of 100 mt or more. The Indus, however, is by far the largest such river to be dammed, and also has a very high energy coastline and shelf (Wells and Coleman, this volume). Thus, cutting off the large

sediment load has particularly great potential for high coastal erosion of the Indus delta. The papers by Schubel and by Wells and Coleman in this volume deal with this problem further.

Note of Additional Evidence

Subsequent to the preparation of this manuscript, one of us (G.S.Q.) obtained a nearly 50-year record of annual discharge and suspended load for the Indus at Kotri, about 250 km upstream from the river mouth. This particularly long record substantiates many of our other data and speculations, and also answers many questions (Fig. 8). Aside from the marked annual variations, three distinct periods are noted in the Kotri records:

1) From 1931 through 1947, discharge generally averaged or exceeded 90 km³/yr, and suspended load exceeded 200 mt. The average for the 17 years prior to 1948 was 225 mt, in reasonable agreement with our previous estimates for the years prior to the late 1940's.

2) From 1948 through 1961, the river was in transition, with continued high discharge but decreasing suspended loads. The measured Indus load at Kotri, for example, fell from 275 mt in 1945 to less than 50 mt in 1948. Other than two years of high loads in 1955-56, the suspended load during this transition period was generally substantially less than 100 mt. Water discharge, however, remained high (Fig. 8). This change in river character probably resulted from construction of

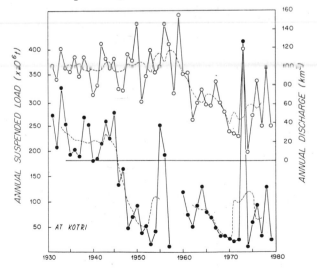

Figure 8. Discharge and sediment loads for the Indus River at Kotri, 1931-1979. Dashed lines represent 5-yr running mean averages. Data from the Indus River Commission.

one or more barrages along the lower course of the river.

3) After 1961, discharge dropped dramatically whereas discharge in the 1950's exceeded 100 km³/yr, it often was less than 60 km³/yr in the 1960's and 70's (Fig. 8). Suspended load dropped slightly, but major yearly fluctuations continued (particularly in the 1970's). As this drop in discharge and the previous drop in sediment load occurred before construction of upstream dams, we assume they resulted from continued barrage construction and corresponding irrigation of the river.

One further indication of change in the river can be seen in the marked change in rating curves for the Indus during these three periods (Fig. 9), with highest competency prior to 1948, and lowest during the transition period (1947-60). The trend of pre-1948 curve parallels and is closest to the Sehwan curve, inferring an approximate continuum with the upstream portion of the river. The strong disparity between the other two curves and the Sehwan curve, however, suggests that the downstream character of the Indus changed before the upstream (1948 versus 1974). Dam construction appears to have strongly altered the carrying capacity of the upper Indus, whereas it appears to have had little impact in the lower Indus, as indicated by the continuity of post '61 Kotri data (Fig. 9). Barrage and irrigation construction presumably was more important in decreasing flow and sediment load in the lower river. Other than the unusually high sediment load in 1976, an average of less than 50 mt was carried past Kotri annually. Presumably the amount presently reaching the ocean in the mid-1980's is even less.

ACKNOWLEDGMENTS

We thank the National Science Foundation for sponsoring the Pakistani workshop that allowed us to discuss this paper. Partial funding for the senior author was from the Office of Naval Research (Contract No. N00014—8IC—0009). We thank Drs. David Aubrey (WHOI) and Bilal Haq (EXXON) for reviewing this paper. This is Wood Hole Oceanographic Institution Contribution Number 5486.

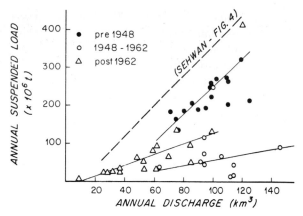

Figure 9. Sediment rating curve at Kotri. Note the marked difference in trends for the 3 periods represented in the 49 year record (Fig. 8). For reference, the Sehwan trend (from Fig. 4). is also shown.

REFERENCES

Beg, M.A.A., 1977. The Indus River basin and risk assessment of the irrigation system. Int. Working Seminar on Environmental Risk Assessment in an International Context. Tihanyi, Hungary, 13 p.

Holeman, J.N., 1968. Sediment yield of major rivers of the world. Water Resources Red., 4: 737-747.

Lisitzin, A.P., 1972. Sedimentation in the world ocean. Soc. Econ. Paleont. Mineral. Spec. Publ., 17: 218 p.

Meybeck, M., 1976. Total mineral dissolved transport by world major rivers. Bull. Sci. Hydrol., 21: 265-284.

Milliman, J.D. and R.H. Meade, 1983. Worldwide delivery of river sediment to the oceans. J. Geol., 91: 1-21.

Pakistan Water and Power Development Authority (WAPDA), 1975. Sediment Appraisal of West Pakistan Rivers, 1960-1972. Surface Water Hydrology Project, Lahore.

Pakistan Water and Power Development Authority (WAPDA), 1978. Sediment Appraisal of West Pakistan Rivers, 1960-1972. Surface Water Hydrology Project, Lahore.

Peshawar University, 1970. The sediment load and measurements for their control in rivers of West Pakistan. Peshawar Univ. Civil Eng. Dept., Water Res. Div., 1: 214 p.; 2: 320 p.

Quraishee, G.S., 1975. Influence of the Indus River on marine environment. Int. Conf. Management of Environment. Pakistan Acad. Sci., Islamabad, p. 111-122.

Strahkov, N.M., 1961. Onekotroykh zakonomernostiakh denudatsii in perenosa osadochnogog materiala na ploschadyakh gymidnykh klimatov, In: N.M. Strahkov, P.L. Bezrykov, and V.S. Yablokov (eds.), Sovremennye osadki moei i oceanov. Moscow, Izdatelstove Akademia Nauk SSSR, p. 5-27.

van Aart, R., 1977. The Indus Delta. ILRI, Netherlands, unpubl. rept.

6

Geology of the Indus Delta

A. H. KAZMI
Gemstone Corporation of Pakistan,
Peshawar, Pakistan

ABSTRACT

The present Indus Delta is located at the head of the Arabian Sea, between Cape Monze and the Rann of Cutch. It covers an area of approximately 1000 sq. miles. However, with the construction of the present canal irrigation system and flood protection levees the river discharge in the deltaic region has been reduced to about one fifth and the river has been confined to a single channel almost upto the coastal area. The present active delta has consequently shrunk to a small triangular area, about 100 sq. miles in extent, in the vicinity of Keti Bundar. East of the present day delta there is the ancient deltaic flood plain and the remnants of ancient tidal deltas.

From inland extending towards the sea, the Indus delta is comprised of deltaic flood plain deposits with an intervening meander belt of deposits from the distributaries, an arcuate zone of older tidal deltaic deposits, followed by more recent deposits of the tidal delta and coastal sand-dunes.

Photogeological studies supplemented with field surveys reveal four ancient courses of the Indus in the lower Indus Plain, the east-west shifts in course being apparently concomitant with shifts in the position of its delta. Thus during the Late Holocene the delta has apparently migrated westward from near the Thar desert to its present position off Karachi.

Geophysical surveys reveal two buried channels in the lower Indus plain which may have been occupied by the ancestral streams of the Indus. One of these is located along the present course of the Indus between Sehwan and Hyderabad and the other follows a NNE-SSW trend east of the East Nara Canal, from near Panjnad, through Cholistan desert up to the mouth of the Rann of Cutch. Subsurface test hole data also reveal a 600 ft. entrenched channel of the Indus, which has been correlated with the lowering of the sea level during the last glacial period. This channel follows the present course of the Indus to Hyderabad, but farther down stream deviates towards Badin. It is filled with sand and gravel deposits of the Tandojam Formation (Late Pleistocene to Early Holocene).

In the south the Tandojam Formation overlies deltaic deposits of the Nabisar Formation (Late Middle Pleistocene). In the northern part between Sehwan and Khairpur the upper part of the Tandojam Formation overlaps or interfingers with the silt and clays of the Larkana Formation (Late Pleistocene) which are probable sub-piedmont deposits.

The earliest geological trace of the Indus delta probably dates back to the early Miocene when it was located in the vicinity of Bugti area, near Sibi ("Bugti Bone Beds", Gaj Formation etc). The Indus Plain had then emerged as a vast monoclinal pediplain consequent to the collision of the Indus and Eurasian plates. Subsequently the delta migrated southward towards Karachi (Manchhar Formation), then eastward in the vicinity of the Rann of Cutch (Nabisar Formation), followed by westward shifts to its present position near Karachi.

INTRODUCTION

The Indus delta is one of the significant geomorphic features of Pakistan and it is located at the head of the Arabian Sea, roughly between Cape Monze and the Runn of Cutch. The present day delta is typically triangular in shape and extends from near the town of Thatta upto the sea. This deltaic complex covers an area of approximately 1,000 square miles (Fig. 1). In recent years construction of three major storage reservoirs in the catchment (Tarbela, Mangla and Bhakra) and an extensive system of irrigation canals and barrages over the Indus has drastically reduced the river discharge in the deltaic region. Consequently the active delta of the Indus has now shrunk into a small 100 square-mile triangular zone near Keti Bundar.

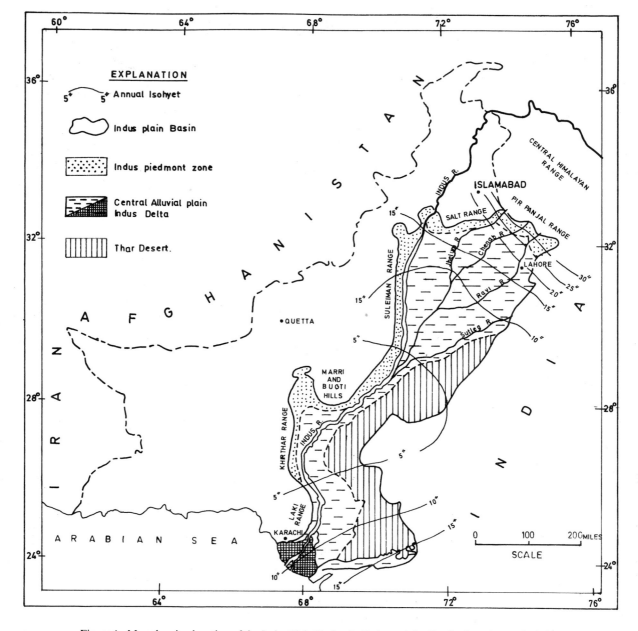

Figure 1. Map showing location of the Indus Plain Basin, the Delta and the distribution of annual rainfall.

The Indus delta comprises an area of considerable economic significance. It constitutes part of the main hinterland bordering the Karachi—Thatta—Hyderabad—Badin urban zone which, besides its trade and industrial complex, contains more than 12% of the population of Pakistan. This region also contains good potential for increase in food production (water resources, agriculture, livestock, fisheries, wildlife) and for development of certain minerals (clays, limestone, coal, and possibly oil and gas). It is, therefore, hoped that apart from scientific and academic interest, a clear understanding of the geology of Indus delta would also help in sound development planning for mobilising the natural resources of this small but important region of Pakistan.

Pakistan's mountainous regions are endowed with some of the most fascinating hard rock geology in the world. It is, therefore, not surprising that until recently most of the geological work has been concentrated in those regions. Only in relatively recent years the search for oil and gas was extended to the off-shore areas and waterlogging and salinity problems necessitated

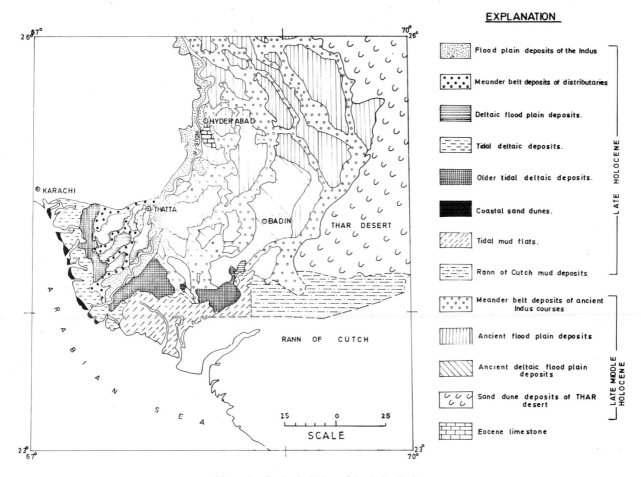

EXPLANATION

Flood plain deposits of the Indus

Meander belt deposits of distributaries

Deltaic flood plain deposits.

Tidal deltaic deposits.

Older tidal deltaic deposits.

Coastal sand dunes.

Tidal mud flats.

Rann of Cutch mud deposits

Meander belt deposits of ancient Indus courses

Ancient flood plain deposits

Ancient deltaic flood plain deposits

Sand dune deposits of THAR desert

Eocene limestone

LATE HOLOCENE

LATE MIDDLE HOLOCENE

Figure 2. Geological Map of the Indus Delta.

geological investigations right upto the deltaic region of the Indus plain. As a result of these activities much useful geological and geophysical information has been collected about the deltaic and off-shore regions of Pakistan. The author has been associated with the ground water investigations in various parts of the Indus plain, particularly in the study of the Quaternary deposits of the Indus Plain. In this paper a summary of the geological information gathered by the author pertaining to the lower Indus plain and the Indus delta is presented.

THE INDUS DELTA

The present day delta of the Indus occupies an area with the shape of an equilateral triangle, each side of which is approximately 60 miles wide. It is the product of the enormous amount of silt and sand brought down annually by the Indus into the sea. Before the construction of the canal system

up-stream it was estimated that the delta was annually growing at the rate of about 113 ft into the sea (Pascoe, 1964). At that time the Indus was transporting nearly 300 million tons or approximately 80,000 acre ft of silt annually to the deltaic region. Since the construction of large dams in the catchment and the canal system, the discharge of the Indus in the deltaic region has been reduced by more than 80 percent. Levees constructed for flood protection have further confined the river to a relatively narrow channel upto Keti Bundar.

The present day deltaic complex is comprised of an upper alluvial part—consisting of the deltaic flood plain and a lower tidal zone, which is the tidal delta. The deltaic flood plain is characterised by meander belt deposits of the distributaries of the Indus which have silted up since the confinement of the Indus through flood protection levees into a single channel. Between the meander belts of the distributing channels and located slightly higher

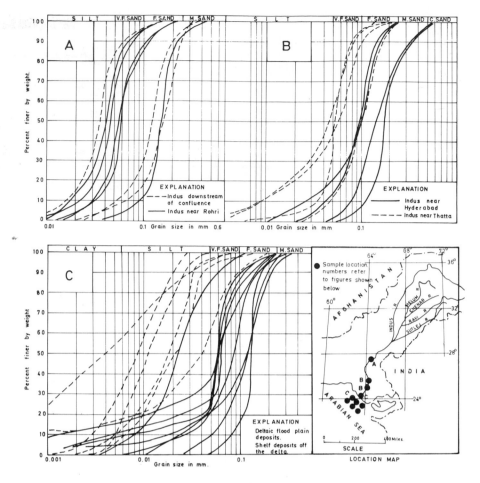

Figure 3. Cumulative curves showing the grain size frequency distribution in the Indus Flood plain, the deltaic flood plain deposits and shelf deposits off the Delta.

are the flood plains formed by the overflow from these distributaries (Fig 2).

The lower margin of the deltaic flood plain is rimmed by the tidal delta. The tidal delta is comprised of three main parts. Bordering the deltaic flood plain is an arcuate zone of older tidal deposits, formed as a result of the silting up of the distributary channels of the Indus. Seaward this zone is followed by the active tidal delta. With the drying and silting up of the major distributaries of the Indus, the tidal delta has been cut off from the alluvial process and has assumed the form of the tidal mud flat. The lower remnants of the distributaries have turned into tidal creeks. Immunity from flood has caused the growth of regular sandy beaches along the seaward edge of the delta. Along the coast there is a narrow zone of coastal sand-dunes, some of which form small islands at the mouths of the major creeks. These sand-dunes and sandy beaches comprise the third component of the tidal delta.

The deltaic deposits mainly comprise interlayered deposits of very fine sand, silt and clay. Pits or shallow wells dug on the deltaic flood plain show the laminated nature of the sediments, which is unlike the common non-bedded or massive or cross bedded character of the flood plain deposits upstream of the delta. Deposits of silt and clay containing abundant mollusc shells underlie the present deltaic flood plain. Whereas the deltaic sediments are significantly finer than the flood plain deposits of the Indus upstream of the delta, physically they resemble the present silty continental shelf deposits at the mouth of the delta (Figs. 3 and 4). The shelf deposits are, however, finer (mainly silt) and contain marine microfossils (mainly foraminifera) which may be one distinguishing feature between these two deposits.

Ancient deltaic deposits, similar to those described above, occur east of the present day

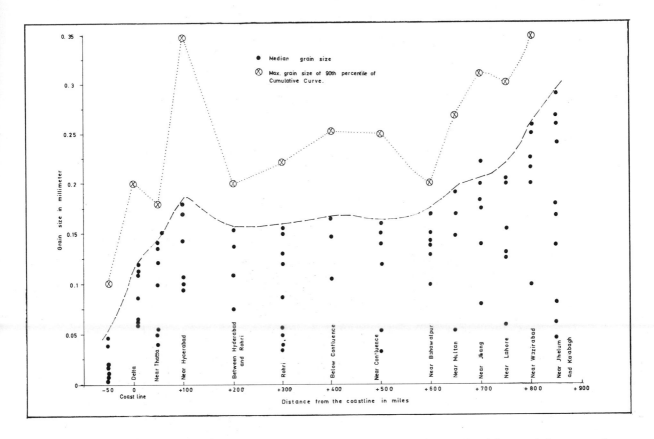

Figure 4. Graph showing the median grain size and the max.grain size at the 90th percentile of the cumulative curves of Indus Plain alluvial sediments (recent) and their variation with distance from the coastline.

delta. This indicates that previously the Indus delta was located farther eastward. The earlier deltas of the Indus and their geological history are discussed in the following pages.

REGIONAL GEOLOGICAL SETTING

Structurally the Indus delta is located south of the Kirthar fold belt bordering the Lyari embayment zone of this belt (Kazmi, 1982). It forms the southwestern part of the Thatta — Hyderabad gravity high and eastward it is followed by the lower Indus trough (Kazmi, 1982; Balkrishan,1977).

A thick sequence of Tertiary sedimentary rocks mainly comprised of limestone, shale, mudstone and sandstone and ranging in age from Eocene to Pleistocene is exposed along the northern margin of the present day delta (Hunting, 1960). In fact at its northern margin the deltaic deposits overlap the gently folded Tertiary rocks. At the upper part of the delta near Thatta there is an outlier of Eocene limestone (Fig 2).

Borehole logs indicate that in the upper part of the delta these Tertiary rocks underlie the delta at shallow depths (at places only about 10 to 100 ft). In the lower part of the delta depth to the bedrock is not known. However, because the present day delta is located on a "gravity high", the bedrock is likely to be at a relatively shallow depth in its lower part also.

Eastwards the unconsolidated Quaternary deposits of the Indus plain complex surround the present day delta. Surficially these sediments mainly comprise flood plain deposits of the present day Indus as well as the flood plain deposits of the extinct courses of the Indus (Fig 2). Linked with these extinct courses of the Indus are the remnants of the ancient deltaic complexes, mainly in the form of extinct deltaic flood plains and tidal deltaic plains. These landforms and geomorphic features are similar to the corresponding forms of the present day deltaic complex except that the surficial features such as traces of meander scars, ox bows, levee remants, distributary channels etc. have become relatively faint, though still

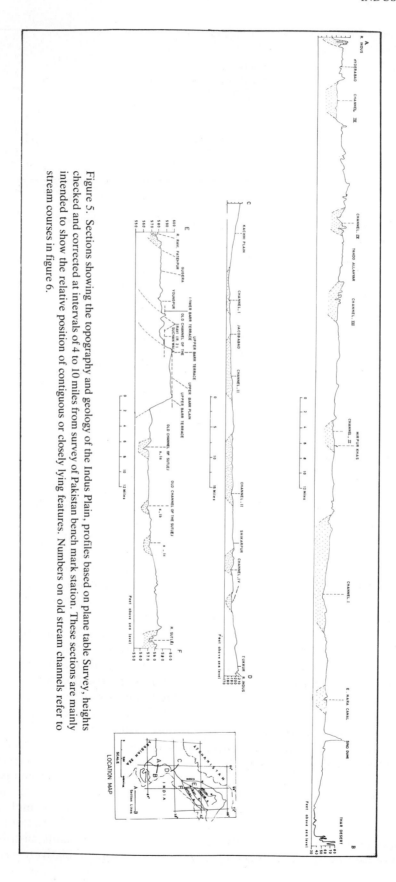

Figure 5. Sections showing the topography and geology of the Indus Plain, profiles based on plane table Survey, heights checked and corrected at intervals of 4 to 10 miles from survey of Pakistan bench mark station. These sections are mainly intended to show the relative position of contiguous or closely lying features. Numbers on old stream channels refer to stream courses in figure 6.

discernible on aerial photographs and landsat imageries (Kazmi, 1966, 1977).

Farther to the east the lower Indus plain is bounded by the Thar desert. In this region the plain slopes eastward as well as southwards. The southward slope conforms to the regional slope of the drainage basin, whereas the eastward slope is mainly due to the gradual but systematic aggradation by the river of its ancient courses followed by consistent westward migration of its channel (Fig 5). This westward migration of the river course was accompanied with the

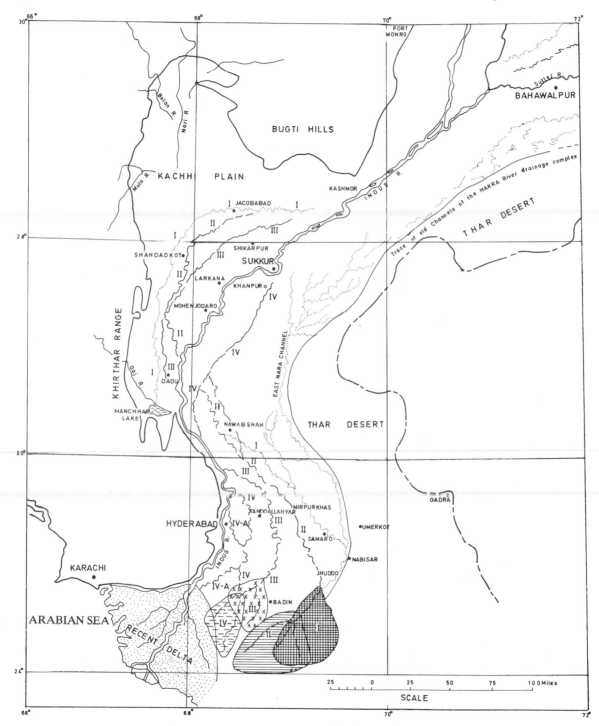

Figure 6. Map showing the old courses of the Indus.

corresponding migration of the delta towards the west.

Whereas in the lower part of the Indus plain, east of the present delta, the Indus shifted its course from the east to the west, further upstream in the vicinity of Jacobabad and Larkana, the evidence indicates that the river aggraded its ancient course and shifted from west to east (Fig 5), until it came to occupy its present course.

Photogeological interpretation accompanied with ground field checks, and evidence from the borehole data reveal that since the earliest Moenjodaro period (about 5000 to 6000 years B.P.), the Indus has made four major shifts in its courses downstream of Sukkur. The position of the Indus delta also shifted east to west at least four times during this period (fig 6).

SUBSURFACE GEOLOGY

Gravity surveys conducted in the Indus Plain region (Glennie, 1955; Balkrishan, 1977) reveal the major features of the basement topography (Fig 7). In the south central part of the Indus plain a major zone of upwarp referred to as the Jacobabad – Khairpur high is indicated. Near Khairpur it contains a large outlier of Eocene limestone. Eastward it is flanked by another zone of upwarp, the Jaisalmer high (Fig 7). In between these two zones of upwarp and the Kirthar Range is a zone of downwarp referred to as the Kachhi foredeep. It has been speculated that such trench-like features may have been occupied by the ancestral streams of the Indus plain during Early and Middle Pleistocene.

Borehole data from oil exploration wells indicate that the basement is overlain by Mesozoic to Tertiary sedimentary rocks which are fairly thick in the downwarp zones (15,000 ft+) and apparently thin out in the zones of upwarp. These rocks are overlain by unconsolidated Quaternary sediments which range in thickness from about 100 to 300 feet in the upwarp zones to more than 600 feet in the downwarped regions.

The examination of the borehole data from a large number of shallow test holes (max. depth about 600 ft) from the south central and lower Indus plain region reveals a 500 to over 600 ft deep and about 30 miles wide trench-like feature roughly following the present course of the Indus between Sukkur and Hyderabad. It is filled with relatively fine to medium alluvial sand referred to by the author as the Tandojam Formation (Kazmi,

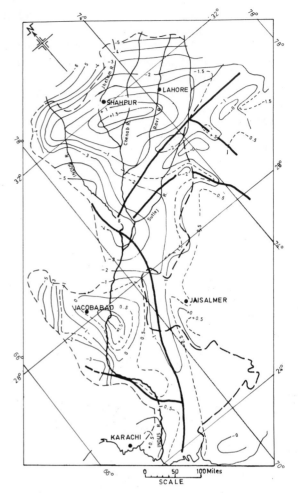

Figure 7. Gravity map of the Indus Plain Basin showing warp contours in kilometers (after Glennie, 1955). Thick lines (added by present author) indicate position of buried trench like features, which may have been occupied by the ancestral streams of the Indus Plain during early and middle Pleistocene.

1966, 1977). Downstream of Hyderabad, this buried channel deviates from the present course of the Indus and follows a course through Mirpurkhas towards Badin (Fig 8).

In the south central part of the Indus plain, between Jacobabad and Moenjodaro, the eastern flank of this channel is underlain by a relatively thin (50 to 100 ft) deposit of alluvial silt and clay (named Larkana Formation by the author), followed by the Eocene limestone (Figs 9, 10 and 11). The western flank of the channel is underlain by gravel, sand and silt of the unconsolidated piedmont and subpiedmont deposits in excess of 200 ft thick. The piedmont deposits comprise conglomerate and sand which extend eastward and interfinger with the silt and clays of the subpiedmont deposits. Near the foothill region of the

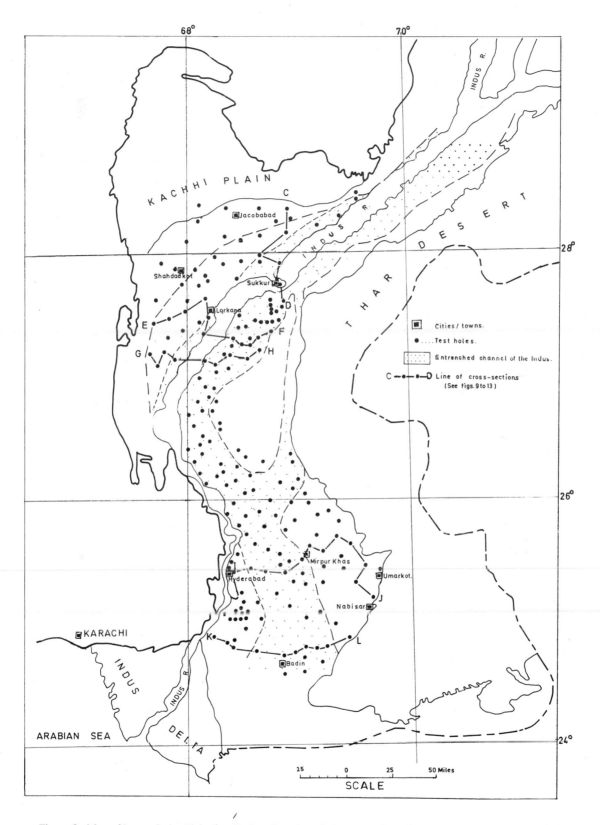

Figure 8. Map of Lower Indus Plain showing location of test-holes and outline of the late Pleistocene entrenched channel of the Indus.

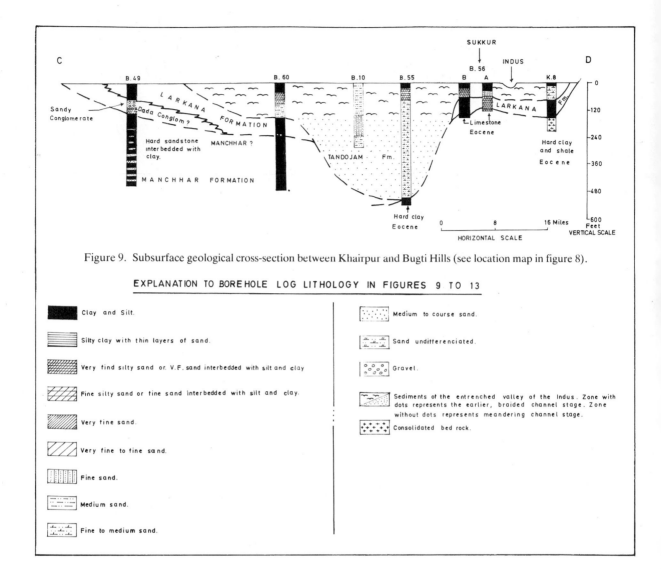

Figure 9. Subsurface geological cross-section between Khairpur and Bugti Hills (see location map in figure 8).

EXPLANATION TO BOREHOLE LOG LITHOLOGY IN FIGURES 9 TO 13

Clay and Silt.

Silty clay with thin layers of sand.

Very find silty sand or V.F. sand interbedded with silt and clay

Fine silty sand or fine sand interbedded with silt and clay.

Very fine sand.

Very fine to fine sand.

Fine sand.

Medium sand.

Fine to medium sand.

Medium to course sand.

Sand undifferenciated.

Gravel.

Sediments of the entrenched valley of the Indus. Zone with dots represents the earlier, braided channel stage. Zone without dots represents meandering channel stage.

Consolidated bed rock.

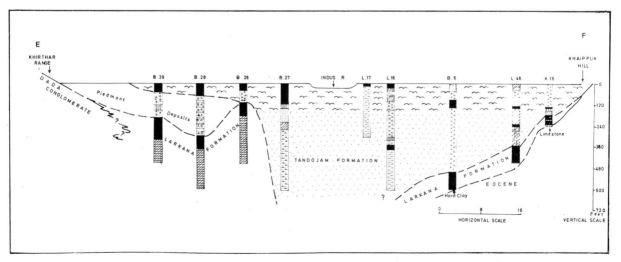

Figure 10. Subsurface geological cross-section from Khirthar Range to Khairpur Hills (see location map in figure 8, for explanation to lithologic symbols see figure 9.)

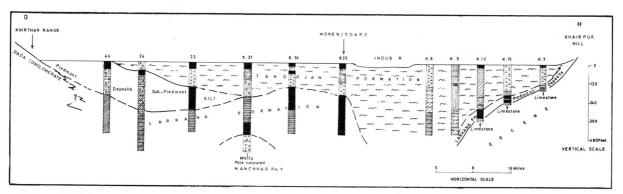

Figure 11. Subsurface geological cross-section between Khairpur Hills, Mohenjodaro and Khirthar Range, (see location map in figure 6, for explanation to lithologic symbols see figure 9).

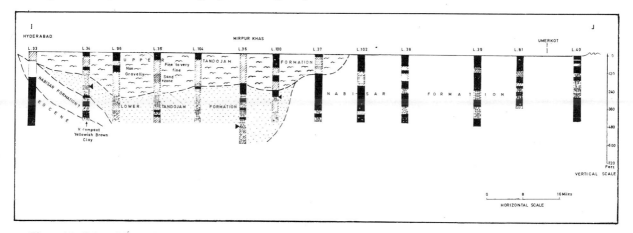

Figure 12. Subsurface geological cross-section between Umerkot and Hyderabad (see location map in figure 8, for explanation of lithologic symbols see figure 9).

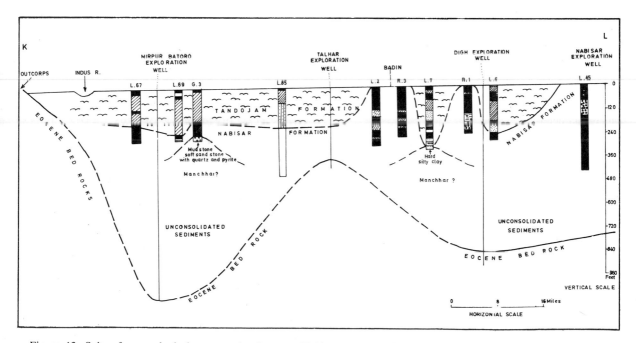

Figure 13. Subsurface geological cross-section between Nabisar, Badin and Thatta. (see location map in figure 8, for explanation to lithologic symbols see figure 9).

Kirthar Range the piedmont deposits are underlain by the Pleistocene Dada Conglomerates (Hunting, 1960). When traced eastward, the Dada Conglomerates interfinger with the Larkana Formation. This structural feature shows that the Dada Conglomerates are essentially piedmont deposits and the silts and clays of the Larkana Formation are the corresponding subpiedmont deposits. Both these formations are underlain by the indurated sandstones, clays, and marls of the Manchhar Formation (Blanford, 1876).

In the lower Indus plain, between Mirpurkhas and Badin, the buried channel of the Indus overlies a relatively thick, hard succession of clay interbedded with thin lenses of very fine sand (Figs 12 and 13). The clay is laminated, ranges in colour from dark grey, brown, greenish brown to greyish white. It is calcareous and at places contains abundant mollusc shells, resembling the present day deltaic deposits named Nabisar Formation by the author on the basis of the borehole data from near the town of Nabisar.

The Nabisar Formation is underlain by sandstones and mudstones of the Manchhar Formation followed by the Eocene limestone.

STRATIGRAPHY AND GEOLOGICAL HISTORY

From the above account of subsurface geology of the lower Indus plain the following stratigraphic sequence may be recognised:

Surficial unconsolidated deposits comprising:- Late Holocene
Flood plain deposits   (3000 to 4000 BP?)
Deltaic deposits
Extinct flood plain
and deltaic deposits.

Tandojam Formation   Early Holocene to Late
                                stage of the last glacial

Subsurface piedmont   Late stage of last glacial
deposits

                    Unconformity

Dada Conglomerate and   Last glacial?
Larkarna Formation

                    Unconformity

Nabisar Formation   Last Interglacial?

                    Unconformity

Manchhar Formation   Early Pleistocene to Pliocene.

                    Unconformity

Eocene limestone,   Eocene
marl and clay

The Eocene limestone underlies the entire lower Indus plain. It was deposited in a shallow sea, the remnant of the Tethys seaway, which was gradually eclipsed as a result of the northward drifting of the Indo-Pakistan plate and its ultimate collision with the Eurasian plate to the north. This collision which probably took place during the Late Eocene resulted in the buckling up of the lower Indus plain into zones of upwarp and downwarp. The Indus plain finally emerged as a land mass and probably went through a cycle of erosion and peneplanation during Oligocene to Pliocene.

During the Oligocene and Miocene the shoreline of the Indus plain apparently had a north-south orientation and extended from the area presently occupied by the Kirthar foredeep upto Karachi and beyond. A relatively shallow sea washed its shores. It is likely that this sea was dotted with a number of offshore islands, more or less parallel to the coastline (Kazmi, 1977; DeJong, 1977). These islands were formed as a result of the earliest phase of the Himalayan Orogeny (Kazmi, 1982). Subsequent phases of orogeny transformed these islands into regular narrow land masses or hill ranges. Between these islands and the Indus coastline the sea formed a relatively shallow and narrow gulf which may be referred to as the Kirthar gulf. The head of this gulf was apparently in the vicinity of what now constitutes the Bolan Pass area, since marine Oligocene—Miocene sediments may be traced upto that region. During this period the north Himalayan drainage or the precursor of the Indus most probably entered the sea in the vicinity of the Bugti area.

In this region the basal part of the 'Bugti Bone Beds' (Gaj Formation) contains a mixed fauna of oysters and mammalian remains (Pascoe, 1965) which suggest an estuarine or deltaic environment. About 500-700 ft above the base there is a fossil bed full of fresh water shells. The vertebrate fossils include *pisces, chelonia, croco- dilia, proboscidea, perissodactyla ancylopoda, artiodactyla* and *carnivora* (Pascoe, 1965). There is also an abundance of fossil wood. These fossil assemblages indicate a locality full of ponds and swamps, leading to the conclusion that the Bugti area represents the first trace of the earliest delta bordering the Indus plain.

There is, as yet, no stratigraphic record to trace a gradual southward migration of this early Miocene delta. In fact, the Middle Miocene phase of Himalayan orogeny suddenly obliterated the Kirthar gulf (Kazmi, 1982), rapidly moving the coastline to the vicinity of Karachi. The Kirthar gulf was replaced by the Kirthar foredeep. This foredeep may have been occupied by one of the major streams of the Indus plain if not by the Indus itself.

In Lower Sind, according to Pascoe (1965), there is a significant intercalation of marine or estuarine beds among the Manchhars and this evidence of deposition in salt water increases in the neighbourhood of the present coast. It can be inferred that during the Pliocene and Early Pleistocene the sea covered a portion of lower Sind, perhaps including a part of the lower Indus plain as well.

In the lower Indus plain there is no record of the events or deposits of the Middle Pleistocene. As may be seen from the stratigraphic columns exposed in a number of wide diameter open wells, the Thar desert is underlain by more than 300 ft thick unconsolidated or semi-consolidated sediments, mainly in the form of interbedded layers of sand and silt. These sediments are presumably Middle Pleistocene in age.

The lower Indus plain contains a good record of the events and sediments of the Late Pleistocene. The Tandojam Formation with coarse gravelly alluvial sand in the lower part and fine to medium alluvial sand in the upper part represents a 400 to 600 ft deep channel filling. This channel was apparently formed as a result of the degradation of its course by the Indus due to the lowering of the sea level during the last glacial period. As large portion of the continental shelf were exposed due to the falling sea level, the Indus may have extended its course across the emerging sea floor right upto the edge of the continental shelf.

Towards the end of the Last Glacial, as the sea level began to rise once again, the Indus aggraded its course and during the Late Glacial to Early Recent deposited the sands and gravels of the Tandojam Formation.

It has been inferred that during the Last Glacial Period (Last Pluvial in the Indus Plain area) the rainfall had apparently increased resulting in more extensive deposition of piedmont (Dada Conglomerates) and subpiedmont deposits (the Larkana Formation) in the Kirthar foothill areas extending upto the slopes of the entrenched valley of the Indus (Kazmi, 1966, 1977).

The entrenched valley of the Indus has apparently dissected the deltaic silt and clays of the Nabisar Formation (Figs. 12 and 13). Nabisar Formation is, therefore, deposited earlier than the Last Glacial. This formation extends for a considerable distance inland even beyond Nabisar. It is unlikely that during a glacial period the delta would have occupied such a deep inland position. On the contrary, during a Glacial period, as in the case of the Last Glacial, due to the lowering of the sea level, the delta may be expected to have migrated seaward. We may, therefore, conclude that the Nabisar Formation represents deltaic deposits formed during the Last Interglacial.

SUMMARY OF CONCLUSIONS

Due to the construction of the canal irrigation system, with its dams, barrages and flood protection levees, the flow of the Indus in the deltaic region has been reduced greatly, resulting in the confinement of the river to a single channel almost to the sea, drying up of most of the distributary channels, shrinking of the active delta from an area of about 1000 sq miles to only about 100 sq. miles, conversion of the entire tidal delta into a tidal marsh and development of sandy beaches and sand dunes along the former deltaic coastline.

The deltaic sediments are laminated and mainly comprise very fine sand in the deltaic flood plain region - and became progressively finer towards the sea. In the tidal delta region and in the shelf area in the immediate vicinity of the delta the sediments are mainly comprised of silt and clays. The shelf deposits contain abundant marine microfossils.

In response to the changes in the geological environment, the Indus has shifted its course frequently accompanied by corresponding changes in the position of its delta. The earliest geological trace of the Indus delta goes back to the Early Miocene when the delta was located in the vicinity of Bugti area near Sibi. At that time the Indus plain had just emerged as vast monoclinal pediplain, consequent to its upwarping due to the collision of the Indian and Eurasian Plates. Subsequently the delta migrated southward until the Pleistocene when it occupied a position in the vicinity of the Rann of Cutch. Apparently, during the Mid. Pleistocene the delta was located north of the Rann of Cutch. In subsequent periods it shifted its position westward.

In the lower Indus Plain, subsurface borehole data reveals a buried, entrenched channel of the Indus, which is linked with the lowering of the sea level during the Last Glacial period, and its subsequent filling up with sand and gravel during the Late Glacial and Early Recent.

One would expect that such cut and fill structures may have formed during each glacial cycle of the Pleistocene. If more such entrenched channels could be located, our inventory of fresh ground water storage would increase considerably. Gravity maps (Fig. 7) show a trench-like feature east of and parallel to the East Nara Canal, traversing Cholistan and Thar desert. No test holes have been drilled in this region. It is likely that this trench may contain buried channels of the Indus dating back to the Mid. Pleistocene Ice Ages. If subsurface exploration fails to locate other major buried channels in this zone of low gravity anomalies of the desert areas, we can conclude that during previous glacials the Indus entrenched itself along its present course, in which case what we see in the geological sections (Figs. 9 to 13) is in effect the major trench of the Indus in which the river was entrenched repeatedly during the last three glacial periods.

REFERENCES

Balkrishan, T.S., 1977. Role of Geophysics in the study and Tectonics. A.E.G., 1:9-27 pp.

Balanford, W.T., 1876. On the geology of Sind. GSI., Rec., 9 (1): 1-15 pp.

DeJong, K.A. and A.M. Subhani, 1979. Note on the Bela Ophiolities with speical reference to the Kanar area. *In:* A. Farah and K.A. DeJong (eds.), *Geodynamics of Pakistan*, 263-269 pp.

Glennie, E.A., 1956. Gravity data and crustal warping in northeast Pakistan and adjacent parts of India. Royal Astron. Soc., Monthly Notices, Geophysics Supp., 7(4): 163-175 pp.

Hunting Survey Corporation Ltd., 1960. Reconnaissance Geology of Part of West Pakistan. (Colombo Plan Coop. Project) Canadian Govt., Toronto, 550 pp.

Kazmi, A.H., 1966. Geology of the Indus Plain. Unpublished Research Paper submitted to the Geology Deptt., Cambridge University, UK, as Nuffield Scholar. 98 pp.

Kazmi, A.H., 1977. Review of the Quaternary Geology of the Indus Plain. Presidential address, Earth Sciences Section, 18th Annual Conference, Multan, Scientific Society of Pakistan (In Urdu): 26 pp.

Kazmi, A.H., 1979. The Bibai and Gogai nappes in the Kach-Ziarat area of northeastern Baluchistan. *In:* A. Farah & K.A. DeJong (eds.) *Geodynamics of Pakistan*, 333-339 pp.

Kazmi, A.H., 1982. Tectonic Map of Pakistan. Published by Geological Survey of Pakistan.

Pascoe, E.H., 1964. A manual of the Geology of India and Burma, Vol.I, Calcutta. Third Edition, 1345-2130 pp.

Pascoe, E.H., 1965. A manual of the Geology of India and Burma. Vol.III, Calcutta. Third Edition 485 pp.

7

Deltaic Morphology and Sedimentology, with special reference to the Indus River Delta

J. T. WELLS and J. M. COLEMAN
Louisiana State University,
Baton Rouge, Louisiana

ABSTRACT

The patterns of sedimentation and morphologic development of a delta result primarily from the interaction of fluvial and marine processes. Historically, the Indus River delta has formed in an arid climate under conditions of high river discharge (400 x 10^6 metric tons of sediment/year), moderate tide range (2.6 m), extremely high wave energy (14 x 10^7 ergs/sec), and strong monsoon winds from the southwest in summer and from the northeast in winter. The resulting sandy, lobate delta, lacking in luxuriant vegetation and dissected by numerous tidal channels, has prograded seaward during the last 5000 years at an average rate of perhaps 30 m/year. Morphology of the Indus Delta lies midway between that of a fluvially dominated delta (elongate, protruding distributaries) and a high-energy wave-dominated delta (beach, beach-ridge, and downdrift deposits).

Whereas sands provide a substrate for the subaerial delta, silts and clays provide material for fill in abandoned channels, delta front outer shelf deposits, and downdrift sedimentation to the east. Coarse sediments from the Indus River generally remain on the inner shelf or undergo transport to deeper water via the Indus submarine canyon. Little of the fine-grained sediment remains within the delta, since maximum river discharge occurs during southwest monsoons, resulting in transport of the muds southeast into the Ranns of Kutch.

Extensive engineering works for irrigation purposes appear to have reduced sediment load to 100 x 10^6 metric tons/year and may reduce it within the next 20 years to virtually zero. This decrease in sediment load, together with the extreme levels of wave energy, will cause rapid wave reworking and transgression of the Indus Delta, not unlike that experienced in other deltas in similar settings, such as the Nile Delta in Egypt and the Tana Delta in Kenya. The end product will be a wave-dominated delta, characterized as a transgressive sand body, capped by extensive eolian dune deposits.

INTRODUCTION

The 30-35 major deltas of the world display a wide range of sedimentary environments and configurations as a result of the complex interactions between marine and fluvial processes. Whereas some deltas experience low wave energy and negligible tides, others are exposed to continuous and severe wave forces or to tide ranges that may exceed 5 m. Many deltas, such as the Mississippi River delta, are dominated by silt- and clay-sized particles, whereas others, such as the Burdekin River delta in Australia, are composed almost exclusively of sand and gravel. The common attribute shared by each of these deltas, regardless of environmental setting, is the ability to accumulate fluvial sediments more rapidly than they can be removed by marine processes.

Previous research has shown that deltaic morphology and sedimentology are a function of numerous processes, most notably climate, sediment yield and type, wave power, tide range, nearshore currents, shelf slope, and tectonic activity (Coleman and Wright, 1975). Attempts to incorporate some or all of these process variables into models for discriminating delta types have resulted in at least three classification schemes. Fisher et al. (1969) proposed high constructive and high destructive delta types based on relative intensity of fluvial and marine processes. Coleman and Wright (1971) and Wright et al. (1974), using a

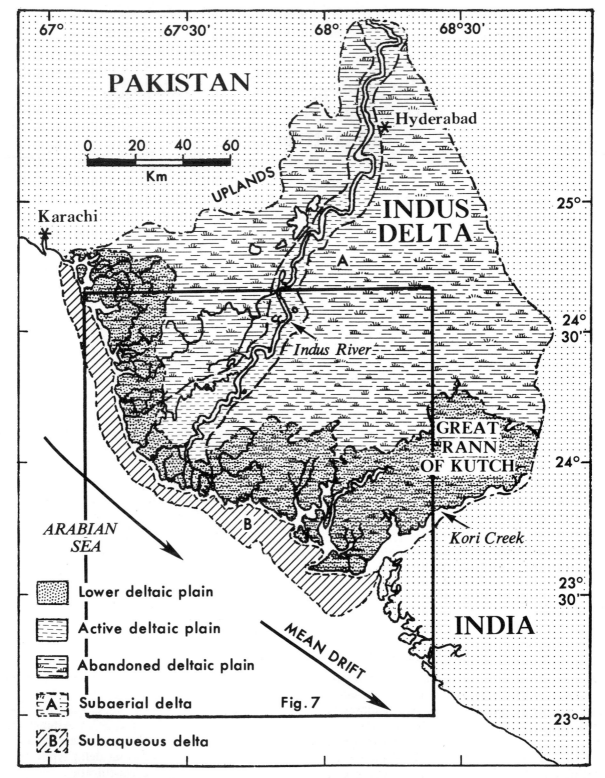

Figure 1. Index map of the Indus River delta in southeastern Pakistan showing the physiographic subdivision of the deltaic plain.

broad range of parameters, quantified the process variables, then clustered deltas into discrete statistical groupings. Most recently, Elliott (1978) proposed a classification scheme based on the earlier work of Galloway (1975) wherein deltas were plotted on a ternary diagram to define general fields of fluvial, wave, and tide dominance.

The most significant aspect of these more recent classification schemes is the recognition of the role of processes in producing responses. In this respect, climate is perhaps the single most important factor in determining fluvial processes and the types of sediment introduced to a given delta. At one end of the spectrum, the Indus River delta in Pakistan can clearly be included in the category of dry (arid) tropical to subtropical deltas. The environmental setting of the Indus is one of low rainfall (35 cm/year), surrounding deserts, erratic discharge, and monsoonal wind patterns. The arid climate, together with a moderate to high tide range along the coast, produces interdistributary evaporites and barren, halite-encrusted mudflats in the adjacent Ranns of Kutch (Glennie and Evans, 1976).

Numerous other deltas occur in similar settings, e.g., the Nile (Egypt), Shatt-al-Arab (Iraq), Senegal (Senegal), Tana (Kenya), and Ebro (Spain). As a generalization, these, too, have deltaic plains largely devoid of vegetation that are characterized by calcretes (Ebro delta) or salinas (Nile delta). Many of the arid deltaic plains are dominated by eolian dune fields of sand that has been eroded from active and abandoned beach ridges of barrier beach shorelines (Elliott, 1978).

It is interesting to note that the Indus River delta, certainly one of the largest in the world, is not included in the classification schemes of Fisher et al. (1969), Galloway (1975), or Elliott (1978), nor does it logically fall into any of the six classification categories of Coleman and Wright (1975). In the following paragraphs we will examine the Indus Delta in more detail and uncover its important and unique characteristics by comparing it to other deltas worldwide.

THE INDUS DELTA

The Indus Delta forms a significant protuberance of clastic sediments, introduced into the northern Arabian Sea by way of the Indus River (Fig. 1). Draining the Himalaya Mountains to the north, the Indus River flows south as a braided stream in its upper valley, but displays a well defined meander belt in the lower reaches of the valley. Sediments of the meander belt consist of river bar deposits and natural levees on a landscape marked by abandoned channels and crescentic meander scars (Holmes, 1968). The complex nature of the river is shown by the numerous tributaries (and distributaries), many from previous river courses such as the Jacobabad, Nawabshah, and Nara.

In tracing the Indus River throughout historic times, Holmes (1968) has shown that the last major change in course occurred in 1758-59 when the river adopted its present course west of Hyderabad (Fig. 1). In 1819, after the lower part of this course became choked with silt, the river began delivering its discharge farther east; each successive shift in the mouth of the river produced a new locus of deposition. The present-day configuration of the delta reflects the extent of these changes in river course.

According to Holmes (1968), the head of the Indus Delta lay 55 km northeast of Hyderabad in historic times; rates of progradation during the last 5000 years appear to have been on the order of 30 m/year. Today, extensive flood protection levees line the banks of the Indus River to prevent overbank flooding and unwanted diversions in the river course. Much of the flood water is diverted through what has become one of the most extensive irrigation systems in the world.

Initial size of the sediment load, amount of discharge, and fluctuations in discharge are determined in the drainage basin to the north of the delta. It is here that the sediment and water originate. Thus the size of the drainage basin provides some measure of the size of a delta. A plot of deltaic plain area versus drainage basin area shows the Indus River to be "average" in that it lies close to the line of least-squares best fit (Fig. 2).*

The deltaic plain alone, that area from the shoreline to the alluvial valley, covers 29,500 km² in the shape of a broad fan (Fig. 1). The plain of the Indus Delta is slightly larger than that of the Mississippi Delta and lies midway on a scale worldwide that ranges from less than 1×10^3 to greater than 4×10^5 km² (Fig. 3). Built by progradation where the river has become a dispersal system rather than a transporting agent, the subaerial delta has been growing most rapidly

*Data used for construction of the scatter plots and bar graphs presented in this paper were compiled by Coleman and Wright (1975). Their values were taken from numerous published and unpublished reports, maps, and data files; definitions and methods can be found in Coleman and Wright (1971) and Wright et al. (1974).

to the east as the river attempts to migrate in this direction. If permitted, all points on the Indus River south of Ganjo Takar, where it makes its last major bend to the west, are potential danger

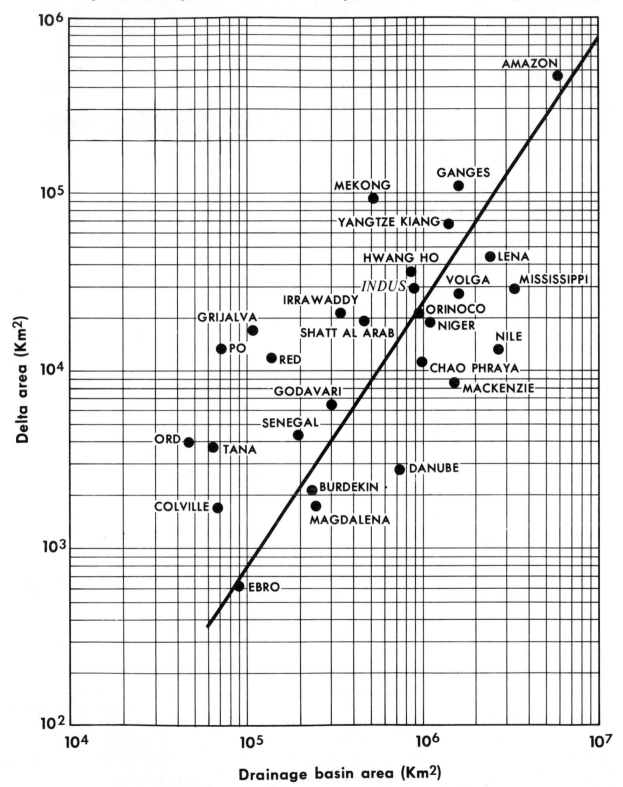

Figure 2. Plot of delta plain area versus drainage basin area for major deltas of the world.

points for a new diversion and future progradation (Holmes, 1968).

Figure 1 shows the subdivision of the Indus Delta into an active and abandoned deltaic plain.

Once the river ceases to deliver sediments as a result of artificial levees or natural diversions, this area of the delta becomes an abandoned deltaic plain. The lower deltaic plain is delineated by the

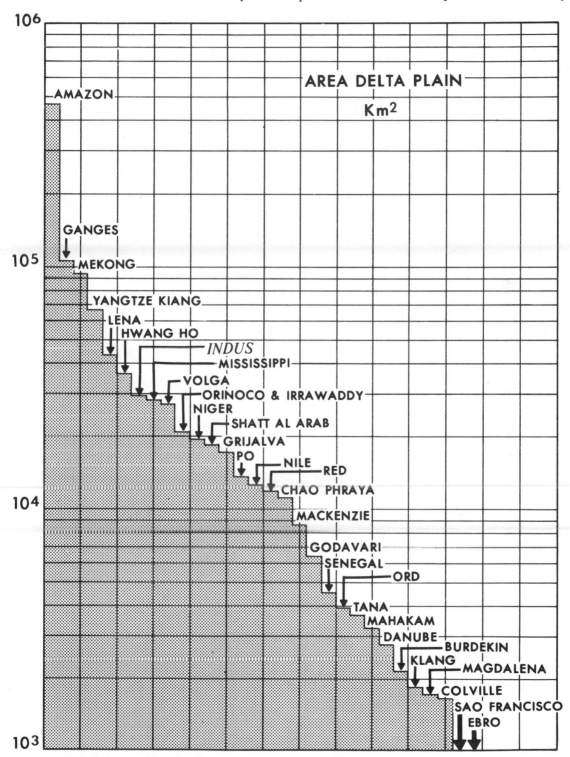

Figure 3. Bar graph of delta plain area for major deltas of the world.

Figure 4. Mangrove-Lined tidal creeks on the lower delta plain of the Indus River.

landward boundary of saltwater intrusion, a boundary that usually does not follow that of the active/abandoned delta plain boundary. In the case of the Indus Delta, substantial storm tides from southwest monsoon winds in summer inundate vast areas of both the active and the abandoned deltaic plains with salt water. This area of lower deltaic plain is characterized by tidal creeks and small overbank splays that are lined with stunted mangroves on a sand/silt substrate (Fig. 4).

An erratic discharge, such as that of the Indus River, is characterized by a predominance of coarse sediments since little opportunity exists in these regimes for sorting of sediments prior to reaching the delta. Although average discharge of the Indus River is nearly an order of magnitude less than that of the Mississippi River (Fig. 5), it is highly erratic, with summer discharge occasionally reaching 30,000 m³/sec, nearly twice the average value for the Mississippi. Sediment discharge is usually reported to be 435-480 x 10⁶ short tons (395-435 metric tons) per year (Holeman, 1968),

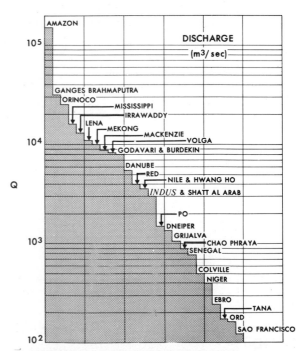

Figure 5. Bar graph of river discharge for major deltas of the world.

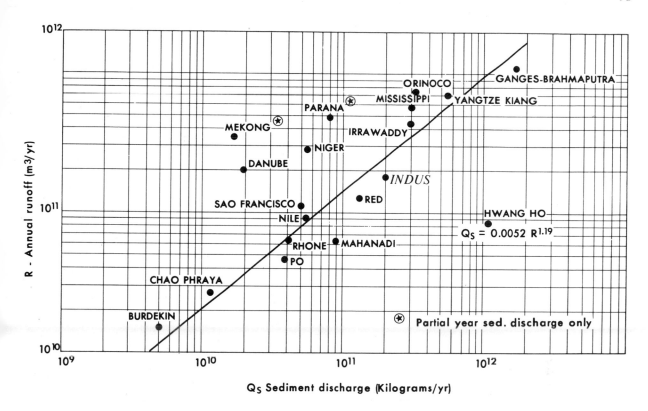

Figure 6. Plot of annual runoff versus sediment discharge for major rivers of the world.

with suspended sediment peaks during August that reach an incredible 3000 mg/*l* (Holmes, 1968). Of particular importance is that, as of the 1960s, sediment discharge of the Indus River was ranked (by most estimates) to be fifth or sixth highest in the world.

Figure 6 shows the relationship between annual river runoff and sediment discharge. Of the 19 data points plotted in Figure 6, nine rivers show greater runoff than the Indus, but only five have greater sediment discharge. This is largely attributable to the erodible sediments of an arid climate and to the fact that the Indus River normally experiences net evaporation over precipitation. Also, it is likely that wind-blown sands from the adjacent desert act as an additional sediment source during northeast monsoon winds

Despite the numerous changes in river course over the last 5,000 years, the Indus Delta has maintained a relatively straight shoreline. The major indentations are now formed by tidal channels, small bays, and, to the east, a large waterway known as Kori Creek (Fig. 7). One measure of crenulation and thus of subaerial complexity of the lower delta plain is the ratio between shoreline length and delta width. With a

ratio of less than two "shoreline kilometers" for each one kilometer of straight coastline, the Indus Delta stands in contrast to highly crenulated deltas such as the Ganges or Mississippi (Fig. 8). Whereas high rates of subsidence (Mississippi Delta) tend to produce a highly crenulated shoreline because of the loss of a coherent delta front, high wave energy tends to produce a straight shoreline because of the strong longshore transport, which spreads sands parallel to depositional strike.

Most deltas extend seaward, well beyond the shoreline, as a platform of sediments that have been deposited subaqueously. As the delta progrades offshore, coarse particles are often deposited over a blanket of delta-front silts and clays laid down under deeper water conditions. Continued progradation produces a coarsening-upward sequence of sediments as shelf-depth waters eventually become subaerial land.

The Indus River has produced a delta that today is largely subaerial (Fig. 9). Similar to the Nile River delta, the ratio of subaerial to subaqueous delta area is nearly 10:1. A relatively coarse sediment load deposited in shallow water has been largely responsible for rapid

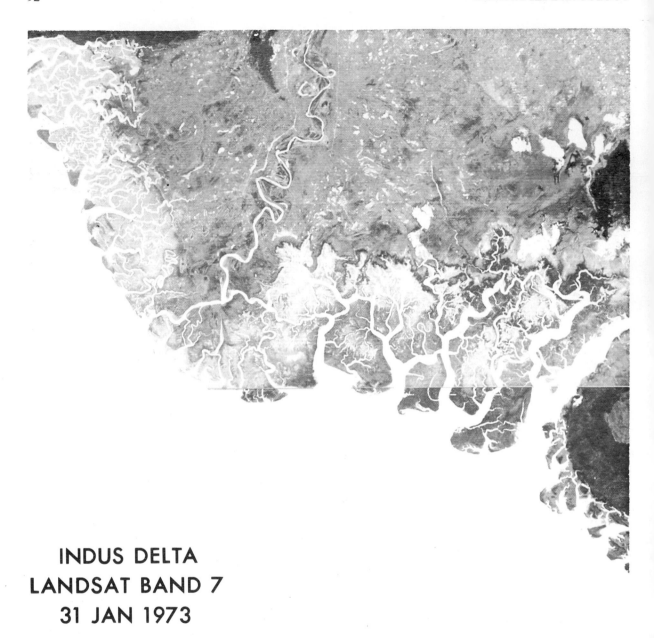

INDUS DELTA
LANDSAT BAND 7
31 JAN 1973

Figure 7. LANDSAT band 7 image of Indus Delta (see Fig. 1 for location).

advancement of the subaerial delta. Although the shelf widens near Karachi to 100-160 km and has a shelf break at 130 m (Closs et al., 1974), the Indus submarine canyon restricts subaqueous delta development by funneling sediment hundreds of kilometers offshore to the Indus cone (Nair et al., 1982).

Tides have been of at least moderate importance in developing the morphology and sedimentology of the Indus Delta. With a tidal excursion of 2.6 m, the Indus Delta can be

included in the 2-3-m mesotide-range category that characterizes many Asian deltas, e.g., the Irrawaddy, Mekong, Shatt-al-Arab, and Chao Phraya (Fig. 10). The role of tides in the Indus Delta has been to produce 1) limited flooding and the formation of evaporites, 2) substantial mixing of riverine and seawater, 3) bell-shaped river mouths and intricate tidal creeks, and 4) bidirectional sediment transport patterns.

Flooded soils and saltwater intrusion have been long-standing problems in the lower Indus

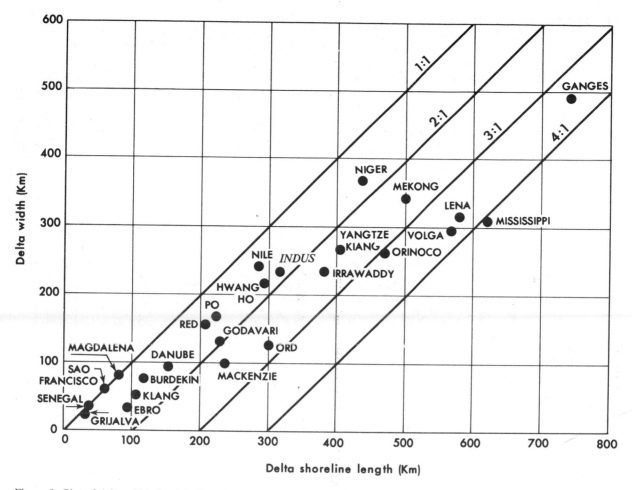

Figure 8. Plot of delta width (straight line distance between lateral extremes) versus shoreline length (digitized) for major deltas of the world.

Delta. Although the area of the lower delta plain (and hence the area of saltwater intrusion) is extensive (Fig. 1), much of this land, such as the Ranns of Kutch, receives its saltwater flooding from sea level elevation associated with the southwest monsoon winds in summer. Likewise, the formation of evaporite deposits is most pronounced in areas subject to the rhythm of annual rather than daily flooding. The degree to which density stratification is affected by tide is unknown. What is known is that tides extend the zone of marine and riverine interactions both vertically and horizontally. With a shelf gradient on the order of 0.1 degree, tides of 2.6 m range would provide an intertidal zone 1,500 m wide. Morphologically, there appears to be a slight tendency toward the formation of bell-shaped river mouths (Fig. 7) that are so common in macrotide-range deltas such as the Ord River delta, and a strong tendency for the development

of intricate tidal creeks (Fig. 4) such as those of the Ganges River delta. Finally, tides are most important sedimentologically in producing reversals in current direction, thus leading to the bidirectional transport of sediment.

Waves have been the single most important process variable in shaping the Indus Delta. Figure 11 shows that, at a water depth of 10 m, the Indus Delta receives the highest average wave energy of any major delta in the world. Intense monsoonal winds arriving from the southwest during May through September are responsible for annual maxima in wave energy that produce such an abnormally high average energy level. Nearshore wave energy, however, shows only weak correlation with offshore wave energy. At the shoreline, for example, the Indus Delta assumes a wave energy ranking of fourth highest, an order of magnitude less than the Senegal and Magdalena deltas (Fig. 12). Nearly three orders of magnitude

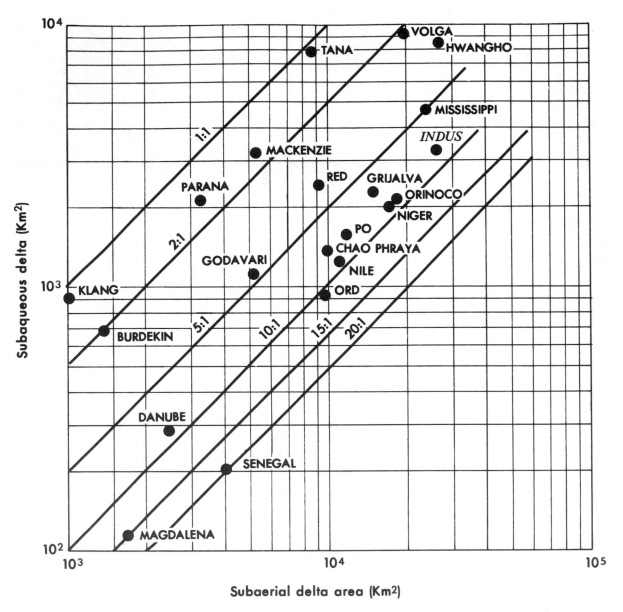

Figure 9. Plot of subaqueous delta area (below mean sea level) versus subaerial delta area (above mean sea level) for major deltas of the world.

greater than the Shatt-al-Arab Delta at the lower end of the scale, the Indus Delta receives more wave energy at the shoreline in a single day than the Mississippi Delta receives in a full year. The change in relative position of the Indus Delta (Fig. 11, 12) is a function of offshore slope, since the amount of wave energy at the shoreline depends mainly on the subaqueous profile; the flatter the slope, the greater the attenuation of deepwater wave energy. Figure 13 shows this relationship in a plot of offshore slope versus wave power at the delta shoreline.

The major effect of incoming wave energy is to sort and redistribute sediments. Waves induce strong longshore currents, straighten shorelines, and produce interdistributary beaches and beach-ridge complexes. In general, an increase in nearshore wave energy (relative to discharge) leads to a shoreline with gentle arcuate protrusions and a uniformly advancing delta front. However, as pointed out by Wright (1978), nearshore wave climate alone is insufficient to explain the degree to which delta morphology is wave dominated. Thus the Indus Delta, even with extreme levels of

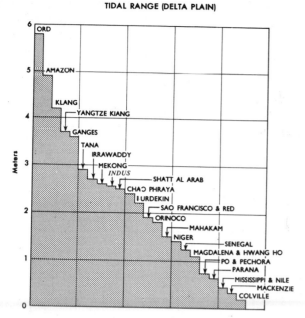

TIDAL RANGE (DELTA PLAIN)

Figure 10. Bar graph of the tide range for major deltas of the world.

wave energy, differs substantially from true wave-dominated deltas such as the Sao Francisco. High discharge from the Indus River has been able to provide sediments to the receiving basin at a rate sufficient to keep pace with the sediment-reworking ability of waves. This is reflected morphologically by the lack of extensive beach and beach ridge deposits and the presence of a broad, arcuate, and moderately indented shoreline.

Dispersal of Sediments

Terrigenous input from the Indus River dominates sediments in the northern Arabian Sea. Sediments are characterized by coarse size, low $CaCO_3$ content, and an abundance of quartz and feldspar. Except for the outer shelf, which has a band of calcium-carbonate-rich sediment, most of the shelf off eastern Pakistan has uniform carbonate content, ranging from 14% to 19% (Stewart et al., 1965). High concentrations of illite and chlorite are common in the clay fraction; toward India, montmorillonite increases, but kaolinite remains virtually absent in northern Arabian Sea sediments (Stewart et al., 1965). The Persian Gulf and Red Sea are not considered to be important sources of sediment to the Indus shelf, but significant contributions may be derived from the dustbearing winds blowing off Africa in summer and northern India in winter. Beyond the

shelf, the median size of sediment decreases to that of silts and clays on the slope, rise, and adjacent sea floor (Kolla et al., 1981).

Sediments discharged by the Indus River into the northern Arabian Sea may be transported farther offshore, accumulate on the continental shelf offshore of the delta, or may be transported by longshore currents to the southeast. Much of the coarse sediment is carried directly offshore to the Indus Fan by way of the Indus submarine canyon (Islam, 1959). The head of the canyon, referred to as The Swatch, is remarkably well aligned with the Kahr distributary of the Indus River and lies only 6.5 km offshore in 35 m of water. As a funnel-shaped feature more than 15 km wide at its head, the canyon provides a direct conduit for offshore transport of sediments. That much of the sediment from the Indus River has followed this route to deeper water is shown by the extent of the Indus Fan, the largest physiographic feature in the Arabian Sea. The Murray Ridge, to the west of the Indus Fan, serves as a sediment dam, thus preventing deposition of Indus River material in the Gulf of Oman.

The size and shape of the Indus Fan suggest that turbidity currents have been important in its formation (Heezen and Laughton, 1963). Transport of Indus River sediments as far as 1500 km into the pelagic regions of the Arabian Sea has resulted in the accumulation of unconsolidated sediments within the Indus Fan (Ewing et al., 1969). Tongues of exceptionally high bottom-water turbidities 1000-1500 km offshore, together with low $CaCO_3$ bands in the easternmost fan (Kolla et al., 1981), support the concept of deep turbid-layer flows. Despite the massive accumulation of terrigenous sediments in the distal fan, the regional increase in $CaCO_3$ at the shelf edge suggests a sedimentary contribution by the Indus River that is highly restricted on some parts of the outer shelf (Nair et al., 1982).

Coarse sediments that are discharged to the northwest or southeast of the Indus canyon will continue to accumulate in the delta or on the inner continental shelf. The protrusion of the Indus Delta and the wide shelf southeast of Karachi are certainly the result of extensive terrigenous sedimentation during the Holocene. Although the percentage of sediment load retained in the delta and inner shelf environments is unknown, seasonal events appear to be quite important in retention of sands because of the in-phase relationship between flood stage discharge and

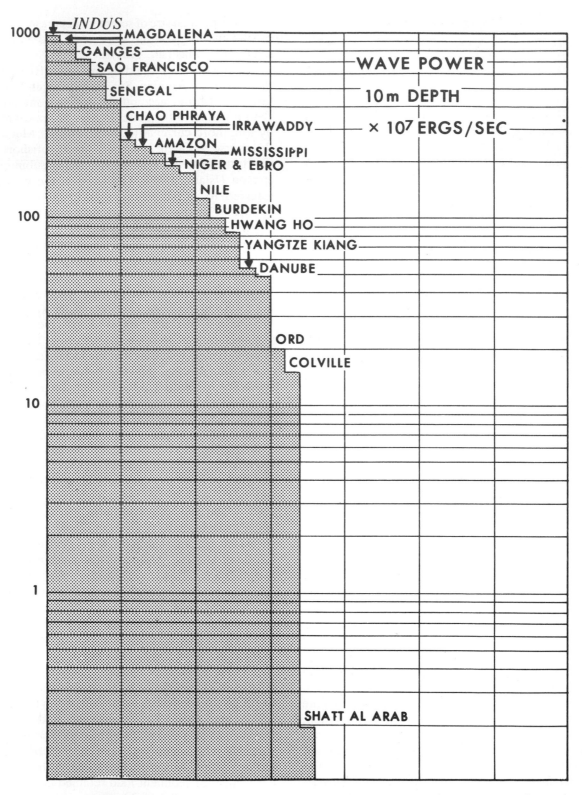

Figure 11. Bar graph of wave power per unit crest width at the 10-m contour for major deltas of the world.

monsoon wind setup. Since maximum wave energy from the receiving basin coincides with maximum sediment discharge from runoff, most of the sediments introduced to the shelf are

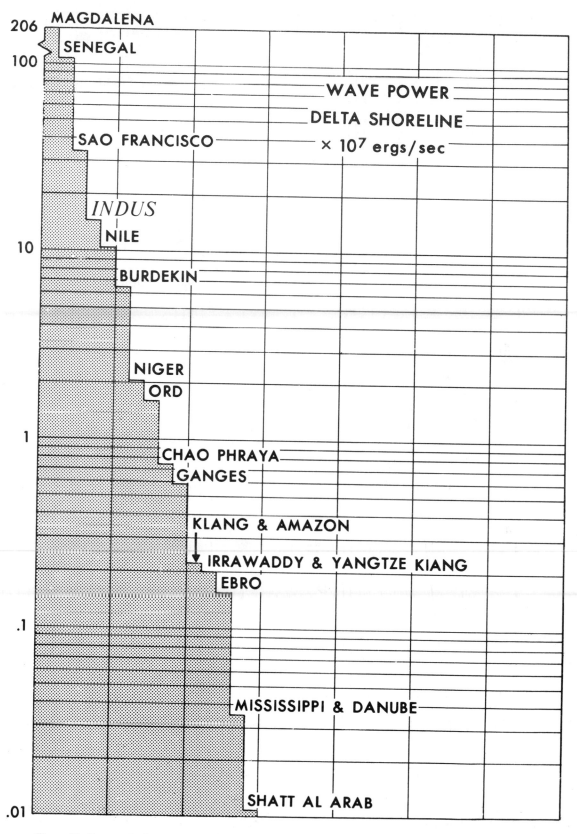

Figure 12. Bar graph of wave power per unit crest width at the delta shoreline for major deltas of the world.

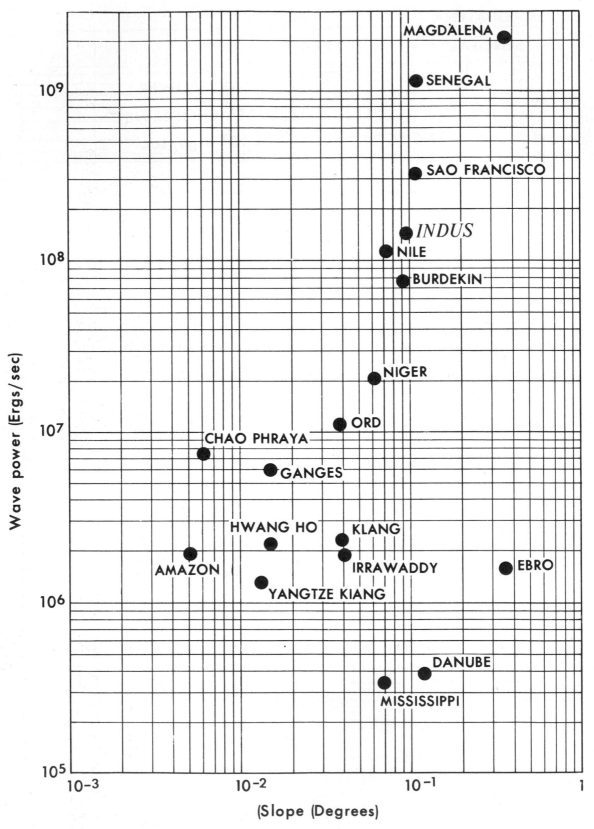

Figure 13. Plot of wave power at the shoreline versus offshore slope for major deltas of the world.

Figure 14. Wind-blown sands along the western margin of the Indus Delta.

immediately reworked and dispersed. Even more important, however, is the fact that the monsoon winds force surges of waters into the lower delta, effectively moving the receiving basin 20-30 km inland, thus promoting retention of sand within the subaerial delta.

Dispersal of sands by longshore currents appears to be relatively minor. As pointed out previously, extensive beach ridges, spits, and other downdrift sand deposits do not have a widespread occurrence. On the other hand, development of extensive mudflats on the north shore of the Gulf of Kutch and within the Ranns of Kutch indicate that transport to the southeast of silts and clays is significant (Glennie and Evans, 1976). The salt-covered deserts of the Ranns of Kutch and the broad Gulf of Kutch embayment receive their fine-grained sediments from storm-tide floods of the southwest monsoon after sediments are transported southeast by wave-and wind-induced longshore currents. Nair et al. (1982) have shown that surface tidal currents of 1.5 – 2.5 knots at the mouth of the Gulf of Kutch

serve as a barrier to the transport of sands as bedload and coarse mica as suspended load, thereby restricting accumulation of nearly all delta-derived sediment to the northwest areas of the India-Pakistan shelf.

Future of the Indus Delta

The future of the Indus Delta is uncertain. Extensive irrigation works during the 20th century have diverted much of the water and sediment away from their natural receiving basin. According to Milliman and Meade (1983), discharge from the Indus River is now only 100×10^6 metric tons/year, a value that reflects a fourfold decrease from previous estimates of discharge. Continued upstream diversions may, in fact, decrease discharge to nearly zero by the turn of the century (S.C. Snedaker, personal communication).

The immediate effect will be an attempt by waves to rework and straighten the shoreline of the existing delta front. As this happens, the "discharge effectiveness" will decrease, i.e., the ratio of discharge per unit width of river mouth to the nearshore wave power per unit width of wave crest (Wright, 1978) will fall from its present value of 1.1×10^{-2}, ultimately reaching a value on the order of 10^{-5} to 10^{-6}. The Indus Delta will then rapidly begin taking on the morphology of a wave-dominated delta.

The loss of discharge is critical for two reasons. First, the Indus Delta, throughout historic times, appears to have retained more sediment for subaerial land-building processes than has been transported by littoral processes. That is, the delta has been able to prograde under conditions of extreme wave energy only because of the high sediment discharge that was deposited rapidly and retained by the delta because of its coarse size. Second, the loss of discharge will cause the delta to become a transgressive sand body, dominated by wind-blown sand deposits. The most likely scenario for the next 25-50 years will be 1) cessation of subaqueous and subaerial delta-front progradation, 2) initial transgression of delta-front sands, 3) increase in wind-blown sands as a result of the loss of vegetation, and 4) establishment of a transgressive beach, dominated by eolian dunes. Figure 14 illustrates that early establishment of eolian sands on the western margin of the Indus Delta near Karachi has already begun.

Similar patterns of deterioration have been observed in other arid deltas subject to loss of discharge and vegetation through activities of man. Perhaps the two best examples are the Nile Delta in Egypt (Academy of Scientific Research and Technology, U.A.R., 1976, 1977; Coleman et al., 1981) and the Tana Delta in Kenya (United Nations Food and Agricultural Organization, 1967), both in advanced stages of deterioration. In the case of the often cited Mississippi Delta, presently experiencing one of the highest rates of land loss of any delta in the world, subsidence, lower overall discharge, and less sand in the sediment load are primarily responsible (Wells et al., 1983). To accurately predict the future of any delta, but particularly a delta such as the Indus, which is undergoing drastic reductions in discharge, will require a better establishment of the sediment budget, littoral processes, and patterns of sediment dispersal.

ACKNOWLEDGMENTS

Much of the long-standing research program on deltas at Coastal Studies Institute has been funded by the Coastal Sciences Program of the Office of Naval Research. Their continued support is gratefully acknowledged. We thank P.A. Byrne and L.J. Rouse, Jr., for reviewing the manuscript and G. Dunn for drafting the figures.

REFERENCES

Academy of Scientific Research and Technology, U.A.R., 1976. Proceedings of Seminar on Nile Delta Sedimentology, Alexandria, April 1975. UNDP/UNESCO/Dept. Geol., Alexandria Univ., Acad. Sci. Res. Technol., U.A.R., 257 pp.

Academy of Scientific Research and Technology, U.A.R., 1977. Proceedings of Seminar on Nile Delta Coastal Processes with Special Emphasis on Hydrodynamical Aspects, Alexandria, 2-9 October 1976. UNDP/UNESCO/Dept. Civ. Engr., Cairo Univ., Acad. Sci. Res. and Technol., U.A.R., 624 pp.

Closs, H., H. Narain, and S. C. Garde, 1974. Continental margins of India. In: C. A. Burk and C. L. Drake (eds.) The Geology of Continental Margins, pp. 629-639, Springer-Verlag, New York.

Coleman, J. M., and L. D. Wright, 1971. Analysis of major river systems and their deltas, procedures and rationale, with two examples. Tech. Rept. 95, Coastal Studies Inst., Louisiana State Univ., Baton Rouge, 125 pp.

Coleman, J. M., and L. D. Wright, 1975. Modern river deltas: variability of processes and sand bodies. In: M. L. Broussard (ed.) Deltas, Models for Exploration, pp. 99-150, Houston Geol. Soc., Houston Texas

Coleman, J. M., H. H. Roberts, S. P. Murray, and M. Salama, 1981. Morphology and dynamic sedimentology of the eastern Nile Delta shelf, Mar. Geol., 42: 301-326.

Elliott, T., 1978. Deltas. In: H. G. Reading, (ed.) Sedimentary Environments and Facies, pp. 97-142, Elsevier, New York.

Ewing, M., S. Eittreim, M. Truckam, and J. I. Ewing, 1969. Sediment distribution in the Indian Ocean. Deep-Sea Res. 16:231-248.

Fisher, W. L., L. F. Brown, A. T. Scott, and J. H. McGowen, 1969. Delta systems in the exploration for oil and gas. Bur. Econ. Geol., Univ. of Texas, Austin, 78 pp.

Galloway, W. E., 1975. Process framework for describing the morphologic and stratigraphic evaluation of deltaic depositional systems. In: M. L. Broussard (ed.) Deltas, Models for Exploration, pp. 87-98, Houston Geol. Soc., Houston, Texas.

Glennie, K. W., and G. Evans, 1976. A reconnaissance of the recent sediments of the Ranns of Kutch, India. Sedimentology, 23:625-647.

Heezen, B. C., and A. S. Laughton, 1963. Abyssal plains. In: M. N. Hill (ed.) The Sea, vol. 3, pp. 312-346, Interscience, New York.

Holeman, J. N., 1968. The sediment yield of major rivers of the world. Water Resources Res., 4:737-747.

Holmes, D. A., 1968. The recent history of the Indus. Geographical Jour., 134: 367-381.

Islam, S. R., 1959. The Indus submarine canyon. Pakistan Geographical Review, XIV:32-34.

Kolla, V., P. K. Ray, and J. A. Kostecki, 1981. Surficial sediments of the Arabian Sea. Mar. Geol., 41: 183-204.

Milliman, J. D., and R. H. Meade, 1983. World-wide delivery of river sediment to the oceans. Jour. Geol., 91: 1-21.

Nair, R. R., N. H. Hashimi, and V. P. Rao, 1982. On the possibility of high-velocity tidal streams as dynamic barriers to longshore sediment transport: evidence from the continental shelf off the Gulf of Kutch, India. Mar. Geol., 47: 77-86.

Stewart, R. A., O. H. Pilkey, and B. W. Nelson, 1965. Sediments of the northern Arabian Sea. Mar. Geol., 3:411-427.

United Nations Food and Agricultural Organization, 1967. Survey of the irrigation potential of the lower Tana River basin. Acres International, Canada, 3 vols.

Wells, J..T., S. J. Chinburg, and J. M. Coleman, 1983. Development of the Atchafalaya River deltas: generic analysis. U. S. Army, Corps of Engineers, Waterways Experiment Station, Final Draft Rept., 90 pp.

Wright, L. D., 1978. River deltas. In: R. A. Davis, Jr. (ed.) Coastal Sedimentary Environments, pp. 5-68, Springer-Verlag, New York.

Wright, L. D., J. M. Coleman, and M. W. Erickson, 1974. Analysis of major river systems and their deltas: morphologic and process comparisons. Tech. Rept. 156, Coastal Studies Inst., Louisiana State Univ., Baton Rouge, 114 pp.

8

Indus Fan: Seismic Structure, Channel Migration and Sediment-Thickness in the Upper Fan

F. COUMES
Elf Aquitaine, Pau, France
and
V. KOLLA
Superior Oil Company,
Houston, Texas

ABSTRACT

Seismic data suggest that the modern Indus Canyon off the Indus River, and atleast three other canyons that existed in the past on the Pakistan-India shelf, fed sediments to several channel-levee systems on the adjacent upper Indus Fan since Oligocene-Miocene uplift of Himalayas. The canyons on the shelf, when they were active, were primarily erosional. The channels were erosional-depositional on the upper fan and may be as much as 10 km wide. The channels in the upper fan as well as the feeder canyons on the shelf migrated extensively in space and time. The channel-levee complexes are stacked up into five or six layers in the upper fan. The uplift of the Murray Ridge, tectonics in the Indus River drainge basin, changes in sediment-input rates / sea level changes, plugging of channels by slumped sediment masses, Coriolis force and channel meander have all been responsible for these migrations.

From the sediment-thickness distribution, several basins or depocenters, separated by basement highs, can be distinguished in the upper fan. One of these basins, the offshore Indus Basin, located on the uppermost sections of the upper Indus Fan and shelf-slope has sediments with thickness exceeding 5.5 sec. (two-way travel time) (>11 km) and reflects sedimentation since rifting of the Indian Margin in late Cretaceous and during earlier times. However, the fan sedimentary sequences have been deposited since about late Oligocene or early Miocene as a consequence of Himalayan uplifts and sea level lowerings.

INTRODUCTION AND REGIONAL SETTING

The Indus Fan with its 1500 km length, 960 km maximum width, and 1.1×10^6 km² area is the most extensive physiographic province in the Arabian Sea. The fan is bounded by the passive continental margin of Pakistan-India, and Chagos-Laccadive Ridge on the north and east, by the Owen-Murray Ridges on the west, and by Carlsberg Ridge on the south (Figs. 1 and 2). The Indus River system draining the Himalayan mountains has been the dominant supplier of sediments to the fan. The Kirthar and Sulaiman mountains in Pakistan bound the Indus River on the west. Physiographically, the Murray Ridge and the Kirthar mountain range appear to be continuous with one another.

The Indus River is about 2900 km long. After leaving the high mountains, the river travels a distance of about 1000-1200 km in the plains before it joins the Arabian Sea. The total drainage basin of the Indus River is 966,000 km² (Krishnan, 1968). The environmental setting of the Indus River is one of mountainous, and tropical to subtropical climate with low rain fall (35 cm/yr), surrounding deserts, erratic discharge and monsoonal wind patterns. Snow-melting from the Himalayas and southwest monsoon winds contribute most waters to the river discharge during the summer months. Although the water discharge of the Indus River is an order of magnitude less than that of the Mississippi River, the sediment load of the former is comparable to that of the latter (Lisitzin, 1972). The total area of both the subaerial and subaqueous portions of the

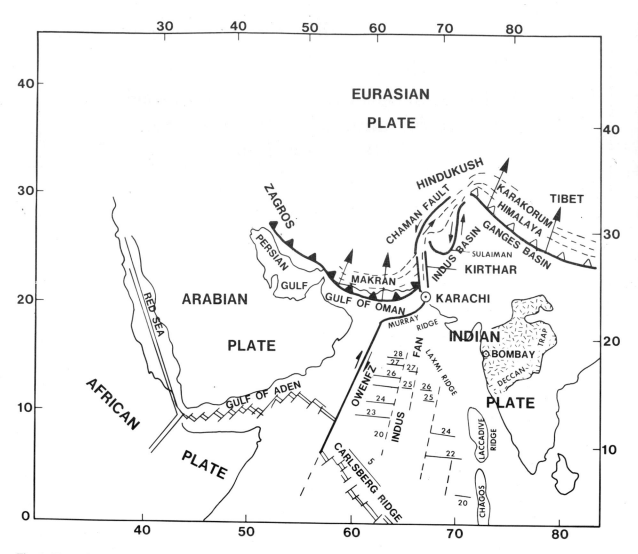

Fig. 1. Tectonic setting of the Indus Fan region (modified from Jacob and Quitmeyer, 1979). The major plates are: Indian, Eurasian, African and Arabian-Plates. The plate boundaries are, open tooth marks: thrusting associated with a continental collision zone; solid tooth marks: thrusting associated with an oceanic subduction zone; double line: spreading on a mid-oceanic ridge or rifting on continents; single solid line: transform fault-Owen fracture zone-Murray Ridge; lines numbered, 20, 22, etc.: Seafloor magnetic anomalies; dashed lines: fracture zones separating magnetic anomalies. Arrows indicate relative sense of plate motions.

Indus Delta is about the same as that of the Mississippi Delta (Wells and Coleman, 1984). However, because of high wave-energy impinging on the Indus Delta, this delta is intermediate in character between the river-dominated (e.g. Mississippi) and wave-dominated (e.g. Sao Francisco) deltas (Wells and Coleman, *op. cit.*).

The Arabian Sea evolved due to rifting of the Indian Margin from Madagascar during the late Cretaceous, and subsequent northward drift of the Indian Plate due to seafloor spreading (Fig. 1; see Naini and Talwani, 1982; Norton and Sclater, 1979). The first land-bridges across portions of the Tethys Sea north of the Indian craton were

established by the end of early Eocene due to the convergence of the Indian and Eurasion Plates, and the water flow in the Tethys Sea was somewhat restricted (Powell, 1978; Sahni and Kumar, 1974; Valdia, 1984; Haq, 1984). By about late Oligocene, the first collision of the Indian Plate with the Eurasion Plate resulted in a significant uplift of the Himalayas. However, it was during the middle to late Miocene, that the major uplift of Himalayas took place (Valdia, 1984; Krishnan, 1968). Another episode of the uplift took place during the late Pliocene to middle Pleistocene time (Valdia, 1984), and the Himalayan uplift continues to the present day.

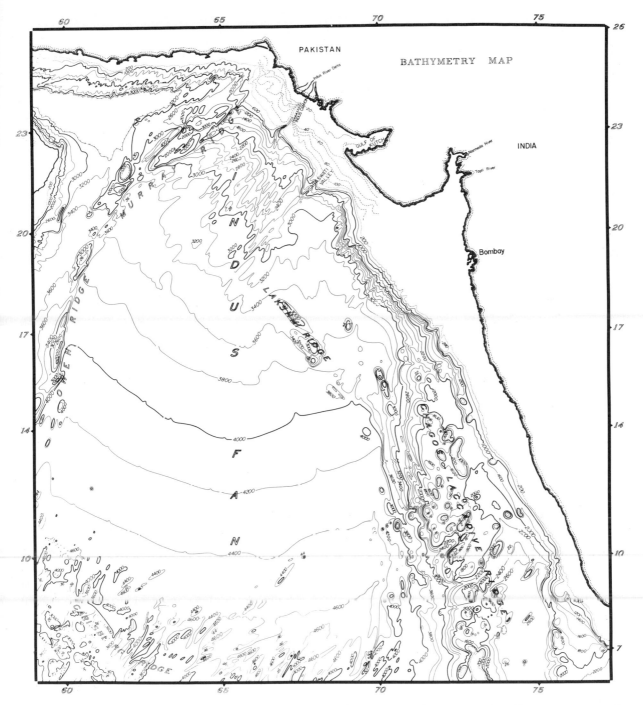

Fig. 2. Bathymetry of the Arabian Sea contours in meters compiled by U.S. Naval Oceanographic Office. Note the irregular topography n the upper Indus Fan resulting from channel-levee build-ups.

The Indus River came in to existence after Eocene, probably during late Oligocene, according the mineralogical data reported on DSDP site 221 (Weser, 1974). The Tethys Sea was gradually closed from north to south both towards the Arabian Sea and Bay of Bengal, probably by late Miocene. Concurrently, the Indus River built a delta advancing southward along the Indus trough bordering the Sulaiman and Kirthar mountain ranges. During the last 5000 years, the delta migrated southward at the rate of 30 m/year (Kazmi, 1984). Although the Indus River has its

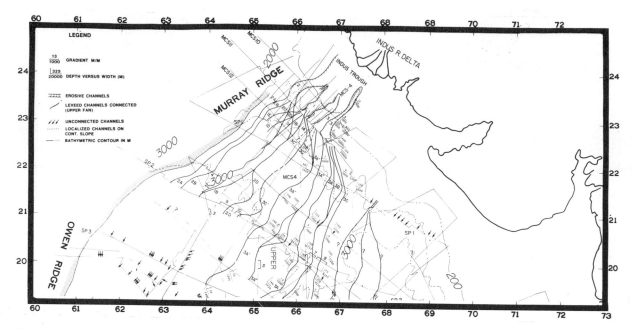

Fig. 3. Patterns of canyons (1, 2, 3) and channels (1A, 1B.., 2A, 2B.., 3A, 3B...) on the Indus Shelf and Fan. The ships' tracks along which multichannel seismic (MCS), sparker (SP) and other seismic data have been collected are shown by thin lines.

origins in active, highly uplifted terranes, the Indus Delta and Fan are located off the passive Pakistan-Indian Margin. However, on several seismic lines (not shown here), we observed channel-levee complexes in the deep Oman Basin west of the Murray Ridge similar to those on the Indus Fan east of the Murray Ridge. These channel-levee complexes in the Oman Basin are seen to be subducted along the Makran Coast. There is reason to believe that these channel-levee complexes in the Oman Basin were also fed by the Indus River. Thus, although the Indus Fan is primarily located off the passive margin, part of it is also located along the active Makran subduction zone (Fig. 1).

In this paper, we discuss bathymetry, seismic structure, channel migration and sediment-thickness distribution of the upper Indus Fan based on multichannel seismic and sparker data, and briefly relate the upper Indus Fan sedimentation to the geologic history of the region. (A detailed paper on seismic stratigraphy and sedimentation of the entire Indus Fan by Kolla et al. is in preparation).

BATHYMETRY

The bathymetry shown in Figure 2 exhibits the typical "fan" shape, although the presence of the Lakshmi Ridge and the northern extension of the Chagos-Laccadive Ridge complicate the topography somewhat in the eastern part of the Indus Fan. The water depths of the Fan range from 1400-1600m at the foot of the continental slope to 4500m at its distal end towards the Carlsberg Ridge.

The shelf-width is about 100-150 km off the Gulf of Kutch-Indus Delta. The shelf-break occurs at an average depth of about 100m along the Pakistan-India Margin. The most pronounced bathymetric feature of the shelf-slope is the Indus Canyon (Fig. 2) with an average width of 8 km and depth of the order of 800 m. The Canyon is 170 km long (Islam, 1959) and apparently commences around 20-30 m water depths on the shelf and ends at 1400 m depth at the foot of the continental slope, where the Canyon widens to 20 km with a relief of about 325 m. The width/depth ratio is around 12 on the shelf and increases to 60 at the foot of the slope. The Canyon appears to exhibit a broad meander on the shelf-slope. There is a pronounced bend in the Canyon at about 1000 m water depth towards east, which resembles channel bends in the northern hemisphere, attributed to Coriolis force. The Canyon gradient on the shelf is around 1/100, and on the slope 1/50 to 1/200. The gradient of the continental slope adjacent to the Canyon is 1/30 to 1/60. Another valley, Saraswati Valley (Fig. 2) of lesser magnitude is present off the Gulf of Kutch, and also appears to make a bend towards east. As

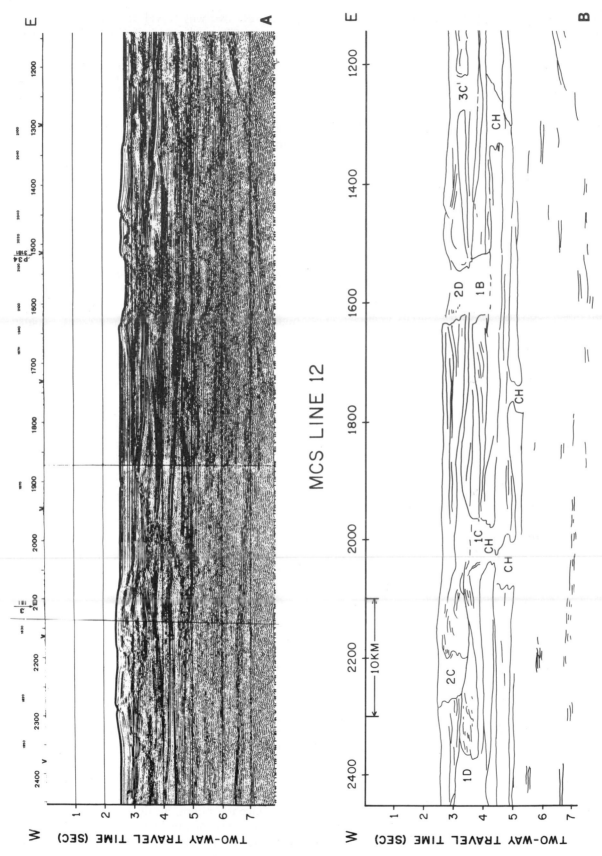

Fig. 4. A: A portion of multi-channel seismic line-12 (MCS-12 in Fig. 3) showing channel-levee complexes stacked into several layers in the upper Indus Fan. B: Line-drawing showing the authors' interpretation of the MCS record.

apparent on the bathymetric map (Fig. 2) and as clearly seen on seismic lines, atleast two canyons existed on the Indus Shelf to the west of the present Canyon.

The seafloor has a relief of several hundred meters in the upper fan and is relatively smooth in the lower fan. The conspicuous seaward bulges in the bathymetric contours of the upper fan reflect the channel-levee built-ups. Opposite the present Canyon, there is a pronounced seaward bulge in the contours down to 3,600-3,800 m and deeper, and results from sediment influx through the Indus Canyon.

SEISMIC STRUCTURE, CHANNEL BEHAVIOR AND MIGRATION

On multi-channel seismic and sparker lines, atleast three canyons on the shelf off Pakistan have been observed (Fig. 3; see McHargue et al., 1978; Coumes and Kolla, 1984). The most recent Indus Canyon is labelled "3" in Figure 3. Two older canyons 1 and 2 exist to the west of 3, and of these, canyon 1 is older than canyon 2. All the canyons essentially lack overbank-levee deposits and were erosional with respect to the surrounding strata when they were active. Available data suggest that these three canyons merged landward into one extensive erosional zone, called Indus Trough (Fig. 3). Seaward, these canyons lead into many channels (connected by continuous lines in Fig. 3) in the upper fan. Channels labelled 1A, 1B, 1C, 1D belong to canyon 1; 2A, 2B, etc. belong to canyon 2, and so on. A represents the oldest channel and other alphabet letters represent sequentially younger channels. More channels exist in the subsurface below the three sets of channels shown in Figure 3. These channels are labelled "CH" on seismic line in Figure 4 and have not yet been related to a specific canyon.

As the Indus Canyon, with an average width of 8 km on the shelf, descends on to the foot of the continental slope at about 1400 m water depth, it becomes very wide (20 km) and has apparently divided in to several channels (Kolla et al., in prep.). The widths of channels in the upper fan originated from the most-recent Indus Canyon as well as other channels originating from the other two canyons vary in width, and may exceed 10 km. The channel-levee heights in the upper fan may exceed 100's of meters. The combined widths of channels and levees may exceed 30-40 km. The thickness of the individual levees may be as much

as 0.6 secs two-way travel time (ca. 0.6 km; see, Fig. 4). The relief of the individual levees associated with each channel burried deeper in the sediment column is significantly higher than that of the levees of the near-surface (Fig. 4). We believe that this is primarily due to overburden and high compaction of very fine-grained sediments on portions of levees distal to channel floors compared to the sediments on portions of levees proximal to channel floors. The size of the channel-levee complexes generally increases upward in the sediment column (Fig. 4). This may reflect progradation of fan sedimentation and/or increase in turbidity current activity in time due to the uplift of the Himalayan mountains.

Both the canyon and channel floors consist of transparent zones and high amplitude reflections locally organized into groups (Figs. 5 and 6). From analogy with river channels, we speculate that the high-amplitude reflections represent coarse sediments and that the transparent zones represent homogenous fine-grained sediments.

TYPES OF CHANNEL MIGRATION

Both the canyons on the shelf and the channels in the upper fan migrated gradually as well as in "jumps" (Figs. 5 and 6). In gradual migration, channels and canyons moved usually in one direction for sometime as the banks on one side of channels eroded and receded, and the banks on the other side advanced due to deposition. At the foot of the continental slope, where the Indus Canyon is very wide (20 km), the gradual eastward migration of the Canyon is more pronounced than that on the shelf, causing a bend in the canyon path towards east. In jump migration, the channels and canyons abandoned their previous courses and jumped entirely to different locations, opening up new channels (Figs. 4, 5 and 6). The gradual and jump migrations may be related to: (1) sea level changes; (2) uplift of Kirthar-Sulaiman mountain ranges and tectonic uplift in the rest of the Indus River drainage basin, causing the eastward migration of the river; (3) uplift of the Murray Ridge; (4) Coriolis force; (5) channel meander (see Komar, 1969); (6) sliding, slumping and channel plugging; (7) sediment-buildups due to deposition at any one location, and as a consequence the tendency of channels to occupy the adjacent topographically low areas.

To the east of the Indus Canyon, there is yet another valley (Saraswati Valley) from which

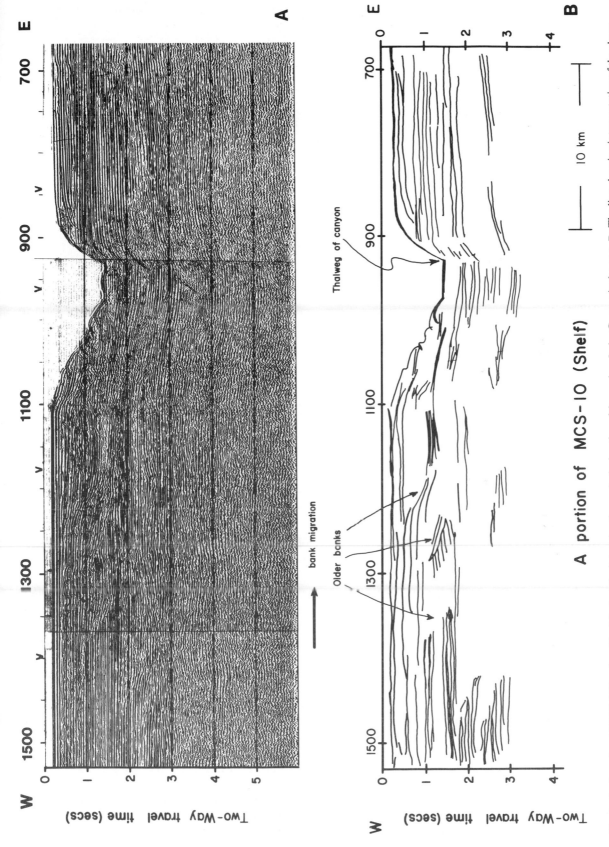

Fig. 5. A: A portion of MCS-10 line (see Fig. 3 for location) showing the gradual migration of the modern Indus Canyon in the past. B: The line-drawing interpretation of the above MCS record. Note the older banks and how they have migrated eastward.

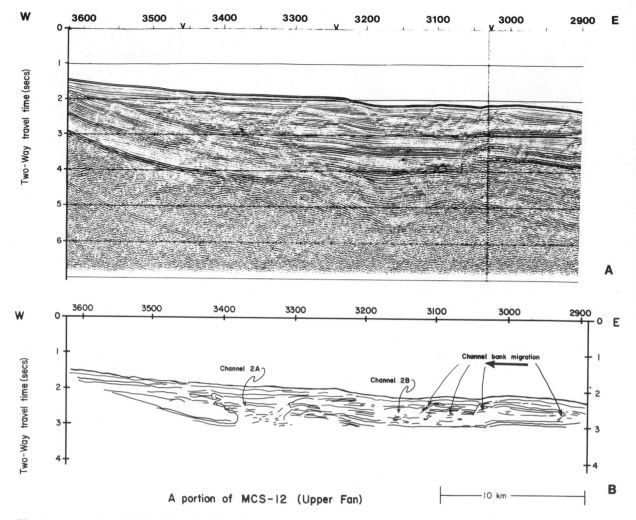

Fig. 6. A: A portion of MCS-12 line (see Fig. 3 for location) showing in detail the gradual and "jump" migrations of channels in the upper fan. B: Line-drawing interpretation of the above MCS record. Note how the banks of each of the channels 2A and 2B have migrated westward. Channel 2A is older than 2B; both of them originating from the canyon 2 (see Fig. 3). After some time, channel 2A was abandoned and channel 2B was opened. This channel abandonment and avulsion is an example of jump migration.

some channels in the upper fan also originated. Presently available data do not permit the assignment of ages to this channel complex relative to the Indus Canyon-channel complex. It is, however, clear that the upper Indus Fan consists of numerous channel-levee complexes which migrated extensively both in space and time. These channel-levee complexes are stacked in up to five or six layers, as can be seen on a portion of multichannel seismic line 12 (Fig. 4)

Apart from the types of shelf-slope and upper fan canyons and channels discussed above, there are several channels locally restricted to continental slope off the Indus Delta. These channels are connected by dotted lines in Figure 3. These channels may be V, U or spoon-shaped

depressions on the seafloor, and may reflect instabilities of the seafloor on the slope. Along the MCS dip-line 4, the continental slope represents a concave shape and the upper fan a pronounced convex shape (Fig. 7). The slope and upper fan morphologies imply sediment flushing or relatively more erosion on the slope, and high rates of sedimentation through channel-levee build-ups in the upper fan, respectively.

SEDIMENT-THICKNESS DISTRIBUTION IN THE UPPER INDUS FAN

The total sediment thickness of the upper fan in seconds of two-way travel time is shown in Figure 8. Three broad sedimentary basins or depocenters

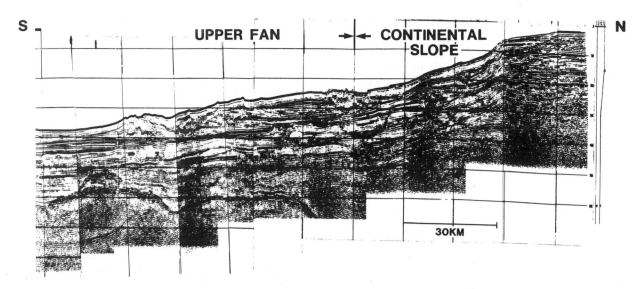

Fig. 7. A portion of dip-line (MCS-4) showing the concave and convex morphologies of continental slope and upper fan, respectively.

separated by a ridge complex can be distinguished in the upper Indus Fan: offshore Indus Basin, the Eastern Basin and the Western Basin (see also, Naini and Kolla, 1982).

The offshore Indus Basin itself can be divided in two smaller basins (Fig. 8). The sediment thickness in these two sub-basins is more than 5.5 seconds of two-way travel time (>11 km). However in the Eastern and Western Basins the sediment thicknesses are only slightly in excess of 2 seconds and 3 seconds, respectively. This total sediment-thickness distribution in the upper fan reflects sedimentation since rifting in late Cretaceous time. However, the sedimentation of the offshore Indus Basin may have begun even before rifting as a part of on-shore Indus Basin at the edge of Indian craton, before the Indian and Eurasian Plates colloided. From DSDP (Weser,

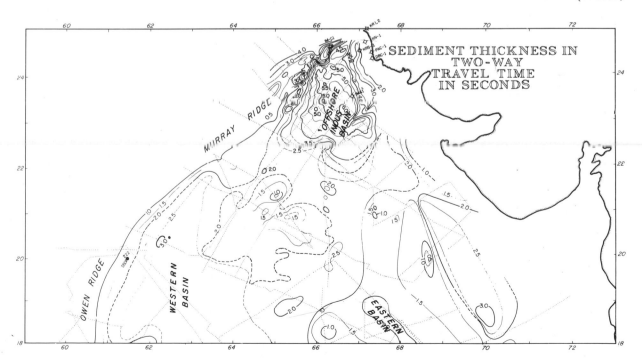

Fig. 8. Sediment-thickness distribution in seconds of two-way travel time. Note three major depocenters: Offshore Indus Basin, Western Basin and Eastern Basin.

1974) and Industry well data on the shelf (Shuaib, 1982), we infer that the fan sedimentary sequences began to be deposited in late Oligocene or early Miocene. The tectonic uplift of Himalayas, combined with the lowering of sea levels in late Oligocene and early Miocene, probably caused the initiation of fan sedimentation. The canyon 1 probably funnelled the sediment to the deep sea at this time. Canyons 2 and 3 came into existence during the later episodes of tectonic uplifts and sea level lowerings. It appears that the canyon 3 was active mostly during the Pleistocene time (Kolla et al., in prep.).

CONCLUDING REMARKS

Identifying areas with sand lithology in terrigenous clastic depositional systems, such as deltas, shelves and deep-sea fans in subsurface is essential to hydrocarbon exploration. Mapping the canyons, channels, etc., in these systems, because of their importance to predicting the lithologies, is a first step in this direction. Our mapping and brief discussions of canyons and channels on the Indus Shelf-Slope and Fan is based only on the seismic data along the ship's tracks shown in Figure 3. Much more data than we have utilized have been collected by the Industry and other agencies on the Indus Shelf and Fan. Seismic sequence and facies analysis of all these data as well as utilization of the data for accurate mapping of the paleo-canyons and channels will allow precise delineation of lithologies and should help identify locales of sand deposits.

The Indus Fan is one of the major deep-sea fans in the world ocean. Deltailed understanding of this fan has immense significance to subsurface fan exploration in general. Side-scan sonar, high resolution seismic and coring studies of the modern Indus Canyon and the deep-sea fan channels which originated from it, will provide a wealth of information on the anatomy, behavior (meander, migration) of channels, and the types of lithologies within these features, that hitherto are not available. Such information is critical in order to develop tools for predicting the lithologies in subsurface fan systems.

ACKNOWLEDGEMENTS

F. Coumes thanks Elf Aquitaine for permission to publish this paper. V. Kolla thanks Superior Oil Company for giving an opportunity to finalize the paper and Chris Owens for typing the manuscript. B. U. Haq has kindly reviewed the manuscript and offered helpful suggestions.

REFERENCES

Coumes, F. and V. Kolla, 1984, Channel migration in upper Indus Fan in relation to geologic history of the region. AAPG Bulletin, 466 p.

Haq, B.U., 1984; Paleoceanography: A Synoptic overview of 200 million years of ocean history. *In.* B.U. Haq and J.D. Milliman (eds.): *Marine Geology and Oceanography of Arabian Sea and Coastal Pakistan.* Van Nostrand Reinhold, New York pp. 201-231 (this volume).

Islam, S.R., 1959; The Indus submarine canyon. Pakistan Geographical Review, XIV: 32-34.

Jacob, K.H. and R.C. Quittmeyer, 1978; The Makran Region of Pakistan and Iran: Trench-arc system with active plate subduction. In. A Farah and K.A. De Jong (eds.): *Geodynamics of Pakistan.* Geological Survey Pakistan, Quetta, pp. 305-315.

Kazmi, A.H., 1984; Geology of the Indus Delta, *In.* B.U. Haq and J.D. Milliman (eds.): *Marine Geology and Oceanography of Arabian Sea and Coastal Pakistan.* Van Nostrand Reinhold, New York. pp. 71-84 (this volume)

Kolla, V., F. Coumes, and A. Lowrie, (in preparation); Internal structure, seismic stratigraphy and sedimentation of the Indus Fan.

Komar, P.D., 1969; The channelized flow of turbidity currents with application to Monterey Deep-Sea Fan Channel. J. Geophy. Res., 74: 4544-4558.

Krishnan, M.S., 1968; Geology of India and Burma. Higginbothams (P) Ltd., Madras, 536 pp.

Listzin, A.P., 1972; Sedimentation in the World Ocean. Society of Economic Paleontologists and Mineralogists, Special Publication 17, 218 pp.

McHargue, T.R., J.E. Webb, and L.G. Kessler, 1978; Ancient Submarine canyon and channel morphology, Indus Cone, offshore Pakistan. AAPG-SEPM, Annual Meeting, 93 p.

Naini, B.R. and M. Talwani, 1982; Structural framework and evolutionary history of continental margin of Western India, *In.* J.S. Watkins and C.L. Drake (eds.): *Studies in Continental Margin Geology,* AAPG Memoir 34, pp. 168-191.

Naini, B.R. and V. Kolla, 1982; Acoustic character and thickness of sediments of the Indus Fan and the continental margin of Western India. Mar. Geol., 47: 181-195.

Norton, I.O. and J.G. Sclater, 1979; A model for the evolution of the Indian Ocean and the break-up of Gondwanaland. J. Geophy. Res., 84: 6803-6830.

Powell, C. McA, 1979; A Speculative tectonic history of Pakistan and surroundings: some constraints from the Indian Ocean. *In.* A. Farah and K.A. De Jong (eds.): *Geodynamics of Pakistan,* Geological Survey of Pakistan, Quetta, pp. 5-24.

Sahni, A., and V. Kumar, 1974; Paleogene paleobiogeography of the Indian Subcontinent. Paleogeogr., Paleoclimat., Paleoecol., 15: 209-226.

Shuaib, S.M., 1982; Geology and hydrocarbon potential of offshore Indus Basin, Pakistan. AAPG Bulletin, 66:940-946.

Valdia, K.S., 1984; Evolution of the Himalaya. Tectonophysics, 105: 229-248.

Wells, J.T., and J.M. Coleman, 1984; Deltaic morphology and sedimentology, with special reference to the Indus River Delta. *In.* B.U. Haq and J.D. Milliman (eds.): *Marine Geology and Oceanography of Arabian Sea and Coastal Pakistan,* Van Nostrand Reinhold, New York, pp. 85-100 (this volume).

Weser, O.E., 1974; Sedimentological aspects of strata encountered on Leg 23 in Northern Arabian Sea. Initial Reports of DSDP, Vol. 23, U.S. Government Printing Office, Washington D.C., pp. 503-520.

C. SEDIMENTARY PROCESSES

9

Estuarine Circulation and Sedimentation:
An overview*

J. R. SCHUBEL
State University of New York,
Stony Brook, New York

ABSTRACT

All of the world's estuaries were formed by the most recent rise in sea level which began approximately 15,000 years ago. All are less than 10,000 years old, and most are only a few thousand years old. Once formed, estuaries are ephemeral features on geological time scales having lifetimes of from a few thousands to at most a few tens of thousands of years. Most estuaries are filled rapidly with sediments. Characteristically, sedimentation rates are highest near the heads of estuaries—near the upstream limit of sea salt penetration. An estuarine delta forms and gradually grows seaward, extending the realm of the river and thereby expelling the intruding sea from the semi-enclosed coastal basin. The basin is transformed from an estuarine basin back into a river valley until finally the river reaches the sea through a broad depositional plain and the transformation is complete. The rate of infilling is a function of the sediment influx, stability of relative sea level, climate, and the estuarine circulation pattern. One can find estuaries around the world in various stages of filling.

It is in the estuary where the mixing of fresh water and sea water produces dynamic conditions leading to the eventual discharge of the river water to the ocean. The mixing may be due primarily to the river, the tide, or the wind. There is a sequence of estuarine circulation types displaying different degrees of mixing. The position that an estuary occupies in this sequence depends primarily upon the relative magnitudes of the river flow and the tidal flow, and upon the geometry of the basin. Changes in any of these factors may produce changes in the estuarine circulation pattern and may thereby alter the resulting sedimentation pattern, sometimes dramatically. The end members of this sequence are the river-dominated (highly stratified) estuary and the tide-dominated (thoroughly mixed) estuary. A decrease in river flow, relative to tidal flow, moves an estuary toward the more thoroughly mixed end of the sequence. A decrease in depth has the same effect. Because of their circulation patterns, river-dominated and partially mixed estuaries are effective sediment traps. More thoroughly mixed estuaries are less effective sediment traps.

Until fairly recently the Indus River had a relatively large, river-dominated estuary. Because of increased diversion of its flow for irrigation, the Indus River now has an estuary only during the monsoon season. During the remainder of the year it has no estuary. This systematic reduction in the flow of the Indus and the resulting elimination of its estuary during most of the year has had a number of dramatic effects on Pakistan's coastal environments: fish stocks have declined; hundreds of thousands of acres of mangrove forests are lost each year; the delta is being eroded; flushing of ports and harbors is sluggish, and as a result concentrations of some contaminants have increased to undesirable levels. An experiment is described in this paper to determine whether maintenance of a small flow of the Indus adequate to sustain an estuarine pattern throughout the year would be desirable.

INTRODUCTION

The term "estuary," derived from the Latin word *aestus* which means tide, has been defined in a variety of ways (Schubel, 1971a). Many marine biologists have considered any body of water to be "estuarine" if the salinity is less than that of average sea water (e.g., Ketchum, 1951) while according to many geologists and physical geographers, estuaries are drowned river valleys (e.g., Howell and Weller, 1960). The definition most useful from the standpoint of understanding

*Contribution 388 of the Marine Sciences Research Center of the State University of New York

processes, particularly circulation and sedimentation processes, is one based on hydrography.

There is a group of coastal embayments in which the mixing of fresh water from the land and salt water from the ocean produces density gradients that drive distinctive circulations leading to the eventual discharge of the fresh water to the sea. It is this group of embayments which we want to characterize with a definition. The definition we will use was proposed by Pritchard (1967):

An estuary is a semi-enclosed coastal body of water which has a free connection with the open sea and within which sea water is measurably diluted with fresh water from land drainage.

According to this definition, there are several characteristics that distinguish estuaries from other bodies of water. First, an estuary is a semi-enclosed coastal body of water. Implicit in this qualification is the restriction that in estuaries the circulation is influenced to a considerable extent by the lateral boundaries of the basin. This imposes an upper size limit on estuaries. Large coastal bodies of water in which the lateral boundaries play a minor role in the kinematics and dynamics of the water movements such as the Baltic and the Gulf of Bothnia are not estuaries. According to Pritchard (1967) these water bodies... "form the coastline rather than being a feature of the coastline." While this restriction is somewhat arbitrary, the distinction is important.

Second, the semi-enclosed coastal body of water must be freely connected with the ocean. Salt water must not only be present for an estuary to exist; it must be present in sufficient quantity and with sufficient regularity to support the characteristic estuarine circulation patterns. Furthermore, a free connection with the open coastal waters is required for the propagation of the ocean tide into the estuary.

The third characteristic that distinguishes estuaries from other bodies of water is that within them sea water must be mixed with and measurably *diluted* by fresh water from the land. In some estuaries submarine flow of groundwater directly through the floor of the basin may be an important source of fresh water, but in most estuaries fresh water from land drainage is the predominant source.

According to the definition we have adopted, we are eliminating from our consideration semi-enclosed coastal embayments, such as Laguna Madre (Tamaulipas, Mexico), in which evaporation exceeds precipitation and runoff. It also follows from this restriction that an estuary extends upstream only to the limit of measurable sea salt. This boundary often is well downstream from the limit of tidal action. For example, in the Potomac River, a tributary to the Chesapeake Bay (USA), tidal action extends to approximately 140 km above the mouth of the estuary where it joins the main body of Chesapeake Bay while the limit of measurable sea salt is only about 75 km above the mouth. In the Chang Jiang (Yangtze), tidal action extends for some eight hundred kilometers above the mouth while the intrusion of sea salt is restricted to a few tens of kilometers during periods of low riverflow. During periods of high flow the sea is expelled completely from the semi-enclosed basin of the Chang Jiang.

The region above the limit of sea salt intrusion but still under the influence of the tide is called the "tidal reaches of the river." The net circulation in the tidal reaches of the river is controlled by the slope of the river surface and not by salt-induced density differences as it is farther seaward in the estuary.

To recap, *every* estuary has *all* of the following characteristics:

(1) It is a semi-enclosed coastal body of water.

(2) It has an unimpeded communication with the ocean at all stages of the tide, and

(3) Within it there is mixing and measurable dilution of sea water by fresh water from the land.

GEOLOGICAL CLASSIFICATION OF ESTUARIES

Most estuaries fill ancestral river valleys, but others occupy fjord basins, bar-built basins, and semi-enclosed coastal basins produced by tectonic processes. In a sense, every estuary is a drowned *river* but the basin that the estuary fills may have been formed, or extensively modified, by processes other than subaerial erosion by rivers. The four geological types of estuaries – (1) submerged river valley estuaries, (2) fjord estuaries, (3) bar-built estuaries, and (4) estuaries formed by tectonic processes – have been used as a classification scheme (Pritchard, 1967). More recently Pritchard (personal communication) has pointed out that the description of the fourth category should be expanded to include all estuaries occupying basins formed by geological processes, other than those responsible for the first three categories.

Drowned River Valley Estuaries

Drowned river valley estuaries fill basins formed by subaerial erosion by rivers (i.e., basins which have retained their dominant ancestral fluvial traits). They usually are relatively shallow, V-shaped in cross-section, and increase in depth and width more-or-less uniformly toward the mouth (Shepard, 1963; Bird, 1969; Schubel, 1971a). Submerged river valley estuaries are found throughout the world. Examples include Chesapeake Bay (USA), Delaware Bay (USA), Mississippi (USA), Chang Jiang (China), Thames (England), Ems (Germany), Seine (France), and the Murray (Australia).

Most of the largest submerged river valley estuaries are incised into low-lying coastal plains, but submerged river valley estuaries are not restricted to coastal plains. Some, such as the steep-sided Broken Bay estuary at the mouth of the Hawkesbury River in New South Wales, are cut into high, rocky plateaus.

Fjord Estuaries

Fjord estuaries fill troughs gouged out by tongues of continental glaciers which moved down pre-existing valleys. Characteristically, they are narrow, steep-sided troughs, U-shaped in cross-section, and relatively straight and deep. Frequently they have shallow sills at their mouths which may be rock or drowned morainic bars (Shepard, 1963; Bird, 1969). To be an estuary, a fjord must have a river entering it. Fjords with sill depths less than about 20 m were lakes until approximately 5,000 years ago. Fjords estuaries with shallow sills often are filled with stagnant, low-oxygen waters below the sill depth. Fjord estuaries are found along formerly glaciated coasts in both the southern and northern hemispheres — in Chile, New Zealand, British Columbia, Alaska, Greenland, Norway, Siberia, Scotland, and many other areas.

Bar-Built Estuaries

Bar-built estuaries occupy basins produced by the formation of barriers and spits which partially enclose reentrants in the coastline. The barriers must be broken by one or more inlets to provide a free connection with the ocean at all stages of the tide. Since the inlets are usually restricted in cross-sectional area, tidal currents in the inlets may be quite strong. But because the cross-sectional area of an inlet is small compared with the total volume of the embayment, tidal action is reduced considerably within the estuary. Bar-built estuaries usually are shallow, and the wind (local and far-field) frequently is the most important driving force for mixing the fresh water and sea water. Bar-built estuaries commonly are found along barrier coasts with large sediment inputs such as the Gulf Coast of the United States. Good examples of bar-built estuaries in the United States are Albemarle and Pamlico Sounds in North Carolina and Great South Bay in New York.

Estuaries Formed by Other Processess

There are several other processes which form the coastal embayments of estuaries. Taken together, there are fewer estuaries in this category than in any of the other three. The several processes forming the estuarine embayments in this category include tectonic processes, volcanic processes, and the partial enclosure of segment of the coast by the terminal moraine of a continental ice sheet.

Estuaries formed by tectonic processes fill basins formed by faulting, folding, or other diastrophic movements. The lower part of San Francisco Bay (USA) was formed, in part, by the slippage of fault blocks. The upper part of the San Francisco Bay estuary was formed by drowning of the lower reaches of the San Joaquin-Sacramento River Valley system, and hence the estuary taken as a whole is a composite of two categories.

At least one embayment on the Pacific Coast of Baja California (Mexico), which at least during a part of the year, satisfies the definition of an estuary, was formed by the construction of a peninsula, running parallel to the coast, by a string of emerging volcanos (Pritchard, personal communication). Long Island Sound is an estuary occupying a relatively large semi-enclosed coastal water body formed by the United States mainland coast on the north and by Long Island on the south. Long Island was formed, in part, as a terminal moraine of a continental ice sheet during the last glacial period.

THE ORIGIN AND DEVELOPMENT OF ESTUARIES

All present day estuaries were formed by the most recent rise in sea level which began 15,000 to

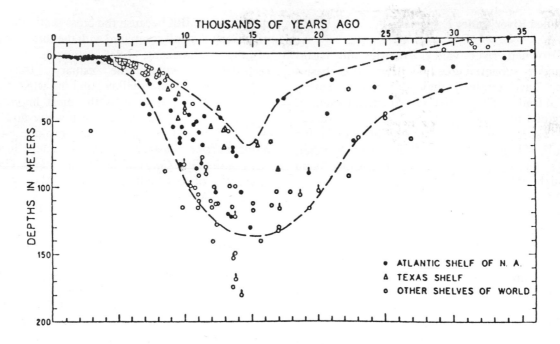

Figure 1. Sea level curve for the past 35,000 years. From Emery and Uchupi (1972).

18,000 years ago. During the last lowstand of sea level when the surface of the sea was approximately 100 – 125 m below its present level (Fig. 1), most of the continental shelves of the world were exposed to the atmosphere (Milliman and Emery, 1968). The coast at that time, formed by the upper continental slope, was steep and made irregular by the mouths of narrow river valleys, incised into the outer shelf and upper slope. Estuaries during that period were confined to these narrow valleys; they were smaller and were not as numerous as today.

Fifteen thousand to 18,000 years ago the Earth's climate began to warm up, glaciers began to retreat, and sea level began to rise. As the sea rose and advanced across the shelf, it progressively drowned the river channels carved into the shelf displacing the estuaries further landward. The general advance of the sea across the shelf was forerun by the estuaries. Sea level rose rapidly until about 5,000 years ago when it reached a position approximately 20 m below its present level. Its rate of rise then decreased substantially. By about 3,000 years ago the level of the sea was within 3 m of its present position. Since then, sea level has risen even more slowly, averaging less than 1 m in every 1000 years.

Estuaries reached their peak in development — in number, size, and complexity – approximately 3,000-5,000 years ago when the rise of sea level

slowed perceptibly and the level of the sea had reached nearly its present position. Since then, sea level changes may be due primarily to the subsidence of continental margins under the increased weight of the added overlying water. According to Bloom (1971), local sea level changes in estuaries and salt marshes may be controlled by the balance between isostatic adjustment and build-up by sedimentation. This isostatic adjustment is greater where shelves are wide than where shelves are narrow.

The rapidity of the rise in sea level was a major factor in the formation of estuaries. Sedimentation could not keep pace with the rising sea that invaded numerous semi-enclosed coastal basins. For the past several thousand years, the relative rates of infilling have been much greater and some estuaries have been nearly or completely filled (Russell, 1938; Meade, 1969). Among them are the Atchafalaya (USA), the Huang Ho (Yellow) (China), and the Chang Jiang (Yangtze) (China).

Once formed, estuaries are filled rapidly with sediments. Characteristically, the sedimentation rates are highest near their heads – near the upstream limit of sea salt penetration. An "estuarine delta" usually forms in the upper reaches of the estuary, near the new river mouth (Russel, 1967). The estuarine delta grows progressively seaward, extending the realm of the river and thereby expelling the intruding sea from the semi-enclosed coastal basin. The basin is transformed from an estuarine basin back into a river valley until finally the river reaches the sea through a depositional plain and the transformation is complete.

The rate of infilling of an estuary is determined by a number of factors including but not necessarily in order of importance: (1) rate of sediment influx, (2) stability of relative sea level, (3) climate, and (4) estuarine circulation pattern.

(1) *Rate of Sediment Influx*. Sediments are introduced into estuaries by rivers and streams, by shore erosion, by primary productivity, by anthropogenic sources, and by the sea. The sources are thus external, internal, and marginal. The greater the inputs, of course, the more rapidly a given estuary is filled (Meade, 1969; Schubel, 1971a). In most estuaries that have been studied in detail, the fluvial inputs of sediment have been shown to be the principal sediment sources. In a few estuaries such as the Ems (Crommelin, 1940; Van Stratten, 1960) and the Seine (Rajcevic, 1957; Vigarie, 1965), the sea has been shown to be the major source of sediment. The importance of the role of the sea as a source of sediment to estuaries probably has been exaggerated, and the importance of shore erosion underestimated (Schubel, 1971a). Man has had a major effect on the rates of input of sediments to estuaries. This is discussed in a later section.

The internal sources of sediment are derived from the remains of the large populations of planktonic and benthic organisms which estuaries support. All contribute to the sedimentation of an estuary, but in the aggregate they rarely, if ever, are the major source of sediment to an estuary as a whole. The most important role of organisms in estuarine sedimentation is the agglomeration and deposition of fine suspended particles by filter-feeding zooplankton and benthos (Haven and Morales-Alamo, 1966; Meade, 1972; Rhoads, 1963; Schubel, 1971a; Schubel and Kana, 1972).

(2) *Stability of Relative Sea Level*. If relative sea level is stable, estuaries evolve rapidly and are filled with sediments. They have lifetimes of from a few thousands of years to at most a few tens of thousands of years. A drop in relative sea level shortens estuarine life spans; a rise in relative sea level rejuvenates estuaries and favors the development of new ones.

(3) *Climate*. Since climate affects soil and rock weathering rates, it plays an important role in determining sediment yields. The kind, amount, and distribution of precipitation are important in determining sediment yields and in determining fresh water inputs and, as a result, estuarine circulation patterns. Climate also affects vegetation which plays an important role in estuarine and tidal flat sedimentation (Bird, 1969; Chapman, 1960). Generally, sediment yields are greater in semi-tropical regions than in temperate regions and at higher latitudes.

(4) *Estuarine Circulation Pattern*. The *tidal* circulation in an estuary is important in the formation of channels, tidal flats, and tidal deltas (Hayes, 1971), but the *net non-tidal* circulation is more important in determining the rates and patterns of sedimentation in most estuaries. The most important role of the tides is in providing the energy to mix the fresh and sea water which produces the density gradients that drive the net non-tidal estuarine circulation. This is the subject of the next section.

ESTUARINE CIRCULATION AND SEDIMENTATION

An estuarine classification scheme based on the mode of basin formation is of limited use. One based on circulation is of much greater value in understanding estuarine processes and in predicting the effects of those processes.

It is in the estuary where the mixing of fresh water from the land and salt water from the ocean produces dynamic conditions leading to the eventual discharge of the river water to the ocean. The mixing may be due primarily to the action of the river, the tide, or the wind. There is a sequence of estuarine circulation types displaying different degrees of mixing of the fresh water and the sea water (Pritchard, 1955). The position that an estuary occupies in this sequence depends primarily upon the relative magnitudes of the river flow and the tidal flow and upon the geometry of the semi-enclosed coastal basin that contains the estuary. Changes in any of these factors may produce changes in the estuarine circulation pattern and may thereby alter the resulting sedimentation pattern, sometimes drastically. One end member of this sequence is the poorly mixed (highly stratified) salt-wedge estuary – the Type A estuary. The other end member is the thoroughly mixed, sectionally homogeneous estuary – the Type D estuary. Two intermediate types which have been described are the partially mixed estuary (Type B), and the vertically homogeneous estuary (Type C).

Estuaries actually vary continuously in their characteristics and may shift from type to type as conditions change. Also, at any given time, different circulation types may be observed at different locations within an estuary, depending upon the relative magnitudes of the tidal flow and the fresh water flow, and upon the local geometry.

HIGHLY STRATIFIED (TYPE A) ESTUARIES

Highly stratified estuaries are river-dominated estuaries. They occur in semi-enclosed coastal basins where the ratio of river flow to tidal flow is relatively large and the ratio of width to depth is relatively small. It is the relative magnitudes of the river flow and the tidal flow, and not their absolute values that are important in determining the estuarine circulation pattern. Highly stratified estuaries also have been called "salt wedge" estuaries because the encroaching sea water is present as a wedge that underlies the less dense, fresher river water, (Fig. 2).

The mixing of sea water and river water is produced primarily by internal, instability waves that develop along the fresh water-salt water interface – the upper surface of the salt wedge (Keulegan, 1949). Above some critical riverflow, the waves break, ejecting small parcels of sea water from the lower layer into the upper layer where mixing occurs. The waves always break upwards, so that fresh water is not transported from the upper layer to the lower layer. Because of this process – called entrainment – the surface layer becomes progressively more saline in a seaward direction. The only mechanism leading to a progressive freshening of the lower layer in a landward (up-estuary) direction is a weak diffusion. Consequently, the salinity of the salt wedge remains very nearly that of full sea water all the way to the landward tip of the wedge.

Although considerably more effective than the diffusive process at the interface, the interfacial waves provide for only a relatively slow transfer of sea water from the lower layer to the upper layer. Since the salt wedge is maintained, there must be a slow movement of salt water up the estuary in the lower layer to replace that which is transferred to the upper layer. The volume rate of upstream flow of the lower layer is of the same order, or less, than the riverflow.

At locations upstream of the tip of the salt wedge, the net flow is downstream at all depths. Seaward of the tip of the wedge, the net flow is also seaward. During the tidal cycle, there may be a short interval of flood (up-estuary directed) flow, but the ebb flow is clearly dominant. In the lower

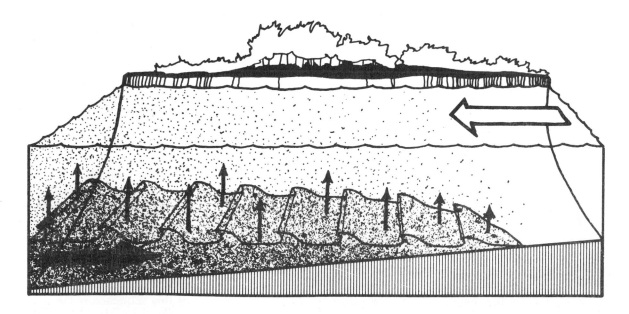

Figure 2. Side view of a highly stratified (salt wedge), Type A, estuary.

layer, the instantaneous flow may be upstream at all times, or it may reverse with the tide, but the *net* flow throughout the lower layer is upstream.

Fine suspended particles that are brought into the estuary by the river and settle into the lower layer are carried back upstream to the tip of the salt wedge by the net upstream flow of the lower layer and accumulate in the vicinity of the tip of the wedge. This fluvial sediment may be supplemented by fine particles from other sources. Heavier particles transported along the riverbed as bedload, by creep and saltation, accumulate upstream of the wedge. The region surrounding the tip of the salt wedge, then, is a zone of rapid shoaling.

The position of the tip of a salt wedge is determined primarily by the riverflow and the channel depth. It migrates upstream and downstream in response to changes in river discharge, in some cases more than 200 km, extending the area of rapid shoaling over a considerable region. With very high riverflow, the salt wedge may be expelled completely from the semi-enclosed coastal basin. During such periods there is no estuary and a large fraction of the river's sediment load may be discharged to the ocean.

A classic example of a saltwedge estuary is the Southwest Pass of the Mississippi River (Wright, 1971). The average flow through Southwest Pass is more than 5,100 $m^3 s^{-1}$, and peak flows may exceed 8,500 $m^3 s^{-1}$. The river completely dominates the circulation. The tidal range in the Gulf of Mexico is only about 35-40 cm. The tip of the wedge migrates more than 235 km in response to changes in the discharge of the Mississippi. During periods of very low flow, salt water may penetrate more than 40 km above New Orleans – nearly 235 km above the mouth of Southwest Pass. During periods of moderately high flow, the tip of the wedge is driven seaward to within a few kilometers of the mouth of the estuary. During such periods the shoaling problem often is so serious in this region that around-the-clock dredging is required to keep the channel open. During a 2-week period when the riverflow through Southwest Pass averaged about 8,500 $m^3 s^{-1}$ and dredges could not operate, shoaling of as much as 8.5 m occurred in one area, and the average fill over the entire region of the tip of the salt wedge was about 2 m (Simmons, 1966).

PARTIALLY MIXED (TYPE B) ESTUARIES

The river dominated, salt-wedge, estuary can exist in the absence of any tide. If the role of the tide is increased to the point where the tide is sufficiently strong to prevent the river from dominating the circulation, the added turbulence that results from the sloshing back and forth of the water by the tides provides the energy for erasing the salt wedge (Fig. 3). This occurs when the volume rate of flow up the estuary during a flood tide is about 10 times the volume rate of inflow of fresh water from the river.

In the partially mixed estuary there is both advection and turbulent mixing across the fresh water-salt water interface. The sharp transition –

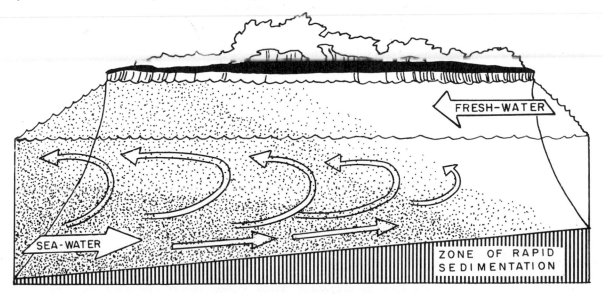

Figure 3. Side view of a partially mixed, Type B, estuary.

the halocline—that separated the fresh water in the upper layer from the sea water in the lower layer of the salt-wedge estuary is replaced by a region of moderate change in salinity. The salinity of the upper layer increases progressively toward the sea because of vertical advection of saltier water from the lower layer and because of vertical mixing. At any point in the estuary, the salinity in the lower layer is always greater than the salinity in the upper layer, and the *difference* in average salinity between the upper and lower layers remains nearly the same over much of the length of the estuary.

Since mixing between the upper and lower layers is much greater than in the Type A estuary, the Type B estuary has much higher volume rates of flow. The net seaward flow of the upper layer may be 10-40 times larger than the total input of fresh water, R, into the estuary above (landward of) that section. Since the estuary as a whole neither is filling nor emptying on the time scales of weeks or months, it must, on the average, move seaward through any cross-section a volume of water equivalent to the total fresh water input to the estuary upstream from that section. Because of the turbulent mixing between the upper and lower layers, the net seaward flow of the upper layer is much greater than the total freshwater input. As a consequence, there must be a concurrent up-estuary flow in the lower layer to satisfy continuity. If the net volume rate of flow moving seaward in the upper layer through a section is $10R-10$ times the total fresh water input, R—then clearly the lower layer must move, on the average, a volume rate of flow of water equivalent to 9R up the estuary through that section; any less and the estuary will drain, any more and it will fill up*. In the Chesapeake Bay, a Type B estuary, the net discharge of the upper layer to the sea at its mouth may at times be 25 times the total fresh water input to the Bay; the corresponding volume rate of flow up the estuary in the lower layer must be 24 times the total fresh water input.

The Coriolis force produces a slight lateral salinity gradient (Fig. 4). The boundary between the seaward-flowing upper layer and the landward-flowing lower layer is tilted slightly. In the Northern Hemisphere the upper layer is

deeper and the flow slightly stronger to the right of a seaward-facing observer. The lower layer is nearer the surface and the flow slightly stronger to the left of the seaward-facing observer. In the Southern Hemisphere the boundary would tilt in the opposite direction. The upstream limit of measurable sea salt penetration moves landward and seaward in response to fluctuations in river flow.

Fine suspended particles which settle into the lower layer are transported upstream by its net landward flow, leading to an accumulation of sediment on the bottom between the upstream and downstream limits of salt intrusion. Usually the most rapid shoaling in partially mixed estuaries is between the flood and ebb positions of the limit of sea salt intrusion. Rapid shoaling may also occur where the net upstream flow of the lower layer is interrupted by tributaries, by abrupt increases in cross-sectional area, or by meandering or splitting of the channel.

Because of turbulent mixing, which is more intense than in a salt-wedge estuary, and because of the well-developed two-layered circulation pattern typical of partially-mixed estuaries, often there is an accumulation of sediment suspended within the water column in the upper reaches of the estuary. Such features, called "turbidity maxima", have been reported in the upper reaches of a large number of partially mixed estuaries throughout the world (Glangeaud, 1938; Postma and Kalle, 1955; Postma, 1967; Schubel, 1968a; Nichols and Poor, 1967; and many others). Characteristically, these turbid zones begin in the

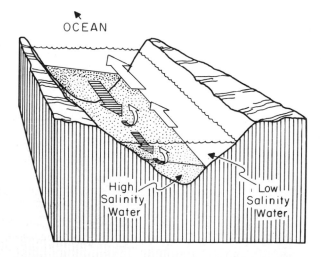

OCEAN

High Salinity Water

Low Salinity Water

Figure 4. Axial view of a partially mixed, Type B, estuary looking seaward in the Northern Hemisphere.

---

*Over periods of a few days estuaries may store excess water or have a net loss of water greater than the total fresh water input as a result of meteorological effects. These are discussed briefly elsewhere in this paper.

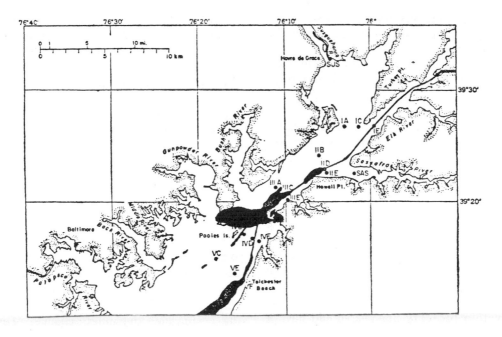

Fig. 5a

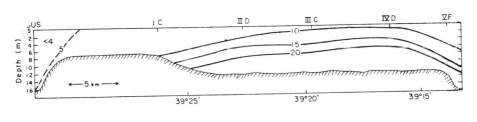

Fig. 5b

Figure 5. Turbidity maximum of northern Chesapeake Bay. (a) Map with stations. All areas deeper than 8m are shown in black. (b) Longitudinal distribution of suspended sediment mg/l typical of periods of low to moderate river flow.

estuary where a vertical gradient of salinity first appears and commonly extends seaward for 20 – 40 km. Within a turbidity maximum the concentrations of suspended sediment and the turbidity, are greater than either farther upstream in the source river or farther seaward in the estuary, (Fig. 5). Turbidity maxima migrate downstream during periods of high riverflow and upstream during periods of low flow. Their formation has been attributed to the flocculation of fluvial sediment (Lüneburg, 1939; Ippen, 1966), to the deflocculation of fluvial sediment (Nelson, 1959), and to hydrodynamic processes (Glangeaud, 1938; Postma and Kalle, 1955; Inglis and Allen, 1957; Postma, 1967; Nichols and Poor, 1967; Schubel, 1968a, b; Allen, 1973; Gallene, 1974). Postma (1967) suggested that the strength

of a turbidity maximum, the excess concentration of suspended matter, depends upon the strengths of the sediment sources at both ends – river and ocean – the settling velocities of the suspended particles, and the strength of the estuarine circulation pattern.

Studies of Chesapeake Bay (USA) and several of its major tributary estuaries – the Rappahannock, York, James, and Potomac – and of the Thames (England), Elbe (Germany), Ems (Germany), Gironde (France), Savannah (USA), and the Chang Jiang (China) estuaries, to name a few, have shown that the turbidity maximum in each of these is produced and sustained by the net non-tidal circulation and is located near the upstream limit of sea salt penetration in the lower layer. This limit corresponds to the "null zone;"

the region where the net upstream flow of the lower layer dissipates and landward of which the residual flow is downstream at all depths.

Therefore, the net non-tidal circulation leads to the formation of an effective "sediment trap" in the upper reaches of the estuary. Many of the suspended particles that settle out of the seaward flowing upper layer into the lower layer are carried back upstream by its net non-tidal upstream flow leading to an accumulation of sediment on the bottom and in the water column – the so-called turbidity maximum – near the head of the estuary. Many of these particles are transported back into the upper layer by vertical mixing and vertical advection and the whole process is repeated many times before the particles either are carried so far downstream that they are not returned, or are deposited permanently.

Postma (1967) hypothesized that suspended particles within turbidity maxima should have a relatively narrow spectrum of particle sizes (settling velocities). Schubel (1968b, 1969, 1971b) pointed out that particles with settling velocities much greater than the mean vertical mixing velocity would be deposited, and once deposited, those particles with threshold velocities greater than the maximum current velocity would not be resuspended; others would be alternately resuspended and deposited. Particles with settling velocities much less than the mean vertical mixing velocity would be concentrated within the upper layer and would eventually be carried far downstream or out to sea by the net non-tidal seaward flow of the upper layer. Between these two "critical settling velocities" there would be a sub-set of particles which would be exchanged repeatedly between the upper and lower layers and which would be trapped effectively within the zone of the turbidity maximum. Schubel (1969) examined this hypothesis in the turbidity maximum of the upper Chesapeake Bay and demonstrated the effectiveness of the sorting mechanism.

Schubel (1969, 1971b) showed that the suspended particle population of the Chesapeake Bay's turbidity maximum is composed of two sub-populations: one, a natural background of suspended particles made up of particles in more or less continual suspension throughout the water column; the other, made up of particles alternately suspended and deposited by tidal scour and fill. Since the latter sub-population is superimposed upon the background sub-population, it cannot be

observed directly. It must be determined by removing by calculation the background sub-population from the total particle population. Schubel (1971b) demonstrated that in the Chesapeake Bay's turbidity maximum the background sub-population of particles had a narrow, stable size distribution with a mean settling velocity of about 10^{-3} cm s^{-1} – the same order as the mean vertical velocity. The sub-population of particles alternately suspended and deposited had a much broader, less stable particle size distribution. The mean settling velocity of these particles at 1.5 m above the bottom ranged from about 1×10^{-3} cm s^{-1} near times of slack water to more than 1×10^{-2} cm s^{-1} near times of maximum ebb and flood currents. Schubel (1971b) demonstrated through actual measurements that the range in the size distribution of particles suspended within the turbidity maxima varies as the estuarine dynamics change. Festa and Hansen's (1978) two-dimensional numerical model of turbidity maxima elucidated the connection between particle size (settling velocity) and estuarine circulation.

Models of suspended sediment transport and turbidity maxima have been developed by Ariathurai and Krone (1976) and Festa and Hansen (1978), and have been reviewed by Owen (1977). Festa and Hansen's (1978) steady-state, two-dimensional numerical model of turbidity maxima in partially mixed estuaries supports the hypothesis that estuarine dynamics are primarily responsible for their occurrence. Their model

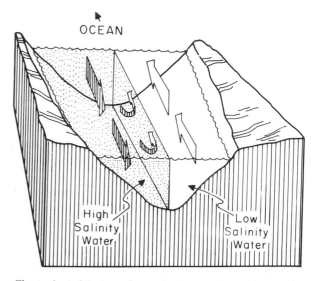

Figure 6. Axial view of a vertically homogeneous, Type C, estuary looking seaward in the Northern Hemisphere.

showed that the magnitude and location of the turbidity maximum depends upon the settling velocity of the sediment, the amount of sediment introduced by the river and the ocean, and the strength of the estuarine circulation.

VERTICALLY HOMOGENEOUS (TYPE C) ESTUARIES

If the role of the tide, relative to that of the river, is increased over that of a partially mixed (Type B) estuary, the turbulent mixing may, theoretically at least, become sufficiently intense to eradicate completely the vertical salinity gradient and produce a vertically homogeneous water column (Fig. 6). The longitudinal salinity gradient would still remain with the salinity increasing seaward. Because of the Coriolis Force, the lateral gradient in salinity also would remain with the higher salinity water to the left of an observer facing seaward in the Northern Hemisphere. The boundary between the lower salinity water flowing seaward and the higher salinity water flowing up the estuary would become nearly vertical and intersect the sea surface. In the Northern Hemisphere, then, the net flow and sediment transport in a Type C estuary generally would be upstream on the left side of the estuary facing seaward and downstream on the right side. Studies in the Thames estuary (England) (Inglis and Allen, 1957) have revealed a sedimentation pattern consistent with such a circulation pattern. Shoaling in Type C estuaries is rapid near the upstream limit of sea salt penetration, in regions of rapid expansion of cross-sectional area, and adjacent to islands and channel bifurcations where the flow is interrupted.

Examples of tidal waterways which approach closely the vertically homogeneous (Type C) estuaries described above include the wider reaches of the Delaware and Raritan estuaries (USA), the Thames (England), and sections of the James River estuary (USA) during periods of very low river flow.

SECTIONALLY HOMOGENEOUS (TYPE D) ESTUARIES

If tidal flow is very large relative to the river flow, it may reach a point, theoretically at least, where it completely overwhelms the effect of the river. Theoretically, the tidal mixing may become so intense that not only is the vertical salinity

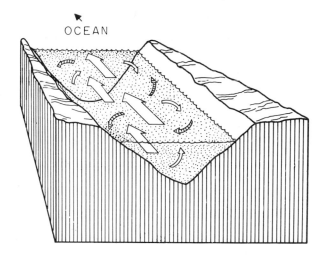

OCEAN

Figure 7. Axial view of sectionally homogeneous, Type D, estuary looking seaward in the Northern Hemisphere.

gradient eradicated, but also the lateral gradient, producing a sectionally homogeneous estuary (Fig. 7). In such a case, the movement of water would be essentially symmetrical about the main axis of the estuary with a slow net seaward flow at all depths throughout the cross section.

In estuaries that are sectionally homogeneous, or nearly so, the most rapid sedimentation would occur in areas where the slow net seaward flow is interrupted by tributaries, by large increases in cross-sectional area, or by obstacles, including islands.

To maintain permanently a salinity distribution which is sectionally homogeneous with the salinity increasing seaward there must be a mechanism capable of moving salt up the estuary from the sea. Clearly, a layered flow with vertical salinity mixing as in Type A and B estuaries is useless since there is no vertical salinity gradient. Further, the river flow must be discharged through the estuary to the sea. Other processes such as longitudinal turbulent diffusion must be at work.

It is conceivable that a longitudinal eddy flux of salt could account for some movement of salt up the estuary. Another possibility is a sort of "advection" that has been observed in the U.S. Army Corps of Engineers' hydraulic model of the Delaware Bay estuary. The model is a relatively realistic one which has along its sides occasional small pockets and embayments. If a blob of dye is released in the model, it spreads rapidly into a band across the estuary and is carried up and down by the tidal current. As the dye passes a pocket some of it enters and is trapped. It then bleeds

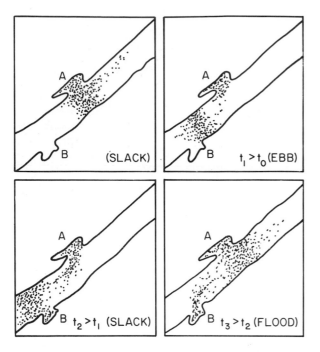

Figure 8. A potential mechanism for the upstream movement of salt in a sectionally homogeneous, Type D, estuary. Figure shows four sketches tracing movement of a blob of dye injected into a model of the Delaware estuary. See text for explanation.

slowly back into the mainstream after the main body of the dye has moved past. This spreads the dye over a larger section of the estuary than it otherwise would occupy (Fig. 8).

A similar phenomenon was observed in aerial photographs of a dye release in the Mattaponi River (a tributary of the York River, Virginia, USA). A band of dye laid across the river moved with the tidal current but left tails of dye along both

banks. These tails diffused back toward the center of the river after the main dye patch had moved away. On one bank the dye tail even moved appreciably in a direction opposite to that of the tidal current. Okubo (1973) presented a theoretical analysis of the effects of shoreline irregularities on longitudinal dispersion in estuaries. While processes such as these undoubtedly are effective in mixing salt over short distances, it is doubtful that they could make a major contribution to mixing over distances as large as the entire length of an estuary.

The strictly homogeneous estuary may be an extreme idealization and not exist in the real world. The estuaries that have been studied so far all have at least slight vertical gradients on the average. Such slight vertical gradients may be all that is necessary to sustain a gravitational circulation. For example, the estuary of the Mersey (England) long was thought to be vertically homogeneous until Bowden reported his salinity measurements and their long-term averages. He found that slight vertical salinity gradients with a maximum salinity difference of 1‰ existed at one stage of the tide (Bowden, Fairbairn and Hughes, 1959). At other stages of the tide the measurable differences vanished.

THE SEQUENCE OF ESTUARINE TYPES

It was found only after examining a relatively large number of estuaries that the theme unifying the classification scheme described above was the salt balance equation, Equation 1.

Equation (1) says that the salinity at any position in

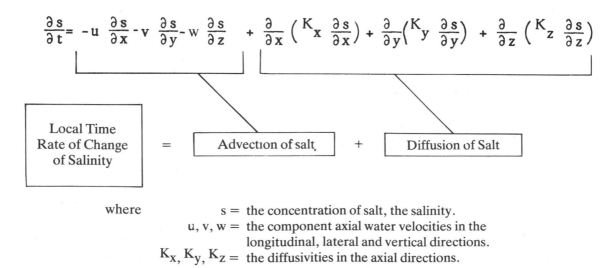

$$\frac{\partial s}{\partial t} = -u\frac{\partial s}{\partial x} - v\frac{\partial s}{\partial y} - w\frac{\partial s}{\partial z} + \frac{\partial}{\partial x}\left(K_x\frac{\partial s}{\partial x}\right) + \frac{\partial}{\partial y}\left(K_y\frac{\partial s}{\partial y}\right) + \frac{\partial}{\partial z}\left(K_z\frac{\partial s}{\partial z}\right)$$

Local Time Rate of Change of Salinity = Advection of salt + Diffusion of Salt

where
s = the concentration of salt, the salinity.
u, v, w = the component axial water velocities in the longitudinal, lateral and vertical directions.
K_x, K_y, K_z = the diffusivities in the axial directions.

an estuary will change with time and that the rate of change is caused by two different physical processes: advection and diffusion. The terms on the right side of Equation (1) involving x are advection and diffusion along the axis of the estuary, those involving y are horizontal across the estuary, and those involving z are vertical. The first three terms on the right side describe advections, the last three terms turbulent diffusions. In any particular estuary you will find that not all six of these processes that move salt around are of equal importance.

Usually, the changes in salinity at any fixed position are quite small. The changes brought about by advection and diffusion are not. Consequently, at any position in the estuary a near dynamic balance must exist between the relatively large advection and diffusion processes.

The classification scheme described above groups together all estuaries in which the changes in salinity (δ_s/δ_t) are produced mainly by the same processes, i.e., by the same terms in Equation (1), (Table 1). The character of the net circulation pattern within an estuary exerts a marked control on the salt balance. When approached in this way, one sees that there is a sequence of estuarine types, each with a different circulation pattern, and each shading gradually from one type to the next as conditions change. Although estuarine circulation patterns form a continuous spectrum, it is useful to identify the end members in this spectrum, Types A and D, and two intermediate Types B and C. While the types, as defined here, are separate and distinct, any estuary you encounter may lie somewhere between the types given. Further, different reaches of the same estuary may be of different types. Estuaries are constantly varying from type to type, in both space and time, as conditions change.

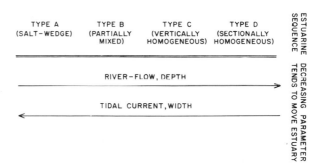

Figure 9. Effects of changes in river flow and geometry on estuarine circulation.

We already have identified the "estuarine driving forces" as the wind, the tide, and the river. In addition, the geometry of the estuary, represented at zero order by width and depth, must play an important role in conditioning the flow and the circulation patterns. If we neglect the wind, we can examine the effects of the river, the tide, the width and the depth on the estuarine circulation pattern. A decrease in river flow or in the depth of the estuary each will tend to move the estuarine circulation type through the sequence in the direction from A toward D. A decrease in tidal flow or in the width of the estuary each will tend to move circulation pattern in the direction from D toward A (Fig. 9).

AN EXAMPLE OF HOW MAN ALTERED AN ESTUARY'S CIRCULATION AND SEDIMENTATION BY ALTERING ITS FRESH WATER INPUT

Charleston Harbor is an interesting example of an estuary whose circulation and sedimentation were altered markedly by changing the freshwater input to the estuary, and to a lesser extent by changing its geometry. The Charleston Harbor estuary, located on the coast of South Carolina

TABLE 1. Summary of the most important processes in determining the salt balance in different types of estuaries.

Estuarine Type	Advection	Diffusion
A	vertical and longitudinal	—
B	vertical and longitudinal	vertical
C	longitudinal and lateral	lateral
D	longitudinal	longitudinal

(USA) is formed by the confluence of the Ashley, Cooper, and Wando Rivers. The mouth of the estuary is restricted, and entrance from the Atlantic Ocean is gained through a single jettied channel. Prior to about 1940, the total freshwater input was very small, averaging only about 2.8 m³ s⁻¹ and the Harbor was somewhere between a vertically homogeneous and sectionally homogeneous estuary. Fine-grained suspended sediment was moved slowly through the estuary to the ocean, and little dredging was necessary. The dredging required to maintain the main navigation channel at a depth of 9 m was only about 61,200 m³ y⁻¹ at a cost of approximately $11,600 y⁻¹.

In late 1941, a dam was completed which diverted water from the Santee River into the upper Cooper River which flows into Charleston Harbor. The average fresh water input to Charleston Harbor rose from only 2.8 m³ s⁻¹ to about 425 m³s⁻¹ — an increase of more than 150 times. The marked increase in the fresh water discharge shifted the circulation pattern in the Harbor from that of a very well-mixed estuary to a two-layered circulation pattern characteristic of a partially mixed estuary. Fine sedimentary particles which previously would have been carried completely through the estuary to the Ocean were now entrapped in the estuary by the net upstream flow of the lower layer and were accumulated in the inner Harbor — in the upper reaches of the net non-tidal estuarine circulation regime. Shoaling became a serious problem. The dredging required to maintain the inner Harbor channels increased to an average of approximately 1,758,000 m³ y⁻¹ at a cost of about $380,000 y⁻¹ during the 9-year period from 1944 to 1952 (Schultz and Simmons, 1957). Since 1952 the required dredging rate has increased further; over the 10-year period from 1960 to 1970 it averaged over 3,800,000 m⁻³ y⁻¹ at a cost of some six million dollars per year. The marked increase in the shoaling rate resulted in part from the addition of a new source of sediment, but the most important factor was the change in circulation produced by the increased river discharge. This was demonstrated conclusively by hydraulic-model studies.

Because of the enormous increased costs of dredging, the Charleston District of the U.S. Army Corps of Engineers developed a plan to redivert most of the Cooper River flow back into the Santee River. Implementation of that plan is now well along at a cost of approximately $100 million.

DEPARTURES FROM AVERAGE CONDITIONS

Two assumptions implicit in the discussion to this point are that wind is of relatively little importance in determining the net estuarine circulation pattern, and that lateral variations in the velocity and density fields are small. One of the most extensive and most thoroughly and thoughtfully analyzed sets of observations for any estuary was that taken in the James River estuary (Virginia, USA) and described by Pritchard in the early 1950's (Pritchard, 1952b, 1954, 1956). The data consisted of three separate subsets of observations of 4, 5, and 11 days, all taken during relatively calm weather. This set of measurements and their analysis laid the foundation for much of our understanding of the physics of estuaries, but as Carter et al. (1979) point out, it so influenced our thinking of how estuaries should operate — particularly partially mixed estuaries — that observations of circulation patterns which were at variance with the classical two-layered circulation pattern were considered to be measurement artifacts or to represent anomalous conditions. Most investigators clung to the belief that observations over several tidal cycles provided a sufficiently long averaging period for representative non-tidal flow estimates (Dyer, 1973). It was not until 1975 that Weisberg (Weisberg, 1976; Weisberg and Sturges, 1976) pointed out that the non-tidal estuarine flow also was highly variable and that the customary rule of thumb of averaging over a few tidal cycles could lead to serious errors.

Prompted by significant advances during the 1960's in instruments to sense and internally record and store large quantities of current speeds and directions, conductivities and temperatures, Pritchard and co-workers set out in 1974 to repeat the "James River experiment" but over a longer period and to do it in the lower Potomac River estuary (a tributary to the Chesapeake Bay, USA). As part of this study, they maintained a reference mooring of three current meters for a full year. Elliott (1976, 1978) analyzed the records from this vertical array. The records were filtered to remove the major tidal components, and averaged over each calendar day to produce daily estimates of the residual flows. The frequencies of occurrence of the six circulation patterns observed are summarized in Table 2. The classical, two-layered circulation pattern occurred a surprisingly

TABLE 2. Circulation Patterns observed in Lower Potomac Estuary (USA) Over One-year and Their Frequencies of Occurrence (Data from Elliott, 1978).

Circulation Pattern		Frequency of Occurrence (% of time)
Classical estuary		43
Reverse estuary		21
Three-Layered		1
Reverse Three-Layered		7
Discharge: flow out at all depths		6
Storage: flow in at all depths		22
		100%

small 43% of the time. However, for averaging intervals of ten days or longer, the classical two-layered pattern always emerged.

Analysis of a three-month record from a two current meter mooring in the upper reaches of the main body of Chesapeake Bay revealed similar variability in the residual, estuarine circulation (Elliott et al., 1978).

The variability in both estuaries was produced in part by local wind forcing associated with the passage of storms along the coast and in part by far-field wind forcing (Elliott, 1978; Elliott et al., 1978; Wang and Elliott, 1978). Their results showed that it was possible to separate variations in current and sea level forced by local wind stress which respond at approximately the seiche period of the Bay from the longer period variations forced non-locally by coastal winds which produced large sea level fluctuations at the mouth of the Bay with a period of about 20 days. These large sea level fluctuations were produced by an onshore or offshore Ekman transport associated with longshore winds.

The significance of the variability of estuarine circulation patterns to estuarine sedimentation has not been established. The effects of these fluctuations on sediment transport and on patterns of sediment accumulation in an estuary will depend strongly upon when they occur.

Dyer (1977) has shown that transverse effects can be important in estuarine circulation. The circulation often is not evenly distributed across the estuary and the water tends to flow in spiral fashion creating "secondary flows." Changes in geometry of the estuary are the principal cause of these transverse effects, but the Coriolis effect and the lateral density field are also important.

SOURCES OF INORGANIC PARTICLES TO THE COASTAL OCEAN

Compared to the total amount of inorganic material discharged by rivers to the Coastal Ocean, the amounts introduced by the atmosphere, by shore erosion, and by industrial activities are relatively small. The latter sources may dominate locally, but their collective contribution to the total mass of new inorganic material added to the Coastal Ocean pales in comparison to that of rivers. In this section, we concentrate on the additions of fluvial sediments to the Coastal Ocean and on the partitioning of these sediments between estuaries and open coastal waters.

TABLE 3a. Holeman's (1968) "Summary of Measured Annual Sediment Yields of Rivers to Oceans (tributaries deleted)." [Holeman's data have been converted to metric units].

Continent	Measured Drainage Area, km²	Annual Suspended Sediment Discharge metric tons (1000 tons)	Tons km⁻²
North America	6,383,416	547,899	85.8
South America	9,894,720	552,863	56.0
Africa	8.149,870	196,191	24.5
Australia	1,073,836	42,955	40.3
Europe	3,515,541	110,620	31.5
Asia	10,911,188	5,819,205	535.9
Total	39,928,571	7,269,733	Average 182 tons km⁻²

TABLE 3b. Holeman's (1968) "Total Sediment Yield to Oceans Extrapolated from Above Data." [Holeman's data have been converted to metric units].

Continent	Total Area Draining to Oceans, km²	Annual Suspended Sediment Discharge — Annual Suspended metric tons km⁻²	Sediment Discharge 10⁹ metric tons
North America	20,719,920	85.8	1.78
South America	19,424,925	56.0	1.09
Africa	19,942,923	24.5	0.49
Australia	5,179,980	40.3	0.21
Europe	9,323,964	31.5	0.29
Asia	26,935,896	535.9	14.43
Total:	101,527,608		18.29

Fluvial Suspended Sediment Input to the Marine Environment

There are few independent estimates of the total discharge of suspended sediment by the World's rivers to the marine environment, and all of these estimates are of the discharge to the coastal zone and not to the "ocean." Estimates range from 58 billion tons per year (Fournier, 1960) to 13 billion tons per year with most estimates falling between 20 and 30 billion tons (Milliman, 1980).

The most frequently quoted world summary is that of Holeman (1968) who estimated that the total discharge of suspended sediment by the World's rivers is about 18 billion tons per year, Table 3. More recently, Milliman (1980) put the total at 16 billion tons, but cautioned that his estimate could be off by as much as 50 percent.

Milliman (1980) pointed out a number of problems associated with such estimates.

1. Rivers with highest sediment loads (e.g., the Huang Ho, (Yellow), Chang Jiang (Yangtze) Ganges/Brahmaputra, Indus, and Amazon) are in developing countries and are poorly documented. Estimates of annual discharges by different investigators often vary by more than a factor of two. Often, measurements are made during dry seasons when river discharge and sediment loads are low. Improper sampling methods and techniques can lead to gross errors in estimating sediment concentrations and sediment discharges.

2. In the absence of adequate data, investigators often have used rating curves to estimate suspended sediment concentrations for different river discharges. Errors of at least 50% can be expected in sediment discharges based on rating curves.

3. In most estimates, bed load is ignored completely, even though in some rivers it may be of great importance.

4. Most measurements are made during average or typical periods. These "average" values of sediment discharge do not reflect catastrophic events such as extreme floods.

5. Average sediment loads of many rivers have been changed by man's activities.

6. A substantial portion of sediment carried by many rivers is stored temporarily along the river as channel deposits or permanently in levees or flood-plain deposits.

In his summary, Milliman (1980) pointed out that in spite of the paucity of data several conclusions can be drawn regarding the transfer of fluvial suspended sediment to the "ocean." These include:

(1) Most *major* rivers have average suspended sediment concentrations between 100 and 1000 mg 1^{-1}. Together the World's *major* rivers carry about 8 billion tons of sediment annually, nearly 50 percent of which comes from two Asian rivers – the Huang Ho (Yellow) and Ganges/Brahmaputra. Other Asian rivers including the Indus, Chang Jiang (Yangtze), Red and Mekong bring the figure to 75 percent of the total for the major rivers. In contrast, North American and European Rivers collectively contribute less than 7 percent of the total discharge of the world's major rivers.

(2) Rivers draining humid areas (e.g., Ganges/Brahmaputra) and desert areas (e.g., Colorado and Orange) tend to have very high suspended sediment concentrations. Rivers draining low-lying areas, regardless of the latitude, have low sediment concentrations. Lowest concentrations of suspended sediment are observed in rivers that empty into lakes before reaching the ocean (e.g., St. Lawrence and Congo).

(3) Perhaps a reasonable figure for the total discharge of suspended sedimentary rivers is 16 billion tons per year assuming that small rivers collectively transport about the same as major rivers. The error in this estimate could, however, easily be 50 percent.

Milliman and Meade (in press) have prepared a new world-wide summary of the discharge of fluvial sediment to the coastal zone incorporating new data that have not been used in previous summaries. Meade (personal communication, November 1981) believes that the contribution of suspended sediment from the smaller rivers may have been underestimated substantially, particularly for those bordering the Pacific.

If the quantities of suspended particulates discharged by rivers are poorly known, even less is known about the composition of these materials. The composition of suspended particulate load of rivers is less well known than that of the dissolved load. According to a study of 20 major rivers (Martin and Meybeck, 1978, 1979) the composition of fluvial suspended particulate matter depends on the ratio of mechanical erosion to chemical denudation, and on the weathering types. In temperate and arctic zones the composition of fluvial suspended matter reflects closely the composition of average fresh rocks, consisting mainly of rock debris and poorly weathered soil particles. In tropical zones, fluvial suspended matter originates mainly from highly developed soils and is strongly depleted in Ca, Na, and Mg and enriched in Al and Fe. For most rivers one is fortunate to know the percentage of the total load accounted for by combustible organic matter, let alone the mineralogy of the inorganic fraction and associated contaminant burdens.

Partitioning of Fluvial Sediments between Estuaries and Open Coastal Waters

There are few data on the partitioning of fluvial suspended sediment discharges between estuaries and coastal marshes on the one hand and the open continental margin and the deep sea floor on the other. Most of the estimates of partitioning that are available are for rivers with relatively small sediment discharges and large estuaries. Drake (1976) estimated that less than 10% of modern river-borne suspended sediment reaches the deep sea (basin and ocean ridge systems) and that most inorganic sediment is deposited on continental margins (including shelf, slope, and rise), or in marginal seas. There probably is general agreement that little fluvial sediment reaches the deep sea, but less agreement on the partitioning between the coastal zone and the shelf.

Nearly all of the sediment that reaches the Atlantic coastal zone of the United States is trapped in estuaries and coastal marshes (Meade, 1969, 1972). Budget calculations indicate that virtually all of the fluvial sediment discharged into large east coast estuaries such as Long Island Sound (Bokuniewicz et al., 1976) and Chesapeake Bay (Schubel and Carter, 1977) must be deposited within these estuaries. In addition, sediments from the inner continental shelf and the ocean shoreline are carried into the lower reaches of estuaries along the east coast of the United States (Firek et al., 1977; Hathaway, 1972; Meade, 1969, 1972; Van Nieuwenhuise et al., 1978; Schubel and Carter, 1977).

Along the southern Atlantic seaboard where the major rivers debouch into the ocean without flowing through large estuaries, the identifiable river-borne sediment is restricted to narrow zone near the coast (Bigham, 1973; Meade, 1980; Milliman et al., 1972). According to Meade (1972) the principal sinks for river sediment in this area are the extensive marshes. Probably less than 5 percent of all river sediment discharged into tidal waters of the U.S. Atlantic seaboard is deposited on the floor of the continental shelf or the deep sea. The fate of suspended sediment discharged by other rivers with major estuaries is similar.

Most of the fluvial sediment discharged by the Parana-Uruguay River system which flows into the large estuary of the Rio de la Plata is entrapped within the estuary (Urien, 1972). Duinker and Nolting (1976) reported that 50 percent of the suspended sediment discharged by the Rhine River is deposited in the Wadden Sea, and Bewers and Yeats (1977) estimated that 93 percent of the suspended sediment entering the Gulf of St. Lawrence is deposited within the marginal sea.

These rivers combined with all the world's other small to medium-sized rivers collectively account for only a relatively small percentage (<10%) of the World's total fluvial discharge of suspended sediment to the coastal zone. The bulk of the total (>75%) comes from a few large rivers that no longer have significant estuaries. The partitioning of the sediment discharged by these large rivers between their estuaries and the adjacent shelf is much different than that for rivers which still have large estuaries. Probably more than 50% of the suspended sediment discharged

by the Chang Jiang (Yangtze), for example, is deposited on the continental shelf (Chen Ji-Yu, personal communication, 1981), and more than 50% (perhaps as much as 90%) of the suspended sediment discharged by the Amazon is deposited on the adjacent continental margin (R. Meade, personal communication, 1981).

Data from the Mississippi indicate that more than 75% of its suspended sediment load near the River's mouth escapes the estuary and is deposited on the delta and in the Gulf of Mexico. For the Huang Ho (Yellow) River, more than 70% of its sediment load reaches the Yellow Sea (Chen Ji-Yu, personal communication, 1981).

Clearly, large estuaries are effective "filters" for removing suspended particles discharged into them by their tributary rivers. As the volume of an estuary decreases, its filtering efficiency for particulate matter decreases also, and at a more rapid rate than its volume. While all modern estuaries were formed at approximately the same time, the stages of filling vary markedly. Estuaries of rivers with large sediment discharges such as the Huang Ho (Yellow), Chang Jiang (Yangtze), and Atchafalaya all have been largely filled with sediments and have low filtering efficiencies relative to estuaries of rivers with smaller sediment inputs such as Chesapeake Bay, Long Island Sound, and the Rio de la Plata.

In summary, the best estimates of the total discharge of suspended sediment to the coastal zone place the value at 16 to 18 billion tons per year. New data may increase this estimate (Meade, personal communication, 1981). There are no world-wide estimates on what fraction of this total escapes estuaries and coastal marshes to reach the open shelf and slope and the deep sea floor. The author's own Mark Twain* estimate places the mass that escapes estuaries between 6 and 10 billion tons per year and rising. One would expect the fraction of fluvial sediment that reaches the sea to increase as estuaries are filled.

MAN'S ACTIVITIES AND SEDIMENT INPUTS TO THE ESTUARIES

Although sediment in estuaries comes from many sources – including erosion of the margins and of the sea floor, and from biological

---

*Mark Twain once pointed out that "There is something fascinating about science. One gets such a wholesome return of conjecture out of such a trifling investment of fact."

productivity — the sources most affected by man are the rivers that carry sediment from upland areas into estuaries. This discussion focuses on the sediment loads of rivers which are increased by activities such as farming, mining, deforestation, and urbanization; and which are decreased by activities such as construction of reservoirs and other protective works.

People's activities have affected not only the amounts of suspended particulate matter added to the Coastal Ocean, but the size distribution and composition of these materials as well. There has been a shift to smaller particles and an increase in the amount of organic matter (Schubel, 1976).

Man's Activities That Increase River Sediment Loads

Soil erosion is the ultimate source of most fluvial sediment. Ever since the first European settlers landed in North America, man has affected the amount of sediment in streams draining this region. The influence of man on sedimentation is especially well documented in the Chesapeake Bay region, where clearing of forests and wasteful farming practices (especially those used in raising tobacco) contributed enormous loads of sediment to the rivers. Clear streams became muddy and once relatively deep harbors at the heads of a number of the tributaries to Chesapeake Bay were filled rapidly with sediment (Gottschalk, 1945). The Potomac River, whose waters were already somewhat turbid but which were still suitable for municipal use in 1853, had become so muddy by 1905 that the city of Washington had to install its first filtration plant. A comparison of the 1792 and 1947 shorelines of the upper Potomac shows that large areas of the Potomac near Washington had been filled with sediments stripped from farmland further upstream. The Lincoln and Jefferson Memorials now stand on what was described in 1711 as a harbor suitable for great merchant vessels. Even today, an average of about 2 million m³ of sediment is deposited every year near the head of tide in the Potomac; not all of this sediment is the result of agriculture, as we shall see. There are other former seaport towns on the western shore of northern Chesapeake Bay where decaying docking facilities now are separated from navigable water by several kilometers of sediment-filled lowland.

Streams that drain modern day farmlands in many of the mid-Atlantic states carry about 10 times as much sediment as streams that drain equivalent areas of forest land. And this relationship is by no means unique. In the Coastal Plain of northern Mississippi, sediment yields from cultivated lands are 10 to 100 times the yields from equivalent areas of forested lands (Gottschalk, 1945; Trimble, 1974). In two other areas where studies have been made — the Tobacco River Valley of Michigan and the Willamette Valley of Oregon — streams draining farmland carry two to four times as much sediment as streams draining equal areas of forested land.

Mining is another activity that has increased the sediment loads of some rivers that flow into estuaries. San Francisco Bay, for example, contains nearly a billion cubic meters of sediment washed from the Sierra Nevada during the approximately 30 years of intensive hydraulic mining for gold in that Range. Even after the hydraulic processing was stopped in 1884, the mining debris continued for many decades to choke the valleys of the Sacramento River and some of its tributaries. Gradually, over the years, the debris was moved downriver to be deposited more permanently in the marshes and shallow areas surrounding San Francisco Bay. The mining debris released in only three decades is more than the total sediment from all other sources (including farmland) that the Sacramento River has carried in the twelve-and-a-half decades since 1850 (Gilbert, 1917). It has been shown that this sediment had an important effect on the circulation of San Francisco Bay; the tidal prism was decreased, and the flushing regime changed significantly.

The high soil erosion rates prevalent in many unglaciated areas during the 19th and early 20th centuries now have been reduced by soil conservation practices and by reversion of uplands to pastures and woodlands (Meade, 1980). But, the effects of the earlier higher erosion rates are still being experienced strongly downstream in the lower reaches of rivers and particularly in their estuaries. "Much of the soil material that was eroded off the uplands since 1700 is stored on hillslopes and on the floors of stream valleys" (Meade, 1980). Many alluvial valleys in the southern and middle-Atlantic Piedmont of the United States are lined with a layer of sediment a meter or more in thickness that has accumulated since European settlers arrived (Costa, 1975). Since the time when upland erosion was curtailed by appropriate soil conservation practices, sediment has been supplied to streams from intermediate storage sites between the uplands and the river (Meade, 1980; Meade and Trimble, 1974; Trimble, 1977). Trimble (1975) has estimated that more than 90 percent of the sediment eroded off the uplands of the southern Piedmont of the United States since 1700 still remains above the Fall line – the boundary between the Piedmont and the Atlantic Coastal Plain. The implication is that soil material removed during an erosional episode can be expected to be released from intermediate storage sites over a period of decades to centuries and move downstream as a wave.

In glaciated areas, the influence of man on soil erosion rates has been less marked than in unglaciated areas (Gordon, 1979; Meade, 1980; Williams and George, 1968). In a study of the Connecticut River valley, Gordon (1979) found little evidence that sediment yields had changed since pre-colonial days.

Urbanization is the most recent of man's activities to contribute large amounts of sediment to streams. Sediment loads derived from land being cleared or filled for the building of houses, roads, and other facilities are best documented in the United States in the area between Washington, D.C. and Baltimore, Maryland. During periods when housing developments, shopping centers, and highways are being built, the soil is disturbed and left exposed to wind and rain. The concentration of sediment in storm runoff from construction sites often is a 100 to 1,000 times what it would be if the soil had been left in its natural vegetated state. Even though the soil is left exposed to erosion of this intensity for only a short time – a few years at most – the amount of land cleared for new housing and ancillary uses in the Washington-Baltimore area was so great in recent years the contribution of sediment has been significantly large. Harold Guy of the U.S. Geological Survey has estimated that the Potomac River receives about a million tons of sediment per year from streams that drain the metropolitan Washington area. This is about the same amount of sediment that the Potomac River brings into the Washington areas from all its other upland sources.

In some areas of the World, lumbering has increased dramatically sediment yields and sedimentation rates in rivers and estuaries. Large areas of forest have been stripped in southern Chile without replanting.

Another of man's activities that increases the sedimentation rates of estuaries is the discharge of dissolved phosphorus, nitrogen, and other plant nutrients into rivers and estuaries. Municipal sewage effluents, including effluents that have received secondary treatment – the highest degree of conventional treatment – contain high concentrations of nutrients. In some areas, agricultural runoff from fertilized croplands and animal feedlots also contributes nutrients to river waters and estuaries. These nutrients promote the growth of diatoms and other microscopic plants (phytoplankton) both in the rivers and in the estuaries into which the rivers flow. The mineral structures formed by many of these organisms persist after the organisms die and become part of the sediment loads of rivers and the sedimentary deposits of estuaries. The U.S. Army Corps of Engineers estimated, for example, that diatom frustules produced in the Delaware River and the Delaware Bay contribute about the same amount of sediment (a million-and-a-half tons per year) to the Delaware estuary as all upland river sources.

The effects of nutrient loading from municipal wastes on primary productivity are readily observable in the Potomac estuary, in Baltimore Harbor and the Back River estuary (Maryland); in Raritan Bay, in the Arthur Kill estuary, in the Hudson estuary, in the Delaware estuary, in San Francisco Bay, and in many other estuaries around the United States, and indeed throughout the World. Stimulation of plant growth by nutrient-enriched runoff from agricultural areas is apparent in the upper Chesapeake Bay, the estuary of the Susquehanna River.

Man's Activities That Decrease River Sediment Loads

Reservoirs probably cause the most significant interruptions in the natural movement of sediment to estuaries by rivers. Reservoirs are built on rivers for a number of purposes: hydroelectric power generation, flood control, water supply, and recreation. Regardless of their purpose, reservoirs share in common the ability to trap sediment (Schubel and Meade, 1977). Even small reservoirs can trap significant proportions of river sediment. For example, a reservoir that can hold only one percent of the annual inflow of river water is capable of trapping nearly half the river's total sediment load. A reservoir whose capacity is 10 percent of the annual river water inflow can trap about 85 percent of the incoming sediment (Meade, 1976; Meade and Trimble, 1974). Although a river will tend to erode its own bed downstream of a reservoir to partly compensate for sediment it has lost, the net effect of the reservoir is to decrease the overall amount of sediment carried by the river. In the larger river basins of Georgia and the Carolinas in the United States, the sediment loads delivered to the estuaries are now something like one-third of what they were about 1910, mainly because of the large number of reservoirs that have been built since then for hydroelectric power and, to a lesser extent, for flood control (Schubel and Meade, 1977).

The trapping, however, cannot always be considered permanent, not even on time scales smaller than the life span of the reservoir. The sediment held behind some dams can be mobilized by extreme flood events (Meade, 1980). Flooding of the Susquehanna River in Pennsylvania and Maryland following passage of Tropical Storm Agnes in June 1972 purged 10-20 years' accumulation of sediment from reservoirs of the lower river (Schubel, 1974; Zabawa and Schubel, 1974; Gross et al., 1978).

On some rivers, settling basins and reservoirs have been built specifically as sediment traps to improve the quality of water farther downstream. In 1951, three desilting basins were constructed on the Schuylkill River of Pennsylvania to remove the excessive sediment that resulted from anthracite coal mining in the upper river basin. The basins are dredged every few years, and the dredged material is placed far enough from the river to be out of reach of floods. As a result of these basins, the sediment load carried by the Schuylkill into the Delaware estuary has been reduced from nearly a million tons per year to about 200,000 tons per year.

Return of cultivated lands to forests can significantly reduce sediment yields. In the last 50 years, the average suspended load of South African rivers has decreased by 50 percent largely because of the stabilization of river banks by introduced vegetation (Rooseboom, 1978).

Net Effect of Man's Activities on Sources of Sediment

The net effect of man's activities no doubt has been an increase in the sediment supplied to most estuaries, but we cannot say by how much. Although reservoirs and other controls have reduced the sediment in rivers in recent years, they have only partly offset the influences that caused the increases in the first place.

Added to this is the fact that sediment takes decades to move through a river system. Much of the sediments released by past mistakes – such as by poor mining practices and by poor soil conservation practices associated with agriculture – are still in the river valleys in transit storage between their sources and the estuaries. Even if the active supply of sediment to rivers were completely checked today, many decades would pass before the sediment loads would drop to their natural levels.

The fight against erosion has been more successful in developed countries than in developing countries because of better regulation of land use. But, deterioration of water quality in coastal areas is a much more serious problem in developed – industrialized – countries than in most developing countries.

A RECOMMENDATION FOR A STUDY OF THE INDUS ESTUARY: AN UNUSUAL OPPORTUNITY FOR PAKISTAN AND FOR ESTUARINE OCEANOGRAPHY

Because of increased utilization of the flow of the Indus River for irrigation for agriculture, the Indus now discharges water to the Indian Ocean only during the Monsoon season. It is only during that period that the Indus has an estuary. During the other 9 to 10 months of the year when the river discharge is eliminated, the Indus has *no* estuary. This systematic reduction in river discharge and the elimination of the estuary has had a number of dramatic effects on Pakistan's Coastal Ocean: fish stocks have declined; hundreds of thousands of acres of mangrove forests have been lost each year; the Delta is being eroded; and flushing of ports and harbors has languished, and concentrations of pollutants in these water bodies have increased to undesirable levels. The losses in coastal resources may be more than offset – at least in the short term – by the gains in food production made possible by irrigation, but an estuarine oceanographer cannot help but wonder whether it would be possible to have both. It is worth investigating.

The systematic and intentional reduction of the

flow of the Indus offers Pakistan an unusual opportunity to conduct an experiment of enormous scientific and societal value; an experiment that would attract attention world-wide. Pakistan is not alone in its intentional elimination of the discharge of one of the World's rivers. The Indus is among the larger rivers whose discharges have been eliminatd, and the largest in terms of sediment load. The impact on coastal resources of eliminating a river is a function not only of the size of that river, but of a number of complex and interrelated environmental, socio-political, and cultural factors. For countries which have little dependence on the Coastal Ocean for food, the impacts of eliminating coastal resources may be unimportant. For countries with a strong dependency on the sea, the impacts of such losses could be debilitating.

Pakistan perhaps has the greatest opportunity to assess the relative gains and losses of the water management strategy it has selected and to assess whether, or not, an alternative strategy, one that permitted à small discharge of the Indus throughout the year, would be practical and more desirable. I urge you to give it careful consideration. I have outlined below a few features of such an experiment. I have made no attempt to design the experiment.

Step 1. Design and conduct a sampling program during the Monsoon season to characterize the velocity and salinity fields to permit model construction.

Step 2. Develop appropriate computer models of the Indus estuarine system using data collected during the Monsoon season when the estuary exists.

Step 3. Using these models, predict the flow field, the salinity field, and distribution of other appropriate parameters one would expect for some reasonable combination of river discharge and oceanographic and meteorological conditions.

Step 4. Regulate the flow of the Indus to provide the discharge used in Step 2 and conduct an appropriate oceanographic and meteorological sampling program.

Step 5. Using data collected in Step 3, adjust the model to make the predictions conform to the observations.

Step 6. Using the adjusted model, predict the flow field, the salinity field and distributions of other appropriate parameters for a second combination of river discharge and oceanographic and meteorological conditions.

Step 7. Regulate the flow of the Indus to provide a second discharge scenario consistent with Step 5 and conduct an appropriate sampling program to verify the adjusted model.

Step 8. Use the adjusted and verified models to assess what minimum river discharge would be required to sustain a sufficiently strong estuarine circulation and rapid enough flushing rates to maintain living marine resources.

The best – most effective – management strategies are those selected from a formulation of objectives and goals, an identification of the full range of alternatives for attaining these objectives and goals, and a rigorous assessment of the advantages and disadvantages associated with each alternative. Only after this has been done is one in a position to select the most appropriate strategy, one that has predictable and acceptable consequences. Maintaining a small discharge of the Indus River throughout the year to sustain an Indus estuary appears to be one management alternative deserving further attention.

REFERENCES

Allen, G.P., 1973. Etude des processus sédimentaires dans l'estuaire de la Girande. Mémoires de l'Institut, de Géologie du Bassin d'Aquitaine, No. 5.

Ariathurai, R. and R.B. Krone, 1976. Finite element model for cohesive sediment transport. Proc. Amer. Soc. Civil Eng., 102:328-338.

Bewers, J.M. and P.A. Yeats, 1977. Oceanic residence times of trace metals. Nature, 268:595-598.

Bigham, G.N., 1973. Zone of influence-inner continental shelf of Georgia. J. Sediment. Petrol., 43:207-214.

Bird, E.C.F., 1969. *Coasts.* The M.I.T. Press, Cambridge. *An Introduction to Systematic Geomorphology,* vol. 4; 246 pp.

Bloom, A.L., 1971. Glacial-eustatic and isostatic controls of sea level since the last glaciation. Pages 355-380, *In* K.K. Turekian (ed.), Late *Cenozoic Glacial Ages.* Yale Univ. press, New Haven, Conn.

Bokuniewicz, H.J., J. Gebert and R.B. Gordon, 1976. Sediment mass balance in a large estuary. Estuarine Coastal Mar. Sci., 4:523-536.

Bowden, K.F., L.A. Fairbairn and P. Hughes, 1959. The distribution of shearing stresses in a tidal current. Geophys. J.R. Astr. Soc., 2:288-305.

Carter, H.H., T.O. Najarian, D.W. Pritchard, and R.E. Wilson, 1979. The dynamics of motion in estuaries and other coastal water bodies. Reviews of Geophysics and Space Physics, 17(7): 1585-1590.

Chapman, V.J., 1960. *Salt marshes and salt deserts of the world.* Interscience, New York., 392 pp.

Coasta, J.E., 1975. Effects of agriculture on erosion and sedimentation in the Piedmont Province, Maryland. Geol. Soc. Am. Bull., 86:1281-2186.

Crommelin, R.O., 1940. De herkomst van het zand van de Waddenzee: Koninkl. Nederlands Aardrijksk. Geroot. Tijdschr. (Leiden), ser. Z, 57:347-361.

Drake, D.E., 1976. Suspended Sediment Transport and Mud Deposition on Continental Shelves. Pages 127-158, *In* D.J. Stanley and D.J.P. Swift (eds.), *Marine Sediment Transport and Environmental Management.* John Wiley & Sons, New York. 602 pp.

Dyer, K.R., 1973. Estuaries: *A physical introduction.* John Wiley and Sons, New York. 140 pp.

Dyer, K.R., 1977. Lateral circulation effects in estuaries. Pages 22-29, *In*

Estuaries, Geophysics and the environment. National Academy of Sciences, Washington, D.C.

Duinker, J.C. and R.E. Nolting, 1976. Distribution model for particulate trace metals in the Rhine estuary, Southern Bight and Dutch Wadden Sea. Neth. J. Sea Res., 10:71-102.

Elliott, A.J., 1976. A study of the effect of meteorological forcing on the circulation in the Potomac estuary. Chesapeake Bay Ins., The Johns Hopkins University, Spec. Rept. No. 55, Ref. 76-8, 35 pp.

Elliott, A.J., 1978. Observations of the Meteorologically Induced Circulation in the Potomac Estuary. Estuarine Coastal Mar. Sci., 6:285-299.

Elliott, A.J. and D-P Wang, 1978. The effect of meteorological forcing on the Chesapeake Bay: the coupling between an estuarine system and its adjacent coastal waters. Pages 127-145, In J.C.J. Nihoul (ed), Hydrodynamics of estuaries and fjords, Elsevier, Amsterdam.

Elliott, A.J. and D-P Wang, 1978. The circulation near the head of Chesapeake Bay. J. Mar. Res., 36(4): 643-655.

Emery, K.O. and E. Uchupi, 1972. Western North Atlantic Ocean: Topography, rocks structure, water, life, and sediments. Amer. Assoc. of Petroleum Geologists Mem. 17, Tulsa, Oklahoma, 532 pp.

Festa, J.F. and D.V. Hansen, 1978. Turbidity maxima in partially mixed estuaries: a two dimensional numerical model. Estuarine Coastal Mar. Sci., 7:347-359.

Firek, F., G.L. Shideler and P. Fleischer, 1977. Heavy-mineral variability in bottom sediments of the lower Chesapeake Bay, Virginia. Mar. Geol., 23:217-235.

Fournier, F., 1960. Climat et erosion. Presses Universitaires de France, Paris.

Gallene, B., 1974. Study of Fine material in Suspension in the Estuary of the Loire and its dynamic Grading. Estuarine Coastal Mar. Sci., 2:261-272.

Gilbert, G.K., 1917. Hydraulic-Mining Debris in the Sierra Nevada. U.S. Geol. Survey Water Supply Paper 236.

Glangeaud, L., 1938 Transport et sedimentation dans l'estuaire et a l'embouchure de la Gironde. Caracteres petrographiques des formations fluviatiles, saumatres, littorales, et neritiques: Soc. Geol. France Bull., ser. 5., 8:599-630.

Gordon, R.B., 1979. Denudation rate of central New England determined from estuarine sedimentation. Am. J. Sci., 279:632-642.

Gottschalk, L.D., 1945. Effects of soil erosion on navigation in upper Chesapeake Bay. Geogr. Rev., 35:319-338.

Gross, M.G., M. Karweit, W.B. Cronin, and J.R. Schubel, 1978. Suspended sediment discharge of the Susquehanna River to northern Chesapeake Bay, 1966 to 1976. Estuaries, 1(2): 106-110.

Hathaway, J.C., 1972. Regional Clay mineral facies in the estuaries and continental margin of the United States East Coast. Pages 293-316, In B.W. Nelson, (ed.) Environmental Framework of Coastal Plain Estuaries. Geol. Soc. Am. Mem. 133, Boulder, Col., 619 pp.

Haven, D.S. and Reinaldo Morales-Alamo, 1966. Aspects of biodeposition by oysters and other invertebrate filter feeders. Limnol. Oceanogr., 11(4):487-498.

Hayes, M.O., 1971. Geomorphology and sedimentation of some New England estuaries. Pages In XII 1—XII 71. J.R. Schubel, (ed.), The estuarine environment: Estuaries and estuarine sedimentation. Am. Geol. Inst. Washington, Short Course Lecture Notes.

Holeman, J.N., 1968. The sediment yield of major rivers of the world. Water Resour. Res., 4:737-747.

Howell, J.V. and J.M. Weller, Chm., 1960. Glossary of geology and related sciences, with supplement. Am. Geol. Inst., Washington, 397 pp. (not numbered consecutively).

Inglis, C.C. and F.H. Allen, 1957. The regimen of the Thames estuary as affected by currents, salinities, and river flow: Inst. Civil Engineers Proc., Maritime Paper no. 38, Maritime and Waterways Engineering Div. Mtg., 7:827-868.

Ippen, A.T., 1966. Sedimentation in Estuaries. Pages 648-672, In A.T. Ippen, (ed.), Estuary and coastline hydrodynamics. McGraw Hill, New York.

Ketchum, B.H., 1951. The exchange of fresh water and salt water in estuaries. J. Mar. Res., 10:18-38.

Keulegan, G.H., 1949. Interfacial instability and mixing in stratified flows. Jour. Research Natl. Bur. Standards., 43:487-500.

Lüneburg. H., 1939. Hydrochemische Untersuchungen in der Elbmundung Mittels Elektrokolorimeter. Arch. Deutsch Seewarte, 59:1-27.

Martin, J.M. and M. Meybeck, 1978. The content of major elements in the dissolved and particulate load of rivers. Pages 95-110, In Biogeochemistry Estuarine sediments. UNESCO, Paris. 293 pp.

Martin, J.M. and M. Meybeck, 1979. Elemental mass-balance of material carried by major world rivers. Mar. Chem., 7:173-206.

Meade, R.H., 1969. Landward transport of bottom sediments in estuaries of the Atlantic Coastal Plain. J. Sediment. Petrol., 39(1): 222-234.

Meade, R.H., 1972. Transport and deposition of sediments in estuaries. Pages 91-120, In B.W. Nelson, (ed.), Environmental Framework of Coastal Plain Estuaries. Geol. Soc. Am. Mem. 133, Boulder, Col., 619 pp.

Meade, R.H., 1976. Sediment Problems in the Savannah River Basin. Pages 105-129, In B.L. Dillman and J.M. Stepp, (eds.), The Future of the Savannah River. Clemson Univ. Water Resources Res. Inst., Clemson, S. Carolina.

Meade, R.H., 1980. Man's Influence on the Discharge of Fresh Water, Dissolved Material, and Sediment by Rivers to the Atlantic Coastal Zone of the United States. Pages 13-17, In River Inputs to Ocean Systems. Proc. of SCOR Workshop, 26-30 March 1979, Rome, Italy. UNESCO, Paris. 384 pp.

Meade, R.H. and S.W. Trimble, 1974. Changes in sediment loads in rivers of the Atlantic drainage since 1900. Pages 99-104, In Effects of Man on the Interface of the Hydrological Cycle with the Physical Environment. Int. Assoc. Sci. Hydrol. Pub. 113.

Milliman, J.D., 1980. Transfer of River-Borne Particulate Material to the Oceans. Pages 5-12 In River Inputs to Ocean Systems. Proc. of SCOR Workshop, 26-30 March 1979. Rome, Italy. UNESCO, Paris. 384 pp.

Milliman, J.D. and K.O. Emery, 1968. Sea levels during the past 35,000 years. Science, 162: 1121-1123.

Milliman, J.D. and R.H. Meade, 1982. World-wide delivery of river sediment to oceans. J. Geol., (In Press).

Milliman, J.D., O.H. Pilkey and D.A. Ross, 1972. Sediments of the Continental Margin off the Eastern United States. Geol. Soc. Am. Bull., 83:1315-1334.

Nelson, B.W., 1959. Transportation of Colloidal Sediment in the Fresh Water Marine Transition Zone (abs.) Pages 640-641, In M. Sears (ed.), 1st Internat. Oceanog. Cong. preprints. Am. Assoc. Adv. Sci., Washington.

Nichols, M.M. and G. Poor, 1967. Sediment transport in a coastal plain estuary. Proc. Amer. Soc. Civil Eng., 93(WW4): 83-95.

Nieuwenhuise, D.S. van, J.M. Yarus, R.S. Przygocki and R. Ehrlich, 1978. Sources of shoaling in Charleston Harbor: Fourier grain shape analysis. J. Sediment. Petrol., 48:373-383.

Okubo, A., 1973. Effect of shoreline irregularities on streamwise dispersion in estuaries and other embayments. Neth. J. Sea Res., 6:213-224.

Owen, M.W., 1977. Problems in the modelling of transport, erosion, and deposition of cohesive sediments. Pages 515-537, In E.D. Goldberg, I.N. McCave, J.J. O'Brien and J.H. Steele (eds.), The Sea, Vol. VI, Marine Modelling, Wiley-Interscience, New York.

Postma, H., 1967. Sediment Transport and Sedimentation in the Estuarine Environment. Pages 158-179, In G.H. Lauff (ed.), Estuaries. Amer. Assoc. Adv. Sci. Pub. No. 83. 757 pp.

Postma, H. and K. Kalle, 1955. Die Entstehung von Trübungszonen im Unterlaufder Flüsse, Speziell im Hinblick auf die Verhältnisse in der Unterelbe. Dtsch. Hydrogr. Z., 8:(4): 137-144.

Pritchard, D.W., 1952. Salinity distribution and circulation in the Chesapeake Bay estuarine system: J. Mar. Res., 11(2): 106-123.

Pritchard, D.W., 1954. A study of the salt balance in a coastal plain estuary. J. Mar. Res., 13:133-144.

Pritchard, D.W., 1955. Estuarine circulation patterns: Am. Soc. Civil Engineers Proc., 81(717)717-1:717-11.

Pritchard, D.W., 1956. The dynamic structure of a coastal plain estuary. J. Mar. Res., 15:33-42.

Pritchard, D.W., 1967. What is an estuary: Physical viewpoint. Pages 3-5, *In* G.H. Lauff (ed.), *Estuaries*. Am. Assoc. Adv. Sci., Washington.

Rajcevic, B.M., 1957. Etude des conditions de sedimentation dans l'estuare de la Seine: Inst. Tech. Batiment et Travaux Publics Annales (Paris), 117:745-775.

Rhoads, D.C., 1963. Rates of sediment reworking by *Yoldia limatula* in Buzzards Bay, Massachusetts, and Long Island Sound: J. Sediment. Petrol., 33(4):723-727.

Rooseboom, A., 1978. Sedimentafvoer in Suider-Afrikaanse Riviere. Water S.A., 4:14-17.

Russell, R.J., 1938. Physiography of Iberville and Ascension parishes. Louisiana Dept. Conserv. Geol. Bull., 13:3-86.

Russell, R.J., 1967. Origin of Estuaries. Pages 93-99, *In* G.H. Lauff, (ed.), *Estuaries*. Amer. Assoc. Adv. Sci. Pub. No. 83, Washington, D.C.

Schubel, J.R., 1968a. Suspended sediment of the Northern Chesapeake Bay. The Johns Hopkins Univ., Chesapeake Bay Inst., Tech. Rept. 35, Ref. 68-2, 264pp.

Schubel, J.R., 1968b. Turbidity Maximum of the Northern Chesapeake Bay. Science, 161(3845):1013-1015.

Schubel, J.R., 1969. Size distributions of the suspended particles of the Chesapeake Bay turbidity maximum. Neth. J. Sea Res., 4:283-309

Schubel, J.R. (ed.), 1971a. The estuarine environment: Estuaries and estuarine sedimentation. Am. Geol. Inst., Washington, Short Course Lecture Notes. 324 pp (not numbered consecutively).

Schubel, J.R., 1971b. Tidal Variation of the Size Distribution of Suspended Sediment at a Station in the Chesapeake Bay Turbidity Maximum. Neth. J. Sea Res., 5:252-266.

Schubel, J.R., 1974. Effects of Tropical Storm Agnes on the suspended solids of the Northern Chesapeake Bay. Pages 113-132 *In* R.J. Gibbs, (ed.), *Suspended Solids in Water,* Marine Science Vol. 4. Plenum Press, N.Y.

Schubel, J.R., 1976. Fine Particles and Water Quality in the Coastal Marine Environment, 34-2 in Internat. Conf. on Environmental Sensing and Assessment, Vol. 2. A joint Conference comprising the Int. Symp. on Environmental monitoring and Third Joint Conference on Sensing of Environmental Pollutants. Inst. of Electrical and Electronics Engineers, Inc., New York. I EEE Catalog ≠ 75-CH 1004-1 II CESA.

Schubel, J.R. and T.W. Kana, 1972. Agglomeration of fine-grained suspended sediment in northern Chesapeake Bay. Powder Technology, 6:9-16.

Schubel, J.R. and H.H. Carter, 1977. Suspended sediment budget for Chesapeake Bay. Pages 48-62 *In* M. Wiley (ed.), *Estuarine Processes,* Vol. II, *Circulation, Sediments and Transfer of Material in the Estuary*. Academic Press, New York. 428pp.

Schubel, J.R. and R.H. Meade, 1977. Man's impact on estuarine sedimentation. Pages 193-209, *In Estuarine Pollution Control,* Proc. of a Conference, Vol. 1. U.S. Env. Prot. Agency, Washington, D.C.

Schultz, E.A. and H.B. Simmons, 1957. Fresh water-salt water density currents, a major cause of siltation in estuaries: U.S. Army Corps of Engineers, Tech. Bull. no. 2. 28pp.

Shepard, F.P., 1963. *Submarine geology,* Second Edition. Harper and Row, New York. 557 pp.

Simmons, H.B., 1966. Field experience in estuaries. Pages 673-690, *In* A.T. Ippen (ed.), *Estuary and coastline hydrodynamics*. McGraw-Hill Book Co., New York.

Trimble, S.W., 1974. Man-induced soil erosion on the Southern Piedmont, 1700-1970. Soil Cons. Soc. Am., Ankeney, IA Viii + 180pp.

Trimble, S.W., 1975. Denudation Studies: Can we assume stream steady state? Science, 188:1207-1208.

Trimble, S.W., 1977. The fallacy of stream equilibrium in contemporary denudation studies. Am. J. Sci., 177:876-887.

Urien, C.M., 1972. Rio de la Plata Estuary Environments. Pages 213-234, *In* B.W. Nelson (ed), *Environmental Framework of Coastal Plain Estuaries*. Geol. Soc. Am. Mem. 133, Boulder, Col., 619pp.

Van Straaten, L.M.J.U., 1960. Transport and composition of sediments. *In* L.M.J.U. Van Straaten (ed.), *Symposim Ems-Estuarium, Nordsee*. Koninkl. Nederlandsch Geol. Mijb. Genoot. Verh., Geol. ser., 19:279-292.

Vigarie, A., 1965. Les modalites du remblaiement alluvial dans l'estuare de la Seine. Cahiers Oceanographiques, 17:301-330.

Weisberg, R.H., 1976. A note on estuarine mean flow estimation. J. Mar. Res., 34(3):387-394.

Weisberg, R.H. and W. Sturges, 1976. Velocity Observations in the West Passage of Narragansett Bay: A partially mixed estuary. J. Phys. Oceanog., 6:345-354.

Williams, K.F. and J.R. George, 1968. Preliminary appraisal of stream sedimentation the Susquehanna River Basin. U.S. Geol. Survey Open-File Rept. 330. 49pp.

Wright, L.D., 1971. Hydrography of South Pass, Mississippi River. Proc. Amer. Soc. Civil Engrs. 97, WW3, 491-504.

Zabawa, C.F. and J.R. Schubel, 1974. Geologic effects of tropical storm Agnes on upper Chesapeake Bay. Marit. Sediments, 10:79-84.

10

Sedimentation Processes on Continental Shelves

R. W. STERNBERG
University of Washington,
Seattle, Washington

ABSTRACT

Shelf sedimentologists are extending the growing body of knowledge accumulated by physical oceanographers and boundary-layer scientists to encompass sediment transport processes on continental shelves. Field studies from various shelf environments have documented significant sediment response to fluid flows generated by wind stress, tides, surface gravity waves, and intruding oceanic currents. Sediment response is measured in terms of frequency of grain motion, mode of transport, particle flux, sediment transport routes, accumulation rates, and total sediment budgets. Research trends, based on recently developed instrumentation and techniques, are expanding shelf sedimentology into important new subject areas such as grain motion under combined oscillatory and steady flows; effects of biological activity on sediment entrainment; flow-sediment-organism interactions; the conditions leading to strata formation and sediment mixing; and computer modeling of sediment transport systems. The present level of understanding of shelf sedimentary processes (physically and to some extent biologically forced) and the perceived importance of combined interactions of processes implies that many major problems of shelf sedimentology would be best studied from a multidisciplinary point of view.

INTRODUCTION

This paper is a review of existing knowledge of sedimentation processes on continental shelves, i.e., a process-oriented approach to the discussion of shelf sedimentation. The particular emphasis is towards a shelf sedimentologist's review of existing knowledge of the major physical and biological processes and of the sediment response in terms of erosion, transport, and deposition of particulate matter on continental shelves.

Results from the scientific literature are used to illustrate various aspects of shelf sediment dynamics. The subject is broad and notably interdisciplinary; therefore it is necessary to draw examples from numerous investigations to characterize the environment. This approach does not provide for coherent discussion of the shelf sedimentology from any particular data base or any region; rather it provides a sense of the complex nature of the continental shelf environment, of the many methodologies being used and developed, and of the general level of understanding of the various sedimentary processes that are being studied around the world.

BACKGROUND

The study of shelf sedimentation processes has always been limited by instrument and measurement capabilities. This is especially evident in the early literature where echo sounders and conventional sediment samplers were the primary instrumentation. Despite these limitations, shelf sedimentologists attempted to deduce the important sedimentary processes from the study of sea floor morphology and grain characteristics. Examples in the literature are numerous; however this discussion will begin with the *Graded Shelf Concept* of Johnson (1919) in which surface gravity waves were considered to be the dominant driving force for sediment motion. The equilibrium shelf was thought to extend to a depth of approximately 200 m, or the seaward limit of gravity wave induced sediment transport. The stirring of sediment by ocean waves was further discussed by Kuenen (1939).

Another concept, the importance of subaerial geological processes in shaping and controlling shelf sediments, was highlighted by Shepard (1932). Subsequently, concepts of relict sediments, the importance of the Holocene marine transgression, and the incorporation of modern fluvial sediments into pre-existing relict sediments have been the source of much discussion (e.g., Stetson, 1939; Curray, 1960; Emery, 1968; Belderson et al., 1971; Swift et al. 1971, 1972; McManus, 1975). On the basis of sediment distribution and hindcast wave statistics, Curray (1960) estimated that fine sands on the edge of the Gulf of Mexico shelf are stirred by hurricane waves approximately once every five years and as frequently as once every two years on other parts of the central shelf.

Interpretations of the sedimentary history of continental shelves applied to shelf evolution following the glacio-eustatic rise in sea level between 18000 and 7000 YBP have resulted in various sedimentation and shelf facies models (e.g., Curray, 1965; Swift, 1970). Figure 1A shows Curray's (1965) neritic sedimentation model and Figure 1B, the resulting Holocene facies model for a broad continental shelf. This facies model typifies a shelf where a moderate input of sediment forms a mid-shelf silt deposit that grades seaward into relict sediment (e.g., Nittrouer, 1978). A

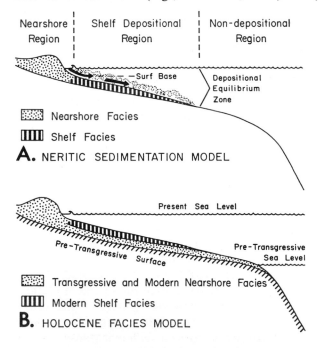

Figure 1. Models: A, neritic terrigenous sedimentation; B, resulting Holocene facies on a broad continental shelf. Nearshore sand facies is contained within a few kilometers of shore. Shelf mud facies is deposited beyond the nearshore facies but not at a high enough rate to cover the underlying transgressive nearshore facies all the way across the shelf (after Curray, 1965).

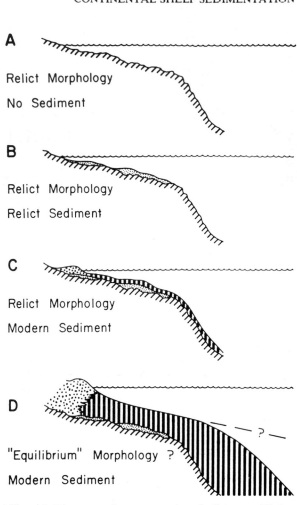

Figure 2. Diagrammatic representation of relict vs. equilibrium sediment and shelf morphology. B, C, and D show a hypothetical sequence of basal transgressive sediments being covered with sediments in equilibrium with the new environment (after Curray, 1965).

hypothetical sequence of the changes that a shelf undergoes with the onset of present sea level is depicted in Figure 2 from Curray (1965). Continental shelves exhibit various stages of evolution as shown in Figure 2, depending on sediment input and oceanic conditions.

Through much of the 1960's, oceanographers began collecting field data that would allow better interpretation of sedimentological processes. Extensive deployment of seabed drifters by physical oceanographers on the northeast (Fig. 3) and northwest (Fig. 4) coasts of the United States showed residual bottom current patterns that were thought to be indicative of sediment dispersal routes (Bumpus, 1965; Gross et al., 1969). At about the same time, decay rate-distribution analysis of short-lived radionuclides from bottom sediments on the continental shelf of Washington

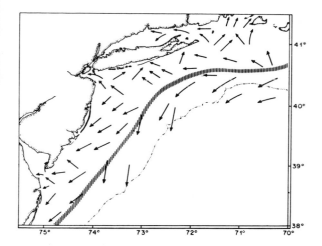

Figure 3. Diagrammatic representation of residual bottom currents on the Atlantic shelf. Dashed line is 200 m contour. Vertically lined strip represents zone of divergence (after Bumpus, 1965).

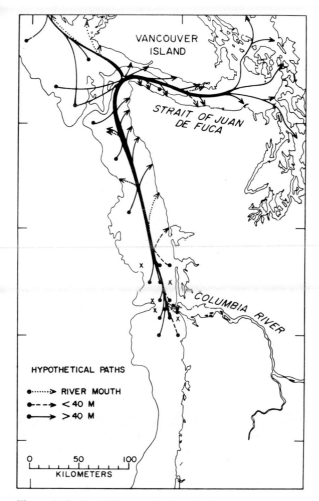

Figure 4. Seabed drifter tracks showing the general northerly flow of the bottom water on the central continental shelf (after Barnes et al., 1972).

and Oregon were used to estimate rates of movement of particulate matter (Gross and Nelson, 1966).

The status of knowledge of shelf sedimentology up to 1969 was reviewed in the American Geological Institute short course "The New Concepts of Continental Margin Sedimentation" (Stanley, 1969). The important physical processes were considered to be: tidal currents (up to 50 cm/sec); wind induced currents; ocean currents (50 cm/sec); and gravity wave induced currents (although there was disagreement about the depth of wave induced sediment transport). Residual bottom flows and net sediment dispersal routes are approximately parallel to the regional isobaths. The components of the shelf velocity field were summarized as shown in Figure 5.

These physical processes discussed above were based almost totally on indirect evidence (e.g., hindcast studies, textural data, drifter trajectory interpretations) as only a few short-term direct measurements of bottom currents or wave motions were available. The resulting concepts were an important early development of the subject of shelf sediment dynamics because correlations were observed between residual flows and suggested sediment dispersal routes, basic techniques to estimate frequency and rate of sediment transport were applied, and a framework of important near-bottom physical processes was outlined. However, shelf sedimentologists were still left with many questions. For example, what are the magnitude and frequency of driving forces on the continental shelf? What is the level of sediment response in terms of the mode of transport, mass particle flux, frequency of sediment transport, and sediment dispersal routes? If sediment is accumulating on continental shelves, what are the mechanisms? These questions can only be addressed by direct measurement of near-bottom conditions on temporal and spatial scales that encompass the important processes and their effect on the seabed.

DISCUSSION

The background section presented selected literature to illustrate the status of our knowledge of shelf sedimentary processes through the late 1960's. Although most of the concepts of shelf sedimentation were based on indirect evidence and limited direct measurements, significant changes in marine instrumentation were occurring

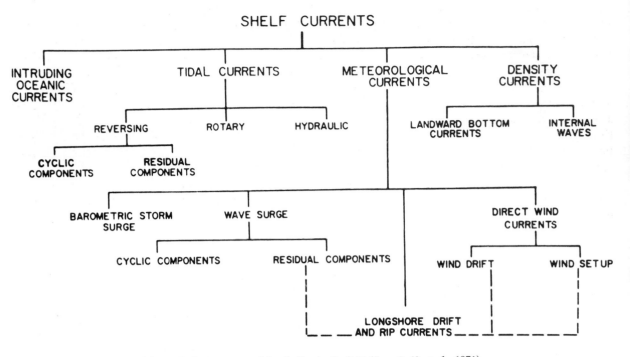

Figure 5. Components of the shelf velocity field (from Swift et al., 1971)

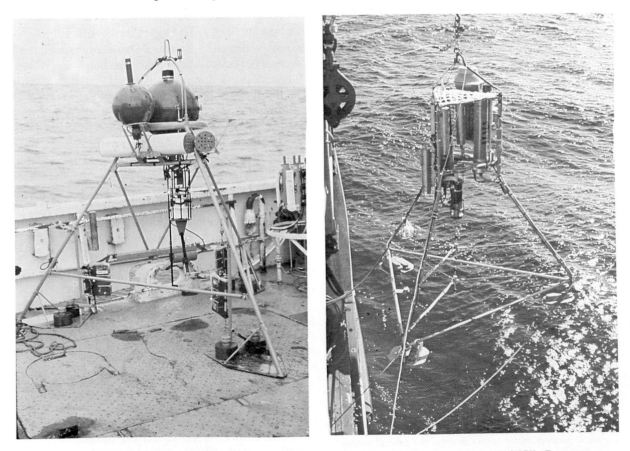

Figure 6. Instrumented bottom tripods ready for deployment. A, system described by Sternberg et al. (1973); B, system described by Cacchione and Drake (1979).

during these years. The application of low-power microelectronics to improve instrument capabilities and the subsequent development of solid state transducers, internally recording instruments, acoustic release mechanisms, and ranging devices vastly improved the measurement and instrument deployment capabilities of marine scientists. It became possible to moor instruments throughout the water column for substantial periods of time so that both mean and fluctuating components of many physical parameters could be monitored. The data collected using these new techniques have significantly improved the physical oceanographer's comprehension of shelf dynamics and have formed the basis for improved understanding of the associated sedimentary response to physical forcing.

During these same years shelf sedimentologists were also developing specialized instrumentation for investigating near-bottom boundary conditions. This included such devices as high resolution geophysical tools (e.g., reflection profiling, side scan sonar) and seafloor monitoring platforms. The primary mode for investigating near-bottom flow and sediment transport conditions was from instrumented tripod platforms that could be placed on the sea bed for extended periods of time. The tripod platforms housed batteries, data recording devices, a variety of sensors to measure flow conditions, wave motions, temperature, suspended sediment concentration, and cameras to take time-lapse photography (Fig. 6). Several research groups in the United States developed various tripod instrumentation systems (e.g., Sternberg et al., 1973; Lesht et al., 1976; Butman et al., 1979; Cacchione and Drake, 1979). These instrumentation systems provide a means to document near-bottom physical and sedimentological conditions on a seasonal basis and during extreme weather conditions. Many of the results presented below were collected from instrumented tripods.

Physical Processes

Following the introduction of internally recording instruments and extended mooring capabilities, a major objective of many physical oceanographic field studies has been the documentation of the spatial, temporal, and amplitude scales of the predominant fluid motions on the continental shelf. Results of long-term observations using moored current meters began

TABLE 1. Physical processes and their scales (from Mooers, 1976)

Process	Time Scale	Horizontal Velocity Scale (cm/sec)	Alongshore* Scale (km)	Transshore Scale (km)
Long-term mean (climatic variability)	Decade	~ 1	$\sim 10^3 - 10^4$	$\sim$ Margin width
Seasonal (year-to-year availability)	Annual and its harmonics	~ 10	$\sim 10^3$	$\sim$ Shelf width
Wind event (season-to-season and year-to-year variability)	Several days to several week	~ 100	$\sim 10^2 - 10^3$	A few horizontal modes, $\sim$ shelf width
Tidal (seasonal variability)	Diurnal and semidiurnal	~ 10	$\sim 10^3 - 10^4$	Entire margin $\sim 1 - 10^2$
nertial (wind event-to-wind event	Latitude dependent variability)	~ 10	(?)10	$\sim (?)10$
Edge wave (day-to-day variability)	Minutes	~ 10	~ 1	$\sim 10^{-1} - 10$
Surface gravity wave (sea and swell) (day-to-day variability)	Seconds	~ 10	$\sim 10^{-1}$	$10^{-3} - 10^{-1}$

*Alongshore scale is an ambiguous quantity; here it has been defined to mean the scale of organized motions, i.e., the scale of coherent motion.

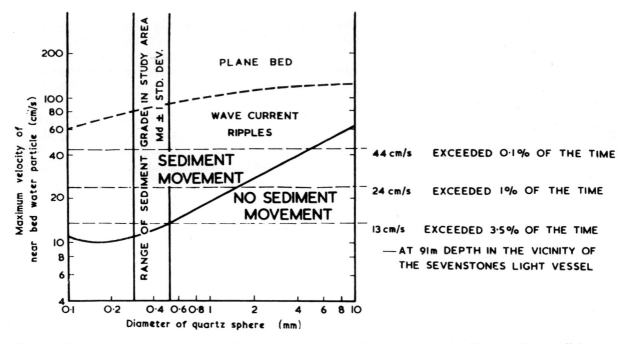

Figure 7. Diagram to show the competence of certain wave-induced oscillatory currents to move bottom sediments off the southwest coast of England (after Channon and Hamilton, 1976).

to appear in the literature in the early 1970's (e.g., Hopkins, 1971; Beardsley and Butman, 1974; Schmitz, 1974; Huyer et al., 1975). These and other observations show that the spectrum of horizontal flows occurs as an ordered sequence of broad-band energy peaks originating from a variety of sources. A summary of this sequence is shown in Table 1.

The physical processes listed in Table 1 suggest a hierarchy of phenomena that may be important driving forces for sediment transport. Wind events are clearly an important process as shown by the magnitude of the horizontal velocity scale. Although most of the other processes listed in Table 1 suggest horizontal velocities that are below the threshold of grain motion, the listed magnitudes do not preclude the potential for grain transport in combination with other processes or under particular conditions. One additional phenomenon (not included in Table 1) that has been investigated by sedimentologists is the impingement of ocean currents on a-shelf region (e.g., see Fig. 15). Although Table 1 is a gross generalization of a mixture of complex phenomena, it provides a guideline for process-oriented shelf sedimentation studies. In general, shelf sedimentologists have attempted, through field experiments, to evaluate the response of shelf sediments over various temporal and spatial scales

commensurate with the processes listed in Table 1.

Sediment Response

Cause and frequency of motion

One of the more obvious applications of near-bottom current meter time-series data has been analyses of the time frequency that near-bottom currents exceed the threshold of grain motion and of the major physical processes that are responsible. Measurements of oscillatory bottom currents from shipborne wave recorders (e.g., Channon and Hamilton, 1976), bottom pressure transducers (e.g., Sternberg and Larsen, 1976), or wave rider buoys (e.g., Jago and Barusseau, 1981) have been used to estimate the frequency of grain motion under surface gravity waves. Figures 7-9 show the frequency in days per month that bottom threshold conditions were exceeded at specific locations on two relatively high energy shelves of southwest England and Washington, and on a low energy shelf, the Roussillan-Longuedoc area of the Mediterranean Sea. The southwest England estimate (Fig. 7) is plotted in terms of a-threshold of grain motion curve for oscillatory flow conditions and the range of seabed particle sizes occurring in the study area (Channon and Hamilton, 1976). These results show that at 91 m depth the threshold of grain motion due to surface

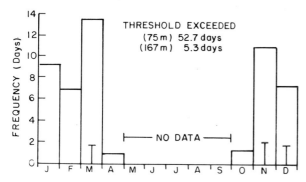

Figure 8. Frequency in days per month in which the threshold of grain motion was exceeded by wave-induced oscillatory flows on the Washington continental shelf. Open bars refer to a mid-shelf station at 75 m depth; line graphs refer to an outer-shelf location at 167 m depth (after Sternberg and Larsen, 1976).

gravity wave activity is 3.5% of the time or approximately 12 days per year The Washington shelf data (Fig. 8) are plotted for 75 m and 167 m depth and show the frequency of sediment movement due to surface wave activity on a monthly basis (Sternberg and Larsen, 1976). Sediment threshold was exceeded most frequently during the winter months and, over the total 203-day wave record that was analyzed, grain motion occurred 57 and 5 days, respectively, at depths of 75 and 167 m.

The data from the Roussillan-Longuedoc studies, a lower energy wave dominated shelf (Jago and Barusseau, 1981), are plotted monthly to show the distribution of threshold conditions for depths of 5, 10, and 20 m (Fig. 9). For example, at 20 m depth, the threshold of grain motion was only exceeded 2 days or less per month between September and May and not at all in July through August. The frequency of exceedence of threshold conditions approached zero at 40 m depth and incipient transport increased to over 70% of the time at a depth of 5 m (Fig. 10). The dominance of wave power as the major sediment transporting and sorting process is suggested in Figure 11 which shows the annual wave power profile and the sediment textural gradient over the inner 6 km of the shelf.

The results of these studies illustrate the order of magnitude of wave generated sediment transport from relatively high and relatively low energy shelf areas. However, the degree to which the observations in each area is representative of "average shelf conditions" cannot be assessed. For example, the frequency of gale force winds in the northeast Pacific over about a 40-year period are

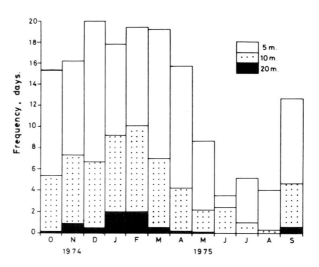

Figure 9. Frequency in days per month (1974/1975) when the maximum near-bed orbital velocity at 5, 10, and 20 m depths exceeded the predicted threshold of grain movement (after Jago and Barusseau, 1981).

known to vary by an order of magnitude (Danielsen et al., 1957), so observations for one year ideally should be placed within a longer term wave climate context.

Efforts have also been made to determine the frequency in time that mean bottom currents exceed the threshold of grain motion at various shelf locations (e.g., Sternberg and McManus, 1972; Gadd et al., 1978; Butman et al., 1979; Vincent et al., 1981). Time-series data were usually collected within 3 m of the seabed and generally consist of time averages of 10 and 20

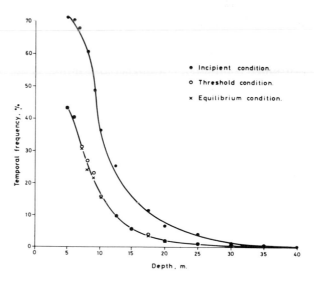

Figure 10. Temporal frequency during 1974/1975 with which incipient, threshold, and equilibrium conditions were surpassed (after Jago and Barusseau, 1981).

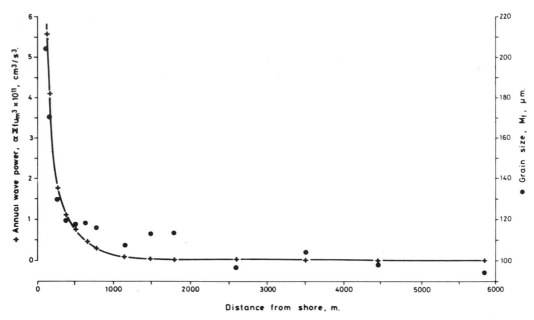

Figure 11. Seaward-attenuating annual wave-power profile and seaward-fining textural gradient of the inner shelf surface (from Jago and Barusseau, 1981).

minute duration. Thus, the observations reflect bottom currents resulting from a combination of wind-stress effects and tidal motions with wave-induced oscillations tending to be averaged out of the records.

Figure 12 summarizes approximately 6200 hours of current meter data collected at 3 m off the seabed at 80 m depth on the Washington continental shelf (Sternberg and McManus, 1972). Analysis of the data on a monthly basis showed that speeds competent to transport bottom sediment occurred for approximately 22 days over the year and varied from several hours per month in the June-September period to 2-14 days per month in November, December, and February. Although competent bottom currents flowed in all directions at various times, monthly and yearly vector averages revealed a shore parallel net transport with a small offshore component similar to the results of the seabed drifter studies of a decade earlier (see Figs. 3, 4).

Figure 13 summarizes over 135,000 hours of observational data collected from 41 stations on the New York Bight continental shelf (Vincent et al., 1981). The dashed line shows the average frequency of time that sediment threshold conditions were exceeded for all stations in <18 m depth, and the solid line is the average exceedence value for all stations located in >18 m depth. In general, sediment threshold is exceeded approximately 1-2% of the time (>18 m depth)

during June-September and as frequently as 15-20% of the time in November-January; at shallower depths (<18 m sediment threshold is exceeded 15% of the time in all but three months (March-May).

The above figures combine time-series data on a monthly basis; however, more detailed analyses show that sediment transporting conditions are associated with individual storm events that sweep the shelf during the winter months and have a duration of one to several days. Tidal currents augment the storm induced events significantly. These physical conditions dictate that grain movement occurs on a storm by storm basis primarily in the winter months. During a winter

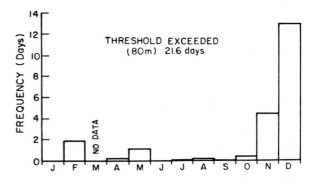

Figure 12. Frequency in days per month in which the mean bottom currents at 80 m depth on the Washington continental shelf exceeded the predicted threshold of grain motion (after Sternberg and McManus, 1972).

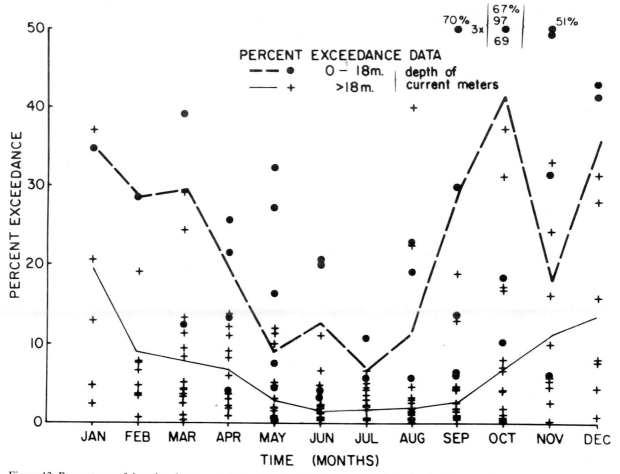

Figure 13. Percentages of time that the currents in the New York Bight apex exceed the threshold value as shown on a monthly basis. Stations in less than 18 m are shown by (•) and in greater than 18 m by (+). The solid line joins the average values of the percent exceedences for the >18 m stations, the dashed line, for the <18 m stations. Exceedences greater than 50% are indicated at the top by the percent exceedence value (after Vincent et al., 1981).

season the shelf may be swept by as many as 5-10 significant storm events; thus bottom sediments are intermittently transported. Estimates from the Washington shelf suggest that sediments being transported as suspended load could experience displacements on the order of 40 km per year, while sediments moving as bedload could be transported on the order of 100 m per year (Smith and Hopkins, 1972).

On many shelf areas tides are the dominant cause of bottom currents, and the magnitude of bottom flows are well in excess of threshold requirements for grain motion (e.g., Stride, 1963; Kenyon and Stride, 1970; McCave, 1971; Hamilton et al., 1980). An example to illustrate the importance that a tidal regime can play in transporting shelf sediments is taken from recent studies in the East China Sea (Larsen, personal communication). During fair weather conditions, the tidal currents at 40 m depth account for over 80% of the velocity variance. The magnitude of U_{100} exceeds 60 cm/sec and, over an annual period, 65% of the maximum values of U_{100} occurring during each flood or ebb cycle exceed sediment threshold requirements, whereas 17% of all tidal generated bottom velocities exceed threshold conditions. Progressive vector plots of U_{100} suggest that tidally induced sediment transport is northward approximately parallel to the isobath trend.

Mass transport and sediment dispersal routes

Several techniques have been used to estimate mass transport of sediment by physical processes. When near-bottom current measurements are used in bedload sediment transport equations, the results are given in terms of potential sediment transport rates, i.e., the quantity of sediment transported, assuming an adequate supply. An

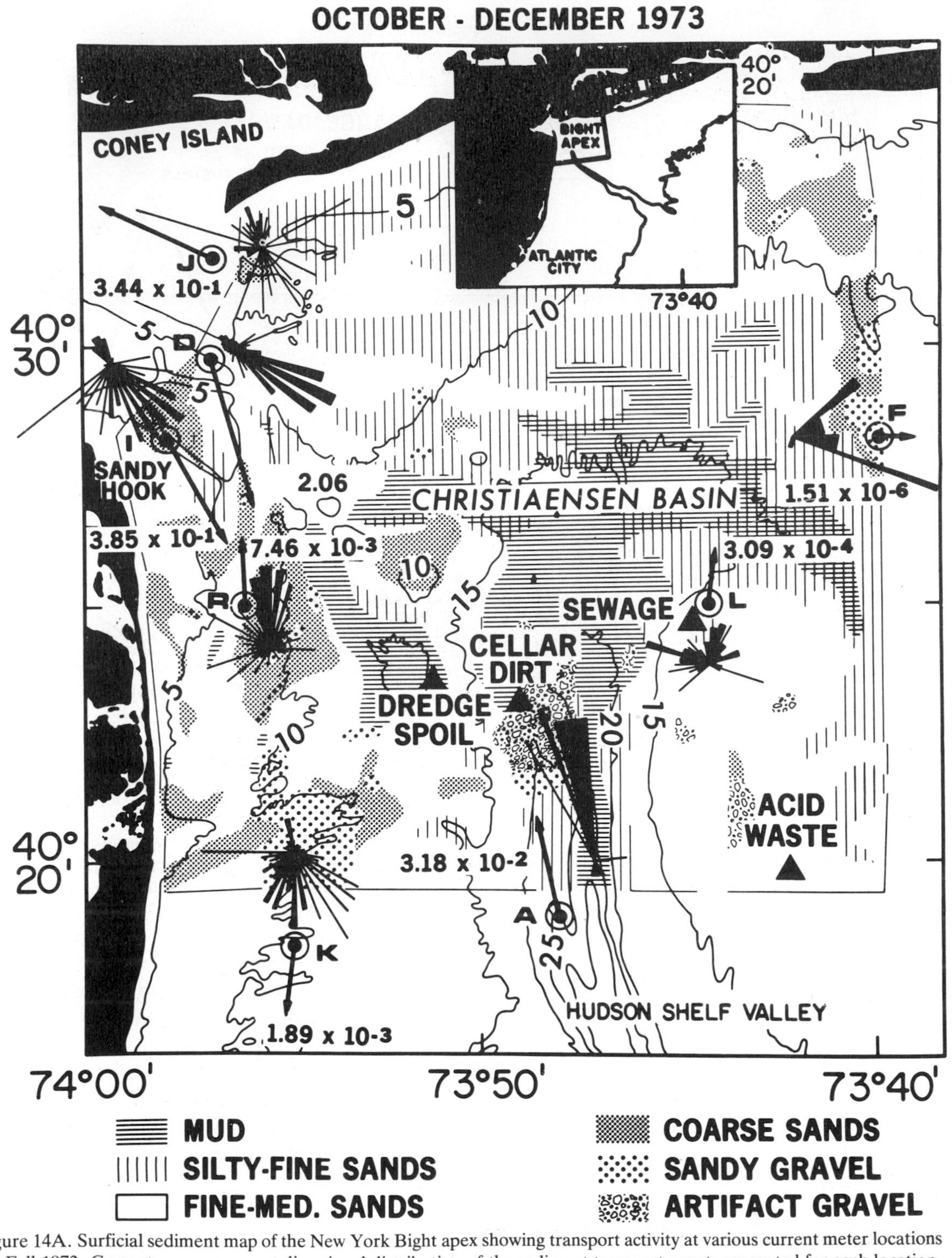

OCTOBER - DECEMBER 1973

Figure 14A. Surficial sediment map of the New York Bight apex showing transport activity at various current meter locations for Fall 1973. Current roses represent directional distribution of the sediment transport events computed for each location. Radial bar length and width represent mean transport rate and frequency, respectively, for each direction class. Each arrow reflects the resultant, or vector sum, of all sediment transport events at each current meter location. Magnitude of net transport rate (gm cm^{-1}s^{-1}) is noted above each vector (after Gadd et al., 1978).

example from the New York Bight apex (Fig. 14A, B) shows the combined results of computations of sediment transport activity in terms of net transport rates (gm cm^{-1}yr^{-1}), frequency of grain motion, and directional distribution of transport for fall and spring (Gadd et al., 1978). Sand transport rates increase towards shore, and the net direction of movement follows the isobath trend;

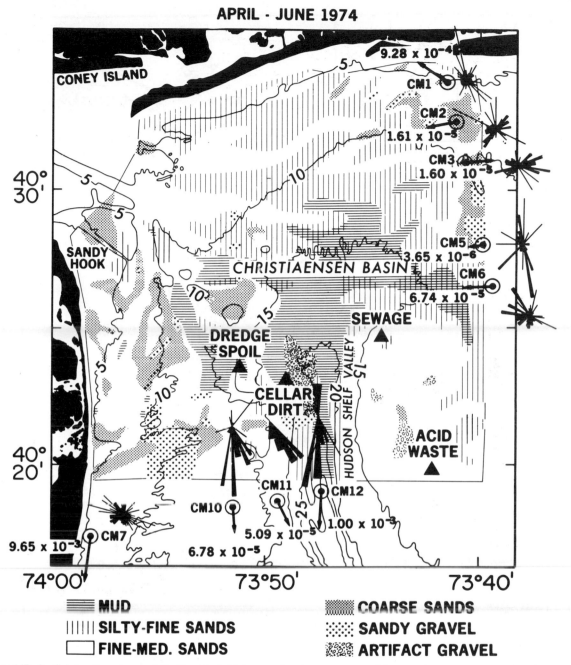

Figure 14B. Surficial sediment map of the New York Bight apex showing transport activity at various current meter locations for Spring 1974. Current roses represent directional distribution of the sediment transport events computed for each location. Radial bar length and width represent mean transport rate and frequency, respectively, for each direction class. Each arrow reflects the resultant, or vector sum, of all sediment transport events at each current meter location. Magnitude of net transport rate (gm cm⁻¹s⁻¹) is noted above each vector (after Gadd et al., 1978).

there is some evidence of seasonal reversals which is consistent with observations of wind stress events in the area.

Another method of estimating mass transport and sediment dispersal direction utilizes a combination of sediment input data, circulation estimates, major bedform analyses, and sediment textural data (Flemming, 1981). A study of the

various dominant parameters from the east coast of South Africa (Flemming, 1981) shows a clearly defined regional pattern of sediment dispersal including estimates of sediment transport rates (Fig. 15). The major parameters used for this sedimentary model are (1) morphology of the continental margin, (2) wave regime and winddriven circulation, (3) influence of the

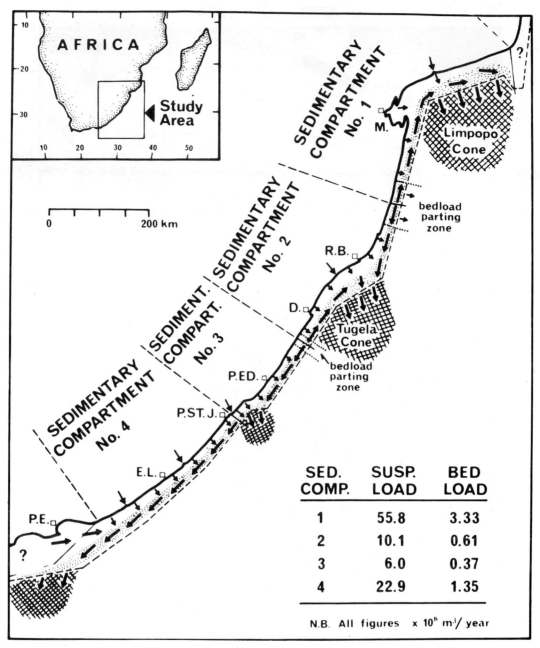

Figure 15. Regional sedimentary model summarizing the dispersal of bedload sediments along the east coast of South Africa (after Fleming, 1981).

Aghulas Current, and (4) sediment supply.

Sediment budgets

Recent development of Pb-210 geochronology has permitted the quantitative investigation of sediment accumulation on time scales of a century (see comprehensive review by Nittrouer et al., 1979). Two shelf areas that have undergone, or are undergoing, extensive investigation of sediment accumulation rates oriented towards constructing a modern sediment budget are the Washington

continental shelf (Nittrouer, 1978) and the Amazon shelf (DeMaster et al., 1980). These studies utilize extensive box coring of shelf sediments with the goals of determining the distribution of sediment accumulation rates and of comparing accumulation with estimates of sediment input from fluvial and other sources.

The Washington shelf represents an excellent example of a shelf associated with a major fluvial sediment source that is undergoing sediment accumulation. The shelf is characterized as a

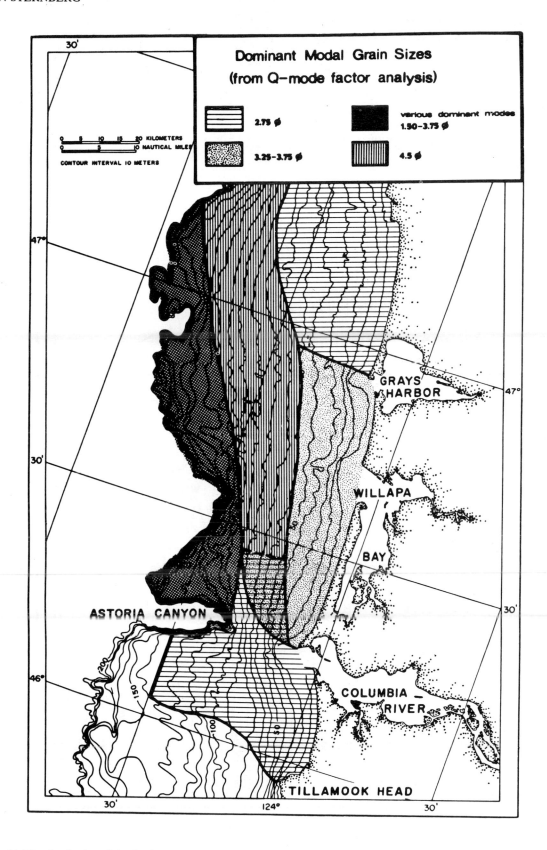

Figure 16. The distribution of the dominant modal grain sizes as delineated by Q-mode factor analysis (after Nittrouer, 1978).

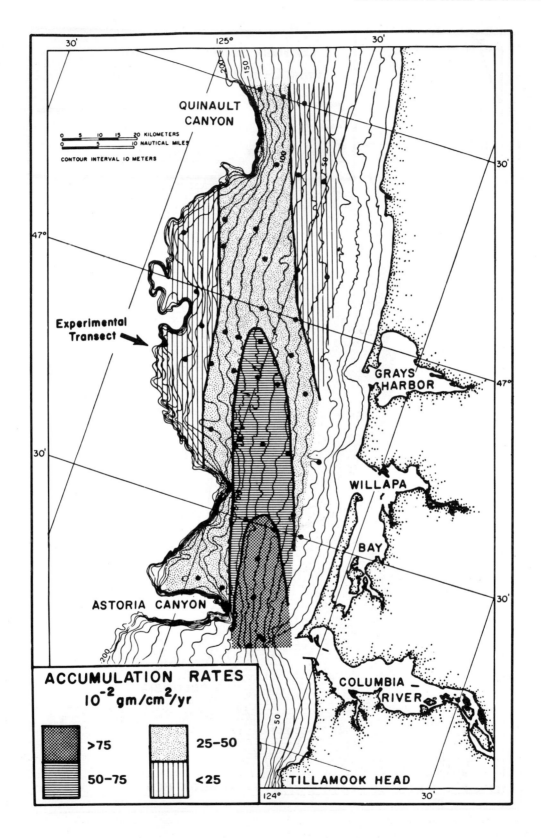

Figure 17. Accumulation rates for the Washington continental shelf (after Nittrouer, 1978).

"classical" pattern of sedimentation (see Fig. 1) with a mid-shelf silt deposit extending northward from the Columbia River, bounded by a modern sand prism on the shoreward side (depth <60 m) and relict sandy sediments on the seaward side (depth >110 m) (Fig. 16). A recent study by Nittrouer (1978) using Pb-210 geochronology shows the distribution of sediment accumulation rates over the study area (Fig. 17). Nittrouer's (1978) conclusions pertinent here are listed below to illustrate the effectiveness of the technique:

a) Rates of modern sediment accumulation progressively decrease with distance from the source, the Columbia River.

b) Approximately two-thirds of the modern fine-grained sediment supplied by the Columbia River accumulates on the shelf; the other third is transported to deeper water. Sediments are transported primarily northward along the shelf; when intercepted by submarine canyons they are routed seaward.

c) The thickness of the mid-shelf silt deposit overlying the basal transgressive sand as determined by acoustic profiling techniques is consistent with the distribution of modern sediment accumulation rates.

d) The sediment within the mid-shelf silt deposit becomes progressively finer with distance from the Columbia River.

e) The portion of the seabed actively stirred by physical and biological processes is approximately the upper 10 cm. Hydraulic sorting of storm deposits and selective feeding by benthic organisms within this zone of the seabed causes the preferential incorporation of coarse sediment within the preserved strata (i.e., below the surface mixed layer).

RESEARCH TRENDS

In reviewing literature published over the past several years, the shelf sedimentologist is made aware of the great variety of interdisciplinary techniques being applied to sedimentological problems. Many of these techniques are recently developed and on the basis of initial results appear to have significant potential for making a major contribution to the study of shelf sediment dynamics. The list below is by no means exhaustive but does provide some indication of recently established or developing research trends.

Formulation and Testing of Sediment Transport Hypotheses

The design of specialized sampling devices to test specific sediment transport hypotheses is an established trend and plays an important role in shelf sedimentology. The instrumented tripods (Fig. 6) discussed above are an example of instrumentation systems specifically designed to test marine boundary layer characteristics. Examples of other systems being used are inverted flumes for in-situ sediment tests (e.g., Young, 1977); acoustic travel time sensors to determine the vector components in turbulent boundary layers (Williams and Tochko, 1977); and heated thermistor probes to investigate viscous sublayer characteristics (Chriss and Caldwell, 1982). Examples of specialized sensors recently developed and being operated in shelf waters include bedload sensors, which use doppler laser (Agrawal et al., 1980) and acoustical (Downing, 1981) techniques as well as acoustic backscatter probes to detect suspended sediment concentration profiles. One type of acoustic backscatter probe scans downward into the lower 1 m of the boundary layer (Young et al., 1982), and another scans upward into the interior flow (Orr and Hess, 1978). Investigations using these new devices uniquely address the specific goals for which they are intended and also provide basic insights into marine boundary layer flows and associated sediment response that advance the whole field.

Geochronology

The use of Pb-210 geochronology discussed above is only one example of a relatively large suite of radioisotopes recently being used to investigate sediment accumulation and reworking processes on the continental shelf. Examples of other radioisotopes, their half-life, sources, and general application are listed in Table 2 (Nittrouer, personnel communication). Application of these techniques in shelf sedimentology is relatively new; however, their importance in investigating sediment accumulation processes, sediment reworking by organisms, physical reworking, and large scale sediment budgets has already been demonstrated. Additionally, the use of the short half-life radioisotopes to estimate

TABLE 2. List of radioisotopes used for geochronology studies (From Nittrouer, personal communication)

Radioisotope	Half-Life	Source	Application
Rn-222	3.8 days	Natural, Ra-226	Irrigation, depth of physical erosion from short term events
Th-234	24 days	Natural, U-238	Particle mixing within the seabed
Be-7	53 days	Cosmic ray produced	Particle mixing within the seabed
Th-228	1.9 years	Ra-228	Particle mixing within the seabed
Pb-210	22 years	Natural, RA-226	Accumulation rates and particle mixing, sediment budgets, modern strata formation
Cs-137	30 years	Bomb produced	Accumulation rates and particle mixing, sediment budgets, modern strata formation
Si-32	276 years	Cosmic ray produced	Accumulation rates
C-14	5,700 years	Cosmic ray produced	Accumulation rates and chronological control for long-term, large-scale stratography

sediment response from short term erosional-depositional events, in conjunction with physical measurements in the bottom boundary layer, has significant potential. Concurrent observations of both process and response conditions may lead, for the first time, to a correlation between strata formed by specific physical events (e.g., storms, wave activity) and subsequent biological reworking rates. Observations of this type could greatly improve our understanding of sediment accumulation, strata development, and facies interpretation.

Sediment-Organism Interactions

The importance of biological organisms in influencing sediment properties and transport characteristics is generally recognized by many sedimentologists but there exist no explicit techniques by which biological effects can be applied to sedimentological problems. In recent years a number of scientists have begun detailed laboratory investigations of a variety of organism-sediment interactions in an attempt to understand these effects in terms of fluid mechanics, sediment

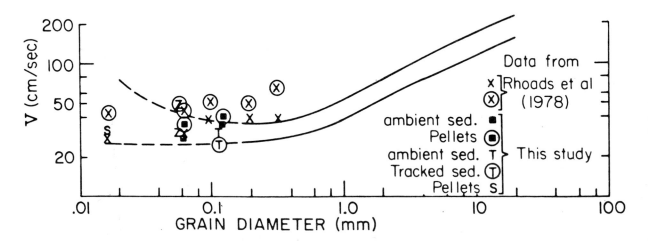

Figure 18. Sundborg's (1967) plot of critical velocity (v) at a height of 100 cm from boundary versus particle diameter, with data superimposed (from Nowell et al., 1981). Data from Rhoads et al. (1978), x and ⊗ abiotic and microbially colonized glass beads, respectively; data from Nowell et al. (1981), ⊙ , ⊗, and s: polychaete fecal pellets.

textural and mass physical property relationships. The results of the investigations include the effects of mucus coatings (Rhodes et al., 1978) and burrows, tracks and fecal pellet mounds (Nowell et al., 1981) on entrainment of marine sediments and the impact of sediment sorting due to particle selection by deposit feeders (Jumars et al., 1981). An example of this influence is illustrated in Figure 18 from Nowell et al. (1981) which shows the effects of tracks, fecal pellets, and mucus coatings on the critical erosion velocity of various sizes of marine sediments. Although studies of this nature are still in their developmental stages, the results call attention to the importance of biota to sedimentological problems and therefore continued investigations of organism-sediment interactions is a requisite.

Sediment Transport Modeling

Numerical modeling of sediment transport systems represents a valuable tool for sedimentological field studies. Large, comprehensive models allow scientists to simulate and combine various boundary layer and sediment transport hypotheses; they also provide a coherent framework for planning field experiments and for handling and interpreting field data. One such numerical model designed by Kachel (1980) to predict sediment transport and strata formation on continental shelves has been further developed by Kachel and Smith (Kachel, personal communication). It is a time dependent model that considers the combined influence of waves and wind-driven currents through the Ekman Boundary Layer, salinity variations in the near bottom region, bottom sediment size distributions with associated settling velocities, critical shear stress values for each size class, and suspended sediment concentrations including sediment induced stratification in the bottom boundary layer. The model output includes computations of

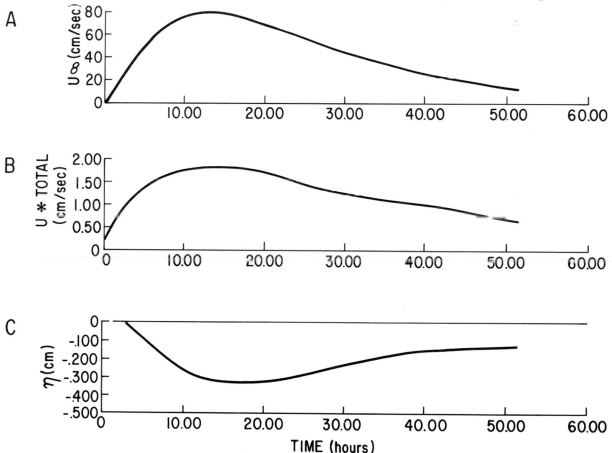

Figure 19. A, interior velocity field (U∞) simulating a storm event on the Washington continental shelf; B, computation of total friction velocity (U* total) resulting from the combined interior flow and wave activity producing maximum bottom oscillatory velocity of 10 cm sec^{-1}; C, predicted seabed erosion curve (after Kachel, 1980).

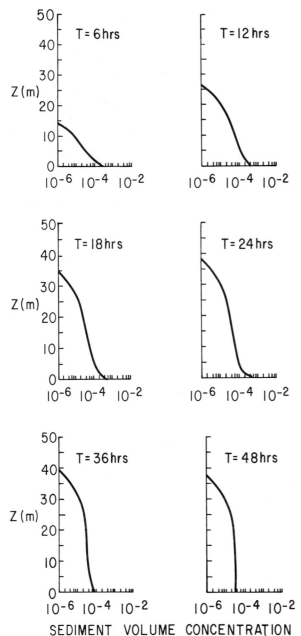

SEDIMENT VOLUME CONCENTRATION

Figure 20. Suspended sediment profiles predicted from the storm event shown in Figure 19. Concentration profiles are shown for the lower 50 m of the water column and are in units of volume concentration (e.g., cm³ cm⁻³) (after Kachel, 1980).

boundary shear stress, vertical profiles of current velocity, eddy diffusivity for mass and momentum, suspended sediment concentration and volume flux over the total water column. Additionally bedload transport rates, depth of erosion, sediment transport distances, and temporal changes in the size composition of surface sediments are estimated. Some aspects of the model are shown in Figures 19-21. In Figure 19, A

presents the wind forced interior velocity input function spanning a 52 hour period similar to flows observed on the Washington continental shelf; B, the magnitude of the total shear velocity which includes the effects of surface gravity waves (12 second period, 10 cm sec⁻¹ maximum bottom oscillatory current); and C, the cumulative erosion curve expected during the 52-hour storm event for a sandy silt bottom sediment characteristic of the Washington mid-shelf silt deposit. Computations of the suspended sediment profiles during the storm period are seen in Figure 20 and predicted variations in surface sediment texture during erosional and depositional phases of the storm in Figure 21.

Sediment transport models utilize the latest theoretical advances (e.g., wave-current interactions for boundary shear stress estimates and sediment induced stratification effects on eddy diffusivity approximations). Measurements required to test these models are within the capabilities of existing instrumentation and field verification can be expected in the future.

SUMMARY

Shelf sedimentologists are extending the body of knowledge accumulated by physical oceanographers and boundary-layer investigators to encompass sediment transport processes. Dominant physical processes that force sediment transport are recognized as surface gravity waves, surface wind stress, tidal motions, and, in some areas, impinging ocean currents. Sediment transport by waves and wind-induced bottom flows tend to occur on a "storm by storm" basis lasting from one to several days. Major transport events primarily occur in the winter months.

Measurements of bottom currents on continental shelves indicate that the frequency of sediment transport is much greater than previously suggested by indirect evidence. For example, in temperate latitudes on the eastern side of ocean basins, waves and wind-induced transport events have been estimated to occur as frequently as 80 days per year on the central shelf and 5 days per year on the outer shelf (Washington shelf). Wave induced transport events on the southwest English shelf at 91 m depth have been estimated to occur as frequently as 12 days per year. On the western side of ocean basins, wind induced currents may move sediment as frequently as 30 days per year (average for all stations with depth <18 m) at

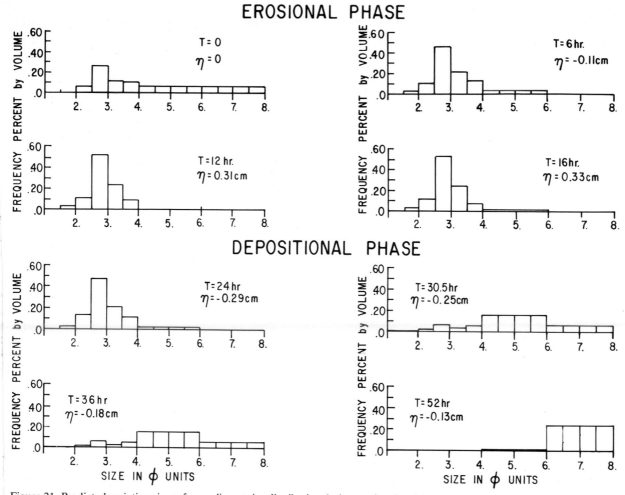

Figure 21. Predicted variations in surface sediment size distribution during erosional and depositional phases of the modelled storm event from Figure 19. T= time in hours; n = surface elevation relatived to initial conditions (T−0) (after Kachel, 1980).

locations off the eastern United States. Sediment transport associated with the tidally dominated shelf of the East China Sea may occur as frequently as 50 days per year (40 m depth).

The movement of sediment tends to be directed by coastal circulation which is predominantly parallel to the shelf with a slight offshore component. Submarine canyons intercept the sediment transport paths and reroute sediment seaward. On shelves receiving sediment from major rivers, sediment accumulation is occurring. A sediment budget estimated for the Washington continental shelf suggests that as much as 67% of the sediment supply is accumulating on the shelf.

Many new research instruments and techniques are being applied to shelf sedimentological problems. Noteworthy examples include solid state acoustic transducers for monitoring suspended sediment concentrations and triaxial velocity components; geochronology using isotopes with half lives ranging from days to decades; sediment-organism interactions expressed in terms of fluid mechanics and the physics of grain motion; and numerical models that incorporate the recent theoretical advances in boundary layer flow-sediment transport relationships.

REFERENCES

Agrawal, Y.C., W.E. Terry, Jr., and A.J. Williams, 3rd 1980. Backscatter laser Doppler velocimeter for energetic deep-sea flows—performance in particulate—dense water. Oceans '80: An international forum on ocean engineering in the '80's, Seattle, Washington, Sept., 8-10: 332-337.

Barnes, C.A., A.C. Duxbury, and B.A. Morse, 1972. Circulation and selected properties of Columbia River effluent at sea. In: A.T. Pruter, and D.L. Alverson (eds.), The Columbia River Estuary and Adjacent Waters. Environmental Studies: 41-80. University of Washington Press, Seattle and London.

Beardsley, R.C. and B. Butman, 1974. Circulation on the New England continental shelf: Response to strong wind storms. Geophysical Research Letters, 1: 181-84.

Belderson, R.H., N.H. Kenyon and A.H. Stride, 1971. Holocene sediments on the continental shelf west of the British Isles. In F.M.

Delaney, (ed.), *The Geology of the East Atlantic Continental Margin.* ICSU/SCOR Working Party 31 Symposium, Cambridge 1970. Institute of Geological Sciences Report No. 70/14: 157-170.

Bumpus, D.F., 1965. Residual drift along the bottom on the continental shelf in the Middle Atlantic Bight area. Limnology and Oceanography, Suppl., 10: R50-53.

Butman, B., M. Noble and D.W. Folger, 1979. Long-term observations of bottom current and bottom sediment movement on the Mid-Atlantic continental shelf. Journal of Geophysical Research, 84: 1187-1205.

Cacchione, D.A. and D.E. Drake, 1979. A new instrument system to investigate sediment dynamics on continental shelves. Marine Geology, 30: 299-312.

Channon, R.D. and D. Hamilton, 1976. Wave and tidal current sorting of shelf sediments southwest of England. Sedimentology, 23: 17-42.

Chriss, T.M. and D.R. Caldwell, 1982. Evidence for the influence of form drag on bottom boundary layer flow. Journal of Geophysical Research, 87: 4148-4154.

Curray, J.R., 1960. Sediments and history of Holocene transgression, continental shelf, northwest Gulf of Mexico. *In:* E.P. Shepard, F.B. Phleger and Tj. H. van Andel (eds.), *Recent Sediments, Northwest Gulf of Mexico:* 221-266. American Association of Petroleum Geologists, Tulsa, Oklahoma, U.S.A.

Curray, J.R., 1965. Late Quaternary history, continental shelves of the United States. *In:* H.E. Wright, Jr. and D.G. Frey (eds.), *The Quaternary of the United States: A Review Volume for the VII Congress of the International Association for Quaternary Research:* 723-735. Princeton University Press, Princeton, New Jersey.

Danielsen, E.F., W.V. Burt, and M. Rattray, Jr., 1957. Intensity and frequency of severe storms in the Gulf of Alaska. Transactions, American Geophysical Union, 38: 44-49.

De Master, D.J., C.A. Nittrouer, N.H. Cutshall, I.L. Larsen, and E.O. Dion, 1980. Short-lived radionuclide profiles and inventories from Amazon continental shelf sediments. EOS, 61: 1004.

Downing, J.P., 1981. Particle counter for sediment transport studies. Journal of the Hydraulics Division ASCE, 107, No. HY11: 1455-1465.

Emery, K.O., 1968. Relict sediments on continental shelves of the world. Bulletin of the American Association of Petroleum Geologists, 52: 445-464.

Fleming, B.W., 1981. Factors controlling shelf sediment dispersal along the southeast African continental margin. *In:* C.A. Nittrouer, (ed.), *Sedimentary Dynamics of Continental Shelves:* 259-277. Marine Geology, 42 (Special Issue).

Gadd, P.E., J.W. Lavelle and D.J.P. Swift, 1978. Calculations of sand transport on the New York shelf using near-bottom current meter observations, Journal of Sedimentary Petrology, 48: 239-252.

Gross, M.G., B.-A. Morse and C.A. Barnes, 1969. Movement of near-bottom waters on the continental shelf off the northwestern United States. Journal of Geophysical Research, 74: 7044-7047.

Gross, M.G. and J.L. Nelson, 1965. Sediment movement on the continental shelf near Washington and Oregon. Science, 154: 879-885.

Hamilton, D., J.H. Sommerville and P.N. Stanford, 1980. Bottom currents and shelf sediments, southwest of Britain. Sedimentary Geology, 26: 115-138.

Hopkins, T.S., 1971. Velocity, temperature, and pressure observations from moored meters on the shelf near the Columbia River mouth, 1967-1968. University of Washington, Department of Oceanography Special Report, 43: 143 pp.

Huyer, A., R.D. Pillsbury, and R.L. Smith, 1975. Seasonal variation of the alongshore velocity field over the continental shelf off Oregon. Limnology and Oceanography, 20: 90-95.

Jago, C.F. and J.P. Barusseau, 1981. Sediment entrainment on a wave-graded shelf. *In:* C.A. Nittrouer, (ed.), *Sedimentary Dynamics of Continental Shelves.* Marine Geology, 42 (Special Issue).

Johnson, D.W., 1919. *Shore Processes and Shoreline Development.* Hafner Publishing Company, New York and London, 1965 (Facsimile of the edition of 1919).

Jumars, P.A., A.R.M. Nowell and R.F.L. Self, 1981. A simple model of flow-sediment-organism interaction. *In:* C.A. Nittrouer, (ed.), *Sedimentary Dynamics of Continental Shelves.* Marine Geology, 42 (Special Issue).

Kachel, N.B., 1980. A time-dependent model of sediment transport and strata formation on a continental shelf. Ph.D. Dissertation, University of Washington, Seattle, Washington.

Kenyon, N.H. and A.H. Stride, 1970. The tide-swept continental shelf sediments between the Shetland Islands and France. Sedimentology, 14: 159-173.

Kuenen, P.H., 1939. The cause of coarse deposits at the outer edge of the shelf. Geologie en mejnbouw, 1: 36-39.

Lesht, B.M., R.V. White and R.C. Miller, 1976. A self-contained facility for analyzing the near-bottom flow and associated sediment transport. EOS, 60: 285.

McCave, I.N., 1971. Sand Waves in the North Sea off the coast of Holland. Marine Geology, 10: 199-225.

McManus, D.A., 1975. Modern versus relict sediment on the continental shelf. Geological Society of America Bulletin, 86: 1154-1160.

Mooers, C.N.K., 1976. Introduction to the physical oceanography and fluid dynamics of continental margins. *In:* D.J. Stanley, and D.J.P. Swift (eds.), *Marine Sediment Transport and Environmental Management:* 7-21. John Wiley and Sons, New York, London, Sydney, Toronto.

Nittrouer, C.A., 1978. The process of detrital sediment accumulation in a continental shelf environment: An examination of the Washington shelf. Ph.D. Dissertation, University of Washington, Seattle, Washington.

Nittrouer, C.A., R.W. Sternberg, R. Carpenter, and J.T. Bennett, 1979. The use of Pb-210 geochronology as a sedimentological tool: Application to the Washington continental shelf. Marine Geology, 31: 297:316.

Nowell, A.R.M., P.A. Jumars and J.E. Eckman, 1981. Effects of biological activity on the entrainment of marine sediments. *In:* C.A. Nittrouer, (ed.), *Sedimentary Dynamics of Continental Shelves:* 133-153. Marine Geology, 42 (Special Issue).

Orr, M.H. and F.R. Hess, 1978. Remote acoustic monitoring of natural suspensate distributions, active suspensate resuspension, and slope-shelf water intrusions. Journal of Geophysical Research, 83: 4062-4068.

Rhoads, D.C., J.Y. Yingst and W. Ullman, 1978. Seafloor stability in central Long Island Sound, 1. Seasonal changes in erodibility of fine-grained sediments. *In:* M. Wiley, (ed.), *Estuarine Interactions:* 221-244. Academic Press, New York, N.Y.

Schmitz, W.J., 1974. Observations of low-frequency current fluctuations on the continental slope and rise near site D. Journal of Marine Research, 32: 233-251.

Shepard, F.P., 1932. Sediments of the continental shelves. Geological Society of America Bulletin, 66: 1480-1498.

Smith, J.D. and T.S. Hopkins, 1972. Sediment transport on the continental shelf of Washington and Oregon in light of recent current measurements. *In:* Swift et. al. (eds.), *Shelf Sediment Transport: Process and Pattern:* 143-180. Dowden, Hutchinson and Ross, Inc., Stroudsburg, Pennsylvania.

Stanley, D.J., 1969. The new concepts of continental margin sedimentation: Application to the geological record. AGU Short Course Lecture Notes. American Geological Institute, Washington, D.C.

Sternberg, R.W. and L.H. Larsen, 1976. Frequency of sediment movement on the Washington continental shelf: A note. Marine Geology, 21: M37-M47.

Sternberg, R.W. and D.A. McManus, 1972. Implications of sediment dispersal from long-term, bottom-current measurements on the continental shelf of Washington. *In:* Swift et al., (eds.), *Shelf Sediment Transport: Process and Pattern:* 181-194. Dowden, Hutchinson and Ross, Inc., Stroudsburg, Pennsylvania.

Sternberg, R.W., D.R. Morrison, and J.A. Trimble, 1973. An instrument system to measure near-bottom conditions on the continental shelf. Marine Geology, 15: 181-189.

Stetson, H.C., 1939. Summary of sedimentary conditions on the continental shelf off the east coast of the United States. *In:* P.D. Trask, (ed.), *Recent Marine Sediments: A Symposium:* 230-244. The

American Association of Petroleum Geologists, Tulsa, Oklahoma, U.S.A.

Stride, A.H., 1963. Current swept sea floors near the southern half of Great Britain. Quarterly Journal of the Geological Society of London, 111: 175-199.

Sundborg, A., 1967. Some aspects of fluvial sediments and fluvial morphology, 1. General views and graphic methods. Geografiska annaler, 49: 333-343.

Swift, D.J.P. 1970. Quaternary shelves and the return to grade. Marine Geology, 8: 5-30.

Swift, D.J.P., D.J. Stanley, and J.R. Curray, 1971. Relict sediments on continental shelves: A reconsideration. The Journal of Geology, 79: 322-346.

Swift, D.J.P., D.B. Duane, and O.H. Pilkey, 1972 *Shelf Sediment Transport: Process and Pattern*. Dowden, Hutchinson and Ross, Inc., Stroudsburg, Pennsylvania.

Vincent, C.E., D.J.P. Swift, and B. Hillard, 1981. Sediment transport in the New York Bight, North American Atlantic shelf. *In:* C.A. Nittrouer, (ed.), *Sedimentary Dynamics of Continental Shelves:* 369-398. Marine Geology, 42 (Special Issue).

Williams, A.J. and J.S. Tochko, 1977. An acoustic sensor of velocity for benthic boundary layer studies. *In:* J.C.J. Nihoul, (ed.), *Bottom Turbulence.* Proceedings of the 8th International Liege Colloquium on Ocean Hydrodynamics: 83-97. Elsevier Scientific Publishing Company, Amsterdam, Oxford, New York.

Young, R.A., 1977. Seaflume: A device for in-situ studies of threshold velocity and erosional behavior of undistrubed marine muds. Marine Geology, 23: M11-M18.

Young, R.A., J.T. Merrill, T.L. Clarke, and J.R. Proni, 1982. Acoustic profiling of suspended sediments in the marine bottom boundary layer. Geophysical Research Letters, 9: 175-178.

D. TECTONICS OF PAKISTAN

11

An Overview of the Tectonics of Pakistan

A. FARAH
Geological Survey of Pakistan,
Quetta, Pakistan,

R. D. LAWRENCE
Oregon State University,
Corvallis, Oregon,
and
K. A. DE JONG
University of Cincinnati,
Cincinnati, Ohio

ABSTRACT

The geological setting of Pakistan in the framework of the modern concept of plate tectonics is unique in the sense that, within an area of about 800,000 sq. km., critical tectonic junctions of different interacting plates and microplates are present in an environment where field exposures are excellent.

Herein we discuss the dynamics of these various plate boundaries. Two types of active plate boundaries are conspicuous: 1) Convergent boundaries characterized by continent-continent collision, obduction and thrusting in the northern region of the Himalaya and by oceanic crust subduction with a volcanic arc and a wide accretionary wedge in the southern region of Chagai and Makran; 2) A transform boundary, the Chaman transform zone, characterized by very large strike-slip and lesser thrusting. The Chaman transform zone connects the Makran convergence zone, where oceanic lithosphere is being subducted beneath the Lut and Afghan microplates, with the Himalayan convergence zone, where the Indo-Pakistan Lithosphere is underthrusting Eurasia. The Chaman zone is at present an intracontinental plate boundary with oblique motion, characterized by north-south strike-slip faults and eastward thrusting and folding in the Kirthar-Sulaiman mountain belt. This mountain belt, the northwestern margin of the Indo-Pak subcontinent, was an Atlantic-type margin from the late Paleozoic until the Cretaceous. In the Cretaceous, the continental margin became a plate boundary; a thrust belt was formed in the Paleocene, and fragments of the oceanic crust were obducted, either as thrust sheets (Muslimbagh) or as an ophiolitic melange (Bela and Waziristan).

INTRODUCTION

Active plate boundaries of various types are exceptionally well exposed in Pakistan (Fig. 1). In the northeast, there is an active continent-island arc-continent collision boundary, the west end of the Himalayan orogen. In the southwest, there is an active boundary of oceanic lithosphere subducting beneath arc-trench gap sediments and continental sediments, the oceanic part of the Arabian plate passing under the Makran arc-trench gap and Afghan microplate. These two convergent boundaries are connected by very large displacement north-south left-lateral strike-slip faults of the Chaman transform zone. The edge of the Indo-Pakistan subcontinent where it abuts the Himalayan convergent zone and the Chaman transform zone is characterized by marginal fold and thrust belts in shelf sediments; these active belts have grown and continue to grow towards the Indian shield. The Makran arc-trench gap is also a fold and thrust belt, but in accretionary wedge sediments, and it grows toward the Arabian plate. The interaction of fault systems and fold belts has imparted a scalloped outline to the tectonic pattern which is distinctive of the geology of Pakistan.

As a result of International Geodynamics Commission projects in Pakistan, much progress has been made toward an improved understanding of these active plate boundaries and their histories in Pakistan. This paper summarizes some of the

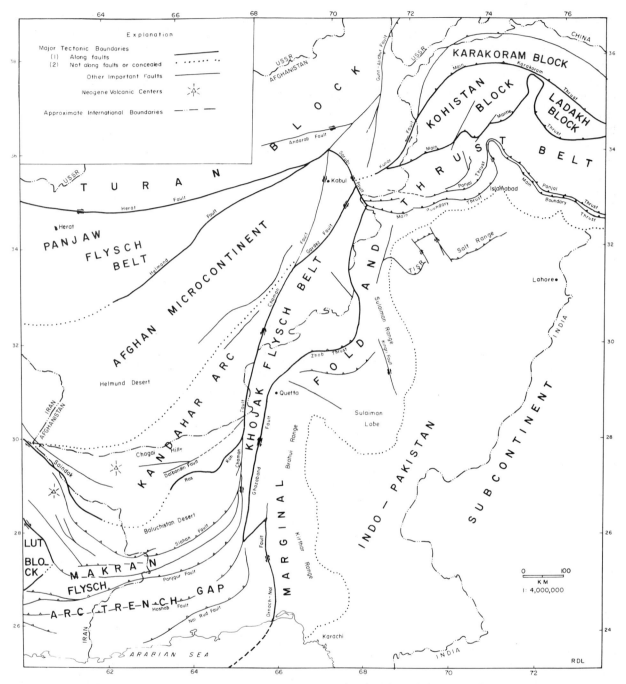

Figure 1. Tectonic boundaries of Pakistan.

major results of this work (see also Farah and De Jong, 1979, and Tahirkheli et al., 1980).

Understanding of the plate-tectonic history of the area has involved study of the magnetic anomaly patterns of the Indian Ocean for the last 80 million years of spreading history culminating in the collision of the Indo-Pakistan subcontinental block with Eurasia (Powell, 1979). Separation of subcontinent from Australia and Antarctica started about 130 million years ago. The spreading rate was high (about 20 cm/yr.) between 80 and 53 Ma. Immediately following collision, the rate of spreading in the Indian Ocean decreased to almost zero, but in the Oligocene it sped up again to about 4 cm/yr. and underthrusting of the Indo-Pakistan plate beneath the Himalayas was initiated (Powell, 1979).

North of the Indo-Pakistan subcontinent, older

oceanic elements are preserved as ophiolites (Stoneley, 1975; Stöcklin, 1977; Mitchell, 1981) which were emplaced as obducted thrust sheets or ophiolite melange on passive continental margins, caught up in suture zones, and/or sliced into the materials of accretionary wedges. All of the major tectonic boundaries of Figure 1 are accompanied by ophiolitic material of one or more of these types. These separate the main pre-Hercynian Eurasian and post-Permian Indo-Pakistan continental blocks from intervening micro-continental blocks that separated from Gondwanaland before the 130 million year breakup. The western Tethys Sea may be separated into two entities, the Paleotethys and Neotethys, respectively north and south of these microcontinental blocks. During the early Mesozoic, the Paleotethys Sea may have resembled the complex of archipelagoes found today between soutwest Asia and Australia. By middle Mesozoic these fragments were swept north into near proximity to Asia as the Neotethys opened. The later history of these archipelagoes in Pakistan is involved in a late Cretaceous andesitic arc. This was an Andean-type marginal arc, the Kandahar arc, along the southern Afghan microcontinent (Bellon et al., 1979; Afzali et al., 1979; Lawrence et al., 1981a; Tapponnier et al., 1981) and an intraoceanic arc, the Kohistan and Ladakh blocks in northern Pakistan (Bard et al., 1980; Jan and Asif, 1981, and Tahirkheli, 1982). Volcanic and plutonic activity has persisted in this arc north of the Makran, in the gap between the Arabian and Indo-Pakistan colliding blocks.

NORTHERN COLLISION BELT

The northern collision belt of Pakistan is a part of the Alpine-Himalayan orogen which, since the work of Argand (1924), has been considered to be the prototype of mountain belts produced by continental collision. There are three major subdivisions to it in this area. North of the Main Karakoram Thrust (MKT) is a terrain of Gondwana affinities that was sutured to Eurasia (the Turan Block) in the Late Triassic to Middle Jurassic (Boulin and Bonyx, 1978; Stöcklin, 1977; Sengor, 1979). The main bulk of the Hindu Kush and Karakoram Ranges form this microcontinent. South of the Main Mantle Thrust (MMT) are the low ranges of Swat, Hazara, and Kashmir and the outlying Salt Range. These are part of the marginal fold and thrust belt of the Indo-Pakistan

subcontinent and pass continuously into the main Himalayan Ranges to the east. Between these two fault-bounded continental blocks are the Kohistan and Ladakh blocks, currently believed to have formed as an intraoceanic island arc (Jan and Asif, 1981; Tahirkheli, 1982).

The geology and tectonics of the Himalaya east of Kashmir are better known and understood than those of northern Pakistan where initial reconnaissance geologic mapping is still in progress. The subdivision of the Himalaya into the Sub-Himalaya, Lesser Himalaya, High Himalaya and Tethys Himalaya and identification of the three great structures, the Main Boundary Thrust (MBT), Main Central Thrust (MCT), and Indus-Tsangpo suture (Gansser, 1981) become unclear in Kashmir. These tectonic features have not yet been satisfactorily traced around the Hazara-Kashmir syntaxis (Yeats and Lawrence, this volume). In Kumaon and farther east, the northern margin of the Indo-Pakistan plate was either doubled tectonically along the MCT (Fig. 2a) or tripled tectonically along the MCT and MBT (Figure 2b) (Powell and Conaghan, 1973; Le Fort, 1975; Powell, 1979). The underthrusting of the Indian plate resulted in crustal telescoping and thickening and the dramatic uplift of the Himalya. As the MBT is younger than the MCT, the thrusting apparently migrated southward toward the Indo-Pakistan subcontinent. The structural equivalent of the MBT can readily be traced around the Hazara-Kashmir syntaxis (Calkins et al., 1975), across the Northwest Frontier Province, and into the Sarubi fault in Afghanistan. It separates the equivalents of the Sub-Himalaya and Lesser Himalaya. On recent small scale maps Gansser (1981) has drawn the MCT into the Panjal thrust of Kashmir. The equivalent of the thrust can be tentatively traced west across the Peshawar Basin toward the Khyber Pass (Fig. 1). However, Valdiya (1980) disputes this designation of the MCT and suggests that it lies farther to the north. Certainly the rocks found above the proposed extension of the Panjal thrust into Hazara, the Precambrian (?) Hazara-Attock slates and overlying lower Paleozoic sediments, do not resemble the crystalline rocks described as above the MCT in the central and eastern Himalayas. Thus the entire area in Pakistan up to the MMT may be the general equivalent of the Lesser Himalaya, and the equivalents of the Great Himalayas and Tethys Himalayas are missing, possibly overthrust by the Kohistan Arc in this region (Fig. 3). Doubling of the Indian plate does not appear to be present.

According to current models, there are two suture zones in northern Pakistan, the MMT and MKT, which join together as the Indus-Tsangpo suture to the east. The map trace of the southern zone, the MMT, comprises two arcs convex toward the south (Fig. 1). These define the Ladakh and Kohistan Blocks, respectively east and west of the Nanga Parbat-Haramosh massif. This is a young uplift (Zeitler et al., 1982) which has deformed the MMT and displaced it to the north. The MMT is marked by mantle-related ultramafics, metavolcanics, metagabbros, phyllitic sediments, blueschists, and garnet granulites (Jan, 1980; Jan and Howie, 1981). In contrast, the northern suture, the MKT, has a map trace that is a smooth arc convex toward the north (Fig. 1). It is marked by flyschoid sediments and shreds of ophiolite (Tahirkheli, 1982).

Between the two sutures is the Kohistan-Ladakh intraoceanic island arc created during subduction of the Tethys Ocean in the Cretaceous.

In Kohistan, current thinking suggests that this arc has been either tilted northward to expose a cross section of an island arc crust (Tahirkheli et al., 1979; Jan et al., 1981; Tahirkheli, 1982) or folded into a tight synclinorium that involves the entire crust (Coward et al., 1982). From south to north, rock types exposed include amphibolite, pyroxene granulite-norite with ultramafic differentiates near the base, a dioritic intrusive complex, and a granitic intrusive complex. Pockets, screens, and pendants of metasediments and metavolcanics are present locally (Tahirkheli, 1979). Interpretation of the structure is as yet very tentative. A north-south geological traverse from Chilas to Gilgit west of the Indus gorge and outside the zone disturbed by the Nanga Parbat-Haramosh massif is needed to test concepts.

Recent fission-track dating (Zeitler et al., 1982) shows that the MMT suture has been "locked" since 15 million years ago. The Kohistan sequence in the north and the Indian plate in the south were apparently uplifted at the same rate over the period. The absence of recent seismicity along the MMT supports the notion that it is

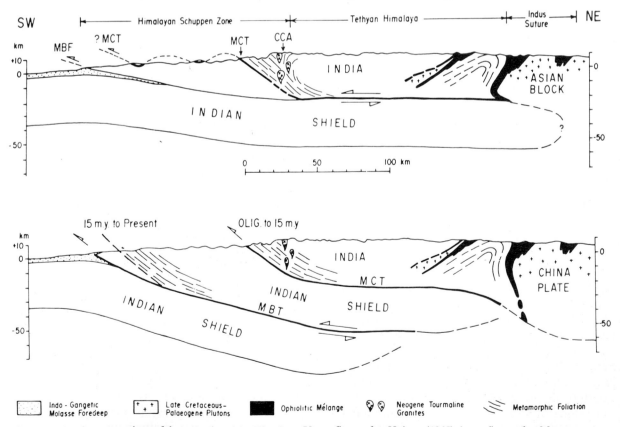

Figure 2. Two interpretations of the tectonics of the Himalaya. Upper figure after Holmes (1965), lower figure after Mattauer (1975); both simplified after Powell (1979).

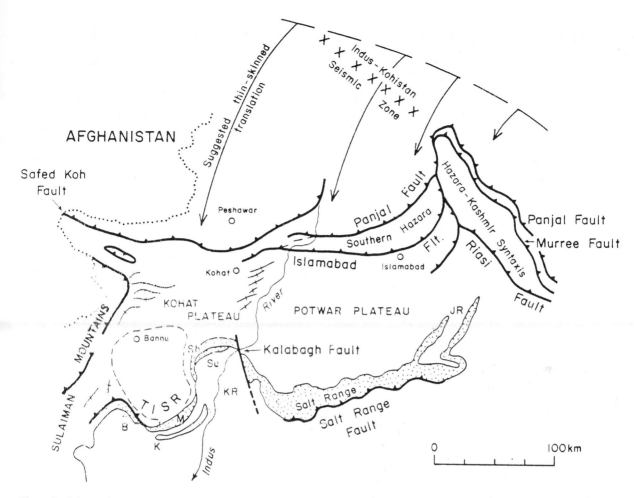

Figure 3. Schematic structural map of the Main Boundary fault and Salt Range zone of northern Pakistan. The arrows suggest a maximum of 250 km of displacement of the southern Himalaya around a pivot point to the east. Abbreviations are: TISR = Trans-Indus Salt Range, B = Bhitanni Range, M = Marwat Range, K = Khisor Range, Sh = Shinghar Range, Su = Surghar Range , KR = Kalabagh Reentrant, and JR = Jhelum Reentrant.

"locked" at the present time. From 25 Ma to 15 Ma, the uplift rate of the block south of the MMT was three to five times greater than the rate for Kohistan. Thrusting of Kohistan over the MMT must have resulted in an uplift of the northern block faster than the southern block. According to Tahirkheli et al. (1979), this thrusting occurred in the Early Tertiary; the time of reversal of this movement is Middle Tertiary (25-15 Ma).

The gravity field across the Karakoram shows a marked symmetrical pattern with negative closure centered over the Karakoram. Marussi (1980) has interpreted this negative axial anomaly bordered by stripes of positive anomalies as "due to the lighter granites that form the Karakoram axial batholith, flanked by metamorphic and basic igneous rocks." A seismic refraction profile between Karakul Lake (northern Pamir in the

USSR), Astor Lake near Nanga Parbat, and Lawrencepur in the Indus plain jointly observed by the Geological Survey of Pakistan and Italian Institutes in 1975 and 1978 indicates a relative thickening of the asthenosphere under the Pamir and Karakoram with a crustal thickness of 60-65 km. The asthenospheric upper layer is recorded at 120 km (7.5 km/sec) and its base at 300 km (8.4 km/sec) in southern Pamir and Karakoram. Under Nanga Parbat, a rise in Moho level is recorded which corresponds to a rise in gravity. As a general surmise, extreme gravity minima and depressions on the geoid surface correspond to the regions with maximum thickness of the earth's crust. On either side of the Karakoram, the M-discontinuity shows a relative rise. Towards the south, under the marginal foredeep, the crustal thickness decreases to 35 km. From south to north, along with an increase in crustal thickness, there is an increase in seismic velocity of the surface layers. The gravity and seismic refraction data may be explained on the basis of the flake tectonic concept of Oxburgh (1972) concerning continental collision. According to this concept, large masses of upper crust of one continental plate may be torn up and driven onto the second one for large distances, thereby accounting for an increase in crustal thickness and in seismic velocity of surface layers from the Indus plain towards the Karakoram in the north.

The gravity field of the Punjab plain suggests a gentle slope of the crystalline basement towards the north and an implication that with increasing sedimentary cover the shield elements can be followed northward beneath the southward thrust of the Himalaya, a result of continent-continent collision (Farah et al., 1977).

The data on earthquake foci and focal mechanism solutions positively define the relative motion of the Indian and Eurasian plates. Molnar and Tapponnier (1975) are of the opinion that central Asia is largely passive tectonically and that the intracontinental seismicity is due to the northward convergence of the Indian plate. On the other hand, Das and Filson (1975) contend that the seismicity of central Asia may be due to the interaction of a number of blocks which move in response to internal and external stresses. Seeber and Armbruster (1979), in their treatment of the Hazara arc in northern Pakistan, a tiny part of the large converging orogen, imply that the tectonics of the region are extremely complex. Based on the data of the telemetered seismic network in northern Pakistan covering the Hazara arc and the foreland thrust and fold belt, they distinguish two northwest-trending seismic zones: the Indus-Kohistan Seismic Zone (IKSZ) and the Hazara Lower Seismic Zone (HLSZ). These seismic zones are well defined by deeper crustal seismicity, and they are discordant to the exposed arcuate structural trends of the Hazara arc. These have been interpreted as buried extensions of the northwesterly trending Himalayan frontal faults beyond their surface termination at the western Himalayan syntaxis. Topographic steps aligned with these fault zones may be reckoned as surface expressions. The IKSZ, by far the more active of the two seismic zones, is associated with underthrusting towards the northeast. In contrast, the HLSZ defines a steeply dipping basement fault (possibly a ramp fault) which probably merges

southward with a shallow dipping surface of detachment between the basement and the sedimentary rock sequence of the Salt Range. Apart from certain conflicting evidence from the eastern portion of the Hazara fault, the general seismicity characteristics of the area indicate vergence towards the Indian shield and a north-south shortening consistent with the convergence of the Indo-Pakistan and Eurasian plates. The emplacement of the Salt Range still remains a problem (Yeats and Lawrence, this volume). The Salt Range is an east-northeast trending rampart which rises abruptly from the Jhelum River plain for a distance of 175 km. The main difference between the Salt Range and the regions to the north is that the Salt Range has undergone strong folding and uplifting in a narrow zone. Crawford (1974), on the basis of the Cambrian Pole positions obtained from the purple sandstone and salt pseudomorph beds, suggests that the Salt Range is a decollement which has been rotated about 75° counterclockwise from an original position in line with the main Himalayan Front, about a pole near the eastern end of the Salt Range. More local is the rotation observed by Opdyke et al. (1981) in the eastern Potwar Plateau; it indicates a maximum 30° counterclockwise movement accompanying southward translation of the Potwar Plateau and the Salt Range.

ARC-TRENCH SYSTEM OF THE CHAGAI-MAKRAN REGION

The area west of the Chaman-Ornach Nal faults is a subduction-convergence zone with an unusually wide and subaerially exposed arc-trench gap (Fig. 1). Data on the area are sparse, and interpretation of the main tectonic features and their history remain speculative. The general geologic setting appears to involve an Andean-type andesitic arc in the Chagai Hills, Ras Koh, and Saindak areas in the north that was most active from the Middle Cretaceous to the early Paleogene but that has had less voluminous igneous activity through the present (Jones, 1961; Arthurton et al., 1979, 1982; Lawrence et al., 1981b). South of this arc belt is a wide area of Quaternary cover in the Baluchistan Desert, where the older history is completely concealed and unknown. South of this desert is the area of the Makran Ranges where arc-trench gap materials are widely exposed. Thick, extensive flysch deposits of late Paleogene/early Neogene age accumulated during the period of slow convergence of the Asian and Indian-Arabian plates in the northern Makran. These may represent an ancient submarine fan similar to the contemporary Indus fan that was developed from rivers draining into the gap between the Indian plate and the Afghan microplate of the Eurasian plate and across the southern edge of the Afghan microplate. The current subduction zone reaches the surface under the ocean in the south (White and Klitgord, 1976; White, 1979; Jacob and Quittmeyer, 1979). Its history may well be explained by the concept of sinking and retreating subduction of Kanamori (1977), who suggests that "after the plates are decoupled, the deformation of the underthrusting lithosphere is possible in several ways. In one case, the subduction zone may retreat in the direction opposite to the direction of the plate motion as the leading edge sinks and falls off. A counterflow may fill the opening between the continental lithosphere and the retreating subduction zone. This counterflow may take place in the form of episodic interarc spreading with upwelling material eventually forming a thin lithosphere between the continental lithosphere and the retreating subduction zone. Once the relatively rigid lithosphere is formed, the coupling between this lithosphere and the oceanic lithosphere may be restored." The history of the area, overall, may thus be one of beginning subduction near the southern margin of the Ras Koh in the late Cretaceous, migrating south toward its present position with a long mid-Cenozoic pause during which a huge submarine fan of flysch was constructed.

VOLCANIC-PLUTONIC ARC OF CHAGAI

The Chagai magmatic belt is convex towards the south and extends for about 500 km in an east-west direction and 150 km in a north-south direction. Arthurton et al. (1979), from field mapping data of the Mashki Chah-Alam Reg area along the southern margin of the central part of the Chagai arc, have worked out a stratigraphy and geologic history that is important for the interpretation of the history of the Chagai arc. They have established the existence of an andesitic volcanic arc in Cretaceous times which was subjected to recurrent uplift through the Paleogene, accompanied by the corresponding formation of the Mirjawa-Dalbandin trough between the Chagai igneous belt and the Ras Koh Range.

Towards the end of the Paleogene, or possibly early in Neogene times, the area suffered a major diastrophism culminating in the establishment of the continental conditions that presently prevail. Two more important phases of volcanism and tectonism in the area have been identified which occurred at about the Miocene-Pliocene boundary and in the early Quaternary. In contrast to the Cretaceous igneous activity, mid-Cenozoic activity was apparently restricted to isolated subaerial centers, one of which, at Saindak, has been radiometrically dated at 20 million years (Sillitoe & Khan, 1977).

The folding episode at about the Miocene-Pliocene boundary has been interpreted as a repercussion of tectonic events of corresponding age in Iran—the collision between the Arabian continental plate and the Iranian component of the Eurasian plate (Lawrence et al., 1981a). Subsequent Pliocene erosional planation was followed by early Quaternary alluvial fan deposition in the Dalbandin trough marking renewed uplift of the Chagai arc. This uplift immediately preceded the extrusion of the Quaternary volcanics of the Koh-i-Sultan suite, members of which intercalate with, and cap, this early Quaternary alluvium.

The arc volcanic and volcaniclastic rocks of the Chagai Hills and Siah Koh are thrust south against and, in part, over the northern Ras Koh, composed of ultramafic rocks of a dismembered ophiolite bounded by faults. The position of the Ras Koh Range in the Chagai arc-trench system is unknown. In the north the Ras Koh Range is in tectonic contact with the gently folded Mesozoic-Tertiary sequence of the Chagai volcanic arc; and, in the south, it is separated by a broad zone of Quaternary deposits from the Makran trench accretionary deposits further south. The Ras Koh Range consists of ophiolites, Paleogene greywacke, volcaniclastics and diabases tightly folded and intruded by calc-alkaline plutons. Basaltic, andesitic, and rhyolitic volcanic rocks of submarine origin are known from northernmost Ras Koh (Bakr, 1964). However, the apparent dominance of volcaniclastic units suggests that they represent a depositional apron south of the main arc.

The aeromagnetic anomaly map of Chagai-Ras Koh shows four distinct east-west trending belts of anomalies (Fig. 4): 1) a northern belt occupied by a cluster of small wavelength positive anomalies over the Chagai volcanic arc, 2) a belt of broad

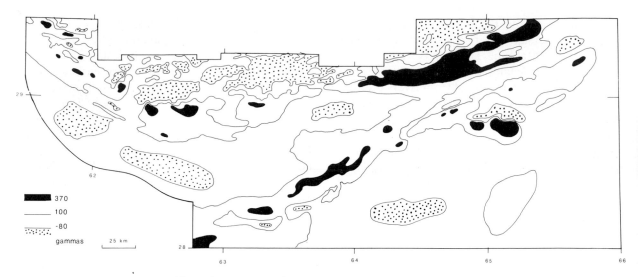

Figure 4. Aeromagnetic anomalies of the Chagai-Ras Koh area.

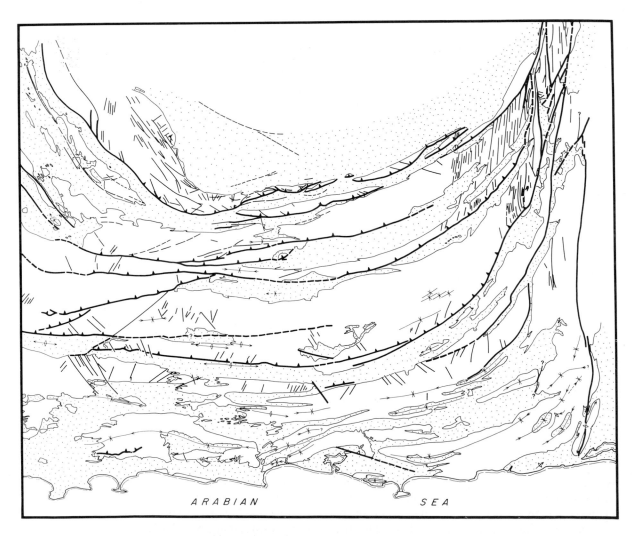

Figure 5. Active faults in the Makran thrust belt, interpreted from Landsat imagery.

negative anomalies over the Dalbandin trough, 3) a belt of elongated positive anomalies over the Ras Koh Range, and 4) a broad belt of low anomalies over the Kharan basin bordering the accretionary wedge. The anomaly belts, which arc fairly distinct and separable in the west, appear to converge and terminate in the east near Nushki, against the Chaman transform fault. A possible geological explanation for these alternating anomaly belts is that the island arc volcanics and volcaniclastics of the Chagai Range were thrust southward, against and, in part, over the northern Ras Koh. The Ras Koh rocks, in turn, have probably been thrust to the south against and, in part, over the northern portion of the Makran accretionary wedge. The three belts terminate to the east, against the Chaman transform boundary.

MAKRAN MARGIN

The Makran margin is formed of an accretionary wedge of sediments developed between a buried offshore trench and Quaternary volcanoes (Faroudi and Karig, 1977). White (1979) finds that gravity modelling supports the hypothesis that the Arabian plate dips at a shallow angle beneath the Makran Ranges. The exposed rocks of the ranges are youngest at the coast and generally become older to the north (Ahmed, 1969; Raza et al., 1981). The major structural features of the Makran are broad synclines separated by sharp, thrust-faulted anticlines. Evidence for Quaternary to Recent activity on many of these thrust faults can be seen on Landsat imagery throughout the Pakistani Makran (Fig. 5).

The structures of the Pakistani Makran are confined between two strike-slip fault systems: (1) to the west the Nch faults of the Hari Rud zone along the eastern boundary of the Lut Block and (2) to the east the Chaman transform zone. The Iranian Makran is directly south of the Lut Block and proves to have very different structures from the Pakistani Makran. Structures extend across the Pakistani Makran with great continuity (Fig. 5), but in Iran, they are broken up and rotated. Thus, the major synclines of the Iranian Makran are broken up into a series of en echelon, slightly elongated basins. While colored melange is known from the northern Iranian Makran (Delaloye and Desmons, 1980), it is not reported from the equivalent position in Pakistan.

White (1979), on the basis of the results of the ship-borne seismic survey off the western extremity of Pakistani Makran, confirms that it is an arc-trench gap. An interesting conclusion from the seismic data is that "only the upper 2½ km of the 5 km thick sediment pile is initially folded." He is of the opinion that the Makran accretionary sediment pile which forms the continental margin is probably emplaced in imbricate thrust sheets in a similar manner to that observed on the inner trench walls of many other subduction zones. This observation is supported by the details of the subduction processes described by Kulm et al. (1977).

Jacob and Quittmeyer (1979) examined the Makran subduction margin on the basis of the seismicity of the region. They noted anomalous features uncommon for a typical trench-arc system. For example, the trench gap measures 400-600 km across, about twice the width of a typical gap, and the shallow dipping seismic zone is weakly developed and extends to a depth of only 80 km. Their analysis of the geological setting and of the earthquake data clearly bring out three points: 1) the Makran margin is an active plate boundary with ongoing subduction at a rate of about 5 cm/year, 2) the subduction model implies that a large part of southern Makran is underlain by a mobile oceanic basement, and 3) the Makran subduction margin is one of the most interesting places to study the subduction processes of accretion and consumption in an arc-trench gap because of its subaerial exposure.

CHAMAN TRANSFROM ZONE

The Chaman transform zone connects the Makran convergent margin in the south with the Himalayan convergence zone in the north. It has an internal convergence zone in the Zhob thrust belt. The Chaman transform zone is of particular significance in the evolution of the structural style observed in Pakistan.

The transform character of the sinistral Chaman strike-slip fault (nearly 900 km long), was first recognized by Wilson in his pioneering paper of 1965. Subsequent studies (Lawrence & Yeats, 1979) have revealed that it is not a simple intra-continental transform. The Chaman fault is the western margin of a large and complex zone comprising three major fault sets and numerous smaller faults. The history of the development of the Chaman transform zone is as complex as that of the San Andreas fault system. The major rock type of this zone is the Khojak Flysch (Oligocene/

Miocene) of Vredenburg (1901). The width of the transform zone varies markedly. In the south it is about 100 km wide; it narrows to 30-40 km between Khuzdar and Quetta; and, in the Zhob thrust belt, it widens abruptly to over 200 km and then narrows gradually towards the Kabul block in the north.

The eastern edge of the zone is marked by continental margin type Mesozoic rocks and the obducted ophiolites of Las Bela and Muslim Bagh/ Fort Sandeman. The western margin is marked by the easternmost outcrops of the Tertiary and Mesozoic rocks of Afghanistan. From Spin Tezha in Pakistan to the Usman fault in Afghanistan, the transform zone is bounded on the west by upper Cretaceous to Eocene andesitic arc material (Jones, 1961; Bakr, 1964; Ahmed, 1964; Lawrence et al., 1981b). The Chaman fault proper is marked by a zone of gouge including an assemblage of red and green clastic sedimentary rock of the Kamerod and Rakhshani Formations, ultramafic rocks, and foraminiferal limestones of the Kharan formation (Lawrence and Yeats, 1979). Most of the material in the fault zone is derived from the west side. These units all first appear at the Usman fault. A minimal left slip of 200 km is recorded by displacement of these units in Pakistan alone (Lawrence and Yeats, 1979). How far this assemblage extends into Afghanistan is not known, but sizeable bodies of ultramafic rocks first appear about 400 km north of the Usman fault. The Spin Tezha crystalline terrain (33° 33'N; 66° 23'E) in Pakistan extends south along the Chaman fault into Afghanistan and is covered by the Helmund desert to the west. It is the eastern continuation of the calc-alkaline arc terrain of the Chagai Hills dragged by oroclinal flexing into the Chaman transform zone (Lawrence et al., 1981b, and Klootwijk et al., 1981). To the north it connects with the Kandahar volcanic arc. The metamorphic complex may represent the basement on which the arc terrain rests, only exposed due to a strong vertical uplift near the Chaman fault.

The displacement along the Chaman fault zone is determined by the motion of the respective plates and the amount of decoupling between the upper and lower layers of the plates near the fault. The decoupling results in 1) less slip than would be expected on the basis of instantaneous plate motions (Minster and Jordan, 1978; Jacob and Quittmeyer, 1979), and 2) thrusting and folding.

Northeast of Quetta, the foldbelt increases its width; the displacement along the Chaman fault decreases, and is finally absorbed in the wide convergence zone of the Himalaya. Thus the displacement increases southward, and may have been 450 km since the Early Miocene (Abdel-Gawad, 1973, and Auden, 1975). It can be inferred that the Makran flysch succession originally extended eastward into a similar flysch sequence which now occupies the area east of the Chaman fault in Katawaz (Afghanistan) and the Kakar Ranges (Pakistan).

The last known large seismic event on the Chaman fault occurred at about latitude 30. 3°N on 20 December 1892. No instrumental or felt earthquakes were reported for the section of the Chaman fault south of this latitude until 1975. Since then, the record of the Chaman fault includes two teleseismic epicenters (3 October 1975, M 6.7, 16 May 1978, M 5.2) and many epicenters of minor earthquakes observed by a seismic network (June to November 1978), covering a 2 degree square centered at Quetta (Armbruster et al., 1979). This increase in seismicity in the southern portion of the Chaman fault warrants careful consideration as a premonitory signal.

The eastern margin of the transform zone is formed by the Quetta Line (Gansser, 1966), along which the sediments of the Khojak flysch belt come into contact with the Mesozoic and Cenozoic platform sediments of the Indian subcontinent. This line is formed by a series of faults of different character. From north to south these include (Fig. 1) the Zhob thrust, the Ghazaband fault, and the Ornach-Nal fault. These faults have a complex Cenozoic history starting with obduction of ophiolite (see below). All show evidence of neotectonic activity. The Zhob fault is currently a thrust that brings flysch belt sediments over the ophiolite complex. The Ghazaband fault probably began as a thrust, but the neotectonic features along its trace strongly suggest strike slip motion. The Ghazaband fault is connected to the Ornach-Nal structural line. This feature has not yet been studied in the field, but appears to be a complex fault set on satellite imagery. The Khuzdar knot, a complex area of folding at the northern end of the Ornach-Nal fault, is also an expression of the structural transition from the Ghazaband to Ornach-Nal fault lines. Neotectonic strike slip motion is also suggested from studies of imagery.

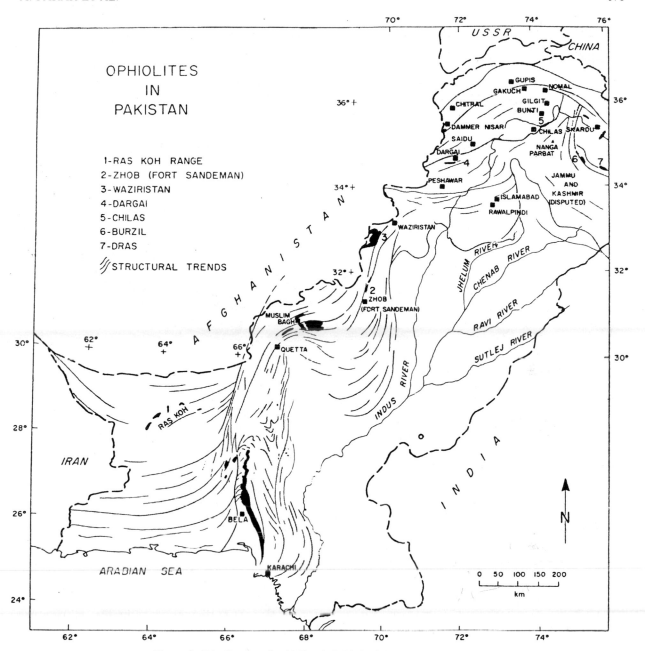

Figure 6. Distribution of ophiolites in Pakistan (after Asrarullah, 1979).

INTERNAL CONVERGENCE ZONE OF QUETTA-LORALAI-BARKHAN

This zone, east of the Chaman transform zone, contains large, upright folds. They resemble folds of the Zagros Foothills (Falcon, 1974) and suggest compression and convergence achieved by southward decollement thrusting of the sedimentary sequence. The thrusting was probably synchronous with the 50° counter-clockwise rotation postulated on the basis of paleomagnetic data by Klootwijk et al., (1981). The seismicity of

Quetta-Loralai-Barkhan zone is of a high level (Quittmeyer et al., 1979), although the largest events have only moderate magnitudes. Focal mechanism solutions determined for several earthquakes (Armbruster et al., 1979) indicate a north-trending P axis which is interpreted as an expression of north-south compression in this zone. The present available geological and geophysical information clearly supports the general statement of Powell (1979) and Sarwar and De Jong (1979) that the Pakistani festoons (including the Quetta-Loralai-Barkhan festoon)

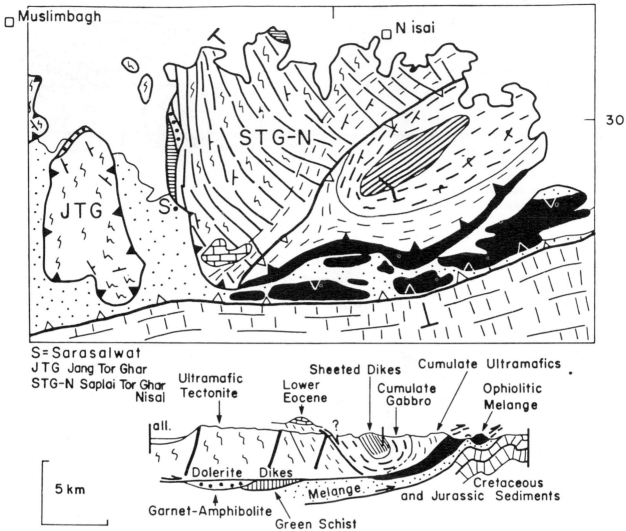

Figure 7. Geologic map and schematic cross-section of the Muslim Bagh ophiolites (after Abbas and Ahmad, 1979).

are the result of a combined northward and westward convergence and counter-clockwise rotation of the Indo-Pakistan subcontinent with Eurasia.

ZONE OF OPHIOLITES AND OPHIOLITIC MELANGES

It is believed that the ultramafic and associated mafic rocks along the axial belt in Pakistan are ophiolites (Fig. 6; Asrarullah and et al., 1979), obducted upon the sediments of the margin of the Indo-Pakistan subcontinent. Of these, the Muslim Bagh and Bela ophiolites have been studied in some detail. The Muslim Bagh ophiolites comprise (1) tectonites (2) ultramafic and mafic cumulates, and (3) sheeted dykes (Abbas and Ahmad, 1979;

Fig. 7). The disjointed ophiolite masses are underlain by ophiolitic melanges which comprise sedimentary, metamorphic, and ophiolitic blocks. The disjointed distribution of the ophiolite bodies is confirmed by the random and restricted aerial extent of the residual gravity anomalies (Farah and Zaigham, 1979). According to Abbas and Ahmad (1979), the Muslim Bagh ophiolites "are not the erosional remnants of a single overthrust sheet, but the fragments of an overthrust sheet which lost its coherence during the final stages of emplacement" (Fig. 7).

A question of significant interest in this area is whether the flysch deposits associated with the ophiolite bodies northwest of Muslim Bagh constitute 1) thrust sheets, resting on continental

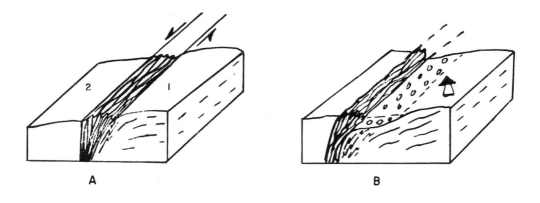

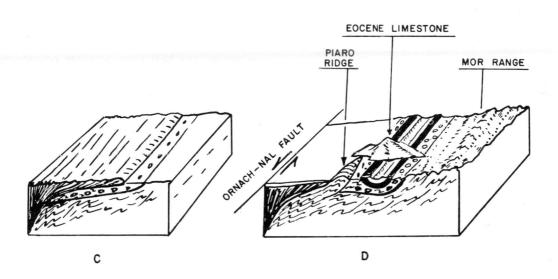

Figure 8. Model for the formation of the Kanar melange. A: Transform faulting between the Indian Plate (1) and the Neo-Tethys (2) creates a tectonic melange (proto-Kanar melange) B: Paleocene oblique convergence causes the emergence of the belt, its erosion, and sedimentary emplacement of the Kanar melange in a tectonic foredeep on the continental margin C: Continuing convergence leads to obduction of the Bela ophiolites on top of the Kanar melange inducing a tectonic fabric. Ophiolite obduction is complete in the Early Eocene. D: Continuing oblique convergence leads to a syncline of the Bela ophiolites. Nal limestone is neo-autochthonous, deposited on the folded and eroded ophiolites in the Middle Eocene-Oligocene (after Sarwar and De Jong, 1982).

margin crust, or 2) an accretionary wedge (to the Indo-Pakistani subcontinent or to Afghanistan), overlying oceanic crust. The Bouguer gravity anomalies of the area indicate thickening of the crust toward the northwest area of Muslim Bagh (Farah and Zaigham, 1979) suggesting that the flysch deposits in the area are underlain by continental crust, not by oceanic crust.

The Bela ophiolites occupy a large area of exposure compared to the Muslim Bagh complex. Pillow lavas dominate in the southern part and mafic and ultramafic rocks in the north. The complex is conspicuous for its extensive melange zones recently described by De Jong and Subhani (1979); Sarwar (1982) and Sarwar and De Jong (1982). The major melange is the Kanar melange which is overlain by a thrust sheet of Cretaceous ophiolites. The melange consists of a chaotic mixture of millimeter- to kilometer-size clasts embedded in a dominantly argillaceous matrix. The clasts include both continental and oceanic rock types. The melange has a complex origin.

Oblique convergence between the Indian plate and the Neo-Tethys resulted in an emergent zone of tectonically mixed continental and oceanic rocks (proto-Kanar melange). Debris from this mixed zone was shed into a tectonic foredeep on the adjacent continental margin (Fig. 8). The presence of sedimentary features in the matrix of the melange and its depositional relationship with the underlying formation clearly indicate that the emplacement of the melange was controlled by sedimentary processes (Sarwar, 1982).

Synclinal folding of the ophiolite nappe resulted from Neogene deformation during which the major tectonic features of the Kirthar-Sulaiman belt were shaped (Sarwar and De Jong, 1979). The folding is visible in the field but can also be deduced from the aeromagnetic anomalies which show a pronounced symmetry when indicating the two limbs of a syncline (Fig. 9).

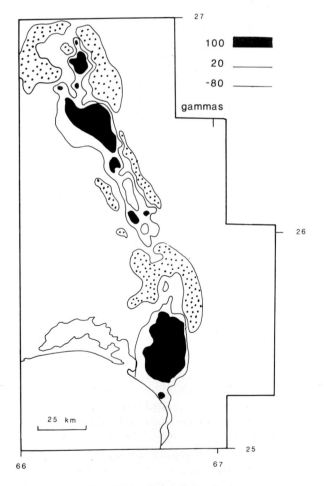

Figure 9. Aeromagnetic anomalies of the Bela area. Deformation of the ophiolites into a fold is indicated by the symmetry of the anomalies.

CONCLUSIONS

It is exciting and informative to note that the piece of the earth's crust, comprising Pakistan, has traversed a circuitous and eventful path and has undergone subtle transformation through geological time. In the process, truly spectacular plate boundaries have developed involving continental collision, oceanic crust subduction and obduction, and transform faulting. Continental collision in the north resulted in the elevation of extremely high mountains including K-2, the second highest mountain in the world.

Subduction and volcanic arc development in the Chagai region bordering Afghanistan and Iran produced a geologically complex scenario that has excellent metallic mineral potential. The wide arc-trench gap of the active Makran subduction margin and the active transform zone of Chaman are among the best exposed plate boundaries in the world. The available geological and geophysical information about the plate boundaries in Pakistan indicate that an Atlantic-type (spreading) boundary was present from the early Mesozoic (or late Paleozoic) until the late Cretaceous, and an Andean-type (subduction) boundary from the late Cretaceous to the present. The Chaman transform zone connects the converging zone of Makran-Chagai in the south with that of the Himalayas in the north. Research involving the geodynamic evolution of the plate boundaries in Pakistan will enable earth scientists and engineers to work out a scientifically sound strategy for mineral and hydrocarbon exploration. It will also allow them to determine the nature and probable recurrence of earthquakes, and prepare effective schemes for minimization of earthquake hazards.

ACKNOWLEDGEMENTS

The encouragement of Asrarullah, Director General, Geological Survey of Pakistan is gratefully acknowledged. Colleagues too numerous to detail from many Pakistani geological organizations generously shared their ideas with us. Partial financial support was provided by NSF grants INT 76-22304, EAR 78-15476, and EAR 81-13158 by NASA Grant NAG 9-2, and by the Geological Survey of Pakistan.

REFERENCES

Abbas, S.G. and Z. Ahmad, 1979. The Muslimbagh Ophiolites. In: A. Farah and K.A. De Jong (eds.), Geodynamics of Pakistan, Geol. Surv. of Pakistan, Quetta, p. 243-249.

Abdel-Gawad, M., 1971. Wrench movements in the Baluchistan Arc and relation to Himalayan-Indian Ocean tectonics. Geol. Soc. of Am. Bull., 82: 1235-1250.

Afzali, H., F. Debron, P. Le Fort, J. Sonet, 1979. Le massif monzo-syenitique de Zarkachan (Afghanistan central): caracteres, age Rb-Sr, et implications tecto-orogeniques., C. R. Acad. Sc. Paris, 288 (D): 287-290.

Ahmed, S. S., 1969. Tertiary geology of part of South Makran, Baluchistan, West Pakistan, Bull. Am. Ass. of Petro. Geo., 53 (7): 1480-1499.

Ahmad, W., 1964. Iron-copper deposits of Bandgan, Kimri, and Jadino; Raskoh Range, Chagai District, West Pakistan. Symp. on Mining Geology and the Base Metals, Ankara, Turkey, p. 181-190.

Argand, E., 1924. La tectonique de l'Asie, C. R. 13e, Congl. Geol. Intern., p. 171-372.

Armbruster, J., L. Seeber, R. C. Quittmeyer, and A. Farah, 1979. Seismic network data from Quetta, Pakistan. Rec. Geol. Surv. Pakistan, 49: 1-5.

Arthurton, R. S., G. S. Alam, S. A. Ahmad, and S. Iqbal, 1979. Geological history of the Alamreg-Mashki area, Chagai District, Baluchistan. In: A. Farah and K.A. De Jong (eds.), Geodynamics of Pakistan, Geol. Surv. of Pakistan, Quetta, p. 325-331.

Arthurton, R. S., A. Farah, and W. Ahmad, 1982. The Late Cretaceous history of NW Pakistani Baluchistan—the northern margin of the Makran subduction complex. In: J.K. Leggett (ed.), Trench-Fore Arc Geology, Geol. Soc. London Spec. Publ. 10: 343-385.

Asrarullah, Z. Ahmad, and S. G. Abbas, 1979. Ophiolites in Pakistan: An introduction. In: A. Farah and K.A. De Jong (eds.), Geodynamics of Pakistan Geol. Surv. of Pakistan, Quetta, p. 181-192.

Auden, J. B., 1975. Afghanistan-West Pakistan. In: A.M. Spender (ed.), Mesozoic-Cenozoic orogenic belts, Geol. Soc. Lond. Spec. Publ. 4: 235-253.

Bakr, M. A., 1964. Geology of the western Ras Koh Range, Chagai and Kharan districts, Quetta and Kalat Divisions, West Pakistan, Rec. Geol. Surv. of Pakistan, 10 (2A): 1-28.

Bard, M. A., H. Maluski, and F. Proust, 1980. The Kohistan sequence: crust and mantle of an obducted island arc. Proc. Intern. Commit. Geodynamics, Grp. 6, Mtg. Peshawar, 23-29 Nov. 1979. Spec. Issue Geol. Bull. Univ. Peshawar, 13: 87-94.

Bellon, H., P. Bordet, and C. Montenant, 1979. Histoire magmatique
· de l'Afghanistan central: nouvelles donees chronometrique K-Ar. C. R. Hebd. Seanc. Sci., Paris, 289: 1113-1116.

Boulin, B. and E., Bonyx, 1978. Orogenese hercyniennc, bordure gondwanienne et tethysian en Afghanistan. Ann. Soc. Geol. N., 97: 297-308.

Calkins, J. A., T. W. Offield, S. K. M. Abdullah, S. T. Ali, 1975. Geology of the southern Himalayas in Hazara, Pakistan, and adjacent areas. U. S. Geol. Survey Prof. Paper, 716-C, 29 p.

Coward, M. P., M. Q. Jan, D. Rex, J. Tarney, M. Thirlwal, and B. F. Windley, 1982. Structural evolution of a crustal section in the western Himalaya, Quat. Jour. Geol. Soc. London, 139: 299-300.

Crawford, A. R., 1974. The Salt Range, the Kashmir syntaxis and the Pamir Arc, Earth Planet. Sci. Letters, 22: 371-379.

Das, S. and J. R. Filson, 1975. On the tectonics of Asia. Earth Planet. Sci. Letters, 28: 241-253.

De Jong, K. A. and A. M. Subhani, 1979. Note on the Bela ophiolites with special reference to the Kanar area, In: A. Farah and K.A. De Jong (eds.), Geodynamics of Pakistan, Geol. Surv. of Pakistan, Quetta, p. 263-269.

Delaloye, M. and J. Desmons, 1980. Ophiolites and melange terrains in Iran: a geochronological study and its paleotectonic implications. Tectonophysics, 68: 83-111.

Farah, A. and K. A. De Jong, (eds), 1979. Geodynamics of Pakistan.
| Geol Surv. Pakistan, Quetta, 361 p.

Farah, A. and N. A. Zaigham, 1979. Gravity anomalies of the ophiolite complex of the Khanozai-Muslim Bagh-Qila Saifullah area, Zhob District, Pakistan. In: A. Farah and K.A. De Jong (eds.), Geodynamics of Pakistan, Geol. Surv. of Pakistan, Quetta, p. 251-262.

Farah, A., M. A. Mirza, M. A. Ahamad, and M. H. Butt, 1977. Gravity field of the buried shield in the Punjab Plain, Pakistan. Geol. Soc.

America Bull., 88: 1147-1155.

Farhoudi, G. and D. E. Karig, 1977. Makran of Iran and Pakistan as an active arc system, Geology, 5: 664-668.

Gansser, A., 1966. The Indian Ocean and the Himalayas, a geological interpretaion. Eclogae Geol. Helvetiae, 67: 497-507.

Gansser, A., 1981. The geodynamic history of the Himalaya: In: H.K. Gupta and F.M. Delaney Zagros Hindu Kush Himalaya Geodynamic Evolution, Am. Geophys. Union Geodynamics Series 3: 111-121.

Jacob, K. H. and R. L. Quittmeyer, 1979. The Makran region of Pakistan and Iran: trench-arc system with active plate subduction. In: A. Farah and K.A. De Jong (eds.), Geodynamics of Pakistan, Geol. Surv. of Pakistan, Quetta, p. 305-317.

Jan, M. Q., 1980. Petrology of the obducted mafic and ultramafic metamorphites from the southern part of the Kohistan island arc sequence. Proc. Intern. Commit. Geodynamics, Group 6, Mtg. Peshawar, 23-29 Nov. 1979, Spec. Issue Geol. Bull. Univ. Peshawar, 13: 95-107.

Jan, M. Q. and M. Asif, 1981. A speculative tectonic model for the evolution of NW Himalaya and Karakoram. Geol. Bull Univ. Peshawar, 14: 199-201.

Jan, M. Q. and R. A. Howie, 1981. The mineralogy and geochemistry of the metamorphosed basic and ultrabasic rocks of the Jijal Complex, Kohistan, NW Pakistan. J. Petrol., 22: 85-126.

Jan, M. Q., M. Asif, T. Tahirkheli, and M. Kamal, 1981. Tectonic subdivision of granitic rocks of north Pakistan. Geol. Bull. Univ. Peshawar, 14: 159-182.

Jones, A. G., (ed.), 1961. Reconnaissance geology of part of West Pakistan, a Colombo Plan Cooperative Project. Government of Canada, Toronto (Hunting Survey Report). 550 p.

Kanamori, H., 1977. Seismic and aseismic slip along subduction zones and their tectonic implications. In: M. Talwani and W.C. Pitman III (eds.), Island Arcs, Deep Sea Trenches, and Back-Arc Basins, Maurice Ewing Series, Am. Geophys. Union, 1: 163-174.

Klootwijk, C. T., R. Nazir-Ullah, K. A. De Jong, and A. Ahmed, 1981. A paleomagnetic reconnaissance of NE Baluchistan, Pakistan. J. Geophys. Res. 86: 289-306.

Kulm, L. D., W. J. Schweller, and A. Masias, 1977. A preliminary analysis of the subduction processes along the Andean continental margin, 6° to 45° S. In: M. Talwani and W.C. Pitman III (eds.), Island Arcs, Deep Sea Trenches, and Back-Arc Basins, Maurice Ewing Ser. Am. Geophys. Union, 1: 285-301.

Lawrence, R. D., and R. S. Yeats, 1979. Geological Reconnaissance of the Chaman fault in Pakistan. In: A. Farah and K.A. De Jong (eds.), Geodynamics of Pakistan Geol. Surv. of Pakistan, Quetta, p. 351-358.

Lawrence, R. D., S. H. Khan, K. A. De Jong, A. Farah, and R. S. Yeats, 1981a. Thrust and strike slip fault interaction along the Chaman transform zone, Pakistan. In: K. McClay and N. Price (eds.), Thrust and Nappe Tectonics Geol. Soc. Long., Spec. Publ. 9: 363-370.

Lawrence, R. D., R. S. Yeats, S. H. Khan, A. M. Subhani, and D. Bonelli, 1981b. Crystalline rocks of the Spintizha area, Pakistan. Jour. Struct. Geol., 3: 449-457.

Le Fort, P., 1975. The collided range. Present knowledge of the continental arc. Am. Jour. Sci., 275-A: 1-44.

Marussi, A., 1980. Gravity, crustal tectonics, and mantle structure in the central Asian syntaxis. Proc. Intern. Commit. Geodynamics, Grp. 6, Mtg. Peshawar, Nov. 23-29, 1979, Spec. Issue Geol. Bull. Univ. Peshawar, 13: 23-27.

Mattauer, M., 1975. Sur le mecanisme de formation de la schistosite dans l'Himalaya, Earth Planet. Sci. Letters, 28: 144-154.

Minster, J. B. and T.H., Jorden, 1978. Present-day plate motion. J. Geophys. Res., 83: 5331-5334.

Mitchell, A. H. G., 1981. Phanerozoic plate boundaries in mainland SE Asia, the Himalayas, and Tibet, J. Geol. Soc. London, 138: 109-122.

Molnar, P. and P. Tapponnier, 1975. Cenozoic tectonics of Asia: Effects of a continental collision, Science, 189 (4201): 419-426.

Opdyke, N. D., N. M. Johnson, G. D. Johnson, E. H. Lindsay, and R. A. K. Tahirkheli, 1981. Paleomagnetism of the Middle Siwalik formations of northern Pakistan and rotation of the Salt Range decollement. Palaeog., Palaeocl., Palaeoec., 37: 1-16.

Oxburgh, E. R., 1972. Flake tectonics and continental collision. Nature, 239: 202-204.

Powell, C. McA., 1979. A speculative tectonic history of Pakistan and surroundings: some constraints from the Indian Ocean. *In:* A. Farah and K.A. De Jong (eds.), *Geodynamics of Pakistan* Geol. Surv. of Pakistan, Quetta, p. 5-24.

Powell, C. McA., and P. J. Conaghan, 1973. Plate tectonics and the Himalayas. Earth Planet. Sc. Letters, 20: 1-12.

Quittmeyer, R. C., A. Farah, and K. H. Jacob, 1979. The seismicity of Pakistan and its relation to surface faults. *In:* A. Farah and K. A. De Jong (eds.), *Geodynamics of Pakistan,* Geol. Surv. Pakistan, Quetta, p. 271-284.

Raza, H. A., S. Alam, S. M. Ali, N. Elahi, and M. Anwar, 1981. Hydrocarbon potential of Makran region of Baluchistan basin. Seminar on Mineral Policy of Baluchistan; Quetta, 29-30 April 1981. Hydrocarbon Develop. Instit. Pakistan.

Roy, S. S., 1976. A possible Himalayan microcontinent, Nature, 263: 117-119.

Sarwar, G., 1982. Geology of the Bela ophiolites in the Wayaro area, Las Bela District, SC Pakistan. Rec. Geol. Surv. Pakistan, in press.

Sarwar, G. and K. A. De Jong, 1979. Arcs, oroclinces, syntaxes: the curvatures of mountain belts in Pakistan. *In:* A. Farah and K.A. De Jong (eds.), *Geodynamics of Pakistan* Geol. Surv. of Pakistan, Quetta, p. 341-350.

Sarwar, G. and K. A. De Jong, 1982. Compostion and origin of the Kanar melange, southern Pakistan. Spec. Publ. Geol. Soc. Am., (in press).

Seeber, L. and J. Armbruster, 1979. Seismicity of the Hazara Arc in northern Pakistan: decollement versus basement faulting. *In:* A. Farah and K.A. De Jong (eds.), *Geodynamics of Pakistan* Geol. Surv. of Pakistan, Quetta, p.

Sengor, A. M. C., 1979. Mid-Mesozoic closure of Permo-Triassic Tethys and its implications. Nature, Lond., 279: 590-593.

Sillitoe, R. H., and S. N. Khan, 1977. Geology of the Saindak porphyry copper deposit. Transactions, Section B, Institue of Mining and Metallurgy, 86: B27-B42.

Stocklin, J., 1977. Structural correlation of the Alpine range between

Iran and Central Asia, Mem. h. ser. Soc. geol. France, (8): 333-353.

Stoneley, R., 1975. On the origin of ophiolite complexes in the southern Tethys region, Tectonophysics, 25: 303-322.

Tahirkheli, R. A. K., 1979. Geology of Kohistan and adjoining Eurasian and Indo-Pakistan continents, Pakistan. *In:* R.A.K. Tahirkheli and M.Q. Jan (eds.) *Geology of Kohistan, Karakoram, Himalaya, Northern Pakistan,* Spec. Issue Geol. Bull. Univ. Peshawar, 11:1-30.

Tahirkheli, R. A. K., 1982. Geology of the Himalaya, Karakoram, and Hindu Kush in Pakistan. Spec. Issue Geol. Bull. Univ. Peshawar, 15:51.

Tahirkheli, R. A. K., M. Q. Jan, and M. Majid, (eds.), 1980. Proceedings of the International Committee on Geodynamics, Group 6 Meeting at Peshawar, 23-29 November 1979. Geol. Bull Univ. Peshawar, 13: 1-213.

Tahirkheli, R. A. K., M. Mattauer, F. Proust, and P. Tappoinier, 1979. The India-Eurasia Suture zone in northern Pakistan: synthesis and interpretaion of recent data at plate scale. *In:* A. Farah and K. A. De Jong, (eds.) *Geodynamics of Pakistan* Geol. Surv. of Pakistan, Quetta, p. 125-130.

Tapponier, P., M. Mattauer, F. Proust, C. Cassaignean, 1981. Mesozoic ophiolites, sutures, and large scale tectonic movements in Afghanistan. Earth Planet. Sci. Lett., 52: 355-371.

Valdiya, K. S., 1980. The two intracrustal boundary thrusts of the Himalaya. Tectonophysics, 66: 323-348

Vredenburg, 1901. Geological sketch of the Baluchistan desert, and part of eastern Persia, Geol. Surv. India Memoir, 31: 179-302.

White, R. S., 1979. Deformation of the Makran continental margin. *In:* A. Farah and K. A. De Jong (eds.), *Geodynamic of Pakistan,* Geol. Surv. of Pakistan, Quetta, p. 295-304.

White, R. S. and K. Klitgord, 1976. Sediment deformation and plate tectonics in the Gulf of Oman. Earth and Planetary Science Letters, 32: 199-209.

Zeitler, P. K., N. M. Johnson, C. W. Naeser, and R. A. K., Tahirkheli, 1982. Fission-track evidence for Quaternary uplift of the Nanga Parbat region, Pakistan. Nature, 298: 255-257.

Zeitler, P. K., R. A. K. Tahirkheli, C. W. Naeser, and N. M. Johnson, 1982. Unroofing history of a suture zone in the Himalaya of Pakistan by means of fission track annealing ages. Earth Planet. Sci. Letters, 57: 227-240.

12

Tectonics of the Himalayan Thrust Belt in Northern Pakistan

R. S. YEATS and R. D. LAWRENCE
Oregon State University, Corvallis, Oregon

ABSTRACT

The major tectonic subdivisions of the Pakistani Himalaya are, from south to north, the Indian foreland, the Salt Ranges, the Potwar and Kohat plateaus, the Hill Ranges, the intermontane basins, the southern Kohistan ranges, the Nanga Parbat-Haramosh massif, the Main Mantle Thrust (MMT), and the Kohistan island arc. The Precambrian foreland is overlain by undeformed platform Cambrian, Permian, Mesozoic and early Tertiary strata, largely marine, and by nonmarine late Cenozoic molasse. Within the foreland is the seismically-active Sargodha gravity high, which is parallel to Indian Himalayan trends and may have been formed by late Cenozoic faulting. The Salt Range and Trans-Indus Salt Range form the leading edge of a decollement within Eocambrian evaporites bringing Phanerozoic strata over crystalline basement. Major thrusting occurred in the last 0.5 m.y. in the Salt Range, but is older northward in the Potwar Plateau. Strata in the Potwar and Kohat plateaus are more strongly deformed northward, and they are in thrust contact with Cretaceous and early Tertiary rocks of the Hill Ranges (Margala, Kala Chitta, Attock, and Kohat Hills and Safed Koh). These ranges contain Paleozoic strata and Precambrian slate in addition to Mesozoic and younger rocks. The Peshawar and Campbellpur intermontane basins began receiving sediments in Pliocene time. They may be the surface expression of low-angle detachment faults which ramp upward beneath the Hill Ranges. Early Tertiary alkali granitic rocks occur in the northern Peshawar basin. The southern Kohistan ranges contain metamorphic rocks associated with granite and gneiss 515 Má. in age. These rocks are in abrupt contact with Devonian marine strata south of Swat and with Cambrian strata in Hazara. The Nanga Parbat-Haramosh massif is a granitic and migmatitic body which yields Tertiary apparent ages; older rocks may be present, but they have not been dated. Farther north and west, the MMT brings the Kohistan island-arc sequence over the crystalline rocks of the southern Kohistan ranges; the Dargai klippe of ultramafic rock at the northern edge of the Peshawar basin indicates that this thrust has large displacement.

The Pakistani Himalaya is divided by the north-south Hazara-Kashmir syntaxis and its northern continuation, the Nanga Parbat-Haramosh massif. Farther west, the Kalabagh reentrant continues northward as the boundary between the Kohat and Potwar plateaus. Even farther west, the Himalayan structure is terminated by great strike-slip faults which mark the modern western edge of the Indo-Pakistan plate. The Pakistani subdivisions do not correlate well with the Indian Himalaya, although the Salt Ranges and the Kohat and Potwar plateaus correspond to the Indian Sub-Himalaya, the Hill Ranges correspond to the Lesser Himalaya, and the Nanga Parbat-Haramosh massif corresponds to the High Himalaya. There appears to be no Tethyan Himalaya in Pakistan, and the High Himalaya ends at the Indus River.

INTRODUCTION AND PLATE TECTONIC SETTING

The Himalayan ranges of southern Asia are a dilemma in modern plate tectonic theory. Subduction of oceanic crust is thought to be accomplished largely by slab pull (Forsyth and Uyeda, 1975), but buoyancy considerations prohibit the Indian continent from being subducted by this mechanism or by a push generated from the ridge crest to the south. As discussed by Alvarez (1982), a new subduction zone south of the Indian continent should have formed after collision of India with Asia, but it did not. Following initial collision, the rate of relative motion slowed from 100-180 mm/y to about 50 mm/y, but even at the slower rate, 300 to 700 km of convergence must be accounted for by underthrusting of the Indian continent beneath the Himalaya and Tibet (Molnar and Tapponnier,

1975) or by some sort of accordion collapse in the wide orogenic belt between cratonic India and Asia (Sengor and Kidd, 1979).

Alvarez (1982) resolved the problem by suggesting that a line of convergence and convective sinking in the lower mantle underlies the Alpine-Himalayan system. Asia and India are thereby pushed together by drag along a pair of converging convection cells in the lower mantle. Some of the convergence is done by continental underthrusting and by tectonic thickening, but much of it occurs along sets of strike-slip faults which push microplates of Asian crust sideways, out of the way of advancing India, in the direction

of the nearest oceanic crust (Tapponnier et al., 1982).

However, the hypothesis of lower-mantle convection is contradicted by the current theory of hotspots, as Alvarez (1982) noted. If hotspots are generated in the lower mantle, and India rides northward on a lower-mantle convection cell, then hotspots should be fixed with respect to the Indian plate. The Chagos-Laccadive and Ninetyeast ridges, located west and east, respectively, of the Indian peninsula (Fig. 1), were attributed to the Reunion and Kerguelen hotspots, respectively, by Duncan (1981). According to the hypothesis that hotspots are fixed with respect to the lower

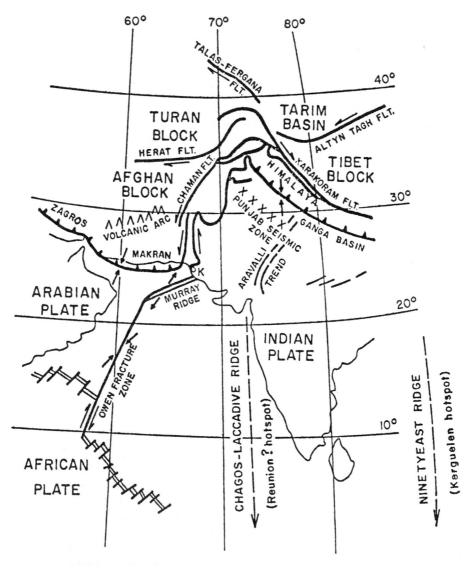

Figure 1. Plate tectonic setting of Pakistan. Chagos-Laccadive and Ninetyeast ridges are either hotspot traces or leaky transform faults. Convergence of Indian plate and various Eurasian blocks accommodated by thrusting in Himalaya and strike-slip faulting farther north. South of Himalaya, Punjab seismic zone is parallel to Himalaya but discordant to Precambrian trends of Aravalli Range. K = Karachi.

mantle, these ridges indicate rapid northward motion of India with respect to the lower mantle. However, these ridges are remarkably parallel to transform faults on the Indian plate, and it may be that volcanism produced along these ridges is related to leaky transform faulting rather than hotspots (Norton and Sclater, 1979).

Regardless of these inconsistencies with present tectonic theory, the Himalaya is a fully developed mountain belt, and it is still actively being deformed. Accordingly it is possible to add geomorphology and those geophysical techniques which reveal the present physical state and dynamics of the crust and mantle to bedrock geology as methods to understand the modern Himalayan system (Lyon-Caen and Molnar, in press). This paper is written at the beginning of a cooperative study of northern Pakistan by the Geological Survey of Pakistan and Oregon State University, assisted by other universities in Pakistan and elsewhere and supported in part by the U.S. National Science Foundation. Therefore, the paper considers some of the problems that we now face in the light of our present knowledge. A companion paper (Farah et al., this volume) discusses the overall tectonic setting of Pakistan and surrounding areas. This paper concentrates on the evolution of the Himalaya south of the Main Mantle Thrust (MMT) and Indus-Tsangpo suture zone, with a special consideration of the neotectonic setting of northern Pakistan.

SUBDIVISIONS OF THE CENTRAL INDIAN HIMALAYA

The subdivision of the central Himalaya generally accepted by most workers was recently summarized by Gansser (1981). From south to north, Gansser's units are (1) the Himalayan foredeep resting on diverse pre-Tertiary elements of peninsular India, (2) the Sub-Himalaya, (3) the Lesser Himalaya, (4) the High Himalaya, (5) the Tethyan Himalaya, and (6) the Indus-Tsangpo suture zone (Fig. 2). The Indian shield consists of Archean granite and gneiss which are overlain by Precambrian strata of various ages, including the older, more metamorphosed Aravalli system and the younger, largely unmetamorphosed Vindhyan system which may extend upsection into the Cambrian (Gansser, 1964). The Malani acidic volcanic rocks and pinkish, medium-grained granite are also younger than the Aravalli System and are dated as 500-700 m.y. (Krishnan, 1968;

Pareek, 1982). Clasts of Malani granite are common in the Permian boulder beds of the Salt Range (Gansser, 1964). Rhyolite and tuff of the Kirana Hills, south of the Salt Range, may be part of the Malani volcanics (Heron, 1923). The Aravalli Range and to a lesser extent the Vindhyan Range trend at an angle to the Himalaya, and it is generally believed that the Aravalli structural belts continue northward at depth beneath the Ganga basin and the Himalayan thrust sheets. In the Aravalli Range, strongly deformed and metamorphosed rocks of the Aravalli System are in fault contact with relatively undeformed Vindhyan sandstone along the Great Boundary Fault of Rajputana. This fault and the internal grain of the Aravalli Range trend northeast, at right angles to the Himalaya (Gansser, 1964).

The Permo-Carboniferous Gondwana formations of peninsular India are also discordant to the Himalaya in many regions. These coal-bearing sequences occupy large faulted intracratonic basins in the Mahanadi and Godavari Valleys, and an east-northeast-trending fault system extending from the Gulf of Cambay up the Narmada Valley is probably also of this same age (Krishnan, 1968). These intracratonic basins contain strata as young as Cretaceous.

In the Himalayan foredeep, the surface of the Indian shield tilts gently northward and is overlain by a northeastward-thickening molasse wedge of Oligo-Miocene (Murree-Dharmsala) and Mio-Pleistocene age (Siwaliks) resting directly on the foreland (Datta and Sastri, 1977). Wells in the Ganga basin show that the pre-Dharmsala geology of the basin floor is a continuation of that of the exposed regions to the south. These wells show that locally the molasse rests upon pre-Vindhyan marble and quartzite, and elsewhere it rests upon Vindhyan strata or upon Permian and Mesozoic rocks similar to those in the Gondwana intracratonic basins to the south (Acharyya and Ray, 1982).

In the Ganga basin, Siwalik strata rest directly on the Indian cratonic assemblage described above. To the northeast, in the Sub-Himalayan Tertiary foothills, the Dharmsala Group wedges in below the Siwaliks. Farther northeast, marine Eocene Subathu beds are present below the Dharmsala, and older rocks wedge in beneath the Subathu (Acharyya and Ray, 1982). In the Simla Hills, the Subathu changes in facies northeastward from fluvio-deltaic to coastal to shallow-marine, with thicknesses controlled by syndepositional

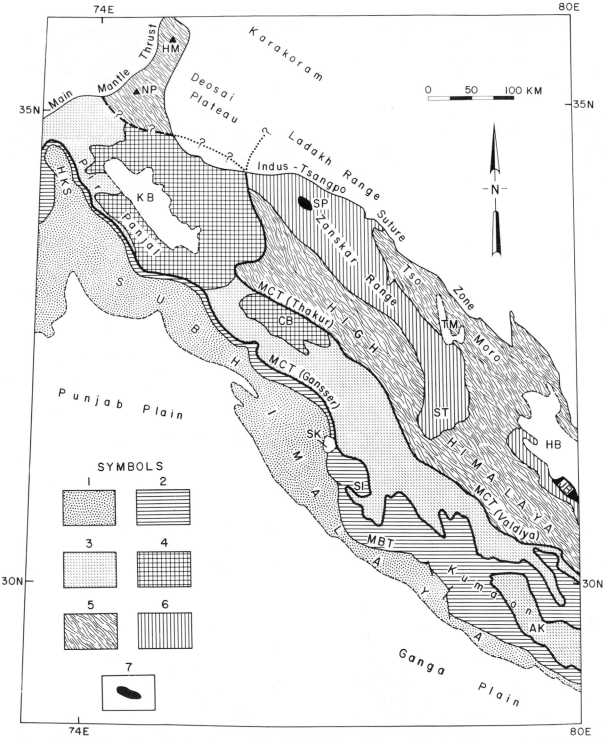

Figure 2. Subdivisions of central and northwestern Himalaya. MCT as extended to northwest by Gansser contrasted with MCT as identified by Valdiya and extended to the northwest by Thakur. Symbols: 1) Deformed Siwalik and Murree sediments of Sub-Himalaya, 2) Sedimentary and low grade metamorphic rocks of the Lesser Himalaya, 3) Medium to high grade, non-migmatized metamorphic rocks of the Lesser Himalaya, 4) Sedimentary rocks of the Kashmir nappe, 5) Crystalline rocks of the High Himalaya, 6) Sedimentary rocks of the Tethyan Himalaya, and 7) Ophiolite. Location: HM = Haramosh, NP = Nanga Parbat, KB = Kashmir Basin, HKS = Hazara-Kashmir syntaxis, SP = Spongtang nappe, CB = Chamba Basin, SK = Suketi Basin, TM = Tso Moro graben, ST = Spiti, SI = Simla, HB = Hundes Basin, JB = Jungbwa nappe, AK = Almora Klippe.

growth faults which are the sites of later thrust faults (Raiverman, 1981). Structurally, the Siwaliks of the Ganga basin are in fault contact with deformed Siwaliks of the foothills along the late Quaternary Himalayan Frontal Fault of Nakata (1972), called the Main Frontal Thrust (MFT) by Gansser (1981). Intensity of folding and faulting increases northward, and the Tertiary sequence is cut by north-dipping thrust faults.

The deformed foredeep strata are terminated on the north by the Main Boundary Thrust (MBT), a set of north-dipping faults that separates the foredeep from nappes of older rocks which comprise the Lesser Himalayan province (Valdiya, 1980a). As summarized by Valdiya (1980b), the Lesser Himalaya in the Kumaon region west of Nepal includes Precambrian and Paleozoic sedimentary and volcanic rocks (Jaunsar Group) overlain by the Blaini conglomerate which may correlate to the Tobra Formation of the Salt Range (Talchir boulder bed of the older literature). The Blaini is overlain by the carbonate-rich Krol and Tal Formations which contain Permian fossils. In a general way, the Lesser Himalayan rocks resemble those of the Indian craton except the sedimentary section is thicker and more complete, undoubtedly due to its proximity to the formerly passive northern margin of the Indian continent. Throughout the western part of the Lesser Himalaya, strata of middle Paleozoic age have not been documented. The Blaini conglomerate rests unconformably upon strata of Precambrian and lower Paleozoic age, much as the Gondwana strata of peninsular India rest unconformably upon the Vindhyan and older systems of peninsular India. The original stratigraphic sequence is commonly distorted by detachment thrusts, leading to problems in correlation. For example, the Simla Slate has long been regarded as Precambrian or Paleozoic, but Acharyya and Ray (1982) reported spores and palynomorphs of Late Paleozoic, Cretaceous, and Paleogene age from these slates. These fossils may indicate the presence of flysch-like Paleocene and Eocene strata associated with the Precambrian Simla Slate in tectonic windows (K. S. Valdiya, pers. com., 1982). This stratigraphic and lithologic similarity between the Simla Slate and younger strata together with the tectonic complications may produce the problems in correlation and dating.

The higher thrust sheets of the Lesser Himalaya, such as the Almora nappe, contain mesograde crystalline rocks of probable Precambrian age. Some of these may originate in schuppen zones below the Main Central (Vaikrita) Thrust (MCT), which marks the northern boundary of the province (Gansser, 1981; Valdiya, 1981). These crystalline thrust sheets may exhibit reversed metamorphism based on the observation that the rocks of ridge crests are of higher grade than those of deep valleys (LeFort, 1975). Alternatively, the appearance of reversed metamorphism may be produced by relatively thin thrust sheets of higher over lower metamorphic grade rocks (K. S. Valdiya, pers. com., 1982). Gansser (1964, 1981) recognized two metamorphic episodes of Precambrian age in these sheets, and he suggested that the young Himalayan orogenic overprint is of relatively little significance in their metamorphic history. The Precambrian section is intruded by S-type granites which are 510-520 m.y. old (Jaeger et al., 1971; LeFort et al., 1979); these include the Almora granite of the Almora nappe in Kumaon (K.S. Valdiya, pers. com., 1982) and the Mansehra granite of northern Pakistan (LeFort et al., 1979).

Transverse folds and faults oriented NE-SW and NW-SE cut across all structures including the MBT and Siwalik belt, which show left- and right-lateral strike separations of up to 10-12 km. Some of these appear to be related to the discordant Aravalli structures of the Ganga plain and the Indian peninsula (Valdiya, 1976).

According to Gansser (1981), the MCT marks the base of a huge intracrustal thrust sheet bringing a 10-15 km thick slab of Precambrian crystalline rocks over the Lesser Himalaya. This slab, comprising the high Himalaya, begins with gneiss followed by quartzite which is more argillaceous and calcareous in its upper part; this latter section alone is locally more than 10 km thick. Because K-Ar ages of granite and gneiss from the High Himalaya are generally Cenozoic, a controversy has arisen over how much of the metamorphism is Precambrian and how much is Tertiary. Precambrian whole-rock ages of 1800 to 1500 m.y. mask still older metamorphic events (Bhanot et al., 1977). Gansser (1981) suggested that the Precambrian phase was much more important than indicated by the K-Ar ages; that is, that most of the slab is made up of Precambrian Indian crust that was variously reworked in the Cenozoic. However, Middle Jurassic corals and cephalopods have been recovered from metamorphosed limestones which underwent polyphase deformation which also included the gneiss of the High Himalaya (Powell and Conaghan, 1973a; Pickett et al., 1975). Rb-Sr ages of whole-rock samples from the gneiss indicated to Powell et al. (1979) that the gneiss was formed during the middle Cenozoic from older Indian crust, including Precambrian crystalline rocks. Most workers recognize a Himalayan metamorphic event of 10-20 m.y. age, including Neogene tourmaline granites (Gansser, 1981), but there is disagreement about whether this is the dominant metamorphic event or is relatively minor (low grade) compared to the Precambrian metamorphism. The MCT carries these Cenozoic metamorphic rocks over the Precambrian metamorphic rocks of the upper thrust sheets of the Lesser Himalaya.

Overlying the crystalline slab of the High Himalaya are the sedimentary rocks of the Tethyan or Tibetan Himalaya. In contrast to the sequence of the Lesser Himalaya, the sedimentary sequence of the Tethyan Himalaya is relatively complete, ranging from Cambrian to early Eocene. Facies changes are from platform marine strata in the south to deeper-water, more clastic deposits in the north, including flysch of late Mesozoic age (Powell and Conaghan, 1973b; Gansser, 1977). Domal granite and gneiss uplifts south of the Indus-Tsangpo Suture Zone are of late Mesozoic but pre-Late Cretaceous age (Chang and Chen, 1973; Chang et al., 1977).

Some common features of the Himalaya south of the Indus-Tsangpo Suture Zone include the following: (1) The transverse structures, some of which appear to be related to the Aravalli trends and others comprising tear faults in thrust sheets continue across the foredeep, Lesser Himalaya, High Himalaya, and Tethyan Himalaya (Valdiya, 1976). (2) The sedimentary rocks of the entire Himalaya correlate with peninsular India and are clearly part of Gondwanaland. (3) The sedimentary rocks change facies northeastwards from a cratonic, largely discontinuous sequence to more continuous platform marine sequences and ultimately to deeper water sequences including flysch; the Himalayan thrusts telescope the northeast-facing Atlantic-type passive margin of the Indian continent (Zhang et al., in prep.).

The Himalayan Frontal Fault and Main Boundary Thrust show the youngest geomorphic signs of tectonic activity in the range, but their instrumental seismicity is relatively low. In contrast, the highest instrumental seismicity, consisting predominantly of observed moderate magnitude earthquakes with thrust-fault focal-mechanism solutions (Seeber et al., 1981), coincides reasonably well with the topographic front between the Lesser and High Himalaya, the surface trace of the MCT in Nepal and Kumaon. River terraces are tilted to the north in the vicinity of the MCT (Valdiya, 1981). Seeber et al. (1981) note, however, that the great Himalayan earthquakes occur south of the MBT; seismic intensity contours show that individual great earthquakes affected a large area of the Ganga plain. According to them, the Ganga plain is characterized by very large infrequent earthquakes with intervening periods of low instrumental seismicity, whereas the MCT is characterized by high instrumental seismicity from earthquakes of much lower magnitude than those to the south. They suggest that the MCT marks the position of an intracrustal basement thrust which flattens southward into a detachment thrust at the top of the basement. This basal detachment thrust produces infrequent, but very large earthquakes.

The Indus-Tsangpo Suture Zone is an ophiolite belt which follows the Tsangpo River and the upper part of the Indus River for nearly 2000 km, mostly in Tibet. The suture zone terminates the Tethyan Himalaya on the north and marks the boundary between late Mesozoic India and various continental fragments to the north (Zhang et al., in prep.). Some of these fragments were once part of Gondwanaland, and they migrated northward separately from India in the early Mesozoic. The telescoping of the Atlantic-type passive margin of India by Himalayan convergence has already been mentioned. Even though flysch deposits of Early Triassic to early Tertiary age are preserved south of the suture zone, it seems probable that most of the continental slope and continental rise deposits of the former passive margin have disappeared, presumably due to subduction beneath Tibet.

The ophiolite zone is generally steeply dipping and highly disrupted, and much of it is melange. Radiolaria from cherts resting on ophiolite are of middle to late Mesozoic age (Xiao et al,. 1980; cited in Zhang et al., in prep.; Nicolas et al., 1981). Melange south of the ophiolite belt contains exotic blocks of Triassic and permian sedimentary rocks in a matrix of Late Jurassic-Cretaceous age (Zhang et al., in prep.).

Southward-directed thrust plates of ophiolite include the Jungbwa and Kiogarh nappes in the central Himalaya and the Spongtang nappe in Ladakh (Gansser, 1981).

The region north of the Indus-Tsangpo Suture Zone will not be discussed in this paper, but it is important to point out the presence of the Transhimalayan plutons, a belt at least 1500 km long of tonalitic intrusions which range in age from Late Cretaceous to Miocene (Gansser, 1981; Zhang et al., in prep.). These plutons are parallel to the suture zone and are related to it. Their low $^{87}Sr/^{86}Sr$ ratios (0.704) suggest an oceanic origin (W. Frank, cited in Gansser, 1981). East of Nanga Parbat, the suture zone is in fault contact with the Paleogene Kailas molasse which transgresses over the Transhimalayan plutons.

TRANSITION FROM CENTRAL HIMALAYA TO NORTHERN PAKISTAN

The subdivisions of the Himalaya discussed above are most clearly defined in Nepal and Kumaon. They are based on three separate criteria that are all satisfied in the central Himalaya, namely, (1) geomorphic development, (2) major structures, and (3) pre-Cenozoic lithostratigraphic packages. The common assertion of their uniform persistence to the west involves several major controversies. A central issue in these is the dispute over the identity of the MCT. This argument is most clearly displayed in the various westward traces of the structure from the central Himalaya offered by different investigators.

Gansser (1981), like many others, places the MCT at the base of the metamorphic rocks of the Himalaya and considers that the metamorphic slabs at the structural top of the Lesser Himalaya are outlying klippen of the High Himalayan crystalline slab. These klippen have a demonstrable south displacement of over 100km and are characterized by Precambrian mesograde metamorphism. This interpretation results in considering all medium to high grade metamorphic rocks as part of the High Himalayan slab; it also results in tracing the MCT west from Kumaon into the Panjal thrust south of the Pir Panjal Range and thence around the Hazara-Kashmir syntaxis. Thus the MBT and Gansser's MCT closely converge along the south of the Pir Panjal Range. The Paleozoic and Mesozoic sedimentary rocks of Kashmir become a southern outlier of the Tethyan Himalayan section, and the very thick crystalline slab of the High Himalaya has essentially disappeared to be replaced by

medium grade metamorphic rocks (Salkhala Formation) (Fig. 2).

Recent work by K. S. Valdiya has questioned this widely accepted interpretation used in numerous major geodynamic interpretations, regional tectonic summaries, and published maps of the Himalaya. Valdiya (1980a, 1981) has shown that in Kumaon the MCT is more accurately located as the thrust at the base of the psammitic gneisses, the Vaikrita thrust. The crystalline outliers of the Lesser Himalaya are an older, separate thrust sheet. They are distinguished from the crystalline slab of the High Himalaya by lack of migmatization and granitization, lack of Neogene tourmaline granite intrusions, and little or no overprint of Himalayan mid-Cenozoic metamorphism. The Vaikrita thrust follows the base of the topographic step of the High Himalaya through Kumaon. The Vaikrita thrust is the logical extension of the MCT identified in Nepal by French workers (Bordet, 1973). To the west, this thrust has been traced by Thakur (1980) along a line trending toward Kargil. Thus the MCT, as revised by these investigators, passes completely north of Kashmir (Fig. 2). In this interpretation, the MCT remains a consistent distance from the central crystalline axis with its steeply foliated root zone (Pickett et al., 1975; Powell et al., 1979).

The interpretation of Valdiya (1980a) and Thakur (1980), which we believe has considerable merit, is important in emphasizing the separate existence of the higher crystalline nappes of the Lesser Himalaya. This revives the concept of the Kashmir nappe of Wadia (1928) and broadly equates it with the Jutogh and Almora structures. This is quite compatible with the stratigraphic findings of Shah (1978, 1980) that the Paleozoic and Mesozoic section of Kashmir has characteristics related to both the Tethyan and the Lesser Himalayan sections. Marine Paleozoic to Triassic sediments of Tethyan affinities are present, but the upper Paleozoic is represented by continental Gondwana sediments similar to continental India and the Lesser Himalaya. Less well known fragments in the Lesser Himalaya include the Chamba basin, the Godavari-Pulchauki basin in Nepal, and the Tang-Chu basin in central Bhutan (Gansser, 1964, 1981; Stöcklin, 1979). All of these rest on crystalline nappes that structurally overlie the sedimentary Lesser Himalayan rocks.

This new interpretation assists in trying to identify those elements of the subdivisions of the

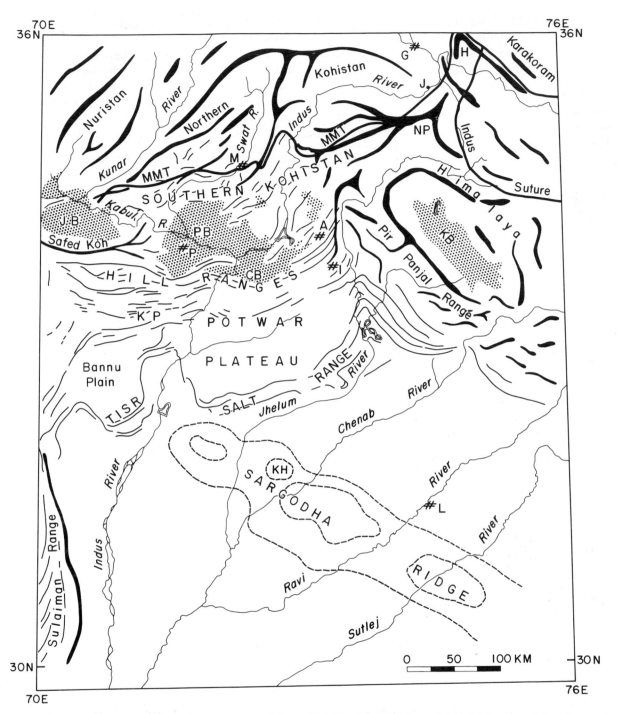

Figure 3. Subdivisions of northern Pakistan. Topographic trend lines shown with higher terrain in thicker lines. Subsurface Sargodha Ridge in dashed lines. Plio-Pleistocene basins are stippled. Locations: JB = Jalalabad Basin, M = Mingora, G = Gilgit, J = Jalipur, H = Haramosh, NP = Nanga Parbat, PB = Peshawar Basin, P = Peshawar, CB = Campbellpur Basin, A = Abbottabad, I = Islamabad, KB = Kashmir Basin, KP = Kohat Plateau, TISR = Trans-Indus Salt Range, KH = Kirana Hills, L = Lahore.

central Himalaya that trace west through Kashmir to Pakistan. The MBT, as the Murree thrust, can be followed around the syntaxis. However, in contrast to the Indian Himalaya farther east, pre-

Murree rocks come to the surface south of the MBT. These are the unfossiliferous Jammu limestone found along the Jammu thrust and considered to be either Carboniferous (Gansser,

1964) or Eocambrian in age (Valdiya, pers. comm., 1982). A more extensive southern appearance of older rocks occurs in the Salt Range of Pakistan. Lithostratigraphically, Kashmir resembles the Lesser Himalaya rather poorly. The Lesser Himalayan sedimentary units are almost pinched out, and instead there is a great development of rocks transitional to the Tethyan Himalayan sedimentary rocks above the upper crystalline nappe. Structurally the MCT does not reach Kashmir. Geomorphically there are two topographic fronts, the Pir Panjal and Zanskar Ranges. Burbank (1982) has presented an interpretation of the recent history of Kashmir which dates thrusting along the north side of the Vale of Kashmir at about 8 to 9 m.y. and along the south side of the Pir Panjal at 2 to 3 m.y. These dates are both younger than the mid-Cenozoic dates usually assigned to the MCT (LeFort, 1975). The classic subdivisions of the central Himalaya do not extend laterally with great neatness and simplicity, so that straightforward correlation to northern Pakistan should not be expected.

Relatively small intermontane basins in the central Himalaya include the Kathmandu basin (Glennie and Zeigler, 1964), Pokhara basin (Freytet and Fort, 1980), and the Suketi basin (Hukku, 1971). The Vale of Kashmir is the easternmost of a group of large intermontane basins of Pliocene to Quaternary age which also include the Campbellpur, Peshawar, Jalalabad, and Kabul basins. The origin of the Kashmir and Peshawar basins has recently been discussed by Burbank (1982) and Burbank and Johnson (1982) who relate the basins to a southward migration of thrust faulting. Whether similar origins can be suggested for the other basins listed above is not certain, but all are located between the MBT and the High Himalaya. Other intermontane basins east of Kashmir are north of the High Himalaya. These include the Hundes basin of the upper Sutlej drainage (Gansser, 1964) and the Thakkhola graben of Nepal (Colchen et al., 1979). It seems less likely that these eastern basins north of the central crystalline axis have the same genetic character as those from Kashmir west. The appearance of these large young basins in the low, southern Himalaya is one of the main contrasts between the western and central Himalaya.

SUBDIVISIONS OF THE HIMALAYA OF NORTHERN PAKISTAN

Since it has been fashionable to carry the subdivisions of the central Himalaya west into Pakistan, a common nomenclature of the natural subdivisions of this area has never been established. We have already seen that serious difficulties lie along this path, and so we offer a purely Pakistani set of subdivisions here that can be compared and contrasted with those of the central Himalaya. From south to north these are (Fig. 3): (1) the Indian shield and Punjab plain foreland, (2) the Salt Range and Trans-Indus Salt Range, (3) the Potwar and Kohat Plateaus, (4) the Hill Ranges including the Margala, Kala Chitta, Attock, and Kohat Hills and the Safed Koh, (5) the Plio-Pleistocene Peshawar and Campbellpur basins, (6) the southern Kohistan Ranges, including the Khyber-Mohmand, Lower Swat, and Hazara Ranges, (7) the Nanga Parbat-Haramosh massif, (8) the Main Mantle Thrust, and (9) the Kohistan island arc. These subdivisions are internally differentiated by major north-south trending transverse structures that separate the individual ranges and basins. These start with the Hazara-Kashmir syntaxis and its northern extension the Nanga Parbat-Haramosh massif in the east. The next is an unnamed alignment of structures starting with the Kalabagh reentrant in the south. It continues north as a set of en echelon folds that separate the Potwar and Kohat Plateaus

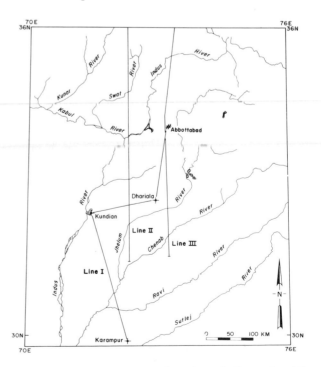

Figure 4. Location of cross-sections shown in figures 6 and 8 plotted on same map area as Figure 3.

and farther north to anomalous features along the Tarbela segment of the Indus River. Farther west, north of the Trans-Indus Salt Range, the eastern margin of the Sulaiman Ranges terminates the Bannu plain. This trend is aligned with the north-south Khyber Mountains that separate the Peshawar and Jalalabad basins. Finally the Himalayan system is cut off in the west by large strike-slip faults such as the Chaman fault of Baluchistan and eastern Afghanistan and the Kunar fault of Nuristan.

From a structural point of view, the broad equivalence between these subdivisions and those of the central Himalaya is as follows: (1) The Sub-Himalayas to the Salt Range and the Potwar and Kohat plateaus, (2) the Lesser Himalaya to the Hill Ranges, Pleistocene basins, and Lower Kohistan, and the High Himalaya to the Nanga Parbat-Haramosh massif. The Tethyan Himalaya is absent, and the High Himalaya present only in the east. There are, however, major contrasts that make any detailed reliance on this equivalence unreliable. The Indian shield does extend into central Pakistan, but the Paleozoic sequence is more complete in the Punjab Plains than it is in the Ganga basin. The Salt Range brings to the surface a thick sequence of strata of Eocambrian to early

Tertiary age within the foredeep. The Siwaliks and Murrees are present over a much broader area in Pakistan than in India, but they are not terminated northward by a discrete MBT. The MBT may be traced around the Hazara-Kashmir syntaxis, but north of Islamabad it consists of several thrust faults, none of which has a particularly large displacement. Farther west, south of Safed Koh, these faults eventually rejoin and connect to the Sarubi tear fault. Nonmarine strata contemporary with the upper part of the Siwaliks in the Peshawar and Campbellpur basins were deposited after the Hill Ranges had isolated these basins from the foredeep to the south. In the Peshawar basin, Murree strata were strongly deformed prior to deposition of these Siwalik-equivalent beds (Burbank, 1982). The strata of the Hill Ranges and Lower Kohistan can be matched very tentatively to those of the Kashmir nappe or the lesser Himalaya, depending on location. The MCT is not recognized in Pakistan. Possibly it will eventually be distinguished in the northern Kaghan area south of Nanga Parbat, an area where the geology is very poorly known. The Indus-Tsangpo Suture Zone bifurcates in Ladakh and across northern Pakistan to rejoin in northeastern Afghanistan. The bifurcation marks the limits of the Kohistan

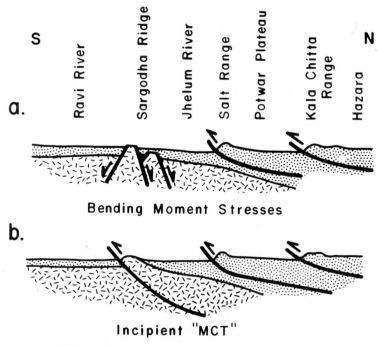

Figure 5. Diagrammatic cross-section across foreland thrust belt of Pakistani Himalaya illustrating two possible origins of a neotectonic Sargodha Ridge. A) Loading of shield by Himalayan thrusts produces bending-moment stresses and an "outer swell" at Sargodha Ridge. B) Buoyancy prevents further subduction of continental crust within Himalaya, and a new, seismically active subduction zone appears within the shield at Sargodha Ridge.

island arc complex (Tahirkheli, 1979, 1982); this ensimatic arc terrain has no exact equivalent in or north of the central Himalaya. Its southern boundary is most nearly equivalent to the Indus-Tsangpo Suture Zone and is known as the Main Mantle Thrust (MMT). The MMT lies successively adjacent to more southerly subdivisions of Pakistan as one passes west from Nanga Parbat to Nuristan. Thus it abuts and overlies very different stratigraphic sequences from east to west across Pakistan.

Indian Foreland

Rocks of the Indian shield are only exposed in Pakistan at Nagar Parkar near the Rann of Cutch and in the Kirana Hills south of the Salt Range. In the Indus basin, only one well, the Shell-Karampur well, has reached basement, which there consists of granitic and metamorphic rock. As discussed above, the rocks of the Kirana Hills were correlated to the Malani volcanics of the Indian shield (Heron, 1923), but this correlation was questioned by Davies and Crawford (1971). A summary of radiometric dates for intrusive and extrusive rocks of western Rajasthan, India and the Kirana Hills indicates that the Kirana rocks are younger than some rocks commonly assigned to the Malani and older than others (Kochhar, 1982; Pareek, 1982).

The Kirana Hills are part of the east-southeast trending Sargodha Ridge of raised continental

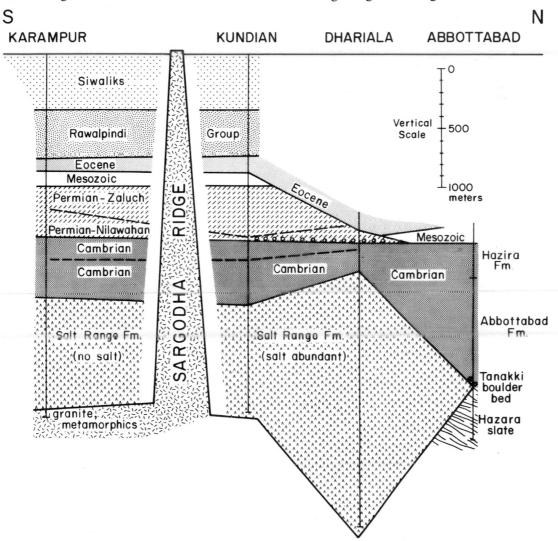

Figure 6. Stratigraphic correlation section from Punjab plains (left) to Hazara region (right) located on figure 4. No horizontal scale. All are exploratory well sections except Abbottabad.

crust which extends from the Indus River at least as far as the Indian border at the Sutlej River (Farah et al., 1977) (Figs. 3 and 5). This ridge is expressed by positive gravity anomalies and is inferred by Farah et al. (1977) to be a horst-like block of continental crust. It is often assumed that this ridge represents an old Precambrian feature which was buried by younger sediments, analogous to the northeast-trending Aravalli Range of India. However, the Sargodha Ridge is seismically active (Punjab seismic zone of Seeber and Armbruster, 1979) (cf. fig. 1), and it may be part of a southeast trending seismic zone which extends as far as Delhi (Menke and Jacob, 1976). The topographic break between the Kirana Hills and the surrounding plains is sharp, suggesting relative youth. The similarity of stratigraphy between the Karampur and Kundian wells drilled south and north, respectively, of the Sargodha Ridge (Figs. 4 and 6) suggests that these wells were drilled in what was originally a single basin. Thus the ridge may post-date rather than pre-date the deposition of strata found in these wells. Finally, the Sargodha Ridge is parallel to the Himalaya and discordant at least to the older Aravalli Range of India.

Two hypotheses may be advanced to explain the Sargodha Ridge and Punjab seismic zone (Fig. 5). The ridge may be analogous to the outer swell commonly developed by buckling of the lithosphere seaward of a subduction zone, in which case the earthquakes may result from bending-moment extensional stresses normal to the Himalaya as the northern edge of the shield is loaded by Himalayan nappes. A fault plane solution for an earthquake near Delhi supports this idea (Molnar et al., 1973), but strike-slip fault-plane solutions obtained by Seeber and Armbruster (1979) do not. LeFort (1975) considered the possibility of a new thrust fault in front of the Himalaya involving basement, a "new MCT". The strike-slip solution could be related to segmentation of this thrust fault. This solution would be more attractive if thrust-fault solutions to earthquakes in this zone were more common than they are and if there were geologic evidence for active faulting at the southern edge of the Kirana Hills. The strike-slip fault plane solutions may indicate that the seismicity is actually controlled by transverse structures (Valdiya, 1976) which themselves localize the positions of reentrants in the frontal fold belts of Pakistan (Seeber and Armbruster, 1979).

Salt Range

The Salt Range and Trans-Indus Salt Range are the surface expressions of the leading edge of a decollement thrust in which the crystalline basement is not involved (Crawford, 1974; Seeber and Armbruster, 1981). The zone of decollement appears to be within evaporites of the Cambrian or Eocambrian Salt Range Formation which underlies the Salt Range and the Potwar Plateau to the north. In the Dhariala well in the Salt Range, the formation is more than 2000 m thick (Shah, 1977). Although the Salt Range Formation is similar in age and lithology to the Hormuz salt of Iran, it is probably that it has only limited distribution in the Indo-Pakistan subcontinent. It is not found in India. The Ganga flood plain of India is the site of great historic earthquakes whereas the Punjab Plain of Pakistan lacks such earthquakes; possibly this is due to the absence of salt under the Indian plains (Seeber et al., 1981). North of the Potwar Plateau, the Hazara region also lacks salt, although other Cambrian facies (red sandstone, glauconitic sandstone, and dolomite) are found in both regions.

Non-marine and shallow-marine Cambrian strata of the Salt Range are overlain with angular unconformity by a Permo-Carboniferous sequence which begins with the Tobra conglomerate and is followed by clastic strata and by a marine Permian carbonate section. The relationship between the Permian and Cambrian sequences is similar to that between the Gondwana beds and the Vindhyan of peninsular India, differing mainly in the more marine aspects of the upper part of the Permian sequence (Zaluch Group) of the Salt Range. The Permian is overlain by Triassic, Jurassic and Cretaceous rocks which are most complete near the Indus River. Farther east, the Mesozoic and the Permo-Carboniferous sequences are overlapped progressively by the Eocene such that the Eocene rests directly upon the Cambrian in the easternmost Salt Range. The Eocene is overlain by the nonmarine Rawalpindi Group (Murree, Kamlial), which is followed by the Siwaliks (Chinji, Nagri, Dhok Pathan, and Soan or "upper Siwaliks").

As mapped by Gee (1980), the Salt Range is deformed in several ways. The sequence is cut by south-verging imbricate thrust faults, the lowermost of which brings the entire sequence over late Quaternary fanglomerate and Jhelum River alluvium (Yeats et al., in press). Near the Indus

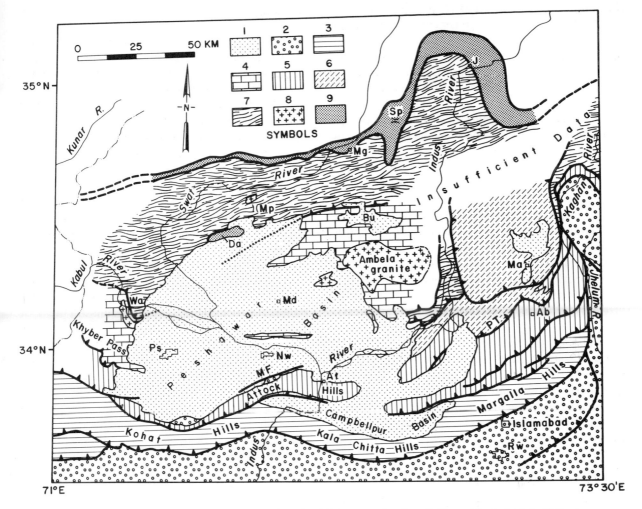

Figure 7. Tectonic sketch of northwestern Pakistan from the Hill Ranges to Southern Kohistan. Geologic Units 1) Quaternary of Plio-Pleistocene basins, 2) Murree and Siwalik strata 3) Mesozoic and Cenozoic rocks with stratigraphic affinities to the Sulaiman Range, 4) Late Precambrian to Paleozoic sedimentary rocks with stratigraphic affinities to Kashmir, 5) Late Precambrian and lower Paleozoic rocks with stratigraphic affinities to the Lesser Himalaya, 6) Tanawal-type metamorphic rocks and included lower Paleozoic gneisses, 7) Salkhala-type metamorphic rocks and included lower Paleozoic gneisses, 8) Alkaline granitic rocks of Ambela, Warsak, and Malakand, 9) Suture zone complex including ophiolite of the Main Mantle Thrust. Locations: Da=Dargai Klippe, Mp=Malakand Pass, Mg=Mingora, Sp=Shangla Pass, J=Jijal, Wa=Warsak, Ps=Peshawar, Md=Mardan, Nw=Nowshera, Bu=Buner, PT=Panjal Thrust, Ma=Mansehra, Ab=Abbottabad, MF=Manki fault, At=Attock, Rw=Rawalpindi.

River, these thrusts give way to right-lateral tear faults which terminate the Salt Range on the west. In addition, high-angle faults bound narrow horsts which trend obliquely into the range and bring Salt Range Formation to the surface. Because the Salt Range Formation is easily eroded, these horsts form deep gorges in which are found some of the classic stratigraphic sections of the Salt Range (Khewra, Nilawahan, and Warchha gorges). In the western Salt Range, the Salt Range Formation forms diapirs which are localized along high-angle faults including the right-lateral tear faults. The largest of these diapirs is found at Kalabagh on the Indus River. The northern Salt Range is a fold belt which includes strata as young as the Siwaliks. The Siwaliks are overlain with angular unconformity by late Quaternary gravels (Kalabagh Formation) which are themselves folded.

Potwar and Kohat Plateau

The Salt Range fold belt gives way northward in the Potwar Plateau to a broad syncline, the axis of which is followed by the Soan River. This syncline is asymmetric and verges south. Farther north, the intensity of folding increases, and north-dipping thrust faults appear (Pinfold, 1918; Gill, 1953), culminating in the faults bounding the Kala Chitta and Margala Hills along the north margin of the Potwar Plateau. In the Kohat Plateau west of the Potwar Plateau, Eocene through Siwalik strata are involved in a complex fold and thrust belt in which Eocene salt occupies the cores of many of the anticlines (DeJong and Qayum 1981). The Kohat Plateau structure differs from that of the Potwar Plateau largely because of this higher salt detachment horizon in the Kohat area.

The age of the major deformation is older northward. In the the Salt Range, a major unconformity occurs between the upper Siwaliks, as young as 0.4 m.y. in age (Johnson et al., 1979), and overlying fanglomerates of Salt Range provenance, but the fanglomerates are themselves strongly faulted and folded (Yeats et al., in press). The presence of conglomerate with sedimentary rock clasts in the Soan Formation (upper Siwaliks) indicates strong uplift of nearby regions to the north (Raynolds et al., 1980). In the Soan Valley south of Rawalpindi, the main deformation occurred during the time of Siwalik deposition. An angular unconformity occurs within the upper Siwalik sequence (Johnson et al., 1982; Raynolds and Johnson, in press).

Hill Ranges

The Hill Ranges, Plio-Pleistocene basins, and Southern Kohistan are probably elements of a single structural block (Figs. 3 and 8). Seeber et al. (1981) have shown a single detachment surface under the entire area from the Salt Range to the MMT on the basis of seismicity. We would interpret the Salt Range and the Hill Ranges as the locations of major ramp faults where rocks from just above the detachment are brought to the surface.

The Hill Ranges consist of a pair of low, interrupted mountain chains extending from the Hazara-Kashmir syntaxis to the Safed Koh. The faults along the more southerly range bring strongly deformed Jurassic, Cretaceous, and Tertiary rocks to the surface, while those along the northern range bring up Precambrian and Paleozoic rocks (Fig. 7). The faults of the Hill Ranges have been considered the Punjabi equivalent of the MBT, although total displacement on the entire fault set is relatively small (a few tens of kilometers at most). Coeval formations of Paleozoic and Mesozoic age are easily correlated from the Potwar Plateau across the Hill Ranges to Hazara, and Eocene strata similar to those of the Kala Chitta and Margala Hills are found in deep wells in the Potwar Plateau.

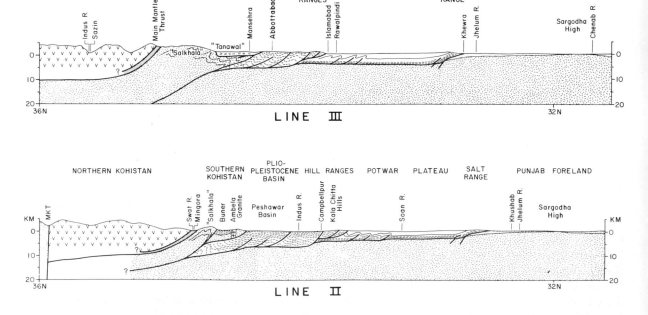

Figure 8.   Structural cross-sections of northwestern Pakistan. Locations shown on figure 4. Vertical exaggeration 2 times.

The several thrusts of the Hill Ranges appear to rejoin to the west in Waziristan along the south margin of the Safed Koh. They pass into the active Sarubi fault, a tear fault with right-lateral displacement (Prevot et al., 1980). No single fault of the Hill Ranges plays the same tectonic role that the MBT has in the central Himalaya.

The Mesozoic and lower Tertiary section of the southern Hill Ranges is fossiliferous and marine. It correlates well with the units of the Sulaiman-Kirthar Ranges to the west (Fig. 7), but does not have a clear equivalent to the east. The Mesozoic portion of the section is similar to that of the Tethyan Himalaya, but there the Tertiary rocks are absent. There are thin Jurassic rocks in Kashmir, but nothing younger until the Plio-Pleistocene. The Lesser Himalaya have nothing equivalent to the fossiliferous marine rocks of the southern Hill Ranges. Thus we see here a major stratigraphic contrast as units from the western Sulaiman marginal belt interleave with the rocks of the Himalaya to the east. The multiple thrust fault traces may eventually be found to reflect this interaction of structure and stratigraphy.

The northern Hill Ranges are clearly defined south of the Plio-Pleistocene basins, but in the east they merge imperceptibly with the Hazara area of southern Kohistan (Fig. 7). In this area, the Hazara Formation, composed of interbedded graywacke and siltstone, unconformably underlies the Tanakki boulder bed at Abbottabad, and both underlie fossiliferous Cambrian strata (Latif, 1974; Ghaznavi et al., 1983). Rb/Sr total-rock ages of 740 ± 20 m.y. and 930 ± 20 m.y. (Crawford and Davies, 1975) confirm the Precambrian age of the Hazara slate. The Cambrian section of Hazara is unconformably overlain by Jurassic, marine Cretaceous, and Eocene marine strata. Both the middle Paleozoic and the Gondwana formations of the upper Paleozoic are missing. Valdiya (pers. comm., 1982) finds stratigraphic correlation between the Precambrian and Cambrian units and the Lesser Himalaya of Kumaon convincing. Farther west, the Attock slate on the Indus River is assumed to be Precambrian, but stratigraphic evidence is lacking due to structural complications. The Attock slate is intercalated, probably tectonically, with quartzite and limestone which differ considerably from the Cambrian sequence of Hazara and the Salt Range. Tahirkheli (1970) distinguished two slate units in the Attock area, one with these inliers and diabase dikes, and another without. Valdiya (1980b)

makes a similar distinction in the Kumaon area. Farther west, in the Khyber Pass area, the Landi Kotal slate shows relations similar to those at Attock (Shah et al., 1980).

Farther north, at Nowshera in the Peshawar basin, limestones yield Silurian and Devonian fossils (Stauffer, 1967, 1968), and are associated along the eastern margin of the Peshawar basin with white quartzites that may correlate with the Muth quartzite that is widespread in Kashmir and the Tethyan Himalaya. The Silurian and Devonian are missing in Hazara and the Salt Range and, indeed, in much of adjacent peninsular India, but they are the only Paleozoic periods represented by fossils in the Peshawar basin. Where is the Cambrian in the Peshawar basin? The Permo-Carboniferous may be represented by the Khyber Limestone which rests on Devonian near the Khyber Pass (Shah et al., 1980), but these strata are not obviously related to the Salt Range Permian. One might speculate that the Peshawar Basin was a different tectonostratigraphic terrane than the Hazara, Salt Range, and peninsular India, and one closely allied with Kashmir. However, both the Peshawar Basin and the Hazara area are underlain in part by similar slates. Detailed mapping of tracts of moderately-deformed carbonate rocks east of Karakar Pass in lower Swat may resolve the problem.

Plio-Pleistocene Basins

Two major basins are currently active geomorphic features in northern Pakistan (Fig. 7). These are the Campbellpur and Peshawar basins, recently studied by Burbank (1982, 1983). Aggradation of fluvial sediment by meandering and braided rivers and accumulation of alluvial fans continues in these basins behind temporary baselevels created and maintained by the rising Hill Ranges. This is in significant contrast to the structurally similar Potwar and Kohat Plateaus which are loci of active degradation at present. On the basis of magnetic-polarity stratigraphy and fission-track dates on enclosed volcanic ashes, Burbank (1982) concluded that sedimentation began around 3 m.y. ago in the Peshawar basin and about 1.8 m.y. ago in the Campbellpur basin. Widespread sedimentation in the southern Peshawar basin terminated about 600,000 years ago due to accelerated uplift of the Attock Range. Sediments along the northern flank of the Attock range are cut by the active Manki fault (R. Yeats and A. Hussain, work in progress). In the Attock Range Paleozoic, Mesozoic, and Eocene marine strata and Murree redbeds are all folded in association with the Hill ranges thrust faults. The Murree is overlain with sharp angular unconformity by gently-dipping strata equivalent in age to part of the Siwaliks (Burbank, 1982). These strata are cut by the Manki fault.

Like Burbank (1982) we consider these basins to be the surface expression of a major low-angle detachment fault. The Hill Ranges mark several ramp structures that bring rocks from just above the basal detachment to the surface (Fig. 8). Thus the structure of the Plio-Pleistocene basins is analogous to the structure of the Potwar Plateau detachment-Salt Range ramp system.

Much more poorly understood is the Pliocene (?) history of the Mansehra and Buner areas (Figs. 7 and 8). These areas have never been investigated from this point of view, but a thick sequence of late Cenozoic sediments is present as raised terraces in the Mansehra area (Pakhli Plain of Shams, 1969) and as valley fills around Buner. The sediments vary from fine to coarse grained and are fluvial and/or lake deposits, presumably similar to the Karewa beds of Kashmir. Thick saprolitic weathering has affected the granitic and metamorphic rocks in much of the Mansehra area. Thus it seems likely that a basin similar to the Peshawar basin existed here in the late Neogene which was destroyed by more recent uplift of the area. This uplift may be reflected in the Indus-Kohistan seismic zone and overlying tectonic step discussed by Gornitz and Seeber (1981).

A zone of alkali granitic rocks of early Tertiary age (41-50 m.y. dates reported by Kempe, 1973) extends from the Afghan border for 200 km eastwards across the Peshawar basin to the Indus River (Kempe and Jan, 1970; Jan et al., 1981). Such rocks are commonly associated with rifting, and there is no obvious counterpart of this zone elsewhere on the subcontinent. Neither is there any other evidence of rifting. These rocks record some as yet unidentified event related to the period immediately after collision, but before substantial underthrusting and associated uplift occurred.

Southern Kohistan

The area north of the Plio-Pleistocene basins is dominated by crystalline metamorphic and intrusive rocks (Figs. 7 and 8). Much of the area is inadequately explored so our knowledge is limited. Local areas that are moderately well known include Hazara-Mansehra (Calkins et al., 1975; Shams, 1969; LeFort et al., 1980), lower Swat-Buner (Martin et al., 1962), and Malakand (Chaudhry et al., 1974; Gansser, 1981). Tentatively these rocks can be considered equivalent to the mesograde metamorphic rocks of the upper thrust units of the Lesser Himalaya. Metamorphic rock types, plutonic sheets and masses, and known stratigraphic and radiometric ages all support this inference. Structural relations are poorly known and prevent either a unique structural interpretation of the area or a clear structural correlation to the better known rocks of the central Himalaya.

In northern Hazara, in the region from Mansehra north, metamorphic and plutonic rocks are widely exposed. The stratigraphic section of lower Hazara, south of the Panjal thrust, has already been discussed. North of the Panjal thrust the Hazara Formation is reported to be unconformably overlain by the late Precambrian Tanawal Formation (Calkins et al., 1975) which is itself overlain by the Abbottabad and Hazara Formations (M. Ghaznavi and T. Karim, pers.comm., 1982). In the Sherwan area west of Abbottabad, the Tanawal Formation is a clastic unit of sandstone, siltstone, and minor conglomerate with well preserved sedimentary structures such as ripple marks and cross-beds. Carbonates are rare or absent. Thus the Panjal thrust is stratigraphically manifested by the absence of the Tanawal Formation south of the fault and its widespread presence north of the fault. In the type area metamorphism of the Tanawals barely reaches chlorite grade. Farther north the Tanawal Formation is overlain by the Mansehra granite along a sharp, arcuate contact (Shams, 1969; Calkins et al., 1975). The gneissic rock is a large sheet-like body folded with the metamorphic rocks. Previous workers considered all gneiss contacts to be intrusive or metasomatic, but we suggest that a thrust fault may be present (Figs. 7 and 8). The tectonic setting and nature of the contact resemble those described in the higher nappes of the Lesser Himalaya (Gansser, 1964; Valdiya, 1981). Within the granitic area middle to high grade, non-migmatized metamorphic rocks are widespread. These are dominantly psammitic units varying gradually or abruptly from impure to pure quartzites. Interbedded pelitic and psammitic units are common. Metaconglomerates are known in two locations (Shams, 1969). Carbonates are again rare. These rocks seem most likely to be stratigraphically equivalent to the Tanawal Formation. As in the Lesser Himalaya, higher grade rocks overlie lower grade rocks.

The Mansehra granitic and gneissic rocks have been carefully studied by Shams and his students and colleagues (Shams, 1969). These workers have distinguished the most abundant unit of the terrain, granite gneisses with huge potash feldspar megacrysts, and younger, relatively small, tourmaline granite bodies. The two kinds of units are considered to be genetically related. The Mansehra granite grades into gneiss (Shams, 1969) and, in part, may be a sheet-like body tightly folded along with country rocks (LeFort et al., 1980). LeFort et al., (1980) have found the Mansehra granite to be an S-type cordierite granite with a Rb/Sr age of 516 m.y. and an initial $^{87}Sr/^{86}Sr$ ratio of 0.7189. Granitic rocks of this age and type have been found in a number of places in the Lesser Himalaya and High Himalaya (Ferrara et al., 1982; Bhanot et al., 1980).

The Salkhala Formation was defined in northwestern Kashmir where it consists of carbonaceous schists, marbles that are commonly dolomitic, calcareous schists, quartz-mica schists and quartzites. Along the Kishenganga River, it is reported to grade upward into Dogra slate(= Hazara Formation) which itself grades upward into fossiliferous Cambrian rocks (Wadia, 1931). Along the Tarbela Reservoir area of the Indus, Salkhala Formation is again mapped. The rocks along the Indus have the same characteristic carbonaceous schist-carbonate lithology, but are overthrust by Chamba quartzite.

Similar rocks appear south of the MMT in lower Swat, where there is a terrain of garnet to kyanite grade dolomitic marbles, graphitic schists, and quartzites which are intruded by sheets and massive bodies of granitic augengneiss containing large potassium feldspar megacrysts. Biotite from this gneiss has an $^{40}Ar/^{39}Ar$ age 515 m.y. (H. Maluski in Jan et al., 1981). Foliated tourmaline granites are found in association with these gneisses. The lower Swat schists probably correlate with the Salkhala Formation, but not with the Tanawal Formation of northern Hazara. In contrast, the augengneisses of Swat and those of Hazara are of similar age and lithology. Both areas show strong similarities to some areas of the Lesser Himalayan crystalline nappes. These rocks have been deformed four times (R. Lawrence, work in

progress; Martin et al., 1962). Metamorphism and schistosity formation predate emplacement of ultramafic bodies directly on top of these rocks during the early stages of development of the MMT. Absence of any regional evidence of Mesozoic or Paleozoic metamorphism strongly supports a Precambrian age of metamorphism. The remaining deformations are subsequent to ultramafic emplacement. The second deformation produced large (10 to 20 km wavelength) folds with generally north-south axial trends and new axial plane cleavage. Very large (50 km wavelength) east-west folds of the third deformation produce strong plunges on the north-south folds. The ultramafic material of the MMT is largely preserved in such plunging synclines. Suturing on the MMT accompanied these east-west folds. The youngest deformation event records drag associated with lateral motion on the MMT.

In the Buner area south of lower Swat these mesograde metamorphic rocks come into abrupt contact with Devonian marine sedimentary rocks (Martin et al., 1962; Stauffer, 1968). The nature of this contact is unclear; it may be a thrust or a steep to overturned fold flank. However the Devonian rocks are not metamorphosed to any significant degree and are only gently folded on east-west axes. These folds are of similar magnitude to the east-west folds of lower Swat. This terrain is intruded by the early Tertiary alkaline Ambela granite batholith and cut by major, post-Ambela, east-northeast shear zones (M. Rafique, work in progress). South of this intrusion more Devonian rocks and the Chamba quartzite which may be correlative with the Muth Quartzite of Kashmir are exposed (Martin et al., 1962). Thus the area from lower Swat to Tarbela may have structure and stratigraphy roughly analogous to that of Kashmir with Paleozoic rocks over Precambrian Salkhalas and all of these riding on a major thrust similar to the Kashmir nappe.

Metamorphic geology similar to that of lower Swat extends from the Besham area on the Indus River (Jan and Tahirkheli, 1969) at least to Malakand Pass (Chaudhry et al., 1974). South of Malakand Pass on the northern edge of the Peshawar plain is the Dargai klippe, an ultramafic body preserved south of the large east-west anticline as a thin thrust sheet (Malinconico, 1982; Uppal, 1972); this is a Pakistani equivalent to the Spongtang klippe. This mesograde terrain probably extends to the Warsak area (Ahmad et al., 1969) and on into Afghanistan.

Nanga Parbat Massif

The principal difference between the Pakistani and the Indian Himalaya is the absence of a thick crystalline slab such as that comprising the High Himalaya of India and Nepal. The only likely representative of this Himalayan feature is the Nanga Parbat massif. Like the crystalline slab of the High Himalaya the rocks of Nanga Parbat are extensively migmatized by Himalayan age metamorphism and cross-cut by small Tertiary tourmaline granite intrusive bodies (Misch, 1949). The Nanga Parbat-Haramosh massif is a narrow northward extension of these crystalline rocks deep into the Kohistan-Ladakh andesitic arc terrain. The rise of this terrain is very recent and continues at a rate of at least 1 cm/yr (Zeitler et al., 1982). Thus the emergence of the Nanga Parbat massif implies that much of the Kohistan arc terrain is allochthonous. More speculative is the possibility that the Nanga Parbat-Haramosh massif actually represents the western end of the High Himalayan crystalline slab (Fig. 2). In this case the MCT must pass south of Nanga Parbat and terminate against the MMT not far to the west. Support for the idea that the Nanga Parbat massif represents the western end of the Himalayan crystalline slab comes from deep seismic sounding profiles between Kashmir and the Pamirs which pass directly under the Nanga Parbat area and show a substantial rise of the Moho under Nanga Parbat (Kaila, 1981; Beloussov et al., 1980). The only known feature in Southern Kohistan that is comparable to the MCT is the northwest trending topographic step extending beyond the Hazara-Kashmir syntaxis across northern Hazara and related to a deep seismic zone, the Indus-Kohistan seismic zone (Gornitz and Seeber, 1981). However, this structure relates most clearly to the thrust along the southern Pir Panjal range which is a younger structure similar to the MCT.

The Indian shield must terminate westward in the region between documented Precambrian rocks in the Sargodha Ridge area and ophiolites in the axial belt of the Sulaiman Range of Baluchistan. In the Himalaya, this western termination must be represented by the western end of the Precambrian slab of the High Himalaya; this slab may extend into the Nanga Parbat-Haramosh massif, but not farther west into Swat or south into Mansehra and Hazara. Instead, there is a sequence of metamorphosed sedimentary rocks (Salkhala, Swat Schists, etc.) which may mark the

point where the Precambrian margin faces west and correlates with the mesograde thrust sheets of the Lesser Himalaya.

Farther north, the Main Mantle Thrust (MMT) which marks the southern boundary of the Kohistan-arc sequence, has not moved in the last 15 m.y., according to fission-track dating of uplift rates on opposite sides of the fault (Zeitler et al., 1982). The change in direction of the MMT from northwest in Ladakh to west-southwest in Kohistan may reflect the edge of the shield. North of the Himalaya the ultramafic rocks of the Indus-Tsangpo Suture Zone overlie Tethyan sediments which are themselves on top of the High Himalayan crystalline slab. Outliers are found as the Spongtang klippe and Jungbwa nappe. The MMT is the western prolongation of the Indus-Tsangpo Suture Zone and in Swat the ultramafic rocks of this zone overlie mesograde metamorphic rocks equivalent to the Lesser Himalayan upper thrust sheets. The Dargai klippe is an outlier analogous to those to the east but lying on a very different terrain.

Neotectonics

The neotectonic setting of northern Pakistan is somewhat better understood, at least as far north as the Peshawar basin. The Salt Range thrust and a similar thrust bounding the Khisor Range west of the Indus River is analogous to the Himalayan Frontal Fault of Nakata (1972), but it is much more prominently developed and brings deeper horizons to the surface. The view of Crawford (1974) that the Salt Range is a decollement riding on a cushion of Precambrian salt is substantiated by the presence of salt in wells in the Potwar Plateau and the relatively minor influence of the Salt Range on the gravity contours of Farah et al. (1977). Paleomagnetic data from the upper Siwaliks show some post-Siwalik counter-clockwise rotation of the eastern Salt Range but not of the western Salt Range (Opdyke et al., 1982). The similarity of stratigraphy between the Salt Range and the Kundian and Karampur wells in the Punjab Plain to the south suggest that the displacement on the Salt Range decollement is measured in tens rather than hundreds of kilometers. The right-slip tear faults near the Indus River may terminate this slab on the west and not involve basement, but the similarity in trend between these tear faults and earthquake focal-mechanism solutions in the Kirana Hills to the south suggests that these tear faults may involve basement. Resolution of this problem requires high-quality seismic reflection lines across this strike-slip zone which will permit mapping of the basement reflector as well as deep sedimentary horizons.

Farther north, the thrusts and folds appear to be older, not involving the late Quaternary. In the southern Peshawar basin, late Quaternary deformation is along east-northeast-trending linear faults. At Manki, south of Nowshera, one such zone deforms loessic silt and fanglomerate as well as unconsolidated river gravels containing a Karakoram clast assemblage (Fig. 7). These gravels probably represent an older course of the Indus River (R. Yeats and A. Hussain, work in progress). In the Indus Valley near Chilas, northwest of Nanga Parbat, the steeply dipping Jalipur sandstone (Misch, 1936) contains a clast assemblage similar to that of the modern Indus River (Olson, 1982). Jalipur deposition may have been localized by a high-angle active fault which controls the east-west course of the Indus River in this region. East of Chilas near the Rakhiot bridge the Indus Valley bends north. Here the course of the valley follows the MMT as it turns north into the Nanga Parbat-Haramosh massif (Tahirkheli, 1979, 1982). The presence of active faulting in this region is indicated by slickensided fan gravels, offset fans, aligned hot springs, and local emplacement of bedrock over alluvium. This faulting may be related to the rapid rise of the Nanga Parbat-Haramosh massif recorded by fission-track data (Zeitler et al., 1982). Rapid neotectonic uplift from Kohistan to Hunza is indicated by dramatically uplifted Pleistocene terraces throughout the Indus, Gilgit, and Hunza Valleys.

The neotectonic setting comprises thrusting and folding only along the southern margin of the Salt Range where the decollement reaches the surface. Farther north, the deformation occurs on high-angle, probably strike-slip, faults. Thrusting appears to be migrating southward such that the structural style at a given locality such as the Peshawar basin progresses from thrusting to high-angle faulting through time. Farther north, vertical displacement without surface faulting appears to be the primary mode of neotectonic behavior. This still appears to be the surface manifestation of deep-seated detachment faulting (Gornitz and Seeber, 1981).

PROBLEMS FOR FUTURE WORK

Rather than summarize the remarks made above, we close by underscoring what we don't know about Pakistan—problems which are important not only for resolving the tectonic history of the area but also to sharpen the focus in exploration for minerals and fuels resources and in evaluating the seismic and ground rupture hazard of the country.

It is important to define the western edge of the Precambrian shield, and this can be done best geophysically by regional gravity and magnetic surveys and by seismic-reflection profiles which map the basement reflector as it tilts westward toward the Sulaiman Range and northward toward the Himalaya. Analysis of magnetic patterns may permit mapping of the Precambrian structural grain into the subsurface. The question of whether the Sargodha Ridge represents an inherited Precambrian trend or a new Himalayan structural zone should then be resolved.

What is the origin of the Hazara-Kashmir syntaxis, the westward change from the Himalayan trend to the arcuate fold belts of the Sulaiman and Kirthar Ranges? Paleomagnetic data show that the Salt Range did not rotate as a single unit from a former Himalayan trend to its present west-southwest orientation. The Salt Range may be controlled by an underlying anisotropy in the basement, either a Precambrian structural grain, which is not supported by gravity data, or by early Paleozoic faults related to rifting of a proto-continent. Oil-industry seismic data show that the basement reflector is stepped down to the north, a sense of throw opposite to that of south-directed Quaternary thrust motion. Evaporites underlying the Cambrian of the Salt Range and Potwar Plateau may be related to rifting, as may the presence of S-type granites of 510-520 m.y. age farther north. Do the north-facing basement steps control the location of ramps in thrusts of the Salt Range and the northern Potwar Plateau? Do the steps in basement comprise zones of weakness which could yield seismically if the basement is under horizontal stress? Do the apparent strike-slip faults in the Mianwali reentrant between the Salt Range and Trans-Indus Salt Range represent "hang-ups" where salt is absent and the Paleozoic is coupled with the basement?

The Paleozoic sequences of the Salt Range, Potwar Plateau, and Hazara appear to correlate well with one another, but their relation to the Paleozoic of the Peshawar plain and ranges to the north and west is unclear. Many of these sequences are involved in imbricate thrusting, and they do not contain lithologies and megafaunas as distinctive as those in the Salt Range. Available biostratigraphic data confirm only middle Paleozoic rocks west of the Indus; this is the very section that is cut out by a major unconformity to the east. The middle Paleozoic is again present in Kashmir. The problem should be resolved by detailed mapping together with conodont biostratigraphy. To this end an acid laboratory should be established in Pakistan for processing large samples of limestone for conodonts. This work should permit the reconstruction of the northern passive margin of the Precambrian shield in Paleozoic and Mesozoic time.

Detailed mapping should continue northward to resolve the structural relations between the Attock and Hazara slates. Mesograde metasediments of Lower Swat, along the Indus River near Besham, through Mansehra, and east to the Kaghan valley need detailed mapping to determine stratigraphic and structural relations within these units and structural relations to other units to the south. Is there a single thrust which brings mesograde metamorphics over Paleozoic sediments from the syntaxis to the Kunar fault in Nuristan as suggested by Lawrence (1982)? The presence of little metamorphosed ophiolite on mesograde metamorphics in Lower Swat raises the question of the time of metamorphism of these rocks. Is there any significant Himalayan age metamorphism in these units? Where is the westernmost extent of the migmatized crystalline slab of the High Himalaya? Why is the ophiolite of the suture zone emplaced over strikingly different tectonostratigraphic units from Nanga Parbat to Waziristan? These are questions that require detailed mapping and careful study of comparative metamorphic petrology coupled with modern isotopic studies.

Finally, an evaluation of the neotectonic setting of the northern areas is critical for future projects to develop the hydroelectric power potential of this region. It is the most seismically active area of Pakistan, yet little is known about the geological expression of this seismic activity. The seismic zone beneath the Pir Panjal of Kashmir continues west as the Indus-Kohistan seismic zone, but aside from a southwest-facing topographic step, this seismic zone has no surface expression. The young deformation along the Indus River near Chilas and Jalipur indicates an active fault zone, but of what magnitude? What is its tectonic significance? Why is the Nanga Parbat massif rising so rapidly? Is the MKT, the Main Karakoram thrust, an active fault as some have suggested?

These problems and others will occupy the best geological minds of Pakistan and other countries for the years to come. The problems are difficult, and they will require a cooperative effort involving the Geological Survey of Pakistan and other governmental agencies with geological expertise, the major universities of Pakistan, and agencies and universities outside the country. The rewards will be a better understanding of the geological history of this complex region and the significance of this to tectonic principles and a better means to explore for mineral wealth and fossil fuels.

ACKNOWLEDGEMENTS

Partial support for this work was provided by NASA grant NAG 9-2, NSF grants EAR 78-15476 and 80-13158 and NSF grant INT 81-18403. Lawrence benefited from a year at the University of Peshawar as a Fulbright professor. Colleagues at the Geological Survey of Pakistan, the University of Peshawar, and the Gemstone Corporation of Pakistan have been very generous with their knowledge of the geology of the country.

REFERENCES

Acharyya, S.K., and K.K. Ray, 1982. Hydrocarbon possibilities of concealed Mesozoic-Paleogene sediments below Himalayan nappes—reappraisal: American Assoc. Petroleum Geologists Bull., 66: 57-70.

Ahmad, M., K.S.S. Ali, B. Khan, M.A. Shah, and I. Ullah, 1969. The geology of the Warsak area, Peshawar, West Pakistan: Geol. Bull. Univ. Peshawar 4: 44-78.

Alvarez, W., 1982. Geological evidence for the geographical pattern of mantle return flow and the driving mechanism of plate tectonics. J. Geophy. Res., 87: 6697-6710.

Beloussov, V.V., N.A. Belyaevsky, A.A. Borisov, B.S. Volvovsky, I.S. Volvosky, D.P. Resvoy, B.B. Tal-Virsky, I. Kh. Khamrabaev, K.L. Kaila, H. Narain, A. Marussi, and Finetti, J., 1980. Structure of the lithosphere along the deep seismic sounding profile: Tien-Shan-Pamirs-Karakoram-Himalayas: Tectonophysics, 70: 193-221.

Bhanot, V.B., V.P. Singh, A.K. Kansal, and V.C. Thakur, 1977. Early Proterozoic Rb-Sr whole-rock age for central crystalline gneiss of Higher Himalaya, Kumaon: J. Geol. Soc. India, 18 (2): 90-91.

Bhanot, V.B., B.K. Pandey, V.P. Singh, and A.K. Kansal, 1980. Rb/Sr ages for some granitic and gneissic rocks of Kumaon and Himachal Himalaya: In, K.S. Valdiya and S.B. Bhatia, (eds.), Stratigraphy and Correlation of Lesser-Himalayan Formations, Hindustan Publ., Delhi, p. 139-142.

Bordet, P., 1973. On the position of the Himalayan Main Central Thrust within Nepal Himalaya: Seminar on Geodynamics of Himalayan Region, National Geophy. Res. Inst., Hyderabad (India), p. 48-55.

Burbank, D.W., 1982. The chronolgic and stratigraphic evolution of the Kashmir and Peshawar intermontane basins, northwestern Himalaya: Unpubl. PhD. thesis, Dartmouth College, 291 p.

Burbank, D.W. 1983. The chronology of intermontane-basin

development in the northwest Himalaya and the evolution of the northwest syntaxis: Earth Planet Sci. Letts., 64: 77-92.

Burbank, D.W., and G.D. Johnson, 1982. Intermontane-basin development in the past 4 Myr in the northwest Himalaya: Nature, 298: 432-436.

Calkins, J.A., T.W. Offield, S.K.M. Abdullah, and S.T. Ali, 1975. Geology of the Southern Himalayas in Hazara, Pakistan, and adjacent areas: U.S. Geol. Sur. Prof. Paper, 716-C, 29 p.

Chang, C., and H. Chen, 1973. Some tectonic features of the Mt. Jomo Lungma area, southern Tibet: Sci. Sinica, 16: 257-265.

Chang, C., X. Zheng, and Y. Pan, 1977. The geological history, tectonic zonation and origin of uplifting of the Himalayas: Peking, Inst. of Geology, Acad. Siniea, 17 p.

Chaudhry, M.N., S.A. Jafferi, and B.A. Saleemi, 1974. Geology and petrology of the Malakand granite and its environs: Geol. Bull. Punjab Univ., 10: 43-58.

Colchen, M., M. Fort and P. Freytet, 1979. Sédimentation et tectonique plio-quaternaire dans le Haut-Himalaya: l'exemple du fossé de la Thakkhola (Himalaya du Nepal): 7ème Réunion annuelle Sci. de la Terre, Lyon, Soc. Géol. France éd., p. 121.

Crawford, A., 1974. The Salt Range, the Kashmir Syntaxis and the Pamir Arc: Earth Planet. Sci. Lett., 22: 371-379.

Crawford, A.R. and R.G. Davies, 1975. Ages of pre-Mesozoic formations of the Lesser Himalaya, Hazara District, northern Pakistan: Geol. Mag., 112: 509-514.

Datta, A.K., and V.V. Sastri, 1977. Tectonic evolution of the Himalaya and the evaluation of the petroleum prospects of the Punjab and Ganga Basins and the Foot Hill Belt: Himalayan Geology, 7: 296-325.

Davies, R.G., and A. R. Crawford, 1971. Petrography and age of the rocks of the Bulland Hill, Kirana Hills, Sargodha District, West Pakistan: Geol. Mag., 108: 235-246.

DeJong, K.A., and A. Qayum, 1981. Collapsed anticline in Kohat, SW Himalaya foothills, Pakistan: Geol. Soc. Am. Abstr. with Programs 13: 437.

Duncan, R.A., 1981. Hotspots in the southern oceans—an absolute frame of reference for motion of the Gondwana continents Tectonophysics, 74: 29-42.

Farah, A., M.A. Mirza, M.A. Ahmad, and M.H. Butt, 1977. Gravity field of the buried shield in the Punjab Plain, Pakistan: Geol. Soc. America Bull., 88: 1147-1155

Farah, A., R.D. Lawrence, and K.A. DeJong, (this volume). Tectonics of Pakistan.

Farah, A., and R.D. Lawrence (in press). Landsat geophysical map, Punjab Plain, Pakistan: In: K.A. De Jong, and R.D. Lawrence, (eds.). Landsat Atlas of Pakistan. Geol. Surv. Pakistan.

Ferrara, G., B. Lombardo, and S. Tonarini, 1983. Rb/Sr geochronology of granites and gneisses from Mt. Everest region, Nepal Himalaya: Geol. Rundsc., 72: 119-136.

Forsyth, D., and S. Uyeda, 1975. On the relative importance of the driving forces of plate motion: Geophys. J.R. Astron. Soc., 43: 163-200.

Freytet, P., and M. Fort, 1980. Les formation Plio-Quaternaires de la Kali Gandaki et da Bassin da Pokhara (Himalaya du Nepal): Bull. Assoc. Geogr. France, 471: 249-257.

Gansser, A., 1964. Geology of the Himalayas: Wiley Interscience, New York, 289 p.

Gansser, A., 1979. Reconnaissance visit to the ophiolites in Baluchistan and the Himalaya, In: A. Farah and K.A. DeJong (eds.), Geodynamics of Pakistan. Geol. Survey of Pakistan, p. 193-213.

Gansser, A., 1981. The geodynamic history of the Himalaya, In: H. Gupta and F. Delany, (eds.), Zagros-Hindu Kush-Himalaya Geodynamic Evolution. Am. Geophys. Union Geodynamics Series, 3: 111-121.

Gee, E.R., 1980. Pakistan geological Salt Range series, 1:50,000, 6 sheets: Directorate of Overseas Surveys, United Kingdom, for Government of Pakistan, and Geol. Survey of Pakistan.

Ghaznavi, M.I., T. Karim, J.B. Maynard, 1983. A bauxitic paleosol in phosphorite-bearing strata of northern Pakistan: Econ. Geol. 78: 144-147.

Gill, W.D., 1953. The stratigraphy of Siwalik series in the northern Potwar, Punjab, Pakistan: Quart. Jour. Geol. Soc. London, 107 (4): 375-394.

Glennie, K.W. and M.A. Zeigler, 1964, The Siwalik Formation in Nepal:

Proc. 22nd Intern. Geol. Congr. (New Delhi, 1964), 14: 82-95.

Gornitz., V., and L. Seeber, 1981. Morphotectonic analysis of the Hazara arc region of the Himalayas, north Pakistan and northwest India: Tectonophysics, 74: 263-282.

Heron, A.M., 1923. The Kirana and other hills in the Jech and Rechna Doabs: Geol. Survey India Recs., 43 (3): 229-236.

Hukku, B.M., 1971. A note on the probable origin and structure of the Suketi Basin, Mandi District, Himachal Pradesh: Geol. Sur. India Misc. Publ. 15: 163-166.

Jaeger, E., A.K. Bhandari, and V.B. Bhanot, 1971. Rb-Sr age determinations on biotites and whole rock samples from the Mandi and Chor granites Himachal Pradesh, India: Eclogae geol. Helv., 64: 521-527.

Jan, M.Q., and R.A.K. Tahirkheli, 1969. The geology of the lower part of Indus Kohistan (Swat), West Pakistan: Geol. Bull. Univ. Peshawar, 4: 1-13.

Jan, M.Q., M. Asif, T. Tahirkheli, and M. Kamal, 1981. Tectonic subdivision of granite rocks of north Pakistan: Geol. Bull. Univ. Peshawar, 14: 159-182.

Johnson, G.D., N.M. Johnson, N.D. Opdyke, and R.A.K. Tahirkheli, 1979. Magnetic reversal stratigraphy and sedimentary tectonic history of the upper Siwalik Group, eastern Salt Range and southwestern Kashmir, In: A. Farah, and K. DeJong (eds.), Geodynamics of Pakistan. Geol. Survey of Pakistan, p. 149-165.

Johnson, G.D., P. Zeitler, C.W. Naeser, N.M. Johnson, D.M. Summers, C.D. Frost, N.D. Opdyke, and R.A.K. Tahirkheli, 1982. The occurrence and fission-track age of late Neogene and Quaternary volcanic sediments, Siwalik Group, northern Pakistan: Paleogeog., Paleoclimat., Palcoccology, 37. 63-93.

Kaila, K.L., 1981. Structure and seismotectonics of the Himalaya-Pamir-Hindu Kush region and the Indian plate boundary: In: H. Gupta and F. Delany, (eds.), Zagros-Hindu Kush-Himalaya Geodynamic Evolution. Am. Geophys. Union Geodynamics Series, 3: 272-293.

Kempe, D.R.C., 1973. The petrology of the Warsak alkaline granites, Pakistan, and their relationship to other alkaline rocks of the region: Geol. Mag. 110: 385-495.

Kempe, D.R.C., and M.Q. Jan, 1970. An alkaline igneous province in the North-West Frontier Province, West Pakistan: Geol. Mag., 107: 395-398.

Khan, A.B., Z.H. Shah and S.M. Naeem, 1970. Geology of the Ghondai Sar and vicinity Jamrud, Khyber Agency: Geol. Bull. Univ. Peshawar, 5: 115-130.

Kochhar, N., 1982. Petrochemistry and petrogenesis of the Malani igneous suite, India: Discussion. Geol. Soc. America Bull. 93: 926-927.

Krishnan, M.S., 1968. Geology of India and Burma: Higginbothams, Ltd. Madras, 536 p.

Latif, M.A., 1974. A Cambrian age for the Abbottabad group of Hazara, Pakistan: Geal. Bull Punjab Univ., 10: 1-20.

LeFort, P., 1975. Himalayas: The collided range. Present knowledge of the continental arc. Am. Jour. Sci., 275-A. 1-44.

LeFort, P., F. Debon, and J. Sonet, 1980. The "Lesser Himalayan" cordierite granite belt: Typology and age of the pluton of Mansehra (Pakistan): Geol. Bull., Univ. Peshawar, 13: 51-61.

Lyon-Caen, H., and P. Molnar, 1983. Constraints on the structure of the Himalaya from an analysis of gravity anomalies and a flexural model of the lithosphere: J. Geophys. Res., 88: 8171-8191.

Malinconico, L.L., 1982. Structure of the Himalayan suture zone of Pakistan interpreted from gravity and magnetic data: Hanover, N.H., Darmouth College unpublished Ph.D. thesis, 128pp.

Martin, N.R., S.F.A. Siddiqui, and B.H. King, 1962. A Geological reconnaissance of the region between the Lower Swat and Indus Rivers of Pakistan: Geol. Bull. Punjab Univ., 2: 1-14.

Menke, W.H., and K.H. Jacob, 1976. Seismicity patterns in Pakistan and north-western India associated with continental collision: Seismol. Soc. America Bull., 66: 1695-1711.

Misch, P., 1936. Ein gefalteter junger Sandstein im Nordwest-Himalaya und sein Gefuge: Festschrift zum 60. Geburtstag von Hans Stille, Enke, Stuttgart, p. 259-276.

Misch, P., 1949. Metasomatic granitization of batholithic dimensions: Am. Jour. of Sci., 247. 204-245.

Molnar, P., T. Fitch, and F.T. Wu, 1973. Fault-plane solutions of shallow earthquakes and contemporary tectonics in Asia: Earth Planet. Sci. Lett., 19: 101-112.

Molnar, P., and P. Tapponnier, 1975. Cenozoic tectonics of Asia: Effects of a continental collision: Science, 189: 419-426.

Nakata, T., 1972. Geomorphic history and crustal movement of the foothills of the Himalaya: Inst. Geogr., Tohoku Univ., Sendai, p. 39-77.

Nicolas, A., J. Giradeau, J. Marcoux, B. Dupre, X. B. Wang, Y. G. Cao, H. Z. Zheng, and X. Ch. Xiao, 1981. The Xigaze ophiolite (Tibet): a peculiar oceanic lithosphere: Nature: 294, (5840): 414-417.

Norton, I.O., and J.G. Sclater, 1979. A model for the evolution of the Indian Ocean and the breakup of Gondwanaland: J. Geophys. Res., 84: 6803-6830.

Olson, T.M., 1982. Sedimentary tectonics of the Jalipur sequence, northwest Himalaya, Pakistan: unpub. M.A. thesis, Dartmouth College, Hanover, N.H., 152 p.

Opdyke, N.D., N.M. Johnson, G.D. Johnson, E.H. Lindsay, and R.A.K. Tahirkheli, 1982. Paleomagnetism of the middle Siwalik formations of northern Pakistan and rotation of the Salt Range decollement: Paleogeog., Paleoclimat., Paleoecology, 37: 1-15.

Pareek, H.S., 1982. Petrochemistry and petrogenesis of the Malani igneous suite, India: Reply: Geol. Soc. America Bull., 93: 927-928.

Pickett, J., J. Jell, P. Conaghan, and C. Powell, 1975. Jurassic invertebrates from the Himalayan central gneiss: Alcheringa 1: 71-85.

Pinfold, E.S., 1918. Notes on structure and stratigraphy in the northwest Punjab: Recs. Geol. Survey India, 3: 137-160.

Powell, C.M., and P.J. Conaghan, 1973. A Polyphase deformation in Phanerozoic rocks of the central Himalayan gneiss, northwest India: Jour. Geol. 81, 127-143.

Powell, C.M., and P.J. Conaghan, 1973b. Plate Tectonics and the Himalayas: Earth Planet Sci. Lett., 20: 1-12.

Powell, C.M., A.R.. Lrawford, R. L. Armstrong, R. Prakash, and H.R. Wynne-Edwards, 1979. Reconnaissance Rb-Sr. dates for the Himalayan central gneiss, northwest India: Indian J. Earth Sci., 6: 139-151.

Prevot, R., D. Hatzfeld, S.W. Roecker, and P. Molnar, 1980. Shallow earthquakes and active tectonics in eastern Afghanistan: J. Geophys. Res., 85: 1347-1357.

Raiverman, V., 1981. Tectonic control in facies distribution, Subathu sediments, Simla Hills, northwestern Himalaya, In: A. K. Sinha, (ed.), Contemporary Geoscientific Researches in Himalaya 1: 207-211.

Raynolds, R.G.H., 1980. The Plio-Pleistocene structural and stratigraphic evolutior. of the eastern Potwar Plateau, Pakistan: Unpubl. PhD. thesis, Dartmouth College, 265 p

Raynolds, R.G.H., G.D. Johnson, N.M. Johnson, and N.D. Opdyke, 1980. The Siwalik molasse: A sedimentary record of orogeny: Geol. Bull. Univ. Peshawar, 13: 47-50.

Raynolds, R.G.H., and G.D., Johnson, (in press). Rates of Neogene depositional and deformational process in the Himalayan foredeep, Pakistan: Geol. Soc. London Spec. Publ.

Seeber, L., and J. Armbruster, 1979. Seismicity of the Hazara arc in northern Pakistan: decollement vs. basement faulting, In: A. Farah and K.A. DeJong, (eds.), Geodynamics of Pakistan. Geol. Survey of Pakistan, p. 131-142.

Seeber, L., J. G. Armbruster, and R.C. Quittmeyer, 1981. Seismicity and continental subduction in the Himalayan arc: In: H. Gupta and F. Delany, (eds.), Zagros-Hindu Kush-Himalaya Geodynamic Evolution. Am. Geophys. Union Geodynamics Series, 3: 215-242.

Sengor, A.M.C., and W.S.F., Kidd, 1979. Post-collisional tectonics of the Turkish-Iranian Plateau and a Comparison with Tibet: Tectonophysics, 55: 361-376.

Shah, S.K., 1978. Facies pattern of Kashmir within the tectonic framework of Himalaya: In: P.S. Saklani, (ed.), Tectonic Geology of the Himalaya. Today and Tomorrow's Printers and Publishers, p. 63-78.

Shah, S.K., 1980. Stratigraphy and tectonic setting of the Lesser Himalayan belt of Jammu: In Valdiya K.S. and S.B. Bhatia, (eds.) Stratigraphy and Correlations of Lesser Himalayan Formations, Hindustan Publ. Corp., Delhi, p. 152-160.

Shah, S.M.I., (ed.) 1977. Stratigraphy of Pakistan: Geol. Survey Pakistan Mem., 12: 1-138.

Shah, S.M.I., R.A. Siddiqi, and J.A. Talent, 1980. Geology of the

Eastern Khyber Agency, North Western Frontier Province, Pakistan: Records. Geol. Survey Pakistan, 44: 1-31.

Shams, F.A., 1969. Geology of the Mansehra-Amb State area, NW Pakistan: Geol. Bull. Punjab Univ., 8: 1-31.

Stauffer, K.W., 1967. Devonian of India and Pakistan: In: (D.H. Oswald, ed.), International Symposium on the Devonian System Alberta Soc. petrol. Geol. p. 545-656.

Stauffer, K.W., 1968. Silurian-Devonian reef complex near Nowshera, West Pakistan: Geol. Soc. America Bull., 79: 1331-1350.

Stöcklin, J., 1980. Geology of Nepal and its regional frame: J. Geol. Soc. London, 137: 1-34.

Tahirkheli, R.A.K., 1970. The geology of the Attock-Cherat Range, West Pakistan: Geol. Bull. Univ. Peshawar, 5: 1-26.

Tahirkheli, R.A.K., 1979. Geology of Kohistan and adjoining Eurasian and Indo-Pakistan continents, Pakistan: Geol. Bull. Univ. Peshawar, 11 (1): 1-30.

Tahirkheli, R.A.K., 1980. Major tectonic scars of Peshawar Vale and adjoining areas, and associated magmatism: Geol. Bull. Univ. Peshawar, 13:39-46.

Tahirkheli, R.A.K., 1982. Geology of the Himalaya, Karakoram, and Hindu Kush in Pakistan: Geol. Bull. Univ. Peshawar, 15: 1-51.

Tahirkheli, R.A.K., M. Mattauer, F. Proust, and P., Tapponnier, 1979. The India-Eurasia suture zone in northern Pakistan: synthesis and interpretation recent data at plate scale: In: A. Farah and K.A. DeJong, (eds.), Geodynamics of Pakistan. Geol. Surv. of Pakistan, Quetta, p. 125-130.

Tapponnier, P., G. Peltzer, A.Y. LeDain, and R. Armijo, 1982. Propagating extrusion tectonics in Asia: New insights from simple experiments with plasticine. Geology, 10: 611-616.

Thakur, V.C., 1980. Tectonics of the central crystallines of western Himalaya: Tectonophysics, 62: 141-154.

Thakur, V.C., 1981. An overview of thrusts and nappes of western Himalaya: In K.R. McClay and N.J. Price, (eds.), Thrust and Nappe Tectonics. Geol. Soc. London Spec. Publ. 9: 381-392.

Uppal, I.H., 1972. Preliminary account of the Harichand Ultramafic Complex, Malakand Agency, N.W.F.P., Pakistan: Geol. Bull. Punjab Univ., 9: 55-63.

Valdiya, K.S., 1976. Himalayan transverse faults and folds and their parallelism with subsurface structures of north Indian plains: Tectonophycis, 32: 353-386.

Valdiya, K.S., 1980a. Intracrustal boundary thrust of the Himalaya: Tectonophysics, 55: 323-348.

Valdiya, K.S., 1980b. Stratigraphic scheme of the sedimentary units of the Kumaon Lesser Himalaya: In: K. S. Valdiya and S.B. Bhatia, (eds.) Stratigraphy and correlations of Lesser Himalayan formations, Hindustan Publ. Corp., Delhi, p. 7-48.

Valdiya, K.S., 1981. Tectonics of the central sector of the Himalaya: In H. Gupta and F. Delany, (eds.), Zagros-Hindu Kush-Himalaya Geodynamic Evolution. Am. Geophys. Union Geodynamics Series, 3: 87-110.

Wadia, D.N., 1928, The geology of Poonch State (Kashmir) and adjacent portions of the Punjab: Geol. Surv. India, Mem. 51: 2-155

Wadia, D.N., 1934. The Cambrian-Trias sequence of NW Kashmir (parts of Muzaffarabad and Baramula District): Rec. Geol. Surv. Ind. 68: 121-176.

Xiao, X.Ch., J. Ch. Quo, G.M. Chen, and Zh. Zh. Zhu, 1980. Ophiolites of the Tethys-Himalayas of China and their tectonic significance: In "Scientific Papers on Geology for International Exchange," Publishing House of Geology, Beijing, China, p. 143-162.

Yeats, R.S., S.H. Khan, and M. Akhtar, (in press). Late Quaternary deformation of the Salt Range of Pakistan: Bull. Geol. Soc. America.

Zeitler, P.K., R.A.K. Tahirkheli, C.W. Naeser, and N.M. Johnson, 1982a. Unroofing history of a suture zone in the Himalaya of Pakistan by means of fission-track annealing ages: Earth Planet. Sci. Lett.,57: 227-240.

Zeitler, P.K., N.M. Johnson, C.W. Naeser, R.A.K., Tahirkheli, 1982.b Fission-track evidence for Quaternary uplift of the Nanga Parbat region, Pakistan: Nature, 298: 255-257.

Zhang, Zh. M., J.G. Liou, and R.G. Coleman, (in prep.). An outline of plate tectonics of China.

E. PALEOCEANS

13

Paleoceanography:
A Synoptic Overview of 200 Million Years
of Ocean History

B. U. HAQ
Exxon Production Research Company,
Houston, Texas

ABSTRACT

This paper outlines the development of ocean basins and their hydrographic-climatic conditions from the Triassic to the end of Pliocene. The narrative traces the chain of paleogeographic events in the Mesozoic and Cenozoic and their inferred influence on the changes in surface and deep water circulation and global climates. The marked stratigraphic event at the Cretaceous/Cenozoic boundary is reviewed in more detail. The topic of sea level change is discussed separately because of its importance in climatic change and biogenic productivity, as well as its role in determining the sedimentary patterns along the ocean margins. Inferred paleocirculation scenaria are presented for six time-slices; mid Jurassic (ca 178-173 Ma), mid Cretaceous (ca 105-100 Ma), late Paleocene (ca 60-58 Ma), mid Oligocene (ca 31-30 Ma), late early Miocene (ca 17-16 Ma), and near the Miocene/Pliocene boundary (5.5-5.0 Ma).

INTRODUCTION

Paleoceanography is a relatively new field of earth sciences that aims at learning about the evolution of the ocean basins and their hydrographic, climatic and biotic patterns. The methodology of paleoceanography is by necessity very diverse in approach and includes input from many of the traditional fields of marine geology, geophysics and stratigraphy. Numerous methodologies developed only recently have also enhanced our ability to reconstruct the paleoenvironments of the past. Perhaps the most important factor that has accelerated the advancements in paleoceanography has been the availability of the vast amount of deep sea core material recovered by the Deep Sea Drilling Project (DSDP).

The new analytical and statistical techniques are helping to unravel the hydrographic mysteries of the past, which include our first insights into the fertility and dissolution histories of the paleoceans. A new appreciation of the manner in which paleoceanographic changes took place is gradually emerging. These insights indicate that seafloor spreading and plate motions, particularly as they effect the opening and closing of shallow and deep passages, may have been important determinants of the past oceanographic conditions, and the role of oceanic gateways is seen as fundamental in recreating the scenaria of circulation of ancient oceans. Widespread, and at times global (eustatic), sea level changes, which in turn may or may not be related to plate motions, may have had significant influence in modulating past global climates, and thus indirectly also influence the distribution and evolution of biota, as well as the supply and accumulation of organic matter. The distribution and preservation of the latter along the continental margins and in enclosed basins is in itself an important factor for of the accumulation of hydrocarbons. Significant oceanwide anoxic episodes have been detected, during the course of which rich organic deposits were preferentially preserved in isolated euxinic basins. The documentation of the timing and extent of such events is of obvious importance to the search for fossil-fuel.

It is also becoming apparent that the tempo of most oceanographic changes of the past was not gradual. Instead, short-term dramatic shifts between metastable states are common. Catastrophic oceanographic-climatic scenaria have been conceived to explain some of the more marked changes at certain stratigraphic boundaries. For example, both terrestrial and extra-terrestrial causes have been suggested for the relatively rapid and marked changes in the stratigraphic record at the Cretaceous/Tertiary boundary.

The purpose of this paper is to briefly outline our knowledge of the development of the ocean basins, their inferred circulation history, and the influence of oceanographic changes on the world climates. The first section of this overview is essentially a running commentary on the chain of paleogeographic events that conceivably provided the impetus for the major oceanographic-climatic changes in the Mesozoic. The second section includes a brief discussion of the probable causes of the stratigraphic changes at the Cretaceous/Tertiary boundary, some 66 million years ago – a subject that is being hotly debated at the present time. This is followed by a commentary on the oceanographic-climatic changes and the probable causal relationships of events in the Paleogene and Neogene oceans. Because of the importance of the subject and it's relevance to paleoceanography, the final section in the paper is devoted entirely to the discussion of sea level changes and the environmental response to such fluctuations. Much of the relevant literature has been cited in the text which should help the reader pursue the issues in greater detail. Other details on various aspects of paleoceanography can be found in numerous symposia volumes published recently (see Table 1 for selected references recommended for further reading). Several paleoceanographic summaries have also been given in recent synthesis-papers, including: Berggren and Hollister (1977), Schnitker (1980), van Andel (1979), Haq (1981), Kennett (1981, 1983), Berger (1982) and Berggren (1982). In addition, numerous, more recent, DSDP Initial Report volumes contain regional syntheses. An excellent historical review of the development of ideas that led to the revolution in earth sciences during the last two decades, especially the contribution of marine geosciences, has been given by Emiliani (1981). In the same volume Berger (1981) has also presented a succinct outline of the various approaches that are in common use in paleoceanographic research.

MESOZOIC OCEANOGRAPHIC SCENARIA

Much of the Jurassic and older oceanic crust has since been subducted and only a very small percentage of the remaining crust is early Jurassic in age [i.e. older than 190 Ma (Ma=millions of years B.P.)]. Consequently, little is known of the pre-middle Jurassic superocean. Our ability to reconstruct the pre-Mesozoic continental geographies, that are a prerequisite for meaningful paleoenvironmental interpretations of those times, is severly restricted due to lack of reliable data. Attempts at Paleozoic reconstructions have met with difficulties because paleogeographies based on remanent magnetism data are often inconsistent with lithologic and biogeographic patterns; inconsistencies which need to be reconciled before such reconstructions can be considered reliable (Boucout and Gray, 1983). Here we will restrict ourselves to the broad outline of the global circulation and climatic histories since the last breakup of the supercontinent of Pangaea, in the late Triassic. [See Worsley et al., 1984 (this volume) for some ideas about periodic convergence and breakup of supercontinents and their implications].

In the Mesozoic (ca 250-66 Ma) the configuration of the Pangaean supercontinent strongly influenced the global climates. In the Triassic (ca 250-210 Ma) Pangaea was extensively rimmed by evaporites, and to a lesser degree, by coal and bauxite deposits (Frakes, 1979). Mid Triassic time, in particular, saw wide latitudinal expansion of evaporitic deposits and there was significant reef building in the middle and late Triassic. Climates of those times are inferred to have been extremely arid, perhaps the most arid ever in the earth's history (Frakes, op. cit.). The latitudinal temperature gradient was low, as higher latitude temperatures were little different than those in the equatorial regions. Since atmospheric and oceanic circulation are largely driven by the thermal contrast between the high and low latitudes, this lack of moisture and thermal gradients indicates that oceanic circulation was sluggish, although vertical stratification may have been present. The global equatorial "Tethys" Current would not be initiated until sometime in the middle Jurassic, when Pangaea was breached, connecting the proto-North Atlantic with the eastern ocean of Panthalassa, and through the opening between the North and South American plates with the Pacific.

TABLE: 1. Recent symposia volumes and other texts dealing with Paleoceanography (arranged in chronologic order).

AUTHORS, EDITORS/CONVENORS	YEAR	MAJOR CONTENTS	REFERENCE
1. Von der Borch, C.C. (ed.)	1978	Deep sea drilling results from the Indian Ocean	Mar. geol., 26: 175 pp.
2. Cita, M.B. & Wright, R. (eds.)	1979	Messinian salinity crisis in the Mediterranean & its global implications	Paleogeogr., Paleoclimat., Paleoecol., 29: 222 pp.
3. Frakes, L.A.	1979	A summary of climatic history of the earth	*Climates throughout geologic time.* Elsevier, N.Y., 310 pp.
4. Talwani, M., Hay, W. and Ryan, W.B.F. (eds.)	1979	Deep sea drilling results from the Atlantic Ocean	American Geophy. Un., M. Ewing Ser. 3, 437 pp.
5. Schopf, T.J.M.	1980	Methodological approach to paleoceanography	*Paleoceanography.* Harvard Uni. Press, Cambridge, Mass.
6. Emilini, C. (ed.)	1981	Synthesis papers on tectonics, paleogeography, paleoclimatogy and paleoceanography.	*The Oceanic Lithosphere.* J. Wiley & Sons, N.Y. *The Sea,* Volume 7, 1800 pp.
7. Blanchet, R. & Monterdert L. (conv.)	1981	I.G.C. symposium volume on the geology of continental margins.	Int. Geol. Cong. Sym. C3, Oceanologica Acta, sp. suppl. to vol. 4, 294 pp.
8. Le Pichon, X., Debyser, J and Vine, F. (conv.)	1981	I.G.C. symposium volume on geology of the oceans	Int. Geol. Cong. Sym. C4, Oceanologica Acta, *ibid.,* 154 pp.
9. Pomerol, C. (ed.)	1981	Eocene/Oligocene Boundary event and its implications	Paleogeogr., Paleoclimat., Paleoecol., 36: 155-364.
10. Warme, J.E., Douglas, R.G. and Winterer, E.L (eds.)	1981	Synthesis of results of a decade of drilling in all oceans	Soc. Econ. Paleont. Min., sp. pub. 32, 564 pp.
11. Barron, E.J. (ed.)	1982	Paleogeography, climate & oceanography since Mesozoic	Paleogeogr., Paleoclimat., Paleoecol., 40: 252 pp.
12. Berger, W.H. & Crowell, J.C. (eds.)	1982	A report on paleoclimatic research & new directions	*Climates in Earth History.* Stud. in Geophy., Nat. Res. Council, 198 pp.
13. Einsele, G. & Seilacher, A. (eds.)	1982	Productivity past and present.	*Cyclic and Event Stratification.* Springer Verlag, Berlin, 536 pp.
14. Kennett, J.P.	1982	A modern view of marine geology and its various subfields	*Marine Geology.* Prentice-Hall, Inc., Englewood, N.J., 813 pp.
15. Seibold, E. & Berger, W.H.,	1982	Introductory text in marine geology and paleoceanography	*The Sea Floor. An Introduction to marine Geology.* Springer-Verlag, Berlin, 288 pp.
16. Briskin, M. (ed.)	1983	Paleocliamtology and chronology of the Cenozoic	Paleogeogr., Paleoclimat., Paleoecol., 42: 209 pp.

Figure 1 summarizes the major paleo-geographic events against a Jurassic-Cretaceous time scale. The late Triassic-early Jurassic times (ca 230-190 Ma) saw the initiation of break-up of Pangaea. Continental rifting and crustal thinning between North America and Africa began during this time (Grow and Sheridan, 1981), as did, most likely, the initial rifting in the Gulf of Mexico (Buffler et al., 1981). The nascent North Atlantic rift system was a complex of northeast-southwest trending grabens and half-grabens, and sedimentological evidence suggests increasing aridity towards the northern end of the complex (Hay et al., 1982). Initial rift valleys were presumably the site of stratified fresh water lakes like their modern analogs, where significant amount of organic matter was deposited and later overlain by evaporities (Hay, 1981). The rifting phase may have been accompanied by widespread doming in the Gulf of Mexico (Buffler et al., 1981)

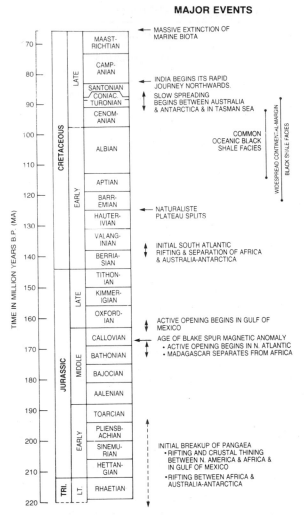

MAJOR EVENTS

Figure 1. The timing of the major paleogeographic events of the latest Triassic through Cretaceous interval, plotted against a time scale in m.y. B.P. (after Harland et al., 1982) and standard European stages. Positions of the widespread mid Cretaceous anoxic events and the massive extinctions at the Cretaceous/Tertiary boundary (some 66 m.y. ago) are also shown.

in anticipation of the spreading between North America and Africa-South America. In the middle and late Jurassic the Gulf area broke up into a series of separate subsiding basins with thick evaporitic deposition. Major drifting and subsidence, however, began in the early late Jurassic (Oxfordian, after 163 Ma), and may have continued until the mid Cretaceous when the early phase of evolution of the Gulf area ceased and is marked by a prominent basin-wide unconformity (Buffler et al., *op. cit.*).

In the Jurassic (212 to 144 Ma) the evaporites were less abundant than in the Triassic but their latitudinal extent was broader (Frakes, 1979).

During the latest Triassic and early Jurassic the global humidity seems to have increased somewhat, although overall conditions were still quite dry. Middle Jurassic once again saw the return of distinctly arid conditions as evidenced by abundant middle and late Jurassic evaporitic deposits. Well developed reefs along the margins of the Tethys, particularly in the Oxfordian (163-156 Ma), and generally warm climates were characteristic of the Jurassic. Callovian (169-163 Ma) and Oxfordian times, may, however, have been relatively cooler (Frakes, *op. cit.*)

In the North Atlantic seafloor spreading began in the early middle Jurassic. The Callovian age of the Blake Spur magnetic anomaly would, however, indicate the initiation of spreading in the southern part of North Atlantic sometime in the early Callovian (Sheridan, 1983) or somewhat earlier (ca 172-167 Ma). Rapid and variable subsidence along the continental margins of the U.S. East Coast was controlled by tansverse fracture zones, segmenting the area into four sedimentary basins (Grow and Sheridan, 1981). Predrift late Triassic and early Jurassic evaporites occur all along the U.S. East Coast. The growth of extensive barrier reefs and associated carbonate platforms along the trailing passive margins persisted after the evaporitic phase (until early Cretaceous), but since then the major site of calcite deposition has shifted to the open ocean, vide calcareous plankton secretion and deposition (Hay, 1981).

Prior to the opening of the Central Atlantic in the middle Jurassic, an epicontinental Central Atlantic seaway existed, but its communication with the open oceans was restricted. Relatively free intermingling may have occurred during times of high seastands such as during the Toarcian (194-188 Ma) and Bajocian (181-175 Ma) (Hallam, 1983). A comparison of bivalve biogeographic data from the western North American, southern Andes and Europe is consistent with the opening of Central Atlantic in the middle Jurassic (Bathonian, 175-169 Ma) times (Hallam, *op. cit.*). In the later middle Jurassic surface water exchange between the Tethys to the east and Pacific to the west through the open Central Atlantic was stabilized and the wind driven surface currents initiated from the east to the west. By Callovian-Oxfordian time the North Atlantic was wide enough to permit equatorial counter current and to create a return flow along the southeast edge of

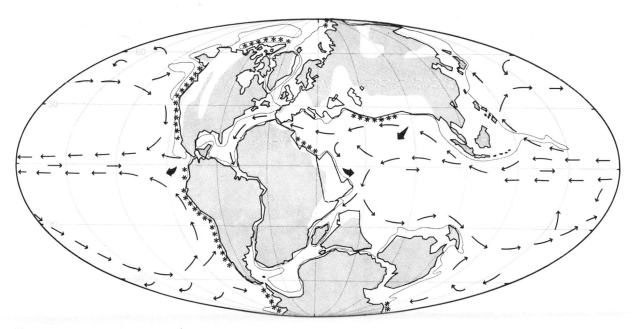

Figure 2. Inferred global paleocirculation scenario during the middle Jurassic (ca 178-173 Ma). In the paleogeographic reconstruction averaged land areas for the time-slice are shown shaded; the epeiric seas unshaded and adapted from Barron al. (1981). 1000m depth contours are also drawn around the continents. The potential source areas for the warm saline bottom waters (WSBW) in the low and middle latitudes are indicated with larger arrows. Inferred surface circulation features are shown with smaller arrows. Asterisks show the predicted upwelling sites (along the East Pacific and Tethys coasts and in the polar regions) and are based on the simulations by Parrish and Curtis (1982) for the late early Jurassic (northern summers).

the North Atlantic (Gradstein and Sheridan, 1983).

Recent interpretations of the magnetic anomaly data between Madagascar and Africa (Rabinowitz et al., 1983) suggested that the motion of Madagascar relative to Africa was from north to south and predates anomaly M25, beginning in later part of middle Jurassic and continuing until anomaly M9 time (early Cretaceous) when it reached its present position relative to Africa. This study puts the age of the separation between Africa and Madagascar at the same time as the beginning of active spreading in the North Atlantic. Mesozoic sediments in the eastern Tethyan (Indian Ocean) region were characterized by relatively rapid accumulation of clays in restricted basins, with calcareous and terrigenous sediments accumulating along the margins (Kidd and Davies, 1978).

Figure 2 shows a mid Jurassic global paleo-geographic reconstruction with superimposed inferred surface circulation patterns (shown by thin arrows), and probable sites of origination of deeper waters from the epeiric seas of the lower latitudes (shown by thicker arrows). As discussed later on, in the Mesozoic the locus of intermediate and deeper waters was most likely to have been in

the low and middle latitude shelves and epicontinental seas, where excess evaporation would make the waters more saline and dense, and more likely to sink to form deep waters.

The reconstruction in Figure 2 shows conditions prior to the opening of the Central Atlantic passageway. The circum-global Tethys Current had not yet developed, although communication of the nascent North Atlantic with the open oceans may have been possible during highstands of sea level. The eventual opening of Central America and the initiation of the Tethys Current had important repercussions for the wide dispersal of tropical marine faunas (Berggren and Hollister, 1974, 1977).

Figure 2 also depicts the areas of probable upwelling (shown by asterisks) along continental margins as determined by Parrish and Curtis (1982). These authors used present day analogies and the Mesozoic-Cenozoic wind patterns predicted by the atmospheric models of Barron and Washington (1982) to infer qualitative upwelling patterns of the past. Such reconstructions show significant correspondence between potential upwelling areas and occurrence of organic-rich rocks in Triassic, early to late Cretaceous and middle Paleogene to Neogene.

Parrish and Curtis (*op. cit.*) ascribe the lack of correspondence for the intervening intervals to differing preservational preferences, controlled by such factors as transgressions and anoxia. They, nevertheless, maintain that upwelling contributed significantly to the patterns of organic-rich facies.

Gradstein and Sheridan (1983) propose that the proto-Gulf Stream may have been initiated in the early Cretaceous by which time Central Atlantic margins would have drifted into the middle latitudes and the effect of westerlies would have developed the North Atlantic gyre. The presence of the Nicaraguan Block would have closed the access from the Caribbean to the Pacific, thereby backing up the trade winds and initiating the proto-Gulf Stream.

The early Cretaceous also saw the separation between Africa and Australia-Antarctica (Norton and Sclater, 1979) where rifting had occurred earlier in the latest Triassic and early Jurassic (McKenzie and Sclater, 1971). Along Naturaliste Plateau crustal splitting had begun prior to magnetic anomaly M4 (ca 125 Ma), which is the oldest anomaly identified in this region (Markl, 1978). A recent study of the magnetic anomalies between Australia and Antarctica puts the breakup of the two continents in the late Cretaceous (ca 90 Ma) (Cande and Mutter, 1982).

In the South Atlantic rifts had formed by the latest Jurassic-early Cretaceous time (ca 145-135 Ma) and a thick series of continental deposits filled the deeper parts of the rifts (Reyment, 1980a). An onshore transgression of southern Africa and Argentina in the Valanginian (138-131 Ma) was followed by the formation of two largely closed northern and southern basins in the South Atlantic. The southernmost basin first opened towards the north in the Albian (113-97 Ma) and the northernmost basin opened in the Santonian (88-83 Ma), but Falkland Plateau high still hindered the inflow of deeper waters (Reyment, *op. cit.*). The pre-Albian South Atlantic would have been much like an isolated Mediterranean, with relatively harsh and arid climates in the low latitudes and cool and dry in the temperate latitudes (Johnson, 1983). At the Aptian/Albian boundary the sudden changes in floral and faunal compositions in the Falkland Plateau cores signal the end of stagnant conditions and the beginning of ventilation of the southwestern Atlantic basin (Wise et al., 1982).

Analyses of stable isotopes of oxygen and carbon in marine calcareous microplankton and benthos have been a major breakthrough in the interpretation of past climatic and environmental conditions (see e.g., Savin, 1977, for a discussion). Analyses of oxygen isotopes of calcareous microfossils can provide valuable clues about such paleoceanographic variables as: the surface water temperature through analysis of surface-dwelling planktonic Forminifera and nannoplankton, bottom water temperature through analyses of benthic Foraminifera, and the structure of the paleo-thermocline through the analyses of deep-dwelling planktonic species. Bottom water paleotemperature values can also be used to estimate the surface water temperatures that prevailed in areas where deep waters originated, i.e., mostly in the low and mid latitudes in the Mesozoic and earliest Tertiary, and in the high latitudes after mid Tertiary. Carbon isotopic ratios have been used as indicators of marine fertility and paleo-productivity, as they are dependent on the supply and storage of organic carbon in the marine realm (see e.g., Kroopnick et al., 1977). However, both oxygen and carbon isotopic composition of marine calcareous microplankton and benthos are subject to biological, and pre- and post-depositional, chemical and preservational factors, necessitating numerous constraints to the interpretations based on stable isotopic data (see Arthur, 1979, for a summary discussion).

Oxygen isotopic paleotemperature data for the pre-Cretaceous is lacking and even for the Cretaceous it is very sparse. Few available data points for the Cretaceous (see, e.g., Douglas and Savin, 1975) indicate the global climates were warm and equable (see Fig. 3) and the thermal gradients were low. The vertical thermal structure of the water column was relatively homogeneous (Douglas and Savin, 1978). Evaporites and coal measures extended up to 45 to 55° latitude in the Cretaceous, and although the total volume of evaporites was smaller, their latitudinal extent was greater than in the Triassic and Jurassic (Frakes, 1979). In the late Cretaceous the sea level highstands may have contributed to greater spread of evaporites and reefs to their widest extent since the Devonian (Frakes, *op. cit.*) The existence of forests and large terrestrial vertebrates in the polar latitudes also testify to the generally warm and equable climate of the Cretaceous (Axelrod, 1984). Both isotopic data and the widespread distribution of reefs suggest a climatic optimum

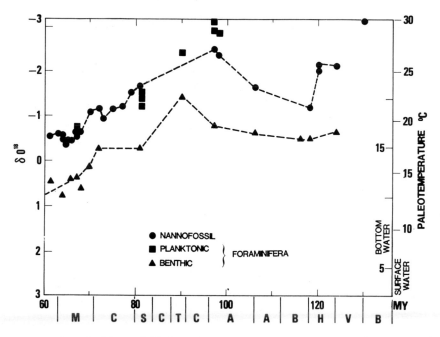

Figure 3. The sparse oxygen isotopic data available from the Cretaceous calcareous nannoplankton and planktonic and benthic foraminifera from DSDP sites on or near the Shatsky Rise in northwestern Pacific Ocean (after Douglas and Savin, 1979 and in COSOD Report, 1982). Paleotemperature scale assumes ice-free polar conditions and salinities similar to today. Letters along the time axis represent the stages of the Cretaceous (see Fig. 1). Notice the peak in both surface (nannofossil) and bottom (benthic foram) temperatures in the mid-Cretaceous (late Albian through Turonian) when the oceans were characteristically warm and less oxygenated. The late Cretaceous (Santonian through Maastrichtian) was characterized by cooler surface and bottom water temperatures and well-oxygenated oceans (see text for details).

from Albian to Santonian times. The Aptian-Albian (119-97 Ma) and Campanian-Maastrichtian (83-66 Ma) sea level highstands may have contributed to the extremely high sedimentation rates during these times (Hay et al., 1981; Southam and hay, 1982).

Global atmospheric modelling of the mid Cretaceous by Barron and Washington (1982) has predicted warm polar surface temperatures ranging form 15° to 19°C, and tropical temperatures only 1-2°C above present. These authors contend that mid Cretaceous atmospheric circulation cannot be assumed to be sluggish because of the low surface temperature gradients. Paleogeography played an important role in influencing these circulation patterns. Cretaceous oceanic thermal patterns may have been more homogenous, but the climates would have been quite variable. Barron and Washington suggest major thermal contrasts between continental margins and their interiors, particularly in Asia, where there is supportive lithologic and floral evidence. Their simulations show a belt of variable high and low pressure zones in the mid latitudes along the edge of the continents, related to the

land-sea thermal contrast. Such conditions would have produced strong moonsonal circulation and high evaporation over the continental shelves and epicontinental seas. These mid and low latitude shallow seas were the sites of the warm, saline and dense bottom waters (WSBW) that characterized the Cretaceous deep water regime (Brass et al., 1981). The mid Cretaceous atmospheric simulations of Barron and Washington (op. cit.) show that the northern margins of Tethys may have received extensive rainfall during that time.

The mid Cretaceous record reveals widespread organic-rich deposits, preferentially preserved in anoxic environments. Carbon-rich sediments deposited between 120 and 90 Ma have been recovered in all ocean basins (Schlanger and Jenkyns, 1976; Jenkyns, 1980; Weissert, 1981) and point to a major mid Cretaceous anoxic phase in ocean history. In the South and Central Atlantic dense brines were formed in the early Cretaceous rifted basins. Roth (1978) suggested that these dense brines spilled over the sills from the evaporite basins and flowed down the continental slopes to form dense saline deep waters which were depleted in oxygen. Sequestering of oxygen

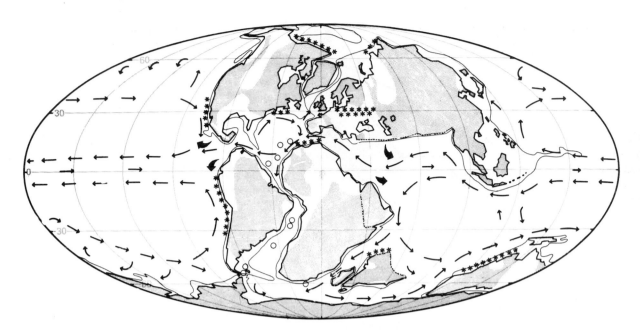

Figure 4. Inferred global paleocirculation patterns during the middle Cretaceous (ca 105-100 Ma). For explanation of symbols see the caption of Fig. 2. Open circles in the North and South Atlantic basins represent deep sea sites where early to middle Cretaceous black shales have been recovered. Upwelling features as shown here are from the Cenomanian (northern summers) patterns of Parrish & Curtis (1982). The mode of deep water formation is still in low and middle latitudes. The circum-global equatorial Tethys Current was operative by this time.

in gypsum may have led to a widespread oxygen minimum zone between 3 and 5 km depth in the Atlantic and Indian Oceans (Roth, *op. cit.*). The well-developed oxygen minimum zone may have persisted throughut the remainder of the Cretaceous (Douglas, 1982). However, Ceno-manian (97-91 Ma) varigated clays deposited over the black shales in the North Atlantic signal the return of oxygenated conditions in that basin (Cool, 1982). Reyment (1980a, b) pointed out that during a major mid Cenomanian to Turonian (ca 95-88 Ma) tectonoeustatic highstand of sea level, Africa was invaded by advancing seas both from the South Atlantic and the Tethys sides through the Benue Trough, the two limbs becoming connected briefly during early Turonian time. True oceanic conditions in the South Atlantic, however, may not have existed until Santonian time (Reyment, 1980a; Barron and Whitman, 1981). Rand and Mabesoone (1982) postulated that the connection between the Brazilia Bulge and Nigeria was not severed until the Maastrichtian (ca 74-66 Ma). They based their conclusions on the peculiarities of Bouguer gravity and vertical field magnetic anomalies. They also contended that there is no evidence of persistent faunal exchange between the North and South Atlantic until late Maastrichtian.

Thierstein (1979) has suggested that in the open ocean the timing of organic-rich intervals appears to be highly variable. This indicates oxygen-poor deep waters on a global scale through most of the Cretaceous, leading to enhanced organic carbon deposition only where local conditions permitted (Thierstein, 1983).

Figure 4 shows a mid Cretaceous (pre-Cenomanian) global reconstruction with inferred surface circulation patterns and potential areas of origination of WSBW in the low to mid latitudes and the inferred upwelling sites. The sites of early to middle Cretaceous organic-rich black shales in the Atlantic basins are also shown (open circles). By this time the open Central American seaway had led to the development of a vigorous circum-global Tethys Current and the North Atlantic gyre (proto-Gulf Stream) was also in place. The southern South Atlantic was connected to the Indian Ocean but the deeper parts still were isolated due to the blocking feature of the Falkland Plateau.

In the Pacific Ocean the northern and southern hemispheric gyres would have been well established, as was probably also the equatorial counter flow. Since the Indian Plate had not yet drifted north, a largely southern hemispheric anti-clockwise gyre is inferred to have characterized the unhindered eastern Tethys. India did not start it's active drift northwards until sometime in the Campanian (ca 83-72 Ma). Similarly, Australia and Antarctica did not begin separating until sometime before anomaly 34 time.

A reinterpretation of the older magnetic anomalies between Australia and Antarctica by Cande and Mutter (1982) has suggested that the anomaly formerly identified as 22 (Weissel and Hayes, 1972; Weissel et al., 1977) is actually anomaly 34. This would agree better with the timing of the initiation of spreading in the Tasman Sea (Hayes and Ringis, 1973), and between New Zealand and Antarctica (Weissel et al., 1977). Thus active spreading between Australia and Antarctica is more likely to have begun in the late Cretaceous (ca 90-85 Ma). The initial spreading phase was characterized by very slow rates. The spreading accelerated only after anomaly 19 time (ca 43 Ma) and Cande and Mutter (*op. cit.*) have pointed out that this timing corresponds to the timing of a major reorganization of plate boundaries in the Indian Ocean (Norton and Sclater, 1979), and to changes in the direction and rate of spreading in the North Pacific where the bend in the Emperor-Hawaiian chain has been dated at about 43 Ma (Clague and Jarrard, 1973).

Saltzman and Barron (1982) have suggested that in the late Cretaceous the deep water circulation may have been highly variable. Isotopic paleotemperature values of about 5-7°C for the deep water in the equatorial Pacific suggest a probable polar source for the bottom water in that basin. In other basins, however, high paleotemperature values (e.g., about 22°C for deep water in the Angola Basin) indicate that bottom water was still originating as WSBW in the temperate latitudes. The relatively more vigorous nature of the late Cretaceous bottom waters is also indicated by the dramatic overall decrease in organic-rich deposits testifying to a more oxygenated and dynamic bottom regime. In the Atlantic the widening of deep passages and the progressively improved connection of the southern basin to the north and to the Indian Ocean led to increased meridional circulation, which would eventually help induce the general cooling of the higher latitudes and enhanced vertical and horizontal thermal gradients (Roth, 1978).

THE CRETACEOUS/TERTIARY BOUNDARY EVENT

The end of the Mesozoic is marked by a major threshold event in earth history. At this dramatic boundary (between the Cretaceous and Paleogene periods, some 66 m.y. ago) a large percentage of the world's biota became extinct in a relatively short time. In the oceanic sections the Cretaceous/Tertiary (K/T) boundary is usually marked by the presence of clays or hard-grounds, and nearshore sections often show a prominent hiatus at the boundary.

In the oceans the groups that were significantly depleted by the K/T boundary event were: the phyto- and zooplankton, the ammonites and belemnites, bryozoans, shallow-dwelling echinoderms, corals, bivalves, larger benthic foraminifera and marine reptiles. (See also Thierstein, 1983, for a summary of the biotic reductions at the K/T boundary). Among the benthos, 20 species of brachiopods became extinct out of a late Maastrichtian assemblage of 26 species. Surlyk and Johansen (1984) noted that the surviving species included forms that could survive in well-aerated shallow water on substrate other than chalk. Deep water benthos were apparently not much affected at the K/T boundary. The nannoplankton species that survive into the Cenozoic are the ones that are adapted to stressful conditions and capable of building fortified skeletons (Percival and Fischer, 1977).

The land record has been recently re-interpreted, and shows a dramatic turnover in the terrestrial fauna (Smith and van der Kaars, 1984). Dinosaurs and flying reptiles that are commonly cited as the prominent casualties of the terminal Cretaceous event, on the other hand, were on a general decline through the late Cretaceous. The K/T boundary episode may have delivered a final *coup de grace* to an already waning lineage. Palynological data (Krassilov, 1979, vide Hallam, 1980) also suggests an abrupt change in the floras across the K/T boundary. There is no evidence, however, of the fresh water fauna being effected at this boundary.

In recent years many explanations have been offered for this catastrophic reduction of biota and the sharp environmental changes at the end of the Cretaceous. Both terrestrial and extra-terrestrial causes have been suggested. Marked climatic changes (both drastic cooling and heating), depletion of nutrients and collapse of the food chain, chemical poisoning of water and atmosphere, physical catastrophies, magnetic reversals, cosmic radiation and impact of large

Apollo objects have all been proposed as possible causes at one time or another (see Beland et al., 1977; Russell, 1977, for recent reviews). Here we will restrict ourselves to the discussion of two of the recently offered terrestrial and extra-terrestrial explanations.

Among the more intriguing terrestrial causes suggested in recent years has been the scenario of "spill-over" of relatively fresh waters from a previously isolated Arctic Ocean (Gartner and Keany, 1978; Thierstein and Berger, 1978). According to this scenario, the less dense, cooler Arctic water would have spread rapidly over the world ocean, causing mass mortalities in the open ocean plankton, disrupting the food chain, and eventually leading to detrimental climatic effects that would trigger a series of ecological disasters. Radical changes in the vegetation and faunal distribution on land are seen as a consequence. The paleontologic (Gartner and Keany, 1978) and isotopic (Thierstein and Berger, 1978) evidence that was brought to bear upon the Arctic spillover model has, however, since been refuted or retracted and the argument for a fresh water injection causing massive extinctions at the K/T boundary has lost much of its appeal.

Since it was first proposed by Alvarez et al. (1979, 1980) the extra-terrestrial cause or the impact hypothesis of the K/T boundary extinctions is becoming steadily more accepted by the earth science community. The discovery of enrichment over background level of Iridium and other noble elements at the boundary sections in Italy (Alvarez et al., 1980), [and later on from Denmark and Spain (Ganapathy, 1980; Smith and Hertogen, 1980)], led Alvarez et al. to propose an asteroidal impact as the probable cause for extinctions in the ocean and on land at the end of the Mesozoic. These authors maintain that the isotopic composition of Iridium suggests an extra-terrestrial source from within the solar system. The scenario proposed a post-impact distribution of pulverized rock in the atmosphere for several years, leading to both heating up and darkening of the atmosphere. This would have resulted in the suppression of photosynthesis and the collapse of the food chain leading to massive extinctions. Emiliani (1980) estimated that an impact of a relatively large Apollo object would trap enough heat in the atmosphere to raise the temperature of the surface of ocean and the lower troposphere by 5 to 10°C. Hsü (1980) favoured a cometary

rendezvous with the earth rather than an Apollo impact, that would have caused cyanide-poisoning of the oceans. [See also the recent symposium volume on geological implications of large body impacts on earth, Silver and Schultz (eds.) 1982].

Officer and Drake (1983) criticized the impact hypothesis on the grounds that a reconsideration of the biological record by them has revealed that faunal and floral changes across the K/T boundary were not abrupt but rather gradual. Alvarez et al., (1984b) refuted these criticisms and maintain that enough evidence has accumulated to suggest that the biotic changes at the boundary are very sharp and apparently synchronous in various parts of the world.

Since its first discovery, the Iridium anomaly has been recorded from the K/T boundary from over 50 sections around the world (Alvarez et al., 1984a), including several non-marine sections (Orth et al., 1981; Pillmore et al., 1984). The ratios among the Platinum-group elements and isotopic composition of the boundary clays (Luck and Turekian, 1983), their association with sanidine sperules that are interpreted as droplets from impact melt, and other theoretical considerations have reinforced the validity of the impact hypothesis (Alvarez et al., 1984a). A recent mineralogical study by Bohor et al. (1984) found quartz in the non-clay fraction at the K/T boundary in eastern Montana to contain features characteristic of shock metamorphism and were interpreted by the authors to be compatible with high velocity impact of a large body.

PALEOGENE PALEOCEANS

The modern vertical structure of the oceanic water masses, the associated circulation, and the predominantly glacial mode of earth history was first initiated in the late Paleogene. These features developed through several step-like transitions (Kennett, 1978; Berger, 1979) into the climatic-oceanographic patterns typical of today. Significant changes occurred during the early Paleogene (ca 66-45 Ma) which was a transitional phase between the predominantly thermospheric circulation of the Mesozoic and the thermohaline circulation which evolved in the late Paleogene and Neogene (see Haq, 1981). In the early Paleogene the higher latitudes began to cool down developing steeper latitudinal thermal gradients, accentuating seasonality, and eventually

MAJOR EVENTS

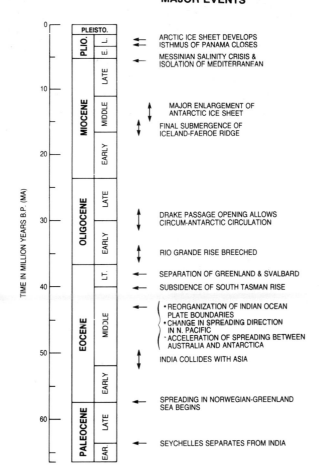

Figure 5. The timing of the major paleogeographic-oceanographic events of the Tertiary, plotted against the Cenozoic time scale recommended by the Decade of North American Geology (DNAG) which was adapted from Berggren et al. (1984).

developing steeper vertical thermal gradients as well when cold bottom waters began forming in the refrigerated higher latitudes in the later Paleogene. The development of the psychrosphere and thermohaline circulation were probably the most significant events in the Cenozoic history of the world ocen (Hsü et al., 1984).

The overall history of the Cenozoic world ocean can be said to have been marked by a long-term regression of the inland and coastal seas (see the section on Sea Level Changes), ice buildup on polar regions and increased albedo that promoted further cooling (Berger, 1982). The ice buildup may have been partly favoured by the more poleward position of the land masses and the thermal isolation of Antarctica (Kennett, 1975).

Figure 5 summarizes the major paleo-geographic-oceanographic events of the Tertiary plotted against a recently revised Cenozoic time scale.

The sedimentary record of the mid Paleocene to early Eocene (ca 60-50 Ma) in the northeastern Atlantic indicates that by this time a relatively calm regime of predominantly calcareous sediments typical of the late Cretaceous was changed, and extensive erosion and redeposition occurred indicating convective overturn and dynamic bottom currents (Berggren and Hollister, 1977). Further north in the Norwegian Sea area there is magnetic evidence of the initiation of seafloor spreading and opening of the Norwegian-Greenland Sea sometime between magnetic anomaly 24 and 25 times (ca 57 Ma) (Talwani and Udintsev, 1976; Talwani and Eldholm, 1977).

In the early Paleogene for the first time a deep connection between the North and South Atlantic may have developed (Berggren and Hollister, 1977). But in the earliest Paleocene the South Atlantic was still characterized by little thermohaline flow at deeper levels and the bottom waters were relatively warm (10-12° C). Surface water temperatures ranged between 12-20°C and the subtropical gyre and associated thermal gradients were prominent features (Johnson, 1983; Hsü et al., 1984). Changes in the deep water structure of the South Atlantic, as indicated by extensive erosion, first occurred in the late Paleocene (ca 60-57 Ma) (Moore et al., 1984).

The Paleocene eastern Tethys was characterized by an active 90-East Ridge, the reorganization of northwestern Indian Ocen when Seychelles separated from India (Northern and Sclater, 1979), and the fast converging Indian Plate towards Asia. A patchy sedimentary record in the area points to considerable bottom water activity (Kidd and Davies 1978).

Figure 6 shows a late Paleocene world ocean with inferred circulation patterns. The most likely source areas of deep waters were probably still in the temperate and low latitudes, although it is likely that even in the late Cretaceous the Pacific deep waters may have been of mixed high and low latitude origin (Saltzman and Barron, 1982). The most prominent feature of surface circulation in the Paleocene was the Tethys Current which dominated the circum-global tropical circulation. Anti-clockwise gyres are inferred to have existed in the southern hemispheric basins. In the Indian

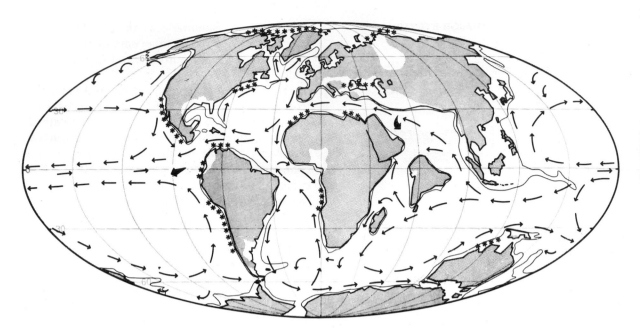

Figure 6. Inferred global paleocirculation in the early late Paleocene (ca 60-58 Ma). For explanation of symbols see the caption of Fig. 2. The upwelling patterns for this time-slice are inferred from the patterns for Maastrichtian (latest Cretaceous) and Lutetian (middle Eocene) of Parrish and Curtis (1982). By this time the mode of deep water formation in the lower latitude was relatively less important than earlier time intervals. The Tethys Current was a dominant feature of the early Paleogene surface circulation.

Ocean, however, south of the still drifting Indian Plate, the cell may have been deflected southwestward, hugging the Antarctica-Australia margins during the return flow eastward. Drake Passage was still closed but Australia had already separated from Antarctica which may have allowed part of the return flow to be deflected into the passage between the two continents, to be stopped by the Tasman Rise, which did not subside till late Eocene (Kennett, 1975, 1977). In the absence of the circum-Antarctic flow a subpolar cell may still have existed in the South Pacific. In the North Pacific both the tropical (and the attendant equatorial counter flow) and subpolar cells are inferred to have been well established. A single gyre analogous to the Gulf Stream dominated the scene in the North Atlantic, with probable inflow towards Labrador and Norwegian Seas. During the Paleocene the southern Labrador passage continued to widen and the northern passage had begun opening (Laughton, 1971; Gradstein and Srivastava, 1980).

The reduction of total continental area in the Pacific and the general increase in pole to pole asymmetry in the early Tertiary may have led to the increased importance of southern high latitudes as the source for deep waters (Schnitker, 1980b). The role of high latitudes of southern

Indian and Atlantic Oceans in the production of deep waters was also enhanced in the early Tertiary, concomittent with the decrease of shallow marginal seas in the low to mid latitude net-evaporation zone following the general regression of this time (Brass et al., 1981).

Both the plankton paleobiogeographic record (Haq et al., 1977; Haq, 1981) and oxygen-isotopic data (see e.g., Savin, 1977; Vergnaud-Grazzini et al., 1978, Buchardt, 1978) are compatible with a general warming trend in global temperatures through the Paleocene (see Fig. 7), culminating in a period of peak warming in the early Eocene (between 57 and 52 Ma), which by all indications was the warmest interval of the Cenozoic (Haq, 1981). Shackleton and Boersma (1981) calculate that the thermal gradient in the later early Eocene was less than half of its present value. Oceanic surface temperatures of 10°C at high latitudes and 20° in the tropics suggest that oceanic circulation contributed a greater proportion of the heat transport to high latitudes (Shackleton and Boersma, *op. cit.*). During this time a widespread transgression apparently enlarged the ecospace for marine organisms, which show generally high diversities in the early and middle Eocene. Increased productivity caused high carbonate and total sediment accumulation rates in the later part

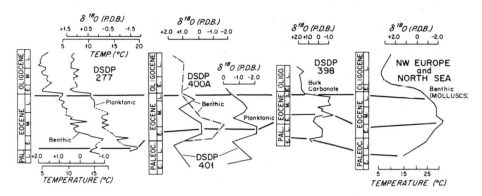

Figure 7. Paleogene oxygen isotopic data from various regions. Planktonic and benthic δ¹⁸O curves from DSDP Site 277 (Campbell Plateau, Southwest Pacific) after Shackleton and Kennett (1975), data from Sites 400, 401 and 398 (North Atlantic Ocean) after Vergnaud-Grazzini et al. (1978); and Mulluscan data from northwestern Europe and the North Sea, after Buchardt (1978). Major climatic events (i.e. cooling in mid-Paleocene; marked warming in late Paleocene-early Eocene; and sharp cooling near the Eocene/Oligocene boundary) are indicated by the connecting horizontal lines. (Figure from Haq, 1981).

of early Eocene and middle Eocene (Davies et al., 1977). Oceanic sections show a low rate of erosion and redeposition in the early Eocene (Moore et al., 1978) and suggest relatively tranquil bottom water regime.

Terrestrial floras and fauna corroborate the peak warming of the early Eocene and also suggest that warm tropical-subtropical belt expanded to double its present latitudinal extent (Wolfe, 1978). The Arctic island of Ellesmere has yielded a rich early Eocene vertebrate fauna, suggesting a temperature in the range of 10-12°C (Estes and Hutchinson, 1980; McKenna, 1980). This presence of temperate floral and faunal elements as far north as 60° N, has been used as an argument to suggest that a very low angle of inclination of the earth's rotational axis in the early Paleogene may have been the causal factor for their existence (Wolfe, 1978). Donn (1982) has suggested that part of the problem of interpreting high latitude climates of the distant past, such as those in the Cretaceous and early Paleogene, may lie in erroneous assignments of high paleolatitude, based on pole positions determined from rock magnetism which may have gross errors due to long-term bias between rotational and diapole locations. Axelrod (1984) has, however, made a fairly convincing case for the non-necessity of invoking the lack of tilt in earth's axis for the existence of rich biota in high Arctic and Antarctic latitudes in the late Cretaceous and early Tertiary. Instead, the mild, equable, polar climates and the well-adapted physiological responses of the plants and animals of those times to the local conditions may have been enough to warrant their existence (Axelrod, op. cit).

The isotopic data shows that after the early Eocene climatic amelioration, there was a long term cooling trend through the remainder of the Eocene, superimposed on marked cooling events in the middle and late Eocene (Fig. 7). A middle Eocene cooling episode is also indicated by paleobiogeographic data (Haq et al., 1977). This is followed by a generally stable climatic phase, before the sharp cooling event in the latest Eocene and early Oligocene.

The presence of a widespread late early Eocene cherts (radiolarite) horizon in the North Atlantic, hich is a part of the "Horizon A-complex" (now identified as Horizon Ac on seismic profiles) has been variously equated with increased silica input through volcanism (Gibson and Towe, 1971), high surface productivity and silica preservation (Herman, 1972) and increased upwelling at low latitudes and increased productivity, as well as increase in the vigor of bottom currents (Berggren and Hollister, 1974). Detailed study of the "Horizon A-complex" by Tucholke (1979) suggested that chertification, volcanism and erosional events in the North Atlantic happened at different times in the middle and late Eocene and can be picked as discrete seismic horizons. Horizon Ac is likely to have been associated with a period of high productivity of siliceous plankton and subsequent preferential diagenesis to chert in the presence of high carbonate content (Lancelot, 1973; Tucholke, 1979).

In the South Atlantic, the middle Eocene (ca 52-40 Ma) is marked by the rise of the calcium carbonate compensation depth (CCD) (Hsü et al., 1984; Moore et al., 1984). The later middle

Eocene and late Eocene of the Pacific also shows extensive carbonate dissolution, high erosion rates, and the narrowing of the equatorial carbonate accumulation zone (van Andel et al., 1976).

In the Indian Ocean there are indications that the Indian Plate may have arrived at a subduction zone in the Tethys Seaway sometime in the latest Paleocene to early Eocene (Laughton et al., 1973; Curray and Moore, 1974). The collision of India with Asia occurred near the early Eocene-middle Eocene boundary (around anomaly 22 time, ca 53-50 Ma) (Johnson et al., 1976; McGowran, 1978). The violent basaltic hydroexplosions along the 90-East Ridge that have been recorded by Fleet and McKelvey (1978) for this time may have been witness to this great tectonic event and its repercussions.

The position of India in the middle Eocene had not yet severely restricted the flow of the Tethys Current north of the Indian Plate, especially during the generally high seastands of late early and early middle Eocene times when extensive epeiric seas were present along the Tethys margins. The Tethys Current continued to use the northern passage until after the early middle Eocene, when the general drop in sea level and further movement of India closer to Asia, reduced the passage more severely and the main flow shifted to the west of the Indian Plate. This is evidenced by the presence of extensive neritic and marginal marine sediments of middle and late Eocene age in Pakistan (McGowran, 1978) and the Middle East. The Tethys flow became sharply reduced and intermittent and may have become restricted to a narrow southwestern passage by the early Oligocene (ca 35 Ma).

In the North Atlantic the widespread erosional/non-depositional interval between middle Eocene and early Miocene suggests that availability of abyssal waters from a northern source and the circulation of the North Atlantic deep water (NADW) may have begun (Ewing and Hollister, 1972) concomitantly with the initial subsidence of Greenland-Faeroe Ridge in the late Eocene (between 40 and 37 Ma). This episode of subsidence may have caused the Ridge to submerge temporarily [where major subsidence did not occur till Miocene (Talwani and Eldholm, 1977)], and resulted in the introduction of higher salinity waters to the intermediate and abyssal depths of the Central Atlantic. As suggested for

the Miocene by Schnitker (1980a), this deep water may have conceivably ended up in the South Atlantic along the Antarctic margins, to be mixed with cold waters to form dense, cold, bottom water (Johnson, 1983), and may have provided the extra moisture needed for the accumulation of ice on Antarctica. By this time the subsidence of the South Tasman Rise permitted a free connection between Indian and Pacific Oceans through the passage between Australia and Antarctica (Kennett et al., 1975). Combined with the greater availability of moisture from North Atlantic sources, the partial isolation of Antarctica may have allowed large scale freezing at sea level and spurred the initiation of Antarctic bottom water (AABW). The onset of cold bottom water is marked by enrichment of benthic $\delta^{18}O$, first documented by Savin et al. (1975) and Shackleton and Kennett (1975). A drop of 4-5° C in bottom water temperature reflects the development of the psychrosphere (Benson, 1975) near the Eocene/Oligocene boundary (at about 36-37 Ma). This event also manifests itself in the presence of widespread erosional hiatuses in the eastern Indian and southwestern Pacific Oceans (van Andel et al., 1976; Moore et al., 1978) where scouring by cold bottom waters has stripped much of the older sediments. The addition of more oxygenated (and thus less acidic) waters to the bottom regime may have been responsible for a lowering of CCD (Berger, 1973) that has been documented for the latest Eocene and Oligocene in the equatorial Pacific (van Andel et al., 1976) and South Atlantic (Hsü et al., 1984; Moore et al., 1984). Overall, there has been a dramatic global increase in the area of deserts over the past 40 million years (i.e. since the late Eocene) and this is ascribed to the cooling higher latitudes and the eventual development of the polar ice caps (Barron & Whitman 1982).

In the North Atlantic the Eocene/Oligocene boundary event is marked by the presence of a widespread hiatus along the basin margins, recognized as the R4 reflector (Miller and Tucholke, 1981). The boundary is also marked by a change form predominantly agglutinated to calcareous benthic foraminiferal assemblages at a site in the southern Labrador Sea (Miller et al., 1980). Miller and Tucholke (1981) consider the Arctic to be the main source of bottom water for this area. By late Eocene (ca 38 Ma) the connection between the Arctic and the Norwegian

Sea was established by the separation of Greenland and Svalbard (Talwani and Eldholm, 1977). Current-controlled sedimentation and erosion continued in the North Atlantic through the Oligocene (Miller and Tucholke, 1981).

The late Eocene also saw rapid declines in both the phyto- and zooplankton diversities (see, Haq, 1973; Fischer and Arthur, 1977; Haq, 1981). Corliss (1979) has indicated a gradual change in the benthic foraminiferal assemblages at a southwestern Pacific DSDP Site during the late Eocene, rather than an abrupt change at Eocene/Oligocene boundary. The same observations and conclusions were drawn by Parker et al. (1984). As at the K/T boundary, anomalously high Iridium concentrations have been recorded from several Eocene/Oligocene boundary sections (Ganapathy, 1982; Alvarez et al., 1982). In some deep sea sections the Iridium anomaly has been reported to correlate with a deep sea microtektite horizon which is considered coeval with North American tektite strewn field (Glass and Crosbie, 1982), leading to the conclusion that the event was caused by a asteroidal impact, analogous to the event at K/T boundary (Alvarez et al., 1982). Recently, Keller et al. (1983) recovered microtektites from three discrete levels in the middle Eocene to middle Oligocene deep sea sections and they found no evidence of mass faunal extinctions associated with any of these horizons. Instead, the authors maintain, all marked faunal changes in this interval can be related to well documented oceanographic phenomena, and no extra-terrestrial impact need be conjured to explain changes in the biologic realm during that time.

There is considerable evidence that the vigorous cold bottom waters that first became important in the late Eocene continued to be strong through most of the Oligocene as well. The separation of Svalbard and Greenland in the late Eocene (Talwani and Eldholm, 1977) ensured the continued availability of higher source of cold, deep water to the North Atlantic. The activity of deep water is indicated by the presence of widespread drift sediments in the North Atlantic. The influence of AABW that flowed through the newly breached Rio Grande Rise in the early Oligocene, is seen through the accumulation of reworked sediments in the South Atlantic (Johnson, 1983). In the Indian Ocean the presence of important Oligocene regional hiatuses can also be ascribed to the activity of AABW which may have developed a strong western boundary undercurrent by that time (Luyendyk and Davies, 1974). In the Pacific the carbonate dissolution at depth increased once again during the later part of early Oligocene, although the carbonate supply in the equatorial Pacific remained high (van Andel et al., 1976).

A paleogeographic event of major significance for the overall Cenozoic oceanographic evolution was the breaching of the straits between Antarctica and South America at the Drake Passage, sometime in the mid Oligocene. The Drake Passage connection was well-established by sometime prior to magnetic anomaly 8 time (ca 29 - 28 Ma) - the oldest coherent anomaly identified in the area (Barker & Burrell, 1977; Kennett, 1977). The cross-section of the Passage grew rapidly after 23.5 Ma (Barker & Burrell, 1977). However, the surface connection at the Passage may conceivably have already been initiated in the early Oligocene. This event led to the enhanced thermal isolation of Antarctica and the further development of the ice-cap, as well as the establishment of circum-Antarctic Current (Kennett, 1975, 1977).

By the mid to late Oligocene (30-24 Ma) time the global surface circulation patterns had essentially evolved the major features of the modern oceans. Figure 8 summarizes the inferred hydrographic patterns of the mid-Oligocene time-slice which look substantially different than those in the Paleocene and much more like the modern ones. The major differences from the earlier patterns is the development of the circum-Antarctic Current and the elimination of the southern hemispheric cyclonic gyres, and the near-cessation of the Tethys Current following the initial uplift of Himalayas. As the stratigraphic record indicates a narrow Tethyan passage may have allowed intermittent connections of the Indian Ocean with the Paratethys (the restricted west-central basin of the old Tethys), particularly during sea level highstands. An open Panama Isthmus still permitted the exchange between Central Atlantic and the Pacific Oceans.

The ice shelves around Antarctica were the dominant sources of cold, dense bottom waters. Northern sources of deep waters may also have been active, as was the exchange North Atlantic surface waters with the Arctic through Norwegian-Greenland and Labrador passages. There may

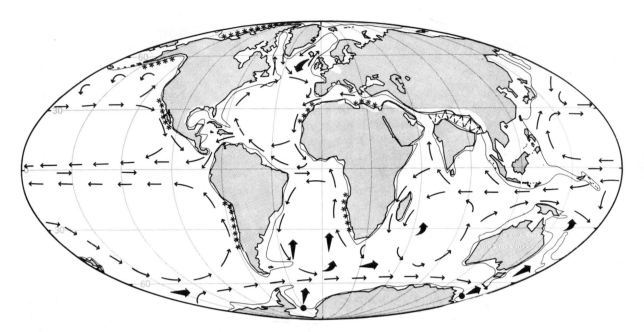

Figure 8. Inferred global paleocirculation during the mid-Oligocene (ca 31-30 Ma). For explanation of symbols see the caption of Fig. 2. Eastern Tethys is now severely restricted following the junction of India with Asia in the Eocene and the Tethys Current has ceased to be circum-global. The circum-global circulation has shifted to southern high latitudes, with opening of the Drake Passage and development of the circum-Antarctic Current. By this time the source areas for deep water formation have shifted completely from lower latitude shallow seas to higher latitude, mostly around Antarctica (Ross and Weddell Sea shelves). The source and initial pathways of deep water are shown with the larger arrows. Upwelling patterns shown here are inferred from the Lutetian and middle Miocene pattern of Parrish and Curtis (1982).

have been some return flow form the Arctic to the Atlantic via the Labrador Sea during late Oligocene (Gradstein & Srivastava, 1980).

Clark (1982) has suggested that indirect evidence indicates initiation of ice-cover in the formerly ice-free Arctic, sometime in the mid Cenozoic (Oligocene). By this time the Central Arctic was nearly land-locked and the connections to the open ocean were restricted enough that the Eurasian river input of fresh water, combined with salinity stratification in the restricted basin may have allowed first winter-ice to form. This, however, did not effect the deeper Arctic waters, as the presence of benthic faunas of up to late Miocene in this basin would indicate (Clark, *op. cit.*)

More evidence for an important mid Oligocene sea level fall (the major 30 Ma sea level drop of Vail et al., 1977) has recently come from the changes in benthic assemblages observed at Walvis Ridge in the South Atlantic (Moore et al., 1984) and from stable isotopic studies (Miller and Fairbanks, 1983; Keigwin and Keller, 1984). The isotopic studies offer the suggestion that only the accumulation of a significant amount of polar ice can explain the prominent benthic $\delta^{18}O$

enrichments in the mid Oligocene, which in turn could explain the major sea level fall indicated by Vail et al. (1977).

Thus it would seem that the Oligocene may have been the most influential epoch in revolutionizing the global oceanographic-climatic conditions. The most prominent change was the shift of the circum-global circulation (and the major means for biotic dispersal) from the equatorial regions to the southern high latitudes. Both the restriction of the main Tethyan flow from the Indian to the Atlantic Ocean and the initiation of the circum-Antarctic flow after the development of the Drake Passage connection occurred during the Oligocene. The repercussion of these changes have already been outlined above, however, the importance of these events to the development of modern oceanographic-climatic conditions and the entry of the earth into a predominantly glacial mode needs to be underscored. The shift from a predominantly non-glacial to glacial mode occurred when the magnitude of transgressions over the continents during high seastands had been considerably reduced, eliminating vast areas of epicontinental seas. The shift accentuated seasonality by

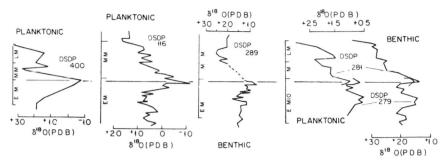

Figure 9. Miocene oxygen isotopic curves from DSDP sites from the north Atlantic (Site 400 after Vergnaud-Grazzini et al, 1978, and Site 116 after Blanc et al., 1980) and the Southwest Pacific (Sites 279 and 281 after Shackleton and Kennett, 1975). The horizontal line connects the late early Miocene warming peak that separates cooler early and middle Miocene intervals.

increasing the land-sea and equator-polar thermal contrasts, eventually leading to the development of a permanent ice-cap on the already thermally isolated Antarctic continent. Prior to the Oligocene the storage of water on the ice-cap may have been more episodic in nature, i.e. significant melting may have occurred during the warmer episodes.

NEOGENE PALEOCEANS

The Neogene is characterized by the accentuation of oceanographic-climatic patterns that had been initiated in the mid Cenozoic. During the early Miocene (ca 24-17 Ma) the convergence of Eurasia and Africa finally interrupted much of the periodic influence of the Indian on the Mediterranean that had continued intermittently through the Oligocene (Rögl and Steininger, 1983). By the early Miocene the major geographic features of the Indian Ocean were already in place and the sedimentary patterns had begun to resemble those of today. The Neogene sedimentary record in the Indian Ocean is essentially complete, indicating relatively stable bottom conditions (Kidd and Davies, 1978).

The final closure of Tethyan flow through the Mediterranean into the Atlantic is inferred to have significantly modified the flow of the eastern limb of proto-Gulf Stream by deflecting it southward (Berggren and Hollister, 1974). Paleogeographically, the shape of the North and South Atlantic has changed little since magnetic anomaly 6 time (early Miocene) (Sclater et al., 1977), with the exception of the Central American connection to the equatorial Pacific that existed until mid Pliocene. The general deep circulation patterns in the Atlantic had already become established in the mid Cenozoic with the initiation of AABW and proto-NADW.

Neogene stratigraphic and oxygen-isotopic record of the east Central Atlantic, off Northwest Africa shows evidence of considerable climatic fluctuations leading to the present (Sarnthein et al., 1982). Each of the six major climatic deteriorations identified in the Neogene of this area have been found to be associated with enhanced surface circulation, deep sea erosion, and glacial-type meriodinal winds that resulted in greater aridity over North Africa. Two of these events (in the middle Miocene and mid Pliocene) are also seen to be associated with intensified river discharge, increased upwelling and enhanced fertility (Sarnthein et al., *op. cit.*).

The available oxygen isotopic data and paleobiogeographic data show a remarkable agreement in the interpretation of Miocene paleoclimates (see Haq, 1981). The earliest Miocene (ca 24-23 Ma) was an interval of generally cool climates, followed by a climatic amelioration between 22 and 21 Ma (Fig. 9). This was once again followed by a decline in temperatures in the late early Miocene (ca 20-17 Ma) A second climate optimum occurred around 16 Ma, that was in turn followed by a well-documented rapid decline in paleotemperatures between 15 and 14 Ma. This climatic deterioration is generally associated with the development, or a major enlargement, of the East Antarctic ice-sheet during the middle Miocene (Shackleton and Kennett, 1975; Savin, 1977; Woodruff et al., 1980).

Savin et al., (1975) have pointed out the characteristic divergence between the tropical planktonic and benthic $\delta^{18}O$ curves since the middle Miocene as signaling the steepening of thermal gradients, enhancement of the tropical thermocline and intensification of the abyssal circulation.

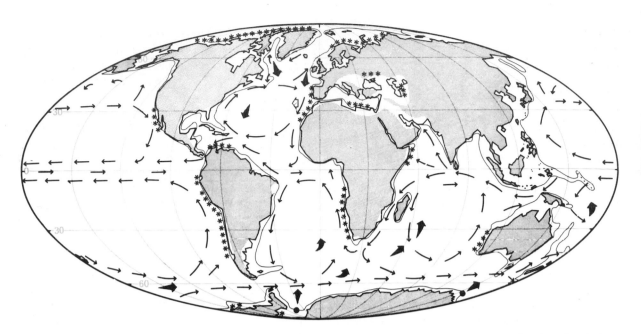

Figure 10. Inferred global paleocirculation patterns for a late early Miocene time-slice (ca 17-16 Ma). For explanation of symbols see the caption of Fig. 2. (The outflow of Mediterranean intermediate water into the Atlantic is shown by the large hatchered arrow). Both surface circulation and loci of formation of polar bottom water look much like their modern counterparts, with the exception of an open Central American Passage. During this time of generally high seastands the Mediterranean may have been temporarily reconnected to the open ocean through the east. The upwelling features shown are those for the middle Miocene (northern summer) by Parrish and Curtis (1982).

The major enlargement of the ice-cap in middle Miocene has been linked to the increased NADW flow (Schnitker, 1980). Schnitker argued that the final submergence of the Iceland-Faeroe Ridge (Talwani and Eldholm, 1977) resulted in the enhanced flow of NADW, which traversed the Atlantic, upwelling off Antarctica and providing the excess moisture needed for an extensive ice accumulation on Antarctica.

The isotopic and paleobiogeographic data show two more episodes of climatic amelioration in the late middle and middle late Miocene (between 13 and 12 Ma and around 8 Ma), which were followed by another decline in isotopic paleotemperatures in the latest Miocene (between 6.5 and 5 Ma) (see Haq, 1980, for further discussion).

The middle Miocene inferred circulation patterns are shown in Figure 10. By this time the eastern limb of the Tethys was closed, but a Langhian (mid Miocene) sea level highstand may have reestablished the Indo-Paratethyan link temporarily (Rögl and Steininger, 1983). The Paratethyan sedimentary record indicates yet another such reconnection possibly in the late Badanian (around 13.5 Ma) time of sea level highstand. The Indo-Pacific link was finally

severed in the late Miocene (between 12 and 11 Ma) following a worldwide sea level fall (Rögl and Steininger, op. cit.).

The sources of NADW from the norhtern high latitude basins and AABW from Ross and Weddell ice-shelves had become well established by the middle Miocene (Fig. 10). The intermittent supply of Mediterranean intermediate water over the Gibralter Straits had probably already begun by middle Miocene. The northern and southern hemisphere gyres in the Atlantic and Pacific looked much like their modern analogs. The Central American opening still provided a connection between the Central Atlantic and the equatorial Pacific.

In the southern high latitudes a major change in oceanic circulation occured in the late Miocene when polar front first migrated to northerly latitudes of the southern ocean and surface temperatures much like today became established (Ciesielski et al., 1983). There is considerable evidence for increased AABW activity and extensive erosion when West Antarctic ice sheet was first formed in late Miocene. Ice-sheets formed in western Antarctic ice shelves were still unstable, however, because they became frequently ungrounded. By late Miocene and early

Pliocene western Antarctic ice-sheet had become stabilized and ice-rafted debris accumulation increased (Ciesielski et al., *op. cit.*).

In the late Miocene (Messinian, ca 5.8 to 5 Ma) the Mediterranean suffered the now well-documented "Salinity Crisis" (Hsü et al., 1973; Cita, 1976) following a sea level fall and the isolation of the Mediterranean basin. This sea level fall is considered a major global event (see e.g., van Couvering et al., 1976; Adams et al., 1977) and may have been caused by yet another major growth in the volume of the Antarctic ice-cap (Shackleton and Kennett, 1975). The isolated Mediterranean developed high salinities and volumnous amounts of evaporites [estimated to be $10^6 km^3$ (Wright and Cita, 1979)] were deposited within a relatively short time. The Mediterranean was reconnected to the open ocean in early Pliocene (ca 5 Ma) when cold deep waters suddenly reentered over the Gibralter Sill (Benson, 1976).

The late Miocene carbon-isotopic shift (around 6.2 Ma) that preceded the Messinian Salinity Crisis and has been recorded from widely different areas of the Indo-Pacific (see e.g., Bender and Keigwin, 1979; Loutit and Kennett, 1979; Vincent et al., 1980; Haq et al., 1980) is now believed to have been caused by an increased flux of organic matter depleted in ^{13}C (Loutit and Keigwin, 1982).

The sea level fall that caused the Messinian Salinity Crisis was also responsible for widespread cooling at the surface and increased bottom water activity, as well as greater fertility (Kennett et al., 1979; Keller 1980, Ciesielski et al., 1982; Vincent et al., 1982). This sea level fall also resulted in increased biogenic productivity and high sediment accumulation rates (Davies et al., 1977).

The Messinian sea level lowering changed the Black Sea into an alkaline lake (Hsü and Giovanoli, 1979). This "Lago Mare" phase was short-lived (some 100 K. yrs.), and connection to the Mediterranean was reestablished in the earliest Pliocene during a transgression. Soon afterwards this connection was once again severed, followed by desalination of the Black Sea which converted it into a fresh water lake. This isolated fresh water lake phase lasted until late Quaternary, when the connection to the Mediterranean was once again established through the Bosporus Straits (Hsü and Giovanoli, *op. cit.*), ushering in the present day conditions.

The inferred paleocirculation patterns near the Miocene and Pliocene boundary (ca 5 Ma) are shown on Figure 11. It is noticeable that the major surface circulation patterns and sources of bottom water supply at this time as well as much of the inferred upwelling sites look much like their modern counterparts. The only major difference is the still-existing connection between the Central Atlantic and the equatorial Pacific at the Panama Straits. Tectonic considerations (Malfait and Dinkelman, 1972) and isotopic studies (Keigwin, 1982) indicate that the connection was operative until mid Pliocene and closed around 3 m.y. ago. This event may have led to a more vigorous Gulf Stream due to deflected energy. Once major development of the Arctic ice cap and the formation of cold Labrador Current following the elevation of the Panama Isthmus was completed, the Gulf Stream may have displaced to its present position (Berggren and Hollister, 1977). The present day circulation patterns in the Caribbean also date back to the closing of the Isthmus of Panama (Keigwin, *op. cit.*).

The last major paleoceanographic-climatic event discussed here is the development of an ice sheet in the Arctic region and its important implications. As mentioned earlier, there is some circumstantial evidence that the Arctic ice cover was already in place in the mid Cenozoic (Clark, 1982). However, there is no unequivocal evidence of any significant ice in the Arctic before mid Pliocene. Recently Backman (1979) has redated the initial ice-rafting event in some of the North Atlantic DSDP sites. His study shows that significant ice-sheet probably did not develop before 2.5 Ma. Although the oxygen-isotopic record would indicate somewhat earlier development of northern hemispheric ice, around 3.2 Ma (Shackleton and Opdyke, 1977; Prell, 1982), the ice-rafted material prior to 2.4 Ma is negligible and may be due to glaciation unrelated to major ice-sheets (Kennett, 1983). Late Pliocene refrigeration and related cooling of surface waters has been reported form wide areas, e.g. in the Mediterranean (Thunell, 1979) in the southern oceans (Kennett, 1979).

A considerable amount of research has gone into the study of Pleistocene biogeographic patterns, paleoclimatology, and paleo-ceanography. Some of these studies include sophisticated reconstructions of the climatic conditions of the past million years and their causal relationships to earth's orbital parameters (see

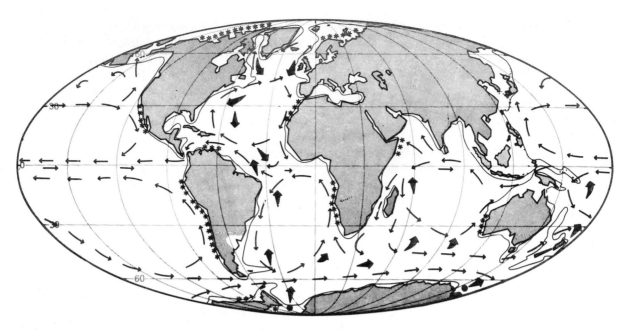

Figure 11. Inferred global paleocirculation just prior to Miocene/Pliocene boundary (ca 5.5-5 Ma). For explanation of symbols see caption of Fig. 2. Patterns are essentially similar to those for the late early Miocene time-slice (compare with Fig. 10). However, during this time of low seastands the Mediterranean was isolated (Messinian Salinity Crisis) and shallow epeiric seas were at their minimum extent. Bottom waters were active and had developed most of the major features of modern circulation. The persistent upwelling loci are simulated form the mid-Miocene patterns (Parrish and Curtis, 1982).

e.g. CLIMAP, 1976; Hays et al., 1976). New insights into the mechanism of glacial and interglacial changes have been provided by detailed biogeographic studies and their spectral analyses (see e.g. Ruddiman and McIntyre, 1979, 1981). The subject of Quaternary paleoceangraphy and climatology is so well-studied and vast that a meaningful summary of the field would be beyond the scope of this overview. The reader is referred to the papers cited above and to the volume edited by and Cline and Hays (1976) for further perusal of the subject.

SEA LEVEL FLUCTUATIONS AND THEIR ENVIRONMENTAL RESPONSE

The stratigraphic record reveals that marine transgressions and regressions have been widespread throughout the Phanerozoic and that many of these events were apparently synchronous over widely separated areas and were therefore, global in extent. Sea level fluctuations have important implications for productivity and sedimentation patterns on the continental margins and the knowledge of their timing, extent and magnitude are of primary interest in hydrocarbon exploration. Paleoceanographic studies suggest

that sea level changes have been one of the more important modulators of earth's climates and past oceanographic conditions, including oceanic sedimentation and dissolution patterns. These environmental parameters in turn could be important in controlling the pathways of biotic evolution. In this section our knowledge of the sea level fluctuations, their probable causal relationships and a historical review of their study is briefly outlined.

Synchroniety over wide areas of episodes of deposition and non-deposition in response to changing sea levels was noted by Suess as early as in 1906. Modern ideas concerning sea level fluctuations as inferred from alternating marine and non-marine lithofacies along the continental margins and interiors can be said to have begun with Hallam (1963), who related Tertiary eustatic fluctuations to changes in the volumetric capacity of the ocean basins. In this paper, written before the concept of plate tectonics, Hallam contended that since the Cretaceous time the average continental relief has undergone a progressive increase due to epeirogenic movements that effected the continents and the ocean margins in the opposite way. In support Hallam cited King (1950) who had inferred a long period of late

Cretaceous and early Tertiary pediplanation, followed by a mid Tertiary uplift producing the striking high plateaux of Africa, South America, India and Australia. This was once again followed by a period of pediplanation and several cycles of uplift and erosion. King (1957) also demonstrated a remarkable parellelism in the Tertiary history of Brazil and southern and central Africa, including an episode of late Oligocene uplift which produced some 3000 feet high surfaces in both regions.

According to Hallam (1963) the progressive overall emergence of the continents during the Tertiary is contrasted by the oceanic subsidence, as evidenced by the presence of guyouts between 500 and 850 fathoms in the Pacific and Atlantic oceans. Hallam also cited the widespread indications of downflexure of the shelf and subsidence of the continental margins along the U.S. East Coast, east coast of Greenland, west coast of Africa, Australia and other regions. He went on to suggest that perhaps the marginal subsidence was episodic, much of it taking place in the late Tertiary and Pleistocene which were also times of intense epeirogeny on the continents.

Using the global paleogeographic atlas of Termier and Termier (1960) and other regional studies, Hallam (1963) estimated that in the late Cretaceous (period of maximum worldwide transgression) some 50% of the present continental area of the U.S. was covered by inland seas, whereas during the Tertiary it never exceeded 10% and averaged only about 3%. Similar major withdrawals of seas since the Cretaceous can also be inferred for other parts of the world. Hallam went on to document the Tertiary transgressive-regressive cycles in the northwestern European basins, the U S Gulf Coast, the Indian subcontinent and elsewhere, and concluded that there was a worldwide synchronism of the major transgressive and regressive events. He made a rough estimate of about 250m of overall sea level fall since the late Cretaceous. He concluded that the volumetric capacity of the world ocean has increased since the Cretaceous, both due to continental emergence and seafloor subsidence, leading to the progressive overall regression of the sea from the continents, interrupted by a series of lesser transgressions resulting from displacement of seawater due to local uplifts of parts of the ocean floor.

Once the implications of plate tectonics and seafloor spreading had become better understood, Hallam (1971) went on to add that the changes in the volume of the ocean basins could be explained by changes in the rates of seafloor spreading. He suggested that the late Cretaceous highstands may have been caused by rapid spreading rates along the mid-ocean ridges. The new hot crust formed at the ridge axes is elevated and as it ages and cools, it subsides, producing a predictable age-depth relationship (Sclater et al., 1971). This means that the volume of the ridge is determined by its spreading rate and its length. Hays and Pitman (1973) computed the volumes of the mid-ocean ridges for several intervals in the Cretaceous and Tertiary, basing their calculations on the standard seafloor age-depth curve of Sclater et al., (1971) and the spreading rates as deciphered from magnetic lineations by Larson and Pitman (1972). They demonstrated the validity of Hallam's hypothesis and concluded that the widespread late Cretaceous transgression was caused by a pulse of rapid spreading and expansion of ridges that led to a marked reduction in the volumetric capacity of the ocean basins. This trend was reversed only after 85 Ma when spreading rates slowed considerably, leading to widespread regression.

Rona (1973) came to a conclusion essentially similar to Hays and Pitman's that the volumetric capacity of the ocean basins was primarily caused by ridge volume changes and orogenic compression of the continental crust. Rona also calculated the sedimentation rates in several late Jurassic to Recent deep wells form the coastal plains and continental margins of North America and northwestern Africa. Average angular sedimentation rates, calculated without correcting for the effect of compaction, revealed three intervals of high sedimentation rates which correlated well with periods of worldwide transgressions, and two intervening intervals of slow sedimentation rates correlated with periods of wide regressions. Rona concluded that to a first approximation sedimentation rate maxima and minima are proportional to seafloor spreading rates. He also suggested that unconformities would develop on sedimentary sequences deposited on the continental shelves during regressions due to nondeposition and bypassing of sediments to the deeper parts of the basins or due to erosion.

Using the paleogeographic facies maps of the Soviet Union and North America, Hallam (1977) calculated the areas inundated by sea in these regions and confirmed the overall trend towards marine regression in the Phanerozoic with the exception of the late Mesozoic highstands. Hallam still maintained that the ocean-volume changes were the primary cause of sea level changes, but he now believed that this was more likely caused by a combination of continental thickening by orogeny (subduction and collision) and changes in the total length of the ridges, rather than changes in the seafloor spreading rates alone, which fail to explain the secular variations (and a long-term regression) of the Tertiary.

Cooper (1977) produced an analysis of the eustatic changes of the Cretaceous and identified thirteen transgressive episodes worldwide. He argued that if the mean sea level is related to volumetric changes in the mid-ocean ridges then it follows that the plate motion must be strongly episodic and the duration of the individual episodes short, rarely exceeding more than 2 m.yrs. Cooper's synthesis is, however, very qualitative and wrought with potential problems of stratigraphic precision and correlation. His cycle chart also does not show the relatively large scale highstands of the late Cretaceous that are widely evidenced by the stratigraphic record.

So far the identification of sea level fluctuations had been based on changes in the lithofacies that characterize transgressions and regressions along the continental margins and in epeiric marine sections. An independent method of determining relative changes in sea level using coastal onlap in seismic sections from the continental margins was published by Vail et al. (1977). This approach combined the study of high quality multi-channel seismic profiles across the margins, as well as well-control and core data, where available. The seismic reflection profile is first subdivided into sequences, or units of genetically related strata, bounded by unconformities or their correlative conformities. The patterns of coastal onlap, toplap and downward shifts of coastal onlap are then used to interpret sea level elevations, and sea level falls, respectively. With this approach Vail et al. first produced regional cycles of sea level change and then a global cycle chart for the Phanerozoic, based on the similarities between the number of cycles in widely separated areas (see Fig. 10). They used the measure of coastal aggradation during each cycle to determine the relative magnitude of the sea level rise during that cycle.

Figure 12 shows the compilation by Vail et al. (1977), based on regional seismic stratigraphic studies of localities in North and South America, western Europe, off West African coast, in southeastern Asia and Australia. The figure succinctly summarizes the Phanerozoic sea level history. Vail et al. recognized two first-order megacycles, the older one extending from the Cambrian to early Triassic (a duration of some 300 m.yrs.), and the younger one starting in the middle Triassic and continuing to the present (a duration of some 225 m.yrs.). Fourteen second-order global cycles have been identified of duration varying between 10 and 80 m.yrs. Third-order cycle (see Vail et al., *op. cit.*, Figs. 2 and 3) have duration of 1 to 10 m.yrs. These authors had greater confidence in the timing of the cycles, than in the amplitude of the fluctuations which are, at best, only approximations determined from grids of seismic sections tied to well-data, where available. Corrections for local and regional subsidence have to be applied before a true picture of the magnitude of such fluctuations can be developed (see Vail et al. *op. cit.*, for a discussion). These curves, however, serve to outline the general sea level history of the earth.

The first-order megacycle curve (Fig. 12) indicates that the early Paleozoic (Cambrian to Devonian) was a period of relatively high seastands, whereas during the interval between the late Mississippian and early Jurassic the seastands were relatively low. Late Jurassic to Eocene is once again a period of high seastands, particularly the late Cretaceous, when the sea levels may have been at their highest. The general withdrawal of the seas from the continental areas began with the Oligocene and has continued to the present. The second-order global curve shows more detail, and especially the abrupt nature of the sea level falls (and relatively gradual rises) becomes obvious. Five prominent falls are seen as the second-order supercycle boundaries within the Cambrian to Triassic megacycle. In the younger first-order megacycle, seven sharp sea level lowerings are seen, with the most prominent falls being in the mid-Oligocene and late Miocene (Vail et al., *op. cit.*).

Cyclic sea level fluctuations from bio- and lithofacies analyses have also been recognized in Australia (Partridge, 1976; Quilty, 1977, 1980; Glenie et al., 1978; and McGowran, 1979) New Zealand (Vella, 1967; Loutit and Kennett, 1981),

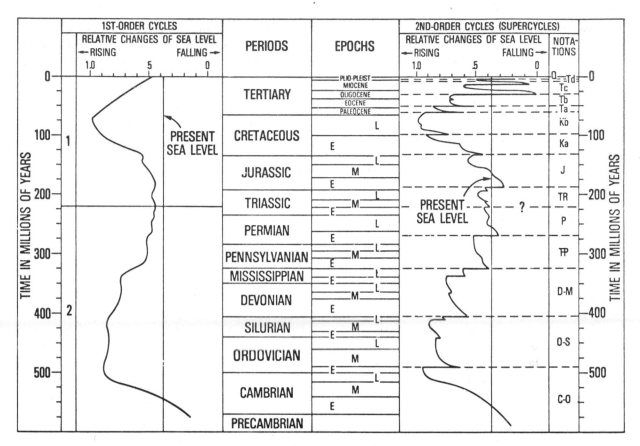

Figure 12. Phanerozoic history of the sea level changes from Cambrian to present day as deciphered by Vail et al. (1977). The first and second-order cycles of global extent are shown against time in million years. Two first-order megacycles (Early and Late Phanerozoic) and fourteen second-order supercycles have been delineated by Vail et al. (1977)

and South Africa (Siesser and Dingle, 1981). Based on rough biostratigraphic estimates these authors suggested some similarities between their regional curves and the global curve of Vail et al. (1977). Loutit and Kennett have suggested that the New Zealand stage boundaries may correlate with the cycle boundaries of Vail et al. The biostratigraphic precision of such correlations is, however, relatively low because of the provincial nature of the austral flora and fauna.

Vail et al. (1977) and Vail and Hardenbol (1979) suggested that the major control on depositional sequences along the continental margins was eustatic and that lowstands produced interregional and global unconformities. They contend that in general, the major unconformities are most easily recognized on seismic data, while medium to smaller events are commonly not detected on seismic lines but become obvious in outcrops, cores and well-logs. As a further documentation of the unconformities, Vail et al. (1980) synthesized the seismic date from the

margin off West Africa and the Blake Escarpment and biostratigraphic data from North Atlantic DSDP sites to map the extent of the Jurassic to Miocene unconformities. In this synthesis they reiterated that in the deep basins periods of extensive erosion seem to coincide with relative lowstands of sea levels. Keller and Barron (1983) and Barron and Keller (1983) correlated Miocene hiatuses in the deep sea with times of strong coastal offlap or lowstands of the Vail et al. curve, and suggested that this was a result of increased intensity of circulation during increased polar refrigerations. However, Loutit and Kennett (1981) argued that during highstands of sea level sediments and organic matter are trapped on the shelves, the possibility of dissolution and hiatuses increases in the deep sea, and not during lowstands when sediments are by-passed to the deeper basins (Rona, 1973). This is also indicated by the estimates of the sediment accumulation rates in the deep sea made by Worsley and Davies (1979) where an overall correspondence between high total and biogenic accumulation and low seastands is seen. Thus apparently the timing of hiatuses is different on the continental margins and the deep sea, as sea level fall may affect the two areas in the opposite way.

With the important assertion of the Vail et al. (1977) method that seismic reflectors represent stratal surfaces and therefore time lines, this approach was seen as a major breakthrough as a correlative tool in exploration. These applications include establishment of regional stratigraphic framework for structural and stratigraphic analysis, prediction of facies distribution, prediction of the geologic age ahead of the drill, and in global stratigraphic correlations (Vail and Mitchum, 1978).

In a discussion of the Atlantic type passive margins and its relationship to eustacy, Pitman (1978 a, b) pointed out that the *rate* of change of sea level may be more important in determining the nature of sequences along the margins than the actual rise or fall of the sea level itself. Pitman ruled out addition or loss of water from the mantle to the lithosphere and sediment influx or removal to or from the oceans as relatively unimportant factors in causing eustatic changes. Crustal shortening due to collision would increase the volume of the basins, whereas hot spots would have the opposite effect. Volume changes in the mid-ocean ridge systems were considered the most important component that influence ocean volumetric changes, as does the creation and destruction of such systems. Pitman's conclusion that transgressive or regressive events recorded at numerous subsiding margins may not be indicative of sea level rise or fall *per se,* but rather the changes in the rate of sea level change has important implications. A decrease in rate of sea level rise would be regressive, as will an increase in the rate of sea level fall. Conversely, a decrease in the rate of sea level fall or an increase in rise would produce transgressive sequences. In a more recent analysis of Jurassic eustatic cycles, Vail et al. (1984) have also cautioned against equating transgressions and regressions of the shoreline with global eustatic sea level rises and falls, as the former are not necessarily globally synchronous. Differing local and regional subsidence and sediment-supply rates can produce apparent transgressions or regressions which may not be synchronous with the global cycles.

Schlanger et al. (1981) have suggested the answers to the late Cretaceous global sea level highstand can be found in the Pacific Ocean where there is evidence of widespread mid-plate volcanism between 110 and 70 Ma. This excessive heating of lithosphere caused a bulge and bathymetric evolution of the area diverged significantly from "normal" subsidence curves of Sclater et al., (1971). This regional shallowing of seafloor and reduction in the ocean volume may have been a major factor in widespread transgressive phase of the late Cretaceous (Schlanger et al., *op. cit.*).

Bond (1978) attempted to separate the sea level effects from the vertical movements of the continents to estimate the "real" elevation of sea level over the continents. Percentages of areas flooded during transgressions were plotted on a hyposometric curve and this was considered a good first-order measure of the sea level rise. Such data can also be used to determine if there has been substantial post-transgressive change in the continental hypsometries. Bond estimated sea levels for Albian, Campanian-Maastrichtian, Eocene and Miocene time intervals. This method may have some promise when used in conjunction with other methods as it can at least provide an outside maximum value of the magnitude of sea level changes. However, as Harrison et al. (1981) pointed out, using present day hypsographic curve to calculate past sea levels can lead to erroneous computations, parcticularly for the Cretaceous, and corrections need to be applied to obtain more realistic values.

Sloss (1978) estimated the volumes of sediments preserved in six Mesozoic-Cenozoic basins, including passive and active margins and the cratonic interiors. He determined that there was a significant correspondence between these basins in episodes of deposition and non-deposition and concluded that cratonic basins and continental margins have roughly synchronous subsidence histories. This suggested to him that a common mechanism must be responsible for both the cratonic and pericratonic subsidence, impling that sea level change was a second order concommitant of a larger scale process.

Hancock and Kauffman (1979) examined the late Cretaceous lithofacies of western interior of the U.S. and northwestern Europe and concluded that major oscillations of sea level were simultaneous on these continents, except in regions of rapid contemporeneous tectonics. With a simplistic model the authors also made crude estimates of rises and falls of sea level on the order of 200 to 400m.

Watts and Steckler (1979) used biostratigraphic data from the COST B-2 well off New York and four deep commercial wells off Nova Scotia to remove the sediment loading component of the subsidence. By assuming that tectonic subsidence is mainly due to the cooling of the lithosphere, the authors also attempted to separate the tectonic component from the eustatic component. They suggest that the total amount of subsidence is a function of initial heating and thinning during rifting and the structural evolution of the margin is mainly controlled by this initial thinning and the subsequent subsidence. It has been suggested that crustal thinning along an Atlantic-type margin is caused by uplift and erosion (Sleep, 1971), but Watts and Steckler argue that this mechanism cannot adequately explain the extensive thinning in the COST B-2 well. Their sea level curve, after the isolation of the eustatic effect, shows a maximum highstand in the late Cretaceous.

Irrespective of what one believes causes sea level change, most workers seem to agree that at least the major sea level changes are eustatic, i.e. global in extent. Mörner (1976), however, introduced another concept capable of explaining time-transgressive sea level changes and implying the non-validity of globally synchronous changes. Mörner argued that the present day geoidal irregularities on the surface of the earth induce sea level difference of as much as 180 m between geoidal peaks and valleys. He asserted that the geoid configuration cannot have remained static in the past. If the assumption of major geoidal changes in the geologic past is valid, then the geoid must have played an important, and at times dominant, role for sea level changes (Reyment and Mörner, 1977). Mörner (1981) analysed the Cretaceous sea level curves of various authors from different parts of the world and inferred that in the Cretaceous there was evidence of large scale latitudinal sea level changes and these represented south to north migrating geoidal highs and lows or gravitational drop motions.

There is some merit to the idea that geoidal irregularities may cause regionally synchronous, but globally diachronous, sea level changes. However, it is difficult, if not impossible, to measure or ascertain the geoidal variations of the past and Mörner's model remains untestable. Mörner's (1981) assertion that bimodal sea level changes in the Cretaceous prove the movement of the geoid, may conceivably be due to problems in precision of stratigraphic correlations. This critique is especially valid in view of the widely different authorships and biostratigraphic methodology used in establishing the regional sea level curves that Mörner used in his analysis. Moreover, Crough and Jurdy (1980) have pointed out that at present the earth's largest geoidal anomalies are associated with the subduction zones and hot spots, implying deep-seated and density related control of the geoid. It follows that as far back in time as subduction zones and hot spots have existed, the geoidal anomalies would be controlled by the position of these features and not move around in a relatively random manner that Mörner's model implies.

Harrison et al. (1981) approached the problem of calculating the magnitude of past sea level changes from two different avenues, i.e. by measuring marine inundations of the present day continents and using present-day hypsographic curves to arrive at the amount of sea level changes, and by estimating the volume of mid-ocean ridge systems. They arrived at a difference of 170m in sea levels by the two methods for a late Cretaceous time slice (80-100 Ma). They explained this discrepancy by two factors. First, the present-day ocean basins have more sediments than the Cretaceous ocean, which would compensate somewhat for the ocean volume changes due to reduction in the ridge volume. Second, and more important factor is that present-day hypsogrphic curve may have less validity for the Cretaceous. Harrison and his colleagues argued that the average continental height is related to its area – the larger a continent, the higher its mean elevation (Hay and Southam, 1977). When the continents are joined together then the hypsographic curve is steeper than when the continents are fragmented. Because of greater elevation in the Cretaceous, a given sea level rise would produce less continental flooding than it would today and a corrected hypsographic curve would be necessary to calculate the "real" sea level changes.

Watts' (1982) modelling studies of the passive continental margins suggested to him that tectonic subsidence is an important control in the development of stratigraphic sequences. Flexural loading produces patterns of onlap that are similar to those used by Vail et al (1977) to infer sea level elevations. Watts contends that although Vail et al. attempted to remove the subsidence effect from their calculations by summing increments of coastal onlap, they did not correct for flexural loading effects. He believes that the patterns of onlap may have been produced by tectonic rather than eustatic control and their widespread coeval occurrence may be because widely separated margins rifted at similar times. Watts show some correspondence between the boundaries of the supercycles of Vail et al. and rifting events. However, the correspondence is good only for the older (pre-Tertiary) events. Younger boundaries show no clear relationship as there are fewer rifting events in the Cenozoic.

Recently, Kominz (1983) attempted a much needed analysis of errors involved in the calculations of sea level fluctuations based on mid-ocean ridge volumes for the past 80 m.y. These errors include inaccurate estimates, such as those for ridge lengths, or partially and completely subducted ridges, problems of time scales, and the lack of information and omission of pre-70 Ma crust in calculations. Kominz found that the main source of reduction in the volume of ridge systems is the Pacific ridges which show a decrease in volume of about 50% from 80 to 15 Ma. Total ridge length, on the other hand, has decreased only slightly through time. Her analysis also showed that the main contribution to the reduction in ridge volumes since the Cretaceous came from reduction in the spreading rates, while ridge length fluctuations may account for variations superimposed on the general trend.

From the preceding discussion it is easy to draw the conclusion that the long term regression from the latest Cretaceous to late Tertiary may have been influenced by several factors working concordantly, including the slowing of the seafloor spreading rates since the late Cretaceous and the resultant changes in the ocean volume, the orogenic movements and the general emergence of the continents, and the development of polar ice-caps in the mid to late Tertiary. The fact that the post-Cretaceous transgressions got progressively smaller in size (i.e., each transgression invaded lesser continental area than the preceding one), so that by Oligocene and Miocene times the transgressions did not invade much of the continental interiors, means that a large amount of ocean water was permanently removed from the ocean basins in the form of polar ice by the mid Tertiary time.

CONCLUDING REMARKS

The brief commentary on the status of paleoceanography provided in the preceding pages illustrates the rapid advancements made in this new field of geoscience during the past decade and a half. However, it also serves to point out the major gaps in our knowledge of ancient oceans and our relative ignorance of the past environmental conditions.

The most important outstanding questions still remain in the Mesozoic superoceans. From the time of the last Pangaean dispersal, some 210 m.y. ago until the mid Cretaceous (ca 100 m. y. ago) little is known of the paleoceans. There are a general dearth of sampling areas and data for the record of those first 110 m. y. of the present megacycle, making reconstructions tenuous at best. Such paleoenvironmental data need to be generated in at least the same kind of detail that is presently available for the Cenozoic. Biogeographic, isotopic, as well as detailed sedimentological data, has to be gathered and evaluated before we can go beyond the present speculative stage and learn more about Mesozoic productivity and hydrographic-climatic conditions.

Some of the most intriguing, and scientifically most promising, problems require drilling in the higher latitudes. In view of the widespread global repercussions that follow the higher latitude environmental changes, drilling in those areas remains a top priority and will require strong governmental committment for the next decade or two.

The importance to hydrocarbon research of the way in which sediments along the continental margins respond to sea level changes has led to a breakthrough in the construction of eustatic cycle charts. However, the all important parameter of the magnitude of such changes still remain an illusive goal. Such estimates need to be carefully constrained by the seafloor spreading and tectonic histories, paleogeographies and sedimentary input factors. In addition, more precise timing and true global extent of each cycle needs to be documented to make them widely applicable in paleoceanographic event-analyses.

ACKNOWLEDGEMENTS

The author is indebted to Ian Norton, Peter R. Vail, and Ramil C. Wright for reviewing this overview. Ian Norton also provided the base-maps on which the paleogeographic reconstructions are based. The author is also indebted to Exxon Production Research Company for permission to participate in the U.S. – Pakistan workshop and for the publication of this paper.

REFERENCES

Adams, C.G., R.H. Benson, R.B. Kidd, W.B.F. Ryan and R.C. Wright, 1977. The Messinian salinity crisis and evidence of Late Miocene eustatic changes in the World Ocean. Nature, 269: 383-386.

Alvarez, W., L.W. Alvarez, F. Asaro and H.V. Michel, 1979. Anomalous iridium levels at the Cretaceous-Tertiary boundary at Gubbio, Italy: negative results of tests for a supernova origin. Geol. Soc. Am. Abstr. Progr., 508, 379.

Alvarez, L.W., W. Alvarez, F. Asaro and H.V. Michel, 1980. Extraterrestrial cause for the Cretaceous-Tertiary extinction. Science, 208:1095-1108.

Alvarez, W., F. Asaro, H.V. Michel et al., 1982. Iridium anomaly approximately synchronous with terminal Eocene extinctions. Science, 216:886-888.

Alvarez, W., E.G. Kaufman, F. Surlyk, L.W. Alvarez, F. Asaro, and H.V. Michel, 1984a. Impact theory of mass extinctions and the invertebrate fossil record. Science, 223: 1135-1141.

Alvarez, W., L.W. Alvarez, F. Asaro and H.V. Michel, 1984b. The end Cretaceous: Sharp boundary or gradual transition? Science, 223; 1183-1186.

Arthur, M.A., 1979. Paleoceanographic events—recognition, resolution and reconsideration. Rev. Geophys., 7:1474-1494.

Axelrod, D.I., 1984. An interpretation of Cretaceous and Tertiary biota in the polar regions. Paleogeogr., paleoclimat., paleoecol., 45:105-147.

Backman, J., 1979. Pliocene biostratigraphy of DSDP sites 111 and 116 from the North Atlantic Ocean and the age of Northern Hemisphere glaciation. Stockholm Contrib. Geol., 32(3): 115-137.

Barker, P.F. and J. Burnell, 1977. The opening of Drake Passage. Mar. Geol., 25:15-34.

Barron, E.J. et al., 1981. Paleogeography, 180 million years ago to present. Eclogae Geol. Helv., 74:443-470.

Barron, E.J., and W.M. Washington, 1982. Cretaceous climate: a comparison of atmospheric simulations with the geologic record. Paleogrogr., paleoclimat., paleocol., 40:103-133.

Barron, E.J., and J.M. Whitman, 1982. Oceanic sediments in space and time In. C. Emiliani (ed.): Oceanic Lithosphere, J. Wiley and Sons, The Sea, vol. 7, pp. 689-731.

Barron, J.A. and G. Keller, 1983. Widespread Miocene deep-sea hiatuses: coincidence with periods of global cooling. Geology, 10:577-581.

Beland, P. et al., 1977. Cretaceous-Tertiary extinctions and possible terrestrial and extaterrestrial causes. Nat. Mus. Can. Bull., 12, 162.

Bender, M.L., and L.D. Keigwin, Jr., 1979. Speculations about the Upper Miocene change in abyssal Pacific dissolved bicarbonate C. Earth Planet. Sci. Letters, 45:383-393.

Benson, R.H., 1975. The origin of the psychrosphere as recorded in changes of deep-sea ostracode assemblages. Lethaia, 8:69-83.

Benson, R.H., 1976. Testing the Messinian salinity crisis biodynamically: an introduction. Paleogeogr., paleoclimat., paleoecol., 20:3-11.

Berger, W.H., 1973. Cenozoic sedimentation in the eastern tropical Pacific. Geol. Soc. Am. Bull., 84: 1941-1954.

Berger, W.H., 1979. Impact of Deep Sea Drilling on paleoceanography. In: M. Talwani et al., (eds.): Results of the Deep Sea Drilling in the Atlantic Ocean. Maurice Ewing ser., Am., Geophys. Union, 3:297-314.

Berger, W.H., 1982a. Paleoceanograpy: The Deep Sea Record. In. C. Emiliani (ed.): Oceanic Lithosphere, J. Wiley and sons, N.Y., The Sea vol. 7, pp. 1437-1519.

Berger, W.H., 1982b. Deep Sea Stratigraphy: Cenozoic climate steps and the search for Chemo-climatic feedback. In. G. Einsele and A. Seilacher (eds.): Cyclic and Event Stratification. Springer-Verlag, Berlin, pp. 121-157.

Berggren, W.A., 1982. Role of ocean gateways in climatic change. In. W.H. Berger and J.C. Crowell (eds.): Climates in Earth History, Studies in Geophysics, National Acad. Sci, Washington, D.C., pp. 118-125.

Berggren, W.A. and C.D. Hollister, 1974. Paleogeography, paleobiogeography and the history of circulation in the Atlantic Ocean. W.W. Hay (ed.). Soc. Econ. Paleont. Mineral. Spec. Pub., 20:12-186.

Berggren, W.A. and C.D. Hollister, 1977. Plate tectonics and paleocirculation—commotion in the ocean. Tectonophysics, 38:11-48.

Berggren, W.A., D.V. Kent, J.J. Flynn and J.A. van Couvering, 1984. Cenozoic Geochronology. DNAG (in press).

Blanc, P.L. et al., 1980. North Atlantic deep water formed by the later middle Miocene. Nature, 283:55-555.

Bohor, B.F. et al., 1984. Mineralogic evidence for the Impact Event at the Cretaceous-Tertiary Boundary. Science, 224:867-869.

Bond, G., 1978. Speculations on real sea-level changes and vertical motions of continents at selected times in the Cretaceous and Tertiary Periods. Geology, 6:247-250.

Boucot, A.J. and J. Gray, 1983. A Paleozoic Pangaea. Science, 222: 571-581.

Brass, G.W., et al., 1980. Ocean circulation, plate tectonics and climate. In. W.H. Berger and J.C. Crowell (eds.): Climates in Earth History, Studies in Geophysics, National Acad. Sci., Washington, D.C., pp. 83-89.

Buchardt, B., 1978. Oxygen isotope paleotemperature from the Tertiary period in the North Sea area. Nature, 275:121-123.

Buffler, R.T. et al., 1981. A model for the early evolution of the Gulf of Mexico Basin. Oceanologica Acta, No. Sp., pp. 129-136.

Cande, S.C. and J.C. Mutter, 1982. A revised identification of the oldest sea-floor spreading anomalies between Australia and Antarctica. Earth Planet. Sci. Lett., 58:151-160.

Ciesielski, P.F., M.T. Ledbetter and B.B. Ellwood, 1982. The development of Antarctic glaciation and the Neogene paleoenvironment of the Maurice Ewing Bank. Mar. Geol., 46:1-51.

Ciesielski, P.F. and F.M. Weaver, 1983. Neogene and Quaternary paleoenvironmental history of Deep Sea Drilling Project Leg 71 sediments, S.W. Atlantic Ocean. Initial Repts. D.S.D.P., vol. 71, U.S. Govt. Printing Office, Washington, D.C., pp. 461-477.

Cita, M.B., 1976. Biodynamic effects of the Messinian salinity crisis on the evolution of planktonic foraminifera in the Mediterranean. Paleogeogr., paleoclimat., paleoecol., 20:23-42.

Clark, D.L. 1982. The Arctic Ocean and Post-Jurassic paleoclimatology. In. W.H. Berger and J.C. Crowell (eds.): Climate in Earth History, Studies Geophysics, National Acad. Sci., Washington, D.C. pp. 133-138.

Clague, D.A. and R.D. Jarrard, 1973. Tertiary Pacific plate motion deduced from the Hawaiian-Emperor chain. Geol. Soc. Am. Bull., 84:1135-1154.

CLIMAP, 1976. The surface of ice-age Earth, Science, 191: 1131-1137.

Cline, R.M. and J.D. Hays (eds.) 1976. Investigation of Late Quaternary Paleoceanography and Paleoclimatology. Geol. Soc. Am. Mem., 145:1-464.

Cool, T.E., 1982. Sedimentological evidence concerning the paleoceanography of the Cretaceous Western North Atlantic Ocean. Paleogeogr., paleoclimat., paleoecol., 39:1-35.

Cooper, M.R., 1977. Eustacy during the Cretaceous: its implications and importance. Paleogeogr., paleoclimat., paleoecol., 22:1-60.

Corliss, B.H., 1981. Deep-sea benthonic foaminiferal faunal turnover near the Eocene/Oligocene boundary. Marine Micropal. 6: 367-384.

Crough, S.T. and D.M. Jurdy, 1980. Subducted lithosphere, hotspots and the geoid. Earth and Planet. Sci. Lett., 48:15-22.

Curray, J.R. and D.G. Moore, 1974. Sedimentary and tectonic processes in the Bay of Bengal Deep-Sea Fan and Geosyncline, In. C.A. Burk and C.L. Drake (eds.): The Geology of continental margins, Springer Verlag, New York, pp. 617-627.

Davies, T.A., et al., 1977. Estimates of Cenozoic oceanic sedimentation rates. Science, 197:53-55.

Donn, W.L., 1982. The enigma of high latitude paleoclimates. Paleogeogr., paleoclimat., paleoecol., 40:199-212.

Douglas, R.G., and S.M. Savin, 1975. Oxygen and carbon isotope analyses of Tertiary from Shatsky Rise and other sites in the North Pacific Ocean. Initial Reports D.S.D.P. vol. 32, U.S. Govt. Print. Off., Washington, D.C., pp. 509-520.

Douglas, R.G., and S.M. Savin,,1978. Oxygen isotopic evidence for the depth stratification of Tertiary and Cretaceous planktic foraminifera. Marine Micropal., 3:175-196.

Douglas, R.G., 1982. Cretaceous benthic foraminiferal paleobathymetry: Quomoda altum id fuit? In. Bottjer et al. (eds.): Late Cretaceous depositional environments and paleogeography, Santa Ana Mountains Southern California. Pacific Sec. S.E.P.M., field trip, vol. and Guidebook, pp. 39-44.

Emiliani, C., 1954. Temperatures of Pacific bottom waters and polar superficial waters during the Tertiary. Science, 119:853-855.

Emiliani, C., 1980. Death and renovation at the end of the Mesozoic. EOS, Trans. Am. Geophys. Union, 61:505-506.

Emiliani, C., 1982. A new global geology. In. C. Emiliani (ed.): Oceanic Lithosphere, J. Wiley and Sons, N.Y., The Sea, volume 7, pp. 687-1728.

Estes,R., and J.H. Hutchinson, 1980. Eocene lower vertebrates from the Ellesmere Island, Canadian Arctic Archipelogo. Paleogeogr. paleoclimatol., paleoecol., 30:325-347.

Ewing, J.I., and C.D. Hollister, 1972. Regional aspects of deep sea drilling in the western North Atlantic. Initial Reports D.S.D.P. vol. 11, U.S. Govt. Print. Off., pp. 951-973.

Fischer, A.G. and M.A. Arthur, 1977, Secular variations in the pelagic realm. In. Cook, H.E. and P. Enos (eds.): Deep water carbonate environments. Soc. Econ. Pal. Mineral., Special Publ., 25:119-150.

Fleet, A.J. and B.C. McKelvey, 1978. Eocene explosive submarine volcanism, Ninety East Ridge, Indian Ocean. Mar. Geol., 26:73-97.

Frakes, L.A., 1979. Climates throughout geologic time. Elsevier Scientific Publishing Co., Amsterdam, 310 pp.

Ganapathy, R., 1980. A major meteoritic impact on the Earth 65 million years ago: evidence from the Cretaceous-Tertiary boundary clays. Science, 209:921-923.

Ganapathy, R., 1982. Evidence for a major meteoritic impact on the Earth 34 million years ago: Implications for Eocene extinctions. Science, 216:885-886.

Gartner, S., and J. Keany, 1978. The terminal Cretaceous event: a geologic problem with an oceanographic solution. Geology, 6:708-712.

Gibson, T.G. and K.M. Towe, 1971. Eocene volcanism and the origin of Horizon A. Science, 172:152-159.

Glass, B.P. and J.R. Crosbie, 1982. Age of Eocene/Oligocene boundary based on extrapolation from North American Microtektite layer. Bull. Am. Assoc. Petrol. Geol., 66:471-476.

Glenie, R.C., J.C. Schofield, and T.C. Ward, 1968. Tertiary sea levels in Australia and New Zealand. Paleogeogr., paleoclimato., paleoecol. 5:141-173.

Gradstein, F.M. and S.P. Srivastava, 1980. Aspects of Cenozoic Stratigraphy and Paleoceanography of Labrador Sea and Baffin Bay. Paelogeogr., paleoclimat., paleoecol., 30:261-295.

Gradstein, F.M. and R.E. Sheridan, 1983. On the Jurassic Atlantic Ocean and a synthesis of results of DSDP leg 76. Initial Reports D.S.D.P., vol. 76, U.S. Govt. Print. Off., Washington, D.C., pp. 913-943.

Grow, J.A. and R.E. Sheridan, 1981. Deep structure and evolution of the continental margin off the estern United States. Oceanologica Acta, No. Sp., pp. 1-19.

Hallam, A., 1963. Major epeirogenic and eustatic changes since the Cretaceous and their possible relationship to crustal structure. Am. J. Sci., 261:397-423.

Hallam, A., 1971. Re-evaluation of the paleogeographic argument for an expanding earth. Nature, 232:180-182.

Hallam, A., 1977. Secular changes in marine inundation of USSR and North America through the Phanerozoic. Nature, 269:769-772.

Hallam, A., 1979. The end of the Cretaceous. Nature, 281:430-431.

Hancock, J.M. and E.G. Kaufman, 1979. The great transgression of the late Cretaceous. J. Geol. Soc. Lond., 136:175-186.

Haq, B.U., 1973. Transgressions, climatic change and the diversity of calcareous nannoplankton. Mar. Geol., 15:25-30.

Haq, B.U., 1980. Biogeographic history of Miocene calcareous nannoplankton and paleoceanography of the Atlantic Ocean. Micropaleont., 26:414-443.

Haq, B.U., 1981. Paleogene paleoceanography: Early Cenozoic oceans revisited. Oceanologica Acta, No. Sp., pp. 71-82.

Haq, B.U., I. Premoli-Silva and G.P. Lohmann, 1977. Calcareous plankton paleobiogeographic evidence for major climatic fluctuations in the Early Cenozoic Atlantic Ocean. J. Geophys. Res., 82:3861-3876.

Haq, B.U. et al. 1980. The Late Miocene marine carbon isotopic shift and the synchroneity of some phytoplanktonic biostratigraphic datums. Geology, 8:427-431.

Harland, W.B. et al, 1982. A geologic time scale. Cambridge Univ. Press, Cambridge. 31 pp.

Harrison, C.G.A. et al., 1981. Sea level variations, global sedimentation rates and the hypsographic curve. Earth Planet, Sci., Lett., 54:1-16.

Hay, W. W., 1981. Sedimentological and geochemical trends results from breakup of Pangaea. Oceanologica Acta, No. Sp., pp. 135-147.

Hay, W.W. and J.R. Southam, 1977. Modulation of marine sedimentation by continental shelves. In. N.R. Anderson and A. Malahoff (eds.): The Role of Fossil Fuel CO_2 in the Oceans, Plenum Press, New York, pp. 569-605.

Hay, W.W. et al., 1982. Late Triassic-Liassic paleoclimatology of proto-Central Atlantic rift system. Paleogeogr., paleoclimat., paleoecolog., 40:13-30.

Hayes, D.E. and J. Ringis, 1973. Sea-floor spreading in the Tasman Sea. Nature, 243:454-458.

Hays, J.D. and W.C. Pitman III, 1973. Lithospheric plate motion, sea level changes and climatic and ecological consequences, Nature, 246: 18-20.

Hays, J.D., J. Imbrie and N.J. Shackleton, 1976. Variations in the earth's orbit: pacemaker for the ice ages. Science, 194:1121-1132.

Hermann, Y., 1972. Origin of deep sea chert in the North Atlantic. Nature, 238:392-393.

Hsü, K.J., 1980. Terrestrial catastrophe caused by cometary impact at the end of Cretaceous. Nature, 285:201-203.

Hsü, K.J., M.B. Cita and W.B.F. Ryan, 1973. Origin of the Mediterranean evaporites. Initial Reports D.S.D.P., vol. 13, U.S. Government Printing Office, Washington D:C. pp. 1203-1231.

Hsü, K.J. and F. Giovanoli, 1979. Messinian event in the Black Sea. Paleogeogr., paleoclimat., paleoecol., 29:75-93.

Hsü, K.J. et al., 1984 South Atlantic Cenozoic paleoceanography. Initial Reports D.S.D.P., vol. 73, U.S. Government Printing Office, Washington D.C., pp. 771-785.

Jenkyns, H.C., 1980. Cretaceous anoxic events: from continents to oceans. Geol. Soc. Lond., 137: 171-188.

Johnson, B.D., C.M. Powell and J.J. Veevers, 1976. Spreading history of the eastern Indian Ocean and Greater India's flight from Antarctica and Australia. Geol. Soc. Am. Bull., 87:1560-1566.

Johnson, D.A., 1982. Abyssal teleconnections: interactive dynamics of the deep ocean circulation. Paleogeogr., paleoclimat., paleoecol., 38:93-128.

Johnson, D.A., 1983. Paleocirculation of the South Atlantic. Initial Repts. D.S.D.P., vol. 72, U.S. Government printing office, Washington D.C., pp. 977-994.

Kcigwin, L.D. Jr., 1980. Oxygen and carbon isotope analyses from Eocene/Oligocene boundary at DSDP Site 277. Nature, 287:722-725.

Keigwin, L.D. Jr., 1982. Isotopic paleoceanography of the Caribbean and East Pacific: Role of the Panama Uplift in Late Neogene. Science, 217: 350-353.

Keigwin, L.D. Jr., and G. Keller, 1984. Middle Oligocene cooling from equatorial Pacific DSDP Site 77B. Geology, 12:16-19.

Keller, G., 1980. Middle to late Miocene planktonic foraminiferal datum levels and paleoceanography of the North and southwestern Pacific Ocean. Mar. Micropal., 5:249-281.

Keller, G. and J.A. Barron, 1983. Paleoceanographic implications of Miocene deep sea hiatuses. Geol. Soc. Am. Bull., 94(5):590-613.

Keller, G., S. D'Hondt, and T.L. Vallier, 1983. Multiple microtektite horizons in the Upper Eocene marine sediments: no evidence for mass extinctions. Science, 221:150-152.

Kennett, J.P., 1977. Cenozoic evolution of Antarctic glaciation, the circum-Antarctic Current and their impact on global paleoceanography. J. Geophys. Res., 82:3843-3860.

Kennett, J.P., 1978. The development of planktonic biogeography in the Southern Ocean during the Cenozoic. Mar. Micropal., 3:310-346.

Kennett, J.P., 1979. Recent zoogeography of Antarctic plankton microfossils. In. S. van der Spoel and A.C. Pierrot-Bults (eds.): Zoogeography and diversity of plankton, Bunge Sci. Pub., Utrecht, pp. 328-355.

Kennett, J.P., 1983. Paleoceanograpy: Global Ocean Evolution. Rev. Geophy. and Space Phy., 21(5):1258-1274.

Kennett, J.P. et al., 1975. Cenozoic paleoceanography in the southwest Pacific Ocean, Antarctic glaciation and the development of the circum-Antarctic current. Initial Reports D.S.D.P. vol. 29, U.S. Govt. Printing Office., Washington D.C., pp. 1155-1166.

Kennett, J.P., et al., 1979. Late Cenozoic oxygen and carbon isotopic history and volcanic ash stratigraphy: DSDP Site 284, South pacific. Am. J. Sci., 279:52-69.

Kidd, R.B. and T.A. Davies, 1978. Indian Ocean sediment distribution since the late Jurassic. In. C.C. von der Borch (ed.): Synthesis of deep-sea drilling results in the Indian Ocean. Mar. geol., 26:49-70.

King, L.C., 1950. The study of the world's plainlands: a new approach in geomorphology. Geol. Soc. Lond. Quat. Jr., 106:101-131.

King, L.C., 1957. A geomorphological comparison between Brazil and Africa (central and southern). Geol. Soc. Lond. Quat. J., 112:445-470.

Kominz, M.A., 1983. Oceanic ridge volumes and sea level change—an error analysis. In. Schlee, J. (ed.): Interregional unconformities and hydrocarbon accumulation. Am. Assoc. Petrol. Geol., Mem. (in press).

Kroopnick, P.M., S.V Margolis, and C.S. Wong, 1977. $\delta^{13}C$ variations in marine carbonate sediments as indicators of the CO_2 balance between the atmosphere and oceans. In. N.R. Anderson, and A. Malahoff, (eds.): The fate of fossil fuel CO_2 in the ocean, Plenum Press, pp. 295-322.

Lancelot, Y., 1973. Chert and silica diagenesis in sediments from the Central Pacific. Initial Repts. D.S.D.P., vol. 17, U.S. Govt. Print, Office, Washington, D.C., pp. 377-405.

Larson, R.L. and W.C. Pitman, 1972. World wide correlation of Mesozoic magnetic anomalies and its implications. Bull. Geol. Soc. Am., 83:3645-3662.

Laughton, A.S., 1971. South Labrador Sea and the evolution of the North Atlantic. Nature, 232-612-627.

Laughton, A.S., D.P. McKenzie and J.G. Sclater, 1973. The structure and evolution of the Indian Ocean, In: D.H. Tarling, S.K. Runcorn, (eds.): Implications of continental drift to the earth sciences, Academic Press, London, pp. 203-212.

Loutit, T.S., and J.P. Kennett, 1979. Application of carbon isotope stratigraphy and correlation to Late Miocene shallow marine sediments, New Zealand. Science, 204:1196-1199.

Loutit, T.S., and J.P. Kennett, 1981. Australasian Cenozoic sedimentary cycles, global sea level changes and the deep sea record. Oceanologica Acta, No. Sp, pp. 45-46.

Loutit, T.S., and L.D. Keigwin Jr., 1982. Stable isotopic evidence for the latest Miocene sea-level fall in the Mediterranean region. Nature, 300:163-166.

Luck, J.M., and K.K. Turekian, 1983. Osmium-187/Osmium-186 in Manganese Nodules and the Cretaceous-Tertiary Boundary. Science, 222:613-615.

Luyendyk, B.P. and T.A. Davies, 1974. Results of DSDP Leg 26 and the geologic history of the southern Indian Ocean. Initial Repts. D.S.D.P. vol. 26, U.S. Govt. Printing Office, Washington D.C., pp. 909-952.

Malfait, B.T., and M.G. Dinkleman, 1972. Circum-caribbean tectonic and igneous activity and the evolution of the Caribbean plate. Geol. Soc. America Bull., 83:251-272.

Markl, R.G., 1978. Further evidence for the early Cretaceous breakup of Gondwanaland, off S.W. Australia. Mar. Geol., 26:41-48.

Matthews, R.K. and R.Z. Poore, 1980. Tertiary $\delta^{18}O$ record and glacio-eustatic sea-level fluctuations. Geology, 8:501-506.

McGowran, B., 1978. Stratigraphic record of Early Tertiary oceanic and continental events in the Indian Ocean region, In. C.C. von der Borch (ed.): Synthesis of deep-sea drilling results in the Indian Ocean. Mar. Geol., 26:1-39.

McGowran, B., 1979. The Tertiary of Australia: Foraminiferal overview. Mar. Micropaleont., 4:235-269.

McKenna, M.C., 1980. Eocene paleolatitude, climate and mammals of Ellesmere Island. Paleogeogr., paleoclimatol., paleoecol., 30:348-362.

McKenzie, D. and J.G. Sclater, 1971. The evolution of Indian Ocean since Late Cretaceous. Geophy. J. Roy. Astron. Soc., 25:437-528.

Miller, K.G., and B.E. Tucholke, 1983. Development of Cenozoic abyssal circulation south of the Greenland-Scotland Ridge: In. M. Bott, S. Saxov, M. Talwani, and J. Theide (eds.): Structure and Development of the Greenland-Scotland Ridge, Plenum Publ., pp. 549-589.

Miller, K.J. and R.G. Fairbanks, 1983. Evidence for Oligocene Middle Miocene abyssal circulation change in western North Atlantic. Nature, 306:250-253.

Moore, T.C., Jr., T.H. van Andel, C. Sancetta and N. Pisias, 1978. Cenozoic hiatuses in the pelagic sediments. Micropaleont., 24:113-138.

Moore, T.C., Jr. et al., 1984. History of the Walvis Ridge. Initial Repts. D.S.D.P., vol. 74, U.S. Govt. Printing Office, Washington, D.C., pp. 873-894.

Mörner, N.A., 1976. Eustacy and geoid changes. J. Geol., 84:123-151.

Mörner, N.A., 1981. Revolution in Cretaceous sea-level analysis. Geology, 344-346.

Norton, I.O. and J.G. Slater, 1979. A model for the evolution of the Indian Ocean and the breakup of Gondwanaland. J. Geophy. Res., 84 (B12): 6803-6830.

Officer, C.B. and C.L. Drake, 1983. The Cretaceous-Tertiary transition. Science, 219.1383-1390.

Orth, C.J., et al., 1981. An iridium abundance anomaly at the palynological Cretaceous-Tertiary boundary in northern New Mexico. Science, 214:1341-1343.

Parish, J. and R.L. Curtis, 1982. Atmospheric circulation, upwelling and organic-rich rocks in the Mesozoic and Cenozoic Eras. Paleoegor., paleoclimat., paleoecol., 40:31-66.

Parker, W.C. et al., 1984. Population dynamics, Paleogene abyssal benthic foraminifers, eastern South Atlantic. Initial Repts. D.S.D.P., vol. 73, U.S. Govt. Print. Off., Washington, D.C., pp. 481-486.

Partridge, A.D., 1976. The geological expression of eustasy in the early Tertiary of the Gippsland Basin. APEA Jour., 16:73-79.

Percival, S.F. and A.G. Fischer, 1977. Changes in calcareous nannoplankton in the Cretaceous-Tertiary biota crisis at Zumaya, Spain. Evol. Theory, 2:1-35.

Pillmore, C.L. et al., 1984. Geologic framework of non-marine Cretaceous-Tertiary Boundary Sites, Raton Basin, New Mexico and Colorado. Science, 223-1180-1183.

Pitman, W.C., III, 1978a. Relationship between eustasy and stratigraphic sequences of passive margins. Geol. Soc. Am. Bull., 89:1389-1403.

Pitman, W.C., III, 1978b. The effects of eustatic changes on stratigraphic sequences at Atlantic margins. Am. Assoc. Petrol. Geol. Mem. 29:453-460.

Prell, W.L., 1982. A re-evaluation of the initiation of northern hemisphere glaciation at 3.2 m.y.: New isotope evidence. Abstacts

95 Ann. Meeting Geol. Soc. Am., p. 592.

Quilty. P.G., 1977. Cenozoic sedimentation cycles in Western Australia. Geology, 5:336-340.

Quilty, P.G., 1980. Sedimentation cycles in the Cretaceous and Cenozoic of Western Australia. Tectonophysics, 63:349-366.

Quilty, P.G., 1984. Cretaceous-Tertiary transgression-regression cycles in Australia (in press).

Rabiwowitz, P.D., M.F. Coffin and D.T. Falvey, 1983. The separation of Madagascar and Africa. Nature, 320:67-69.

Rand, H.M. and J.M. Mabesoone, 1982. Northeastern Brazil and the final separation of South America and Africa. Paleogeogr., paleoclimat., paleoecol., 38:163-183.

Reyment, R.A., 1980a. Paleo-oceanology and paleobiogeography of the Cretaceous South Atlantic Ocean. Oceanologica Acta, 3(1):127-133.

Reyment, R.A., 1980b. Biogeography of the Saharan Cretaceous and Paleocene Epicontental Transgressions. Cret. Res., 1:299-327.

Reyment, R.A. and N.A. Mörner, 1977. Cretaceous transgressions and regressions exemplified by the South Atlantic. Spec. pap. Paleont. Soc., Japan, 21:247-260.

Rona, P.A., 1973. Relations between rates of sediment accumulation on continental shelves, sea-floor spreading and eustasy inferred from Central North Atlantic. Geol. Soc. Am. Bull., 84:2851-2872.

Roth, P.H. 1978. Cretaceous nannoplankton biostratigraphy and oceanography of the Northwest Atlantic Ocean. Initial Repts. D.S.D.P., volume 44, U.S. Govt. Printing Office, Washington, D.C., pp. 731-759.

Rögl, F. and F.F. Steininger, 1983. Vom Zerfall der Tethys zu Mediterran und Paratethys. Ann. Naturlist. Mus. Wien, 85(A): 135-163.

Ruddiman, W.F. and A. McIntrye, 1979. Warmth of the subpolar North Atlantic Ocean during Northern Hemisphere ice-sheet growth. Science, 204:173-175.

Ruddiman, W.F. and A. McIntrye, 1981. Oceanic mechanisms for amplification of the 23,000-year ice volume cycle. Science, 212:617-627.

Russell, D.A., 1977. The biotic crises at the end of the Cretaceous period. Nat. Mus. Can. Bull., 12:11-23.

Saltzman, E.S. and E.J. Barron, 1982. Deep circulation in the Late Cretaceous: Oxygen isotope paleotemperatures from *Inoceramus* remains in D.S.D.P. cores. Paleogeogr., paleoclimat., paleoecol., 40:167-181.

Sarnthein, M., et al., 1982. Atmospheric and oceanic circulation patterns off Northwest Africa during the past 25 m.y. *In*. U. Von Rad et al. (eds.): *Geology of the Northwest African Continental margins*, Springer-Verlag, Berlin, pp. 545-604.

Savin, S.M., 1977. The history of the Earth's surface temperature during the last 100 million years, Ann. Rev. Earth Plant. Sci., 5:319-355.

Savin, S.M., R.G. Douglas, and F.G. Stehli, 1975. Tertiary marine paleotemperatures. Geol. Soc. Am. Bull., 86:1499-1510.

Schlanger, S.O. and H.C. Jenkyns, 1976. Cretaceous oceanic anoxic events: Causes and consequences. Geol. Mijnbouw, 55:179-184.

Schlangér, S.O., H.C. Jenkyns, and I. Premoli-Silva, 1981. Volcanism and vertical tectonics in the Pacific Basin related to global Cretaceous transgressions. Earth Planet. Sci. Lett., 52:435-449.

Schnitker, D. 1980a. North Atlantic oceanography as a possible cause of Antarctic galciation and eutrophication. Nature, 284:615-616.

Schnitker, D., 1980b. Global paeloceanography and its deep water linkage to the Antarctic glaciation. Earth Sci. Rev., 16:1-20.

Sclater, J.G., R.N. Anderson and M.L. Bell, 1971. Elevation of ridges and evolution of the central eastern Pacific. J. Geophy. Res., 76:7888-7915.

Sclater, J.G., S. Hellinger, C. Tapscott, 1977. The paleobathymetry of the Atlantic Ocean from the Jurassic to the present. J. geol., 85:509-522.

Shackleton, N.J. and J.P. Kennett, 1975. Paleotemperature history of the Cenozoic and the initiation of Antarctic glaciation: Oxygen and carbon isotope analyses in DSDP sites 277, 279, and 281. Initial Repts. D.S.D.P. vol. 29. U.S. Govt. Print. Off., Washington, D.C., pp. 743-755.

Shackleton, N.J. and N.D. Opdyke, 1977. Oxygen isotope and paleomagnetic evidence for early northern hemisphere glaciation Nature, 270:216-219.

Shackleton, N.J. and A. Boersma, 1981. The climate of the Eocene Ocean. J. Geol. Soc. Lond., 138:153-157.

Sheridan, R.E., 1983. Phenomena of pulsation tectonics related to the breakup of the eastern North America Continental margin. Initial Repts. D.S.D.P. vol. 76, U.S. Govt. Printing Office, Washington, D.C., pp. 897-909.

Siesser, W.G. and R.V. Dingle, 1981. Tertiary sea-level movements around southern Africa. J. Geol. 89:83-96.

Silver, L.T. and Schultz, P.H., (eds.), 1982. Geological implications of impacts of large asteroids and comets of the Earth. Geol. Soc. Am. Sp. pap. 190, 528 pp.

Sleep, N.H., 1971. Thermal effects of the formation of Atlantic continental margins by continental breakup. Geophys. J. Roy. Astr. Soc., 24:325-350.

Smit, J. and J. Hertogen, 1980. An extraterrestrial event at the Cretaceous-Tertiary oundary. Nature, 285:198-200.

Smith, J. and S. van der Karrs, 1984. Terminal Cretaceous extinctions in Hell Creek, area, Montana: compatible with catastrophic extinction. Science, 223:1177-1179.

Sloss, L.L., 1972. Synchrony of Phanerozoic sedimentary-tectonic events of the North American craton and the Russian Platform. Sect. 6 in Stratigraphie et Sedimentologie: 24th Int. Geol. Cong. (Montreal).

Southam, J.R. and W.W. Hay, 1982. Global sedimentary Mass Balance and sea-level changes. *In*. C. Emiliani (ed.): *The Oceanic Lithosphere*, J. Wiley and Sons, N.Y. *The Sea*, volume 7, pp. 1617-1684.

Suess, E., 1906. *The face of the Earth*. vol. 2, Clarendon Press, Oxford, 556 pp.

Surlyk, F. and M.B. Johansen, 1984. End-Cretaceous branchiopod extinctions in the Chalk of Denmark. Science, 223:1174-1177.

Talwani, M., G. udintsev et al., 1976. Tectonic synthesis. Initial Repts. D.S.D.P., vol. 38, U.S. Govt. Printing Office, Washington, D.C. pp. 1213-1242.

Talwani, M., and O. Eldholm 1977. Evolution of the Norwegian-Greenland Sea. Geol. Soc. Am. Bull. 88:969-999.

Talwani, M., J. Mutter and O. Eldholm, 1981. The initiation of opening of Norwegian Sea. Oceanologica Acta., No. Sp., pp. 23-30.

Termier, H., and G. Termier, 1960. Atlas de Paleogeographie: Paris, Masson and Cie, 99 pp.

Thiede, J., 1980. Pale-oceanography, margin stratigaphy and paleo-physiography the Tertiary North Atlantic and norwegian-Greenland Sea. Phil. Trans. R. Soc., London, A294:177-185.

Thierstein, H.R., 1979. Paleoceanographic implications of organic carbon and carbonate distribution in mesozoic deep sea sediments. *In*. M. Talwani and W.B.F. Ryan (eds.): *Continental margins and paleoenvironment*. M. Ewing Series, vol. 3. Am. Geophys. Un., Washington, D.C. pp. 249-274.

Thierstein, H.R., 1983. Trends and events in Mesozoic oceans. *In*. CNC/SCOR. Proceedings Joint Oceanogr. Assembly 1982. GR. symposia. Canadian National Committee—Scientific Comm. on Oceanic Res., Ottawa, pp. 127-130.

Theirstein, H. and W.H. Berger, 1978. Injection events in Earth history. Nature, 276:461-464.

Thunnell, R.C., 1979. Pliocene-Pliestocene paleotemperatures and paleosalinity history of Mediterranean Sea, results from DSDP Sites 125 and 132. Mar. Micropal., 4:173.

Tucholke, B.E., 1979. Relationship between acoustic stratigraphy and lithostratigraphy in western North Atlantic Basin. Initial Repts. D.S.D.P., vol. 43, U.S. Govt. Printing Office, Washington, D.C., pp. 827-846.

Tucholke, B.E., 1981. Geologic significance of seismic reflectors in the deep western North Atlantic Basin. SEPM Spec. Publ. 32:23-37.

Vail, P.R., and J. Hardenbol 1979. Sea level changes during the Tertiary Oceanus, 22:71-79.

Vail, P.R. and R.M. Mitchum, Jr., 1979. Global cycles of sea-level change and their role in exploration. 10th World Petr. Cong., vol. 2, pp. 95-104, Bucharest, Romania.

Vail, P.R. R.M. Mitchum, Jr. and S. Thompson, III. 1977. Seismic stratigraphy and global changes of sea level, parts 3 and 4: Global cycles of relative changes of sea level. Amer. Assoc. Petrol. Geol. Mem., 26:63-97.

Vail, P.R., R.M. Mitchum, Jr., T.H. Shipley and R.T. Buffler, 1980. Unconformities the North Atlantic. Phil. Trans. Roy. Soc. London, 294:137-155.

Vail, P.R., J. Hardenbol and R.G. Todd, 1984. Jurassic unconformities, chronostratigraphy and sea-level chnges from seismic stratigraphy and biostratigraphy. GCS-SEPM Found. 3rd Annual Res. Conf. Proceed., pp. 347-364.

Van Andel, T.H., 1979. An Eclectic Overview of Plate Tectonics, paleogeography and paleoceanography. *In.* J. Gray, and A.J. Boucont, (eds.): *Historical biogeography, plate tectonics and changing environments.* Oregon State Univ. Press, Corvallis, pp. 9-25.

Van Andel, T.H., G.R. Heath, and T.C. Moore, Jr., 1975. Cenozoic history and paleoceanography of Central Equatorial Pacific Ocean, Geol. Soc. Am. Mem., 143:1-134.

Van Andel, T.H., G.R. Heath and T.C. Moore, Jr., 1976. Cenozoic history and paleoceanography of central Equatorial Pacific ocean: A synthesis based on D.S.D.P. data. A.G.U. Geophy. monograph, 19:281-295.

Van Couvering, J.A., et al., 1976. The terminal Miocene event. Marine Micorpal., 1:263-286.

Vella, P., 1967. Eocene and Oligocene sedimentary cycles in New Zealand. New Zealand J. Geol. and Geophys., 10:119-145 and· 1162-1164.

Vergnaud-Grazzini, C., C. Pierre and R. Letolle, 1978. Paleoenvironment of the Northeast Atlantic during the Cenozoic: oxygen and carbon isotope analyses at DSDP Sites 398, 400 and 401. Oceanol. Acta, 1, (3):381-390.

Vincent, E., J.S. Killingley and W.H. Berger, 1980. The magnetic epoch 6 carbon shift: A change in the ocean's $^{13}C/^{12}C$ ratio, 6.2 million years ago. Marine Micropal., 5(2):185-203.

Watts, A.B., 1982. Tectonic subsidence, flexure and global changes of sea level. Nature, 297:469-474.

Watts, A.B. and M.S. Steckler, 1979. Subsidence and eustasy at the continental margins of eastern North America. *In.* M. Talwani, et al., (eds.): *Deep Sea Drilling Results in the Atlantic Ocean: continental margins and paleoenvironment.* Am. Geophys. Union, Maurice Ewing Ser., 3:218-239.

Weissel, J.K. and D.E. Hayes, 1972. Magnetic anomalies in the southeast Indian Ocean. Am. Geophys. Un., Antarctic Res. Ser. 19:165-196.

Weissel, J.K., Hayes, D.E. and E.M. Herron, 1977. Plate tectonics synthesis: The displacements between Australia, New Zeland, and Antarctica since the late Cretaceous. Mar. Geol. 25, 2312-277.

Weissert, H., 1981. The environment of deposition of black shales in early Cretaceous: an on-going controversy. *In.* J.E. Warme et al. (eds.): *The Deep Sea Drilling Project: A decade of progress.* SEPM Spec. Pub., 2:547-560.

Wise, S.W., et al. 1982. Paleontologic and paleoenvironmental synthesis of the southwest Atlantic Basin based on Jurassic to Holocene faunas and floras from the Falkland Plateau. *In.* C. Craddock (ed.): *Antarctic Geoscience.* Univ. of Wisconsin press, Madison, pp. 155-163.

Wolfe, J.A., 1978. A paleobotanical interpretation of Tertiary climates in the northern Hemisphere. Am.Sci., 66:694-703.

Woodruff, F., S.M. Savin & R.G. Douglas, 1981. Miocene stable isotopic record: a detailed deep Pacific Ocean study and its paleoclimatic implications. Science, 212:665-668.

Worsley, T.R. and I.A. Davies 1979. Sea-level fluctuations and deep-sea sedimentation rates. Science, 203:455-456.

Worsley, T.R., D. Nance and J.B. Moody, 1984. Sea-level and plate dynamics. *In.* B.U. Haq and J.D. Milliman (eds.): *Marine Geology and Oceanography of the Arabian Sea and Coastal Pakistan.* Van Nostrand Reinhold Co.. N.Y., pp. 233-251 (this volume).

Wright, R.C., and M.B. Cita, 1977. Geo- and biodynamic effects of the Messinian Salinity Crisis in the Mediterranean. Paleogeogr., paleoclimat., paleocol., 29:215-222.

14

Sea Level and Plate Dynamics

T. R. WORSLEY
D. NANCE
Ohio University,
Athens. Ohio,
and
J.B. MOODY
Battelle,
Columbus, Ohio

ABSTRACT

Continental freeboard and eustasy, as gauged by the relative position of the world shelf break with respect to sea level, have varied by + 250 m from today's ice-free shelf break depth of ~200 m, during the past 600 Ma.

Assuming constant or uniformly accreting continental crust and ocean water volume in an ice-free world, sea level fluctuations can be attributed to variation in the world ocean basin volume caused by changes in either its area or its depth relative to the world shelf break. An increase in volume and lowering of sea level occur as (1) the world ocean floor ages, cools and subsides; (2) accreting continents collide, thicken and decrease in area; and (3) poorly conductive continental platforms become thermally elevated due to a size-induced stasis over the mantle. Conversely, a decrease in the age of the world ocean floor, attenuation of continental crust during rifting, and an increase in continent number and mobility, will reduce the world ocean basin volume and raise sea level.

Theoretical sea level calculated from these principles correlates well with calibrated, first-order cycles of eustatic sea level change for the Phanerozoic. The record closely fits a simple model of retardation and acceleration of terrestrial heat loss during alternating periods of supercontinent accretion and fragmentation. Calibrated to sea level highstands, successive first-order marine transgressions and orogenic "Pangea" regressions characterize a self-sustaining, ~440 Ma plate tectonic cycle for the late Precambrian and Phanerozoic. The cycle can be recognized as far back as 2 Ga from the tectonic evidence of continental collision and rifting recorded in global orogenic peaks and mafic dike swarms, and may be related to major episodes of glaciation and evolutionary biogenesis.

INTRODUCTION

The growing body of evidence for long-term (i.e., $>10^8$ yr) episodicity in tectonic processes (Stille, 1924; Sloss, 1963; Holmes, 1951; Vinogradov and Tugarinov, 1962; Burwash, 1969; Condie, 1976, 1982; Fischer, 1981 and in press; Anderson, 1981, 1982; Mackenzie and Pigott, 1981) has important consequences for terrestrial heat loss and the differential terrestrial heating that ultimately powers plate motion. As heat production is a continuous process whose rate is smoothly declining through time (Fig. 1), demonstrated plate tectonic episodicity and its effects on the hydrosphere and biosphere must be manifestations of variable terrestrial heat loss. Furthermore, that this episodicity is at least in part a deterministic (nonrandom) process is demonstrated by the periodic recurrence of two and probably three Pangeas in the past one billion years (Windley, 1977; Condie, 1982). Given the insulating nature of continental tectosphere with respect to subcrustal heat flow (Anderson, 1981, 1982), clustered continents (Pangeas) should

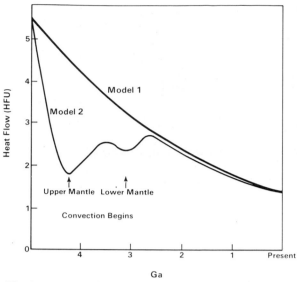

Fig. 1. Heat flow (HFU) at the earth's surface through time (Ga). Model 1 is for an early-convecting earth (McKenzie and Weiss, 1975) and Model 2 for a delayed-convecting earth (after Lambert 1980). Note that both models are virtually identical for the past 2.5 Ga – showing a smooth exponential decline of heat production.

retard and therefore promote hemispherical asymmetry of terrestrial heat loss. Oscillatory terrestrial heat loss would therefore appear fundamental to the repeated assembly and fragmentation of supercontinents.

Figures 2 through 5 illustrate recent attempts at compilation and synthesis of long-term quasi-periodic tectonic episodicity and form in large part the data set upon which we base our model of periodic Pangeas. The authors of these excellent syntheses agree that the driving mechanism of tectonic episodicity and its geological, climatological, and paleontological consequences are linked to episodic changes in heat flow patterns and suggest the cause of the changes to be variations in the intensity and style of convection patterns in the mantle. Anderson (1981, 1982) takes the reasoning one step further and offers evidence that the positions of the continents themselves are the cause, not the result, of changing mantle convection patterns. In this article we use the above information to

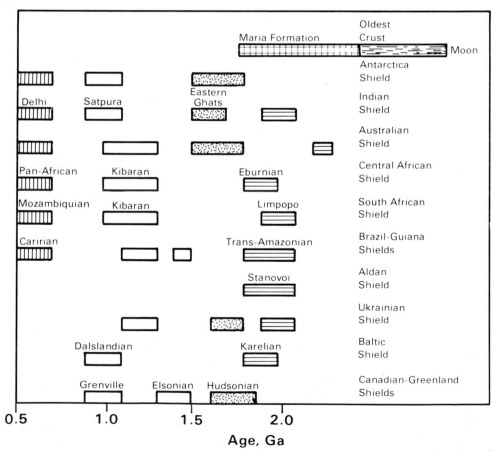

Fig. 2. Major orogenic periods 0.5 – 2.2 Ga after Condie (1976). Note that peaks ending at 1.9 and 1.6 Ga are not clearly resolved.

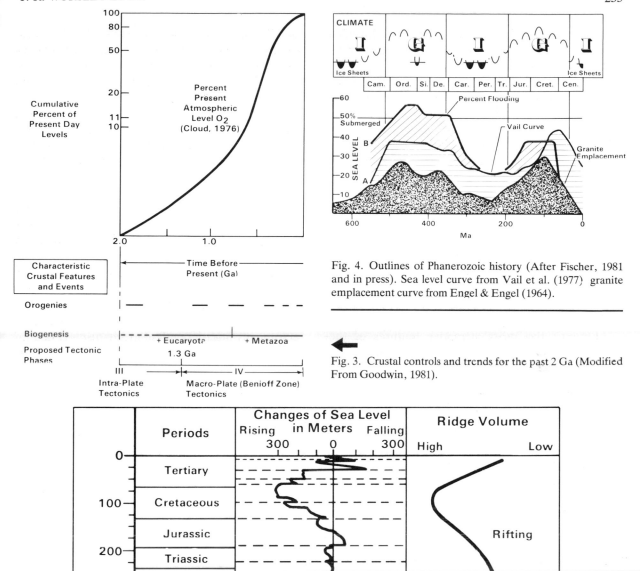

Fig. 4. Outlines of Phanerozoic history (After Fischer, 1981 and in press). Sea level curve from Vail et al. (1977) granite emplacement curve from Engel & Engel (1964).

Fig. 3. Crustal controls and trends for the past 2 Ga (Modified From Goodwin, 1981).

Fig. 5. Comparison of global sea-level changes and mid-oceanic ridge volume. Sea-level curve is after Vail et al. (1977), as calibrated by Pitman (1978) and adopted by Vail and Hardenbol (1979); the Vail et al. first-order curve, which averages out the sharp second-order sea level drops shown in this figure, is plotted on Fig. 14. Ridge volume is from Mackenzie and Pigott (1981) and Shanmugam and Moiola (1982).

demonstrate that relatively simple relationships govern the quasi-periodic behavior of the earth's tectonic history.

To reconcile the repeated formation of Pangeas, we will first summarize the factors that control continental freeboard, or the relative elevation of the continents with respect to sea level (Wise, 1974), and then outline a model that accounts for hemispheric, quasi-periodic acceleration and retardation of heat loss. The model produces a heat-driven, self-replicating cycle of clustered and scattered continents with consequent first-order effects on the freeboard, and hence climate and biogenesis (irreversible jumps in biotic complexity) consistent with available geologic data outlining plate motions for the past 2 Ga. Finally, we calibrate the model to the geologic record using the history of first-order eustatic sea level changes.

One of the most attractive features of the model is its ability to derive global geologic history on a ~50 Ma resolution using only plate tectonic and heat flow information. The averaging techniques used to construct the model recognize local or short-duration events only in proportion to their global contribution to the geologic record. Therefore, the model in its current version cannot be expected to represent, necessarily, either the geologic history of any given crustal block or the detailed history of freeboard in general. However, residual differences between predicted and measured values of eustatic sea level probably represent real second-order geologic events (Fig. 5), such as continental glaciation or continent-to-continent collision.

TECTONIC CONTROLS OF SEA LEVEL

Continental freeboard represents the complement to eustasy and both are most readily defined as average water depths at the world shelf break. The position of the shelf break consequently provides a convenient reference by which to gauge tectonic controls of sea level. In usual convention, a drowned world shelf would have a negative freeboard as the result of a positive eustatic sea level. However, for the purpose of avoiding confusion, we adopt the unorthodox convention of quantifying negative freeboard such that values for both freeboard and eustasy are positive when shelves are flooded and negative when they are emergent. Currently, the negative freeboard for the average world ocean is +130 m

(or 200 m for an ice-free world). However, negative freeboard has oscillated by more than 500 m during the last 600 Ma from deeper than +500 m (300 m below its present ice-free value) during the Late Cambrian/Early Ordovician and Late Cretaceous, to depths less than zero (shelf break emergent) during the latest Precambrian and Early Permian (Fig. 5). Freeboard changes as a function of the volume of water in the world ocean, and the volume of the world ocean basin itself. The main mechanism for changing the volume of water in the ocean is to sequester a portion as ice caps on continents, a process that can lower sea level by as much as 200 m (Pitman, 1978). We will return to water-volume effects when we discuss second-order feedbacks to the primary factor of basin volume.

The volume of the world ocean basin is controlled by three geometrically distinct but tectonically interdependent mechanisms (Fig. 6):

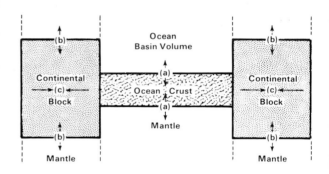

Fig. 6. Controls of ocean basin volume: (a) the thermally controlled bathymetry of the ocean floor, (b) thermally controlled continental elevation; and (c) orogenically controlled continental compression. Continental compression will decrease continental area and hence increase ocean volume. Conversely, continental extension during rifting will decrease ocean volume.

(a) the mean elevation of the world's ocean floor
(b) the continent/ocean basin areal ratio (or simply the area of the continents), and
(c) the mean elevation of the continental blocks.

Each of these controls on volume of the world ocean basin will be discussed in the following sections.

Ocean Floor Elevation

Sclater et al. (1971) have shown that oceanic lithosphere thickens, cools, and subsides with increasing age. Davis and Lister (1974) have demonstrated that the depth increases linearly with square root of age and Parsons and Sclater (1977) have fit the age versus depth relation to the equation:

$$d(t) = 2,500 + 350\ t^{1/2}$$

where d = present day bathymetry in meters and t = age in Ma. As we are concerned with the vertical distance between the ocean crust and the shelf break, we subtract 130 m to yield:

$$d(t) = 2,370 + 350\ t^{1/2}$$

From this equation the seafloor will obviously subside if the world ocean crust ages and shoal if it becomes younger. For example, all other factors held constant, the shelf break to seafloor depth would be 4,845 m $(2,370 + 350\ (50)^{1/2})$ for 50 Ma ocean crust (Fig. 7a) but would subside to 5,081 m

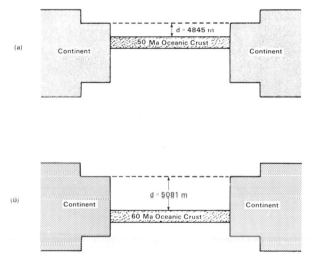

Fig. 7. Bathymetry of (a) 50 Ma and (b) 60 Ma ocean floor (Parson & Sclater, 1977) and consequent increase in ocean volume assuming fixed continental blocks.

as the world ocean floor ages to 60 Ma $(2,370 + 350\ (60)^{1/2}$, yielding a 236 m drop in sea level (Fig. 7b). In practice, numerous assumptions are tied to these numbers and the effects of loading and the extent to which oceanic and continental lithosphere are decoupled, is ignored. Nonetheless, Russell (1968), Larson and Pitman (1972), Hays and Pitman (1973), Turcotte and Burke (1978), and Harrison (1980) have used the

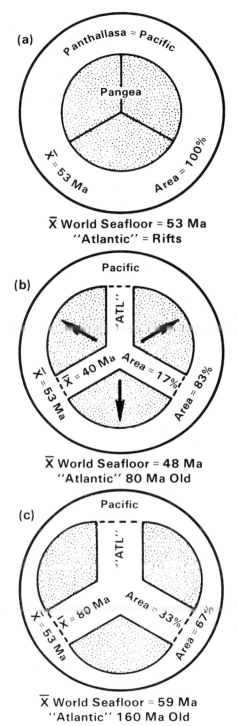

Fig. 8. Schematic representation of the Berger & Winterer (1974) fragmenting Pangea model: (a) represents a Pangea ready to fragment surrounded by 53 Ma Panthalassa; (b) shows the system 80 Ma after rifting. "Atlantic-type" oceans of 40 Ma average age represent 17 percent of the world's sea floor. The weighted average of "Atlantic" and "Pacific-type" ocean floor yields a world's sea floor age of 48 Ma; (c) represents the system 160 Ma after rifting; "Atlantic-type" oceans now average 80 Ma and represent 33 percent of world ocean floor yielding a world seafloor age of 59 Ma; this stage illustrates today's world ocean.

age versus depth relationship to correlate high sea level with young world ocean floor. World ocean floor can become younger either by increasing seafloor spreading rates while retaining constant ridge length or by increasing ridge length while retaining constant spreading rates or a combination of both, as either serves to increase ridge volume and hence raise sea level. As Pangea breakup increases ridge length (see Southam and Hay, (1981) for a review), we start with the assumption (as do Berger and Winterer, 1974) that seafloor spreading rates remain largely constant.

Assuming uniform spreading rates equivalent to today's, Berger and Winterer (1974) have calculated the age of the world ocean crust as a function of the breakup of a Pangea. During such an event, a 100 percent "Pacific-type" world ocean bordered by active margins would evolve into one with a uniformly increasing proportion of "Atlantic-type" ocean bordered by passive

margins. We have adapted their general model here (shown schematically in Figure 8) but have calculated ocean crust bathymetry using our modified $d(t) = 2{,}370 + 350/t^{1/2}$ relationship instead of their empirical age-depth curve. We can see from Figure 9 that the Berger and Winterer world ocean has a mean age of 53 Ma and shelf break to seafloor bathymetry of 4,918 m during Pangea, producing an ice-free shelf break depth of + 340 meters.

However, if the relationship shown in Figure 9 is correct and seafloor age (and therefore elevation) were the only factor controlling sea level, we would expect an ice-free shelf break depth of +340 m during Pangea to deepen to +463 m 80 Ma later when the age of world ocean crust decreases to 48 Ma and achieves a shelf break to seafloor bathymetry of 4,795 meters. This sea level maximum is in phase with empirical eustasy curves but Pangeas are known to have had minimal, if any, shelf break flooding (see Fischer, in press).

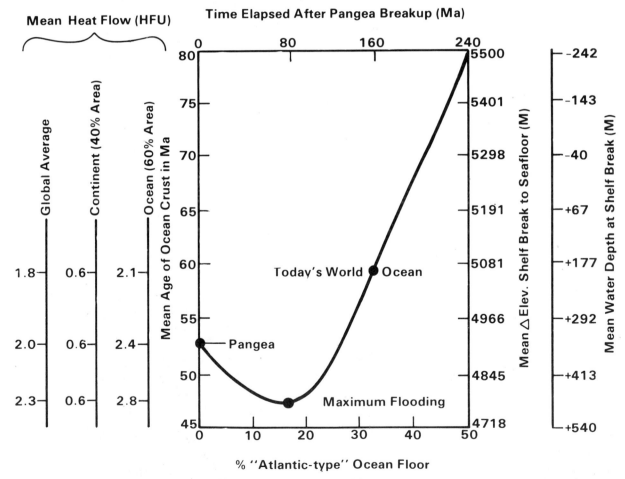

Fig. 9. Age, bathymetry, and mean heat flow of the world ocean as a function of the break-up of Pangea, modified from Berger & Winterer (1974), Parsons & Sclater (1977), and Harrison (1980). The mean shelf break depths assume today's ice-free sea level at + 200 m.

Thus, we conclude that seafloor elevation alone cannot be the sole control over water depths at the shelf break, and that either or both continental area and continental elevation must also influence freeboard. The most likely modulator of continental area is orogeny while that of continental elevation is heat flow retardation and heat storage beneath the continents.

Continental Area

If we assume, as a first approximation, that continental volume is constant and fix all other factors, world continental area will be inversely proportional to world continental elevation (i.e., volume of a slab = base × height). Continental area is reduced and elevation increased during orogeny (Fig. 10) while the reverse is true during rifting.

Orogeny occurs both by ocean crust subduction (i.e., Andes Mountains, Fig. 10b) and continent to continent collision (i.e., India - Asia, Fig. 10c). Ocean crust subduction is a process that can occur more or less continuously (as in the Pacific which has likely been bounded by subduction for most of the Phanerozoic) or discontinuously (as in "Atlantic-type" oceans that are bounded by passive margins during opening, and active margins during closing). The extent of orogeny (and therefore continental compression or ocean basin expansion) is largely dependent upon the rate of subduction and the age of the crust being subducted. Maximum orogeny is commonly attained during "Chilean-type" subduction (Uyeda and Kanamori, 1979) (Fig. 10b) when young, positively buoyant crust is rapidly consumed (Russell, 1968). Minimal orogeny, and perhaps even continental expansion via back-arc basin development (Molnar and Atwater, 1978), tends to occur during "Marianas-type" (Uyeda and Kanamori, 1979) subduction when old, negatively buoyant ocean crust is slowly consumed.

As a first approximation, we can consider the Pacific to be an equilibrium situation in which the average age of crust being consumed does not change through time, only in location (Berger and Winterer, 1974). Consequently, "Pacific-type" oceans will not significantly change continental area or sea level. However, the opening and closing of "Atlantic-type" ocean produces a very different result. Initial rifting during the opening of "Atlantic-type" ocean (Fig. 8), yields

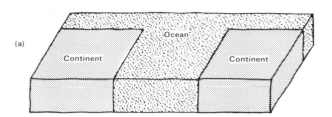

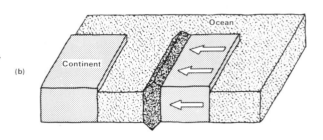

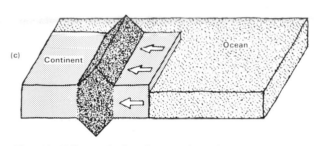

Fig. 10. Effects of changing continental areas on ocean volume, all other variables held constant: (a) static situation in which the shelf break is at sea level; (b) activation orogeny and crustal foreshortening caused by "Chilean" type subduction. Decrease in total continental area increases ocean basin volume and therefore lowers sea level; (c) continent-continent collision results in a sharp decrease in continent-ocean areal ratio and rapid lowering of sea level.

continental extension, which increases continental area, and therefore tends to raise sea level. Continued opening produces no further change in continental area. As an "Atlantic-type" ocean finally stops expanding and begins to close, the old ocean crust that is first subducted is likely to have little effect on continental area except for slight continental stretching via back-arc spreading (Molnar and Atwater, 1978). However, as closing continues, increasingly younger ocean crust must be consumed until finally the ridge itself must be subducted as the converging continents approach collision. Such a situation can only produce continental compression, reduction in area, and a lowering of sea level. In summary, then, ocean margin orogeny of "Pacific-type" ocean appears to occur at a fairly constant rate and is unlikely to change sea level significantly. On the other hand, opening and closing of 'Atlantic-type" ocean should strongly influence sea level through orogenically-driven modulation.

Continental collision is rapid, spectacular, and episodic. In contrast, the fairly slow and even processes of seafloor spreading and subduction can produce only long-term sea level change. As such, continent-continent collision is the one tectonic mechanism that can rapidly alter continent-ocean areal ratio, and therefore produce rapid decreases in sea level. A present-day example of continent-to-continent collision is the $2\text{-}3 \times 10^6$ km² parcel of double thickness (~ 80 km) continental crust of the Tibetan Plateau. Berger and Winterer (1974) estimate that the collision lowered sea level about 40 m, or 13 meters for each 10^6 km² areal reduction. Harrison et al., (1981) calculate the change as 11 meters for each 10^6 km² area reduction. Continent-to-continent collision, therefere, appears to be the only good tectonic candidate for the nonglacial, second-order eustasy cycles of Vail et al. (1977), which number about 25, and involve a 50 to 100 m lowering of sea level (Fig. 5). If collisions occur relatively frequently (i.e.; every 25 to 50 m.), as during a Pangeas assembly, their cumulative effect could sustain an average sea level drop of perhaps 50 m considering that mountain ranges erode exponentially with a half-life of about 10 Ma (see Harrison et al., 1981)

Applying these considerations to the breakup and assembly of Pangeas, we would expect a rise in sea level to accompany initial Pangean fragmentation, as rifting is the dominant process. As "Atlantic-type" oceans begin to grow, continent-to-continent collision should be infrequent as Pangea dispersal is essentially radial (Fig. 8). Orogenic activity should therefore exert minimal influence on sea level. However, as dispersal ceases and a new Pangea begins to accrete, increasingly younger ocean crust is subducted in the closing "Atlantic-type" oceans. The continent-ocean ratio should therefore decrease, thereby lowering sea level. At the same time, the continents are moving inward, and the frequency of continental collision should greatly increase, further decreasing continental area and lowering sea level. Interestingly, orogenic processes that tend to change sea level are exactly in phase with, and therefore reinforce, the modulation of seafloor age discussed previously. In summary, then, Pangea fragmentation initially yield continental extension through rifting, young world ocean crust, and minimal orogeny; all of

which raise sea level. Continued expansion of "Atlantic-type" ocean, however, should begin to age the world ocean floor and lower sea level. Renewed Pangea assembly will reproduce initial conditions, as is explained later in this paper.

Continental Elevation

Unlike the ocean crust, whose elevation can be calculated from thermal considerations alone, the elevations of the continents are governed by the thickness of continental crust and underlying depleted mantle (tectosphere) in addition to temperature (Jordan, 1979; Anderson, 1981, 1982). Continents can be subdivided physiographically into orogenic belts and cratons. Orogenic belts are characterized by compressionally heated and thickened crusts overlying relatively mobile and thin tectospheres (Jordan, 1979). Their high elevation is overwhelmingly controlled by crustal thickening resulting from plate collision (Fig. 10). However, a significant thermal component of elevation is involved, which may migrate laterally to heat and buoy continental platforms and shields adjacent to orogenic belts (Condie, 1982; Dewey and Burke, 1974). The oceanic margins of orogenic belts do not exhibit easily defined shelf breaks and therefore form a poor gauge with which to measure freeboard.

Cratons, on the other hand, are characterized by a fairly uniform crustal thickness of about 35 km (Harrison, 1980; Condie, 1982) underlain by a thick layer of depleted mantle, forming a tectosphere extending to depths in excess of 300 km (Jordan, 1979). As thicknesses are uniform, changes in elevation of continental platforms and shields are governed primarily by changes in tectosphere temperature and consequent changes in phase (Jordan, 1979). In addition, as tectosphere is poorly conductive, thick platform and shield tectospheres are insulators that retard heat flow from the mantle below (Jordan, 1979; Anderson, 1981, 1982; Worsley et al., 1982). We therefore base our calculations of heat controlled freeboard changes upon the stable platform component of continents using today's Africa as an initial model.

Today's Africa differs from the other continents in several significant ways. Unlike all but an ice-free and isostatically compensated Antarctica, Africa is composed almost entirely of stable platform and its negative freeboard is at -200 m (shelf break 200 m above sea level [Harrison et al., 1981; Hay and Southam, 1977]) as opposed to the world average of + 200 m. It has undergone less Phanerozoic flooding than any other continent (Wise, 1974) and is stationary with respect to the mantle within the resolution of plate navigational indicators (McDougall and Duncan, 1980). We propose, following the models of Jordan (1979), Anderson (1981, 1982), and Worsley et al., (1982), that Africa's extreme shelf break elevation, without benefit of orogenic or crustal heating, is a direct consequence of its near stasis over the mantle for perhaps as long as 120 Ma (Burke and Wilson, 1976) or even 300 Ma (Anderson, 1982). As a result, heat buildup at the base of its tectospheric insulating layer has caused it to be thermally buoyed 400 m. Other continental platforms that are in motion over the mantle encounter cool subducted lithosphere beneath their leading edges and are able to dissipate subtectospheric heat buildup through the ocean crust at their trailing edges. When a continent elevates, its tectospheric "roots" deepen and it becomes more difficult to move. Africa's stasis over the mantle therefore may be partly caused by its thick tectospheric anchor. We suggest here that tectospheric thickness is inversely proportional to a continent's tendency to move with respect to the mantle and increases with continental area (Fig. 11). Therefore, the larger the continent, the more likely it is to be stationary. Pangeas (the largest possible continents) should be the most static, and hence the most highly elevated. Geologic evidence (Hay and Southam, 1977; Burke and Wilson, 1976; McDougall and Duncan, 1980) suggest this to be the case

In addition to obvious collision-produced mountains, therefore, the thermal insulation properties of continents (Anderson 1981, 1982) also tend to buoy continents. Reduced heat flow from beneath the continents averages about 0.6 HFU (Sclater et al., 1980), whereas the long-term world average heat flow from beneath ocean crust is about 2.4 HFU (Harrison, 1980), changing with time over 80 Ma period within a total range between 2.1 and 2.8 (Fig. 9). Average heat flow from beneath ocean crust is therefore 4 times greater than that from under continents (Sclater et al., 1980). As oceanic crust represents about 60 percent of the earth's surface (Condie, 1982), total mantle heat dissipated through the world ocean basin is about 6 times greater than that dissipated through the continents.

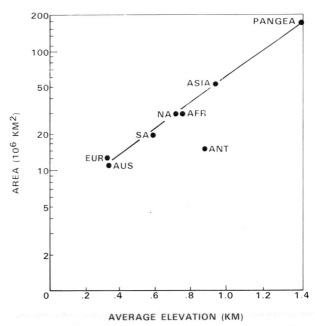

Fig. 11. Continental area versus continental elevation modified from Hay & Southam (1977).

Figure 12 illustrates the effects of a Pangea stasis based mainly on the ideas of Anderson (1982) and Worsley et al. (1982). Pangeas have large areas and hence tend to average out mantle driven tendencies to move in a particular

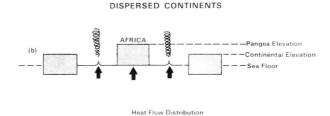

Fig. 12. Schematic thermal subsidence of a rifting Pangea (a) an assembled Pangea, that is stationary over the mantle, retards heat flow and is therefore thermally domed: (b) Pangea breakup with thermal subsidence of dispersing continents accompany gravity sliding off the dome. Thermally elevated Africa remains stationary.

direction. We show a thermally domed Pangea ready to rift and fragment in Figure 12a. The fragments moving outward over the mantle dissipate their subcrustal heat via seafloor spreading processes in the newly-formed "Atlantic-type" oceans (Fig. 12b). The central block (Africa) has not moved and therefore remains thermally domed. Harrison (1980) has shown that oceanic heat loss was about 35 percent greater in the Late Cretaceous than it is today, because the late Cretaceous ocean floor was younger than today's. We plot Harrison's results in modified form in the left-hand column of Figure 9 for Pangea, the Late Cretaceous maximum (80 Ma post-Pangea), and today. Figure 9 indicates that heat flow results are clearly concordant with the Berger and Winterer (1974) model of seafloor age versus time elapsed since breakup of a Pangea (upper scale of Fig. 9). We therefore believe that heat-flow retarding continents tend to thermally dome when they become static over the mantle and/or attain a large area. Stasis and doming would indicate an intermediate global heat flow during Pangea because of an intermediate seafloor age, but a high elevation because of its size. After breakup, thermal relaxation of the continental fragments would ensue. Oceanic heat flow would attain a maximum 80 Ma after breakup and decrease thereafter until Pangea dispersion reverses and a new Pangea begins to assemble, thus initiating a new cycle.

In constructing our model we use today's Africa, whose shelf break is 400 m higher than those of the other continents (Harrison et al., 1981), as a conservative estimate of the shelf break elevation of a Pangea. We do not know whether the 400 m of doming is caused by temperature change (Anderson, 1982) or a volume-increasing phase change (Jordan, 1979; Smith, 1982) but think the latter to be more likely. In any case, we suggest that the continents were at least 400 m more elevated during a Pangea.

We now use the information summarized above to construct a model of a self-sustaining, heat-driven cycle of alternating assembly (retarded heat flow) and fragmentation (accelerated heat flow) of Pangeas. The model can, in turn, be used to predict the tectonic component of eustatic sea level and its effects on climate and the biosphere, including the feedbacks that promote or discourage sequestering of water on ice caps.

THE MODEL

Outline of the Model

We postulate that a strong global asymmetry developed or already existed about 2.5 Ga ago when the first true granitic continental crust formed (Windley, 1977), dividing an oceanic (Panthalassa) from a continental (Pangea) hemisphere. The oceanic crust and its associated hydrothermal system are far more efficient (Sclater et al., 1980) than continental crust at maintaining the heat transfer necessary to preserve the earth's thermal balance. Once hemispheric asymmetry developed in the early Proterozoic, the asymmetry persisted thereafter because the Panthalassan hemisphere always remains an area of general mantle upwelling that pushes continents towards the antipodal Pangean hemisphere. In this model, Panthalassa and its "Ring of Fire" would represent the convective tectonic analog of an ocean gyre, Jupiter's great red spot, the Tharsis region of Mars, or the Hadley cells that dominate atmospheric heat transfer; all of which are thermally self-sustaining convection cells that persist because of the system's configuration, or simply because of their pre-emptive first appearance. In support of a permanent Panthalassa, whose present day manifestation is the Pacific, is the complete lack of geologic evidence that eastern Asia and the western Americas were ever joined, coupled with the widely held view that the present Pacific coastline of North America has been the approximate site of a major continent-ocean boundary for the last billion years (Windley, 1977).

The less efficient heat transfer through a continental Pangean hemisphere would result in a tendency towards thermal doming and perhaps hot spot production (e.g., Burke and Wilson, 1976; Anderson, 1981, 1982). The observed degree of doming should be proportional to the continental area and the absolute continent velocity with respect to the mantle, as well as the heat-producing continental collisions of Pangean assembly. Larger continents would tend to be more stationary than smaller ones because they average tendencies toward convection in a particular direction. Hence Pangeas will tend toward maximum stasis and elevation. In addition, repeated heat-producing continent-to-continent collisions during the assembly of a Pangea, and simultaneous aging and subsidence of the world ocean crust would all argue for maximum marine regression. The anomalous elevation of modern passive-margin-bounded Africa, which is 400 m higher than its crustal thickness and area would predict, may therefore reflect thermal doming resulting from its current stasis with respect to the mantle (Burke and Wilson, 1976; Anderson, 1981, 1982). Furthermore, as a Pangea would cover 40 percent of the globe, the propensity toward continental glaciation during a Pangean episode would be high (see Fischer, in press), further increasing the relative elevation of the supercontinent by storing water on ice caps and hence further lowering sea level.

Pangeas, however, are likely to be relatively short-lived, as prolonged stasis over the mantle will ultimately cause sufficient thermal doming for hot-spot-induced rifting (Burke and Wilson, 1976; Anderson, 1981, 1982) to be initiated, followed by gravitational sliding of the continental fragments into the adjacent, thermally subsided Panthalassa hemisphere. At the same time, the convective conduits of the newly-formed rifts and subsequent young oceans allow for cooling of the Pangea thermal dome, a process we may be witnessing along today's East African Rift and Red Sea. During fragmentation, the average age of the world's ocean crust will become younger and oceanic heat flow will increase, yielding shallower ocean floor and consequently higher sea level as cold, maximally aged Panthalassa crust is replaced with youthful, thermally elevated, "Atlantic-type" oceanic crust. In addition, rifting stretches the continental crust and increases the continental area, which also raises sea level. Decreasing age and areal shrinking of the world ocean coupled with thermal subsidence of Pangea fragments, combine to produce a sea level rise that will continue until both the leading and trailing margins of the fragmented Pangea are finally surrounded by oceanic crust of equivalent age, and the thermal dome formerly under Pangea is convectively dissipated. Following thermal dome dissipation, the impetus of breakup ceases and the convective pressure of Panthalassa reasserts itself, initiating subduction in the oldest parts of the former "Atlantic-type" oceans that leads to their ultimate closure, and a repetition of the cycle.

Calibration of the Model

A crude quantification of the Pangea accretion-fragmentation cycle is most readily achieved through its effect on freeboard with time. Whether one assumes little or no continental accretion and a constant volume of water or continually increasing continent and ocean water volume, freeboard is a two-component system in which the position of sea level with respect to the shelf break is determined by (1) the total volume of the ocean basins as governed by seafloor elevation and (2) the relative elevation of the continental platforms as determined by their areas and subcrustal heating. In this preliminary attempt at quantification, we assume constant continental and ocean water volume in an ice-free world and do not consider the isostatic effects of water or sediment loading. Rather, we simply use today's ice-free water depth at the world shelf break (+200 m) as our initial calibration point.

The above assumptions are admittedly simplistic but they are adequate for calculating relative phase and amplitude of components comprising the cycle. Future formulations allowing for secular continental and seawater accretion, isostatic feedback effects due to water loading, erosion and sedimentation, and sequestering of water on ice caps will only modify amplitudes and introduce secular trends into the cycle. Those effects will not alter the positions of the major peaks and troughs nor their relative amplitudes.

Sea Floor Component

Our modified age-depth equation ($d(t) = 2,370 + 350 t^{1/2}$) for ocean floor with respect to the world shelf break shows a smooth decline from an average shelf-break to ridge-crest elevation (0 Ma) difference of -2,370 m to a difference of about -5,081 m for a crustal age of 60 Ma. "Pacific-type" ocean floor with its rather uniform age of 53 Ma (Berger and Winterer, 1974) corresponds to shelf break-seafloor difference of 4,918 m. "Atlantic-type" oceans, on the other hand, show an elevation inversely proportional to the time elapsed since their inception by continental rifting. The oldest parts of today's marine Atlantic (its margins), for example, average 140 Ma in age. Hence the average age of its seafloor is 70 Ma, corresponding to shelf-break basin-floor difference of 5,298 m (Fig. 9).

Berger and Winterer (1974) point out that a Pangea would be surrounded by 100 percent "Pacific-type" ocean floor of 53 Ma age and 4,918 m shelf break-ocean floor difference, and that Pangea breakup at today's uniform spreading rates would progressively replace "Pacific-type" ocean floor with "Atlantic-type". The globally averaged age of the sea floor according to their model would decrease from 53 Ma to a minimum of 48 Ma in 80 Ma when the ocean floor was 17 percent "Atlantic-type" (Fig. 9). During this period, the average seafloor elevation would rise 123 m from 4,918 m to 4,795 m. However, as the proportion of "Atlantic-type" ocean floor continues to increase beyond 17 percent, the globally-averaged age (and depth) of the seafloor would again begin to increase as "Atlantic-type" floor ages. Today's world ocean comprises about one-third "Atlantic-type" ocean crust (average age 70 Ma) and two-thirds "Pacific-type" ocean crust (average age 53 Ma) yielding an average global age of 59 Ma, equivalent to an average global shelf break-ocean floor difference of 5,058 m. Despite the oversimplifications used in deriving these figures, they compare closely with the measured age and bathymetry of 59 Ma and 5,156 m respectively (Berger and Winterer, 1974; Harrison, 1980). Furthermore, we believed that today's Atlantic and parts of the Indian ocean have achieved maximal crustal age as oceans bounded by passive margins, and must soon begin closure. Our reasoning for this conclusion is discussed next.

Assuming that Africa is stationary (Burke and Wilson, 1976; McDougall and Duncan, 1980) and represents the remains of the former thermal dome of Pangea (Anderson, 1981, 1982), the North and South American Plates have been "pulling" away from Africa by sliding down the periphery of the old thermal dome with the mid-Atlantic Ridge migrating westward at 1/2 the speed of the North and South American Plates to preserve symmetry (Burke and Wilson, 1976). Compressive stress along the North American Atlantic margin (Zoback and Zoback, 1980) but not along the West African Margin (Condie, 1982) suggests that the Americas have now reached the "bottom of the trough" between the former Pangean dome and the Panthalassan convection system now represented by the East Pacific Rise. In addition, plate tectonic considerations (Molnar and Atwater, 1978; Carlson et al., 1982) suggest that seafloor older than 160 Ma forms a thermal "inversion layer" and is thus isostatically unstable and ultimately self-subducting. The margins of the Atlantic have, in places, reached this age and it

seems likely that subduction, already in progress in the Carribean, will ultimately occur along the oldest parts of the compressively stressed West Atlantic. When subduction completely uncouples the American continental margin from the floor of the Western Atlantic via "Marianas-type" subduction, the impetus for the Americas to slide westward ceases, as they are now bounded by a "Chilean-type" subduction system with fast-spreading, young, elevated crust to the west and a "Marianas" or back-arc type with slow-spreading, old, cold, subsided crust to the east. At this point, the young, elevated, fast-spreading Pacific should begin to "push" the Americas back into the old, subsided, slow-spreading Atlantic until the Atlantic narrows and closes. Progressive conversion of the Atlantic into a "Pacific-type" ocean by selectively subducting old, dense lithosphere, will young its ocean crust until the 53 Ma average of "Pacific-type" ocean floor is achieved. At this point, the average global age of oceanic crust will again be 53 Ma and its average elevation -4,918 m.

The upper graph of Figure 13 shows the estimated Phanerozoic seafloor component of negative freeboard corresponding to two Wilson cycles of continental scattering and convergence. The model is based on that of Berger and Winterer (1974) as constrained by the ages and durations of Pangeas (Ziegler et al., 1979; Bambach et al., 1980; Barron et al., 1981). The model is further constrained by the 520 Ma and 80 Ma ages of the two first-order sea level maxima of Vail et al. (1977), as calibrated by Pitman (1978), each of which are taken to correspond to periods of 17 percent "Atlantic-type" ocean. Accordingly, relative to today's ice-free sea level elevation above the shelf break of + 200 m, the seafloor component of sea level is + 340 m during a Pangea, + 463 m 80 Ma after Pangea breakup (at 17 percent "Atlantic-type" ocean), + 200 m 160 Ma after breakup (at the maximal 1/3 "Atlantic-type" ocean, corresponding to today's sea level), and finally + 340 m when a new Pangea accretes and the cycle is completed.

Continental Platform Component

The calculated continental platform component of eustatic sea level for two Wilson cycles is shown on the lower graph (Fig. 13). A Pangea stasis over the mantle should result in at least 400 m of thermal doming based on the

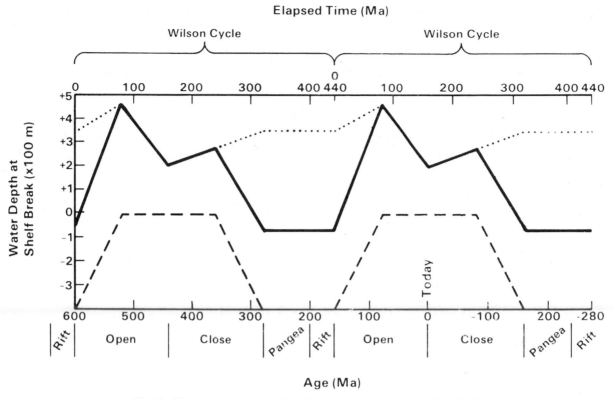

Fig. 13. Tectonic components of sea-level change (see text for explanation).

present elevation of stationary Africa. Pangea breakup, gravitational sliding of continental fragments off the thermal dome, and opening of young "Atlantic-type" oceans with high heat flow would ultimately result in dissipation of the 400 m thermal dome. Thermal relaxation models (Sclater et al., 1971; Sleep, 1971) suggest this dissipation would be largely completed 80 Ma after the initiation of drifting and hence coincident with a 17 percent "Atlantic-type" world ocean (Fig. 9). Aging of "Atlantic-type" oceans to an average age of 80 Ma would yield a maximum age of 160 Ma at their passive margins, making them vulnerable to "auto-subduction", back-arc spreading and eventual closure. Assuming closure to be temporally symmetric as appears to have been the case for Iapetus (Williams, 1980), a new Pangea should accrete after a further 160 Ma. Renewed thermal doming would elevate this Pangea for about 80 Ma (thermal models of Sclater et al., 1971; Sleep, 1971; Jordan, 1979; Anderson, 1982). At this point the rifting and breakup part of the cycle is once again initiated. Using the history of the modern day East African Rift and that of the opening of the Atlantic as our calibration points, we propose (as do Southam and Hay [1981]) that

the "rifting without drifting" phase of a Pangea breakup lasts 40 Ma, (total duration of Pangea 120 Ma) when renewed drifting begins the next 440 Ma cycle.

The middle curve of Figure 13 is our model of ice-free, negative freeboard for the Phanerozoic. This curve represents the algebraic addition of the seafloor (upper curve) and continental platform (lower curve) components. Negative freeboard is -60 m (sea level 60 m below the shelf break) at Pangea breakup, rises 523 m in 80 Ma to a maximum of 463 m, falls 263 m to a +200 m minimum at 160 Ma, rises to a 275 m maximum at 240 Ma before falling to -60 m at 320 Ma. Freeboard remains at about this level for 120 Ma, at which time Pangea rifts and drifts apart, and a new 440 Ma Wilson cycle begins. At present, we are 160 Ma into the current cycle and can expect coalescence of a new Pangea in 280 Ma if our model is correct.

Testing the Model

To test our model, we compiled the Phanerozoic history of three variables of the geologic record for which we believe plate tectonics is a controlling or strongly modulating

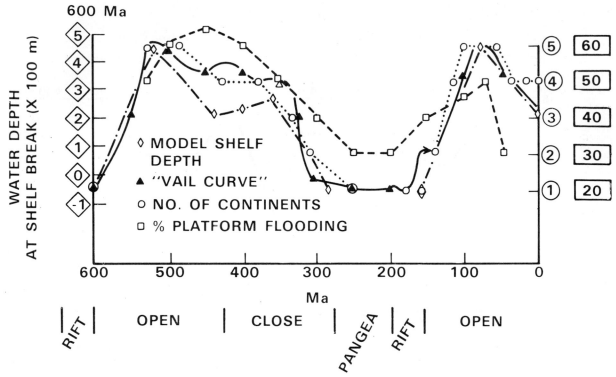

Fig. 14. Comparison of model to three independently derived geological.data sets modified from Worsley et al. (1982).

◇ = model shelf depth as calculated in Fig. 13.
▲ = first-order sea level curve of Vail et. al. (1977), as calibrated by Pitman (1978).
O = number of continents (Ziegler et al., 1979; Bambach et al., 1980; Barron et al., 1981).
□ = % platform flooding, which is an average of the results of Termier and Termier (1952) and Strakhov (1948) as
     presented in Hallam (1981)

factor (Fig. 14). These variables are compared with our calculated freeboard curve. The variables include (1) the number of continents (Ziegler et al.,1979; Bambach et al., 1980; Barron et al., 1981), (2) average continental freeboard as reflected in the Vail et al., (1977) Phanerozoic eustatic sea level curve as calibrated by Pitman (1978), and (3) the Termier and Termier (1952) and the Strakhov (1948) percentage platform flooding curves. As can be seen from Figure 14, a strong covariance exists between our model's predictions and all three variables throughout the Phanerozoic. We now discuss in detail the data sources, modifications, and methods of plotting utilized in deriving Figure 14.

Number of Continents

Data concerning the number of continents during the Phanerozoic (Fig. 14) were derived from Ziegler et al. (1979) and Bambach et al. (1980) for the Paleozoic and Barron et al. (1981) for the Mesozoic-Cenozoic required degradation

to the resolution available for the Paleozoic in order to make the two counts comparable. For example, the periods 0 Ma, 10 Ma, 40 Ma, and 60 Ma are represented by Barron et al. (1981) as having 6, 7, 7, and 6 continents respectively whereas we plot 5, 4, 4, and 4. The excellent seafloor-spreading-based paleogeographic resolution of the past 180 Ma and a knowledge of present continental configuration allows the precise timing of separation of India and Australia from Africa and Antarctica to be known. At the resolution of Ziegler et al. (1979) we cannot clearly differentiate closely adjacent from continuous continents, and as both will show similar isostatic behavior, we adjust the Mesozoic-Cenozoic to the same "lumping" bias as exists for the Paleozoic. In a similar fashion, we do not tally fragments smaller than India.

Regardless of this secular bias, which can only alter the amplitude and not the phase of the correlation, a striking covariance exists between the number of continents and freeboard. Hay and Southam (1977) and Harrison et al. (1981) have demonstrated the existence of a direct exponential relationship between the area and average elevation of today's continents (Fig. 11). Consequently, dispersed continents with relatively small average areas will have low elevations while assembled ones (Pangeas) of maximum area will have the highest elevations. The inverse relationship between number of continents and negative freeboard observed in Figure 14 demonstrates that the empirical area-elevation relationship has persisted throughout at least the Phanerozoic but fails to explain its cause. We believe that the elevation/area relationship is a second order correlation of (1) the inverse relationship between a continent's area and its velocity with respect to the mantle and (2) a shrinking of continental area due to collision (Fischer, in press). The increasing propensity toward stasis with larger continental area may reflect a deeper tectospheric anchor coupled with a trend toward averaging out tendencies to be pulled in any given direction. Subcontinental heat buildup and consequent thermal doming is proportional to residence time over a fixed point in the mantle, which is, in turn, proportional to continental area. Consequently, Pangeas should achieve maximum stasis and area, and hence will attain the highest elevations. In contrast maximum continent numbers having the lowest elevations and highest velocities with respect to the mantle should be achieved as the percentage of "Atlantic-type" ocean increases immediately following Pangea breakup, as illustrated in Figure 14.

Whether the thermal component of Pangea elevation is the result of heat alone or a heat-induced phase change at the base of the tectosphere is uncertain. However, the slight continental response to thermal expansion requires unreasonably high (Smith, 1982) subcrustal temperatures to achieve the required elevations suggesting a phase change is the more likely process (Jordan, 1979; Smith, 1982).

Average Continental Freeboard

Continental freeboard can be predicted from our model and tested against empirically measured sea level curves. The relative sea level curve of Vail et al., (1977) as calibrated by Pitman (1978) and Harrison (1980), and the percent continental flooding curve of Strakhov and Termier and Termier as compiled in Hallam (1981) have been used. As both data sets essentially measure negative continental freeboard, their similarity is not surprising (Hallam 1981). However, the Strakhov-Termier and Termier curves present global averages, whereas Vail et al. (1977) rely mainly on North America (Hallam, 1981) in plotting percentage continental flooding through time. We sampled both the Strakhov (1948) and Termier and Termier (1952) curves as presented by Hallam (1981) at 50 Ma intervals and plotted the average on Figure 14. For the relative curve of Vail et al., (1977) we reproduced its shape and quantified its amplitudes to coincide with the results of Pitman (1978) and Harrison (1980).

If our continental area/velocity/elevation model (Fig. 8) is broadly correct, the elevations of continental platforms will not be constant and the history of freeboard will be unique for each continent. Global freeboard can therefore be obtained only by calculating a weighted average for all continents. It cannot be calculated using any single continent except a Pangea, so that we cannot claim the negative freeboard of our model as absolute. However, we do claim that the timing of the peaks and troughs and relative amplitudes of peaks and troughs are in correct proportion. Thus, the Late Cambrian and Cretaceous were times of maximum and approximately equal drowning while the two Pangeas represent approximately equal emergence. Lesser fluctuations are correctly scaled between these extremes.

EXTRAPOLATING THE MODEL

As the pre-Paleozoic stratigraphic record is so poorly known and its sedimentary history so poorly dated and correlated, Precambrian sedimentological data (i.e., carbonate platforms, banded iron formations or clastic provinces) cannot be used even semi-quantitatively to estimate freeboard. For this vast period of time the only reliably datable events are thermotectonic and equated with orogeny or igneous activity, such as the intrusion of mafic dike swarms. They therefore represent the only points of commonality between the Paleozoic and the Precambrian. These events are plotted on Figure 15 and in our view offer a clear Precambrian extrapolation of our 440 Ma Pangea cycle. Again the correspondence is remarkable. Peaks in worldwide orogenic activity, that we interpret as Pangea buildup, occur at roughly 400 to 500 Ma intervals and are slightly lagged by mafic dike swarms that are suggestive of rifting. Given the uncertainties in Precambrian dating, the cycle we observe appears to have modulated the geologic record for at least the last two Ga and perhaps even longer. Direct evidence(Windley, 1977; Condie, 1982; Anderson, 1982) demonstrates that the two most recent orogenic episodes lagged by mafic dike swarms correspond to assembly and dispersal of supercontinents (Pangeas). Indirect evidence (Windley, 1977; Condie, 1982) suggests the same for the earlier three episodes. However, dispersal of the earlier supercontinents was probably minimal (Kroner, 1981) as it occurred prior to the onset of the present style of Wilson-cycle plate tectonics (fig. 15).

In addition to punctuating the earth's orogenic history, the observed episodicity strongly influences eustasy, which, in turn, affects climate and biota. Fischer (1981 and in press) has made a compelling case for the two Phanerozoic cycles (Fig. 4) and it therefore seems reasonable that similar relationships should hold in the Precambrian if the tectonic model is valid. As Figure 15 shows, important biogenetic events appear to occur (given that they are imprecisely dated by their superpositional and/or cross-cutting relationships with thermotectonic events) about 50 to 100 Ma after fragmentation. Such events are most likely to occur as the result of the rapid drowning of the continental platforms by shallow seas leading to biotic diversity explosion (Valentine and Moores, 1970) and consequent CO_2/O_2, icehouse/greenhouse effects delineated by Fischer (1981 and in press). Furthermore, there appears to be some Precambrian evidence of the correspondence of supercontinents and major glaciation similar to that noted by Fischer for the Phanerozoic (Fig. 15).

Evidence of glaciation is one of the most unambiguous and reliable Precambrian climate indicators (Frakes, 1979). As noted earlier, episodes of glaciation can lower sea level by as much as 200 m, and therefore represent the one major control of freeboard seemingly not subject to plate tectonic influence. However, the prerequisites of prolonged glaciation include a continent over a pole and elevated freeboard for that continent. The argument for polar positioning of a continent is self-evident. The need for elevation is suggested by Antarctica, which has remained largely stationary over the South Pole

since the Permo-Carboniferous glaciations of Gondwanaland yet only acquired its present ice cap following the drop in the high sea level that characterized the Cretaceous. Similar reasoning applies to other nonglaciated continents occupying polar positions. As Fischer notes for the Phanerozoic, Pangea intervals usually satisfy both the polar positioning and high elevation requirements. Assuming a "circular" Pangea of about 7,500 km radius (and area equivalent to today's continental area of 175 x 16^6 km²), a random position of a Pangea would result in about a 75 percent chance of continental crust covering a pole. Consequently, while Pangeas do not mandate glacial episodes or vice versa, there should be a strong correlation between the two. Figure 15 shows the mean ages of major glacial episodes during the past 2 Ga. Other glacial periods, including those of the late Ordovician and Neogene that do not correspond to Pangeas, nevertheless can be attributed to a continent in a polar position (see Ziegler et al., 1979). It is interesting to note that three or perhaps four (if contested evidence of a~1,500 Ma glacial episode is accepted) out of five Pangeas have been glaciated; a percentage in relatively good agreement with the probability of any given Pangea covering a pole. However, Frakes (1979) cautions that, despite their utility as climatic indicators, dating the timing and magnitude of Precambrian glaciations is fraught with pitfalls owing to the difficulties in finding suitable material to date, as well as to the tendency for glaciers to destroy their own records because they must exist above sea level and hence can only leave easily eroded records.

CONCLUSIONS

We have constructed a simple, internally consistent plate tectonic model that explains the quasi-periodic and co-varying history of diverse geologic processes that include orogeny, sea level changes, glaciation and biogenesis as the result of a single fundamental cause. That cause is the periodic acceleration and retardation of the earth's heat loss, assuming that heat production has been smoothly declining since at least 2 Ga ago. The model fits the available data base very well considering the preservation of the geologic record.

The next step in model development will be to quantitatively formulate the interaction of first

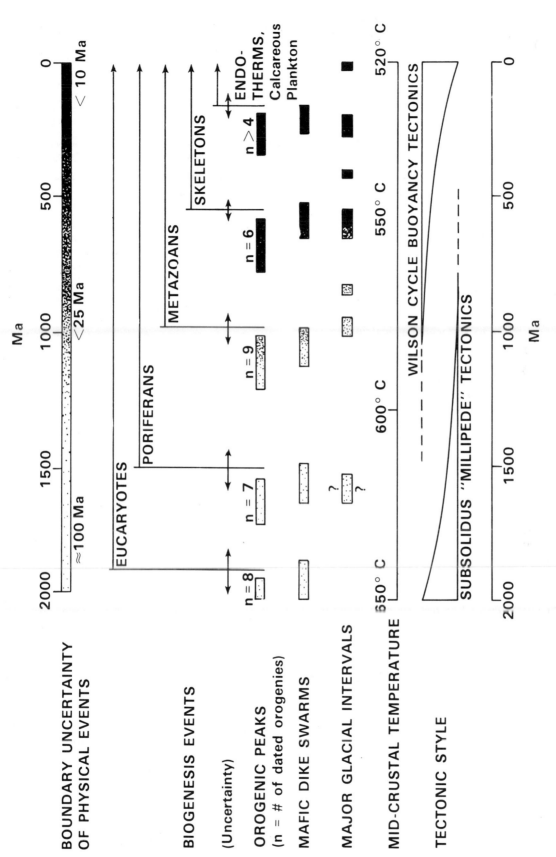

Fig. 15. Summary of quasi-periodic events in earth history for the past 2 Ga. Biogenesis events after Cloud (1968a, b, 1976). Orogenic peaks after Burwash (1969); Condie (1976, 1982). (Note that mafic dike swarms, which represent rifting episodes lag orogenic peaks (Pangea assemblies) by perhaps 50 Ma). Major glacial intervals after Frakes (1979); Cristie-Blick (1982). Mid-crustal temperature after Wynne-Edwards (1976); Tectonic style after Wynne-Edwards (1976); Kroner (1981).

order controls of continental area and elevation in a manner similar to the equation used to calculate the seafloor elevation as a function of its age. When this first order formulation is available, deviations between the predicted sea levels and geologically measured ones should then give us a quantitative estimate of the frequency and intensity of second order controls of eustasy, such as glaciation or continent-continent collision, which could be tested against the geologic record of actual events.

Therefore, the ultimate test of whether our proposed model correlates disparate geologic events or offers an explanation of their underlying fundamental cause depends on the validity of the one critical postulate we used to derive it; namely the present hemispherical asymmetry of the earth's mantle convection system. Arguments that "there has always been a Pacific" and that "all the other planets seem to be asymmetrical in some way", while offering circumstantial support to the postulate, do not prove it. Verification or rejection of the model will only be forthcoming when the detailed effects of continents upon the mantle's heat flow patterns become better understood. This topic is currently being examined by many investigators and hopefully definitive answers will be attained in the near future. We anxiously await those results.

ACKNOWLEDGEMENTS

We thank A.G. Fischer and B.R. Rosendahl for reviewing the manuscript. The members of the Geologic Problem Seminar at Ohio University in the Spring 1982 greatly contributed to focusing our thinking on this problem. Support of this work through National Science Foundation Grant OCE 79-19309 to TRW is gratefully acknowledged. Battelle Memorial Institute is thanked for their generous support of manuscript preparation.

REFERENCES

Anderson, D.L., 1981. Hotspots, basalts, and the evolution of the mantle. Science, 213: 83-89.

Anderson, D.L., 1982. Hotspots, polar wander, Mesozoic convection, and the geoid. Nature, 297: 391-393.

Barron, E.J., C.G.A. Harrison, J.L. Sloan, and W.W. Hay, 1981. Paleogeography, 180 million years ago to the present. Eclogae Geol. Helv., 74: 443-470.

Bambach, R.K., C.R. Scotese, and A.M. Ziegler, 1980. Before Pangea: The geographics of the Paleozoic world. Am. Scientist, 68(1): 26-38.

Berger, W.H. and E.L. Winterer, 1974. Plate stratigraphy and the fluctuating carbonate line. In: K.J. Hsu and H.C. Jenkyns (eds.), Pelagic sediments: On land and under the sea. Spec. Pub. int. Ass. Sediment., 1: 11-98.

Burke, B.C. and T.J. Wilson, 1976. Hot spots on the Earth's surface. Scient. Amer., 235(2): 46-57.

Burwash, R.A., 1969. Comparative Precambrian geochronology of the North American, European, and Siberian shields. Can. J. Earth Sci., 6: 357-365.

Carlson, R.L., T.W.C. Hilde, and S. Uyeda, 1982. A root-t relation between rate of subduction and age of subducted lithosphere. Eos, Trans. Am. Geophys. Union, 68(45): 1112.

Cloud, P.E., 1868a. Atmospheric and hydrospheric evolution on the primitive Earth. Science, 160: 729-736.

Cloud, P.E., 1968b. Pre-metazoan evolution and the origins of the metazoa. In: E.T. Drake (ed.), Evolution and Environment: 1-72. Yale University Press, New Haven.

Cloud, P.E., 1976. Beginnings of biospheric evolution and their biogeochemical consequences. Paleobiology, 2: 351-387.

Condie, K.C., 1976. Plate tectonics and crustal evolution, (1st ed.), Pergamon Press. 288p.

Condie, K.C., 1982. Plate tectonics and crustal evolution, (2nd. ed.), Pergamon Press. 310p.

Cristie-Blick, N., 1982. Pre-Pleistocene glaciation on Earth: Implications for climatic history of Mars. Icarus, 50: 423-443.

Davis, E.E. and C.R.B. Lister, 1974. Fundamentals of ridge crest topography, Earth Planet. Sci. Lett., 21: 405-413.

Dewey, J.F., and K. Burke, 1974. Hotspots and continental break-up: Implications for collisional orogeny. Geology, 2: 57-60.

Engel, A.E.J. and C.B. Engel, 1964. Continental accretion and the evolution of North America. In: A. P. Subramaniam and S. Balakrishna (eds.), Advancing frontiers in geology and geophysics. Indian Geophys. Union, 17-37e.

Fischer, A.G., 1981. Climatic oscillations in the biosphere. In: M.H. Nitecki (ed.), Biotic crises in ecological and evolutionary time: 102-131. Academic Press.

Fischer, A.G., (in press). The two Phanerozoic supercycles. In: W. A. Berggren and J. A. van Couvering (eds.), The new uniformitarianism. Princeton University Press.

Frakes, L.A. 1979. Climates throughout geologic time. Elsevier, Amsterdam. 310p.

Goodwin, A.M., 1981. Precambrian perspectives. Science, 213: 55-61.

Hallam, A., 1981. Facies interpretation and the stratigraphic record. W. H. Freeman and Company, Oxford. 291p.

Harrison, C.G.A., 1980. Spreading rates and heat flow. Geophys. Res. Lett., 7(12): 1041-1044.

Harrison, C.G.A., G.W. Brass, E. Saltzman, J. Sloan, II, J. Southam, and J.M. Whitman, 1981. Sea level variations, global sedimentation rates and the hypographic curve. Earth Planet. Sci. Lett., 54: 1-16.

Hay, W.W. and J.R. Southam, 1977. Modulation of marine sedimentation by the continental shelves. In: N.R. Anderson and A. Malahoff (eds.), The role of fossil fuel CO_2 in the oceans: 569-605. Plenum Press, New York.

Hays, J.D. and W.C. Pitman, III, 1973. Lithospheric plate motion, sea-level changes and climatic and ecological consequences. Nature, 246: 18-22.

Holmes, A., 1951. The sequence of precambrian orogenic belts in south and central Africa. 18th Int. Geol. Congr., London, (1948), 14: 254-269.

Jordan, T.H., 1979. The deep structure of the continents. Scient. Amer., 240(1): 92-107.

Kroner, A., 1981. Precambrian plate tectonics. In: A. Kroner (ed.), Precambrian plate tectonics. Developments in Precambrian geology 4: 781p. Elsevier, Amsterdam.

Lambert, R.S.J., 1980. The thermal history of the Earth in the Archean. Precamb. Res., 11: 199-213.

Larson, R.L. and W.C. Pitman, 1972. Worldwide correlation of Mesozoic magnetic anomalies, and its implication. Geol. Soc. Am. Bull., 83: 3645-3662.

Mackenzie, F.T. and J.D. Pigott, 1981. Tectonic controls of Phanerozoic sedimentary rock cycling. J. Geol. Soc. London, 138: 183-196.

McDougall, I. and R.A. Duncan, 1980. Linear volcanic chains—Recording plate motions? Tectonophys., 63: 275-295.

McKenzie, D.P. and N. Weiss, 1975. Speculations on the thermal and tectonic history of the Earth. Geophys. J. Roy. Astro. Soc., 42: 131-174.

Molnar, P and T. Atwater, 1978. Interarc spreading and cordilleran tectonics as alternates related to the age of subducted oceanic lithosphere. Earth Planet. Sci. Lett., 41: 330-340.

Parsons, B. and J.G. Selater, 1977. An analysis of the variation of ocean floor bathymetry and heat flow with age. J. Geophys. Res., 82: 803-827.

Pitman, III, W.C., 1978. Relationship between eustacy and stratigraphic sequences of passive margins. Geol. Soc. Am. Bull., 89: 1389-1403.

Russell, K.L., 1968. Oceanic ridges and eustatic changes of sea level. Nature, 218: 861-862.

Sclater, J.G., R.N. Anderson, and M.L. Bell, 1971. The elevation of ridges and the evolution of the central eastern Pacific. J. Geophys. Res., 76: 7883-7915.

Sclater, J.G., C. Jaupart, and D. Galson, 1980. The heat flow through oceanic and continental crust and the heat loss of the Earth. Revs. Geophys. Space Phys., 18: 269-311.

Shanmugam, G. and R.J. Moiola, 1982. Eustatic controls of turbidites and winnowed turbidites. Geology, 10:231-235.

Sleep, N.H., 1971. Thermal effects of the formation of Atlantic continental margins by continental break-up. Geophys. Jour., 24: 325-350.

Sloss, L.L., 1963. Sequences in the cratonic interior of North America. Geol. Soc. Am. Bull., 74: 93-114.

Smith, A.G., 1982. Late Cenozoic uplift of stable continents in a reference frame fixed to South America. Nature, 296: 400-404.

Southam, J.R. and W.W. Hay, 1981. Global sedimentary mass balance and sea level changes. In: C. Emiliani (ed.), The sea: The oceanic lithosphere, 7: 1617-1684. John Wiley and Sons, New York.

Stille, H., 1924. Grundfragen der vergleichenden tektonik. Borntraeger, Berlin.

Strakov, N.M., 1948. Fundamentals of historical geology. Gosgeolizdat, Moscow.

Termier, H. and G. Termier, 1952. Histoire geologique de la biosphere. Masson, Paris. 721p.

Turcotte, D.L., and K. Burke, 1978. Global sea-level changes and the thermal structure of the Earth. Earth Planet. Sci. Lett., 41: 341-346

Uyeda, S. and H. Kanamori, 1979. Back-arc opening and the mode of subduction. J. Geophys. Res., 84: 1049-1061.

Vail, P.R., R.M. Mitchum, Jr., and S. Thompson, III, 1977. Global cycles of relative changes of sea level. In: C.E. Payton (ed.), Seismic stratigraphy-Applications to hydrocarbon exploration. Am. Assoc. Petrol. Geol. Mem., 26: 83-97.

Vail, P.R. and J. Hardenbol, 1979. Sea-level changes during the Tertiary. Oceanus, 22: 71-79.

Valentine, J.W., and E.M. Moores, 1970. Plate tectonic regulation of faunal diversity and sea level: A model. Nature, 228: 657-659.

Vinogradov, A.P. and A.I. Tugarinov, 1962. Problemy geokhronologii Dokembriya. Akad. Nauk. S. S. S. R. Kom. Opred. Absolyut Vozrasta Geol. Formatssii, B5: 8-11.

Williams, H., 1980. Structural telescoping across the Appalachian orogeny and the minimum width of the Iapetus ocean. In: D.W. Strangway (ed.), The continental crust and its mineral deposits. Geol. Assoc. Canada Spec. Paper 20: 421-440.

Windley, B.F., 1977. The evolving continents. John Wiley and Sons, New York. 385p.

Wise, D.U., 1974. Continental margins, freeboard and volumes of continents and oceans through time. In: C.A. Burk and C.L. Drake (eds.), The geology of continental margins: 45-58. Springer-Verlag, New York.

Worsley, T.R., D. Nance, and J.B. Moody, 1982. Plate tectonic episodicity: A deterministic model for periodic Pangeas. Eos, Trans. Am. Geophys. Union, 63(45): 1104.

Wynne-Edwards, H.R., 1976. Proterozoic ensialic orogenesis: The millipede model of ductile plate tectonics. Am. Jour. Sci., 276: 927-953.

Ziegler, A.M., C.R. Scotese, W.S. McKerrow, M.E. Johnson, and R.K. Bambach, 1979. Paleozoic paleogeography. Ann. Rev. Earth Planet. Sci., 7: 473-502.

Zoback, M.L. and M. Zoback, 1980. State of stress in the coterminous United States. J. Geophys., Res., 85: 6113-6156.

F. ARABIAN SEA AND COASTAL BIOLOGICAL OCEANOGRAPHY

15

Mangroves: A Summary of Knowledge with Emphasis on Pakistan

S. C. SNEDAKER
University of Miami,
Miami, Florida

ABSTRACT

The global mangrove inventory of 53 species (12 genera in 8 families) is estimated to dominate about 23 million hectares of sheltered coastal intertidal land. Although they are the subject of some 5,700 publications, mangroves remain a poorly understood system compounded by a history of scientific misinterpretation, e.g., mangroves have a growth requirement for seawater and mangroves "build" land by trapping sediments. In the early 1970s, mangroves were recognized to have an important role in the sustenance and maintenance of coastal fisheries, and this perception of ecological value stimulated an expansion of research activity and an improvement in the quality of the research findings. For example, it is now well documented that mangroves have a definite growth requirement for fresh water and that "land building" simply reflects the accretion of sediments followed by mangrove colonization. It is also known that mangrove growth and structural development are positively correlated with annual inputs of terrigenous nutrients and the frequent turnover of interstitial pore water induced by tidal inundation. Conversely, oligotrophic waters, high salinities and inadequate flushing of the sediments impose a chronic stress that is correlated with diminished growth and poor structural development.

The Indus River delta represents a major example of the negative effects of the progressive reduction in freshwater discharge over a period of many years. It is estimated that mangrove forests presently occupy less than 1% of the original quarter-million hectares and that these remnants consist mainly of sparse, small-statured *Avicennia* spp. lining the banks of the better-flushed tidal channels. With the projected further reduction in the fluvial discharge, the remaining mangroves may be expected to experience continued decline with unknown effects on coastal fishery resources. The current and expected phenomena present an opportunity for a long-term scientific study of the variety of changes that are taking place in the delta region. In addition, it may prove beneficial to evaluate a variety of alternative productive uses for the extensive delta area. One possible use is the conversion of suitable areas for shrimp brood and growout ponds.

INTRODUCTION

Mangroves and the mangrove environment represent one of the most intensively studied (Rollet, 1981, indexed 5,653 literature citations), but also one of the least understood, tropical marine ecosystems. Within the past decade, a broad awareness has developed regarding the mangrove environment; its ecological importance in sustaining productive coastal and nearshore fisheries, as well as providing other benefits such as coastal protection and wildlife habitat, is widely recognized. As a result of the recognition of these perceived values, there has been a significant stimulation of interest in mangrove-oriented research throughout the world. The purpose of this paper is to: (1) characterize the mangrove ecosystem with emphasis on the knowledge and status of Pakistan's mangrove environment, and, (2) develop general recommendations for scientific research that may lead to enhanced economic utilization, consistent with conservation in Pakistan.

GENERAL CHARACTERIZATION

Taxonomic Overview

Mangroves are a collection of woody plants, plus associated fauna and flora, that utilize a coastal, saline, depositional environment involving a variety of coastal land forms with typically anaerobic soils (definition adopted by the SCOR Working Group 60 on Mangrove Ecology). The mangroves species are arboreal, woody, spermatophytic halophytes represented by an estimated 53 species (Chapman, 1975) distributed among some 12 genera in 8 families (Lugo and Snedaker, 1974). These are considered to be the "obligate" mangroves in that they are "restricted" to coastal saline intertidal environments whereas other species (e.g., "facultative" mangroves and mangrove "associates"), sometimes associated with mangroves, are not so restricted and can develop in noncoastal environments.

In Pakistan, there are 8 mangrove species in the Indus River delta region and 5 species along the Makran Coast with species of the genus *Avicennia* dominating both areas. The genera and species are: *Bruguiera conjugata, Ceriops tagal, C. roxburghiana, Rhizophora apiculata, R. mucronata, Aegiceras corniculata, Avicennia marina,* and *Sonneratia caseolaris* (Saifullah, 1982, see also Barth, 1982 for taxonomic synonyms); *A. marina* is the dominant species. Compared with the rather luxuriant Asian flora of some 44 species (Chapman, 1975, 1977), the Pakistan component is relatively small. This, in part, may be explained as being a result of the relatively arid environment of the Pakistan Coast, the Indus delta notwithstanding.

Salinity Tolerance

Although mangroves can survive in sea water, it is not a physiological requirement. In contrast, fresh water is required for mangroves to attain their maximum structural development (Cintron, 1981). In fact, mangroves can thrive in a fresh water environment (cf. Snedaker and Brown, 1981a, and citations therein) in the absence of competition from glycophytes (West, 1956; Chapman, 1976). Their dominance in intertidal saline environments is therefore due to their ability to tolerate salinity better than their potential glycophyte competitors. Although mangroves can tolerate saline conditions better than glycophytes, they cannot survive saline concentrations greater than approximately 90 ppt (Scholander et al., 1962) and are progressively "stressed" by increasing salinity through the increased metabolic expenditure of energy to compensate (Waisel, 1972; Queen, 1974; Gale, 1975). Because the individual species differ

significantly in their tolerance of saline conditions Lugo et al., (1975) developed the "metabolic basis for zonation" to explain the reason for the apparent characteristic of mangroves to sometimes occur in predictable zones parallel to the shore (cf. Chapman, 1976). See Snedaker (1982a) for a review of the alternative hypotheses of zonation.

Due to the importance of fresh water in the establishment of mangrove communities, frequently, mangroves are most highly developed in areas of either high-annual direct precipitation and/or surface water runoff from upland watersheds (Pool et al., 1977; Cintron, 1981). Thus, in Pakistan the largest area of greatest mangrove development is in the Indus River delta region located south of Karachi and north of the international border with India. The Indus River is the sixth largest river in the world (see Wells, this volume) and the Indus delta is the second largest in area on the Indian subcontinent (the delta of the combined Ganges-Brahmaputra rivers is the largest). Thus, only in this region is there sufficient fresh water to potentially support a large area in mangrove forest.

Distribution

Intertidal mangrove ecosystems are estimated to cover some 23 million hectares of the world's tropical coastlines with 62% of the area in the Old World and the remainder (38%) in the New World (Snedaker and Brown, in prep.). The influence of fresh water is also apparent in their global distribution, with the largest single areas occurring along coastlines influenced by major rivers (e.g., Amazon, Niger, Orinoco, etc.) and in heavy rainfall areas such as Irian Jaya and along the Pacific Coast of Colombia. In Pakistan, the area of mangrove forest has been estimated at 249,486 ha (Khan, 1966) in the Indus delta with less than 20 ha along the Makran Coast in Baluchistan (Saifullah, pers. comm.). The Indus delta area, thus ranks as the fifth or sixth largest single mangrove area in the world; it ranks second on the Indian subcontinent, deferring to the Sundarbans Forest with some 607,000 ha (Snedaker and Brown, in prep.).

Historical records for British India (prior to 1947) and for Pakistan (cf. Khan, 1966) indicate that the distribution of mangroves in the Indus delta area has significantly changed during the past several hundred years. Earlier, the major spill rivers of the Indus emptied through distributaries close to Karachi. Around that time it is reported that the mangroves were tall and healthy and that the growth was dense and extensive (Khan, 1966). However, it is estimated that at the time of the last major shift in the Indus toward the Rann of Kutch, these mangroves areas began a process of deterioration simultaneous with the rapid development of extensive forests in the delta areas of the new, more southerly, spill rivers. This broad pattern of change is consistent with the observations of geomorphologists (e.g., Thom, 1967; Thom et al., 1975) who have documented the sequential appearance and disappearance of mangroves associated with temporal and spatial changes in coastal fluvial and deltaic processes.

Productivity, Structure, and Litterfall

Tropical forest ecosystems are reported to be among the most productive in the world in terms of gross primary productivity (GPP defined as grams carbon assimilated per unit area of ground surface per unit time) (Rodin et al., 1975). The first empirical field measurements of mangrove productivity were made in the early 1970s in south Florida (Carter et al., 1973) and demonstrated that mangroves are highly productive ecosystems. The data, summarized by Lugo and Snedaker (1974) and Snedaker and Brown (1981b), together with productivity studies underway in Australia (J. S. Bunt, pers. comm.) show, however, that there is considerable variation in GPP with mangroves in "optimum" habitats, which exhibit the highest rates, compared to mangroves in less favorable environments. The values reported range from a high of 13.9 to 1.4 g C m^{-2}. day^{-1} (Lugo and Snedaker, 1974; Snedaker and Brown, in prep.). Although there is no precise quantitative relationship between GPP and standing stock biomass (see Lugo and Snedaker, 1974; Snedaker and Brown, 1981b), there is an apparent positive relationship with community structure, including forest height and basal area (Pool et al., 1977; Cintron, 1981). No measurements of GPP have been made on Indus delta mangroves, but structural comparisons with other world areas suggest that the average values would be relatively low.

Of particular importance is that portion of the GPP known as the net primary productivity (NPP). This represents the portion of the GPP that remains after carbon loss through respiration is deducted (Newbould, 1967) and is accumulated as

biomass, primarily in wood, roots and leaves. The shedding of plant parts, including leaves, floral parts and fruit, twigs and branches, is known as littertall and collectively, is an important source of organic detritus which enters marine foodwcbs (Heald 1970; Odum, 1969, 1971) in particulate form and as dissolved organic matter (Snedaker, 1979). Litterfall, thus represents a fraction of the net primary productivity. To date, no one has measured or estimated litter production in Pakistan mangrove forests, but analogous data from other areas of the world suggest that daily rates of litter production would fall in the range of 1-3 g (dry weight) (summarized in Snedaker and Brown, 1981b).

Uses of the Mangrove Ecosystem

Mangroves and the mangrove environment (i.e., the ecosystem) have a large number of direct and indirect uses and are, in fact, utilized in many parts of the world. These uses fall within the broad categories of food, fuel, fiber and fodder and have been extensively cited by Kunstadter and Snedaker (in press), Walsh (1977), and Saenger et al. (1981). In some countries, the local economy is based on the annual yields of products taken from the mangrove ecosystem. For example, the Khulna paper-pulp mill in southwest Bangladesh is dependent on the annual sustained-yield harvesting of *Excoecaria agallocha* in the Sundarbans forest, which also supplies the local population with timber products, thatch and honey. In addition to these varied kinds of direct uses, harvested yields from nearshore fisheries are reported to be proportional to the coastal area in mangrove forest (Turner, 1977; Martosubroto and Naamin, 1977). The perceived importance of mangroves for the maintenance of nearshore fishery stocks has been a motivating reason for their protection in many parts of the world.

Although there is a variety of potential economic uses of the some 23 million hectares of mangrove ecosystems distributed along pan-tropical coastlines, local utilization and/or use varies along a continuum of nonexistence, to subsistence, to economic management, to potentially-abusive exploitation. In certain extensive regional areas of mangrove forests that have low population densities (i.e., human), for example, along the Pacific Coast of Colombia, parts of north Brazil, or in Irian Jaya, direct utilization of the mangrove ecosystem is minimal to nonexistent. In contrast, in similar coastal areas of high human population densities in developing countries (e.g., India and West Africa), the mangrove ecosystem represents a source of subsistence products that range from gathered firewood to small-scale artisanal capture fisheries. Only a relatively few countries (e.g., Bangladesh, Malaysia, Philippines) presently have sustained yield management plans in force; other countries, particularly in Central and South America, are beginning to recognize the economic benefits to be derived from mangrove timber and wood products. This recognition is based in part on the fact that with suitable modification, mangrove forests can be harvested for commercial purposes as most other forests with minimal environmental degradation (Snedaker, 1982b). At the extreme end of the utilization continuum, are examples of direct and indirect abusive exploitation that include the one-time clearcutting of large areas and the conversion of productive forests to housing sites, agricultural fields and aquaculture ponds (Saenger et al., 1981). Also included among the potentially abusive actions are, the diversion of fresh water and the introduction of wastes and pollutants (see also, Linden and Jernelov, 1980).

In Pakistan, the Indus delta mangrove area supports local fishery enterprises (pomfrets, anchovies, *Hilsa ilisha,* etc.), a commercial shrimp fishery that contributes to the export market, and provides a source of firewood and other forest products for local use (Saifullah, 1982). In addition, the leaves of the dominant genera, *Avicennia,* serve as a dry season fodder for domestic camels. It is also reported that when *Avicennia* leaves are stall-fed to dairy cows, the dietary supplement increases the butterfat content of the milk (Saifullah, in prep.). Although a forest management plan has been developed to utilize mangrove timber and wood products (Khan, 1966), the relatively poor quality and low density of the mangrove trees in the Indus delta appear to have mitigated against implementation of an economically-viable utilization plan. At the present time, the mangrove forests are protected by law, and some revenues, albeit insignificant, are collected from the sale of forest products.

CURRENT STATUS OF INDUS DELTA MANGROVES

There is an absence of reliable reports on the status of Indus delta mangroves, particularly with respect to changes that may have taken place over the past several decades in association with the upstream withdrawal of Indus River water. This is a reasonable expectation due to the great logistic difficulty involved in travelling to, and within, the extensive Indus delta region. Also, until relatively recently, scientific interest in the delta area has been limited to taxonomic and descriptive surveys. As a result, the observations reported here are those of the author, made during visits to the Indus delta in 1977 (as a representative of Unesco's Division of Marine Sciences) and again in 1982 (November) during the U.S.—Pakistan Workshop on Marine Sciences. In 1977, two areas were visited: overland via Thatta to the area around Keti Bunder, and by launch (courtesy of the PNS Zulfiquar) to the seafront portions of Korangi Creek and Karachi Harbour. In 1982, the delta area was again examined during a low-level overflight (visibility, excellent) made possible through the courtesy of the Pakistan Navy. During both visits, reference observations were made relative to Khan's (1966) "Working Plan" for the period 1963-1983 (made accessible to the author through the courtesy of the Sind Department of Forests and Wildlife, Karachi).

It was apparent in 1977 that the tall and extensive forests of the Indus delta (cited in colonial records) were no longer present. Instead, the existing forests generally reflected the imposition of an environmental stress associated with reduced fresh water availability that led to hypersalinity and nutrient impoverishment. Apparently, the best developed trees only occurred along well-flushed tidal channels, whereas in the interiors of the deltaic islands (i.e., poorly flushed areas) were sparsely inhabited by the dwarf form of *Avicennia* spp. The village of Keti Bunder showed dramatic evidence (e.g., surface salt accumulation and corrosive deterioration of domestic structures) of the effects of hypersalinity that presumably led to the associated partial-collapse of a once viable rice and fishing industry. These observations and the anecdotes from informants were consistent with reports on the reduction in freshwater discharge from the Indus due to upstream diversion and use in agricultural irrigation. In this arid part of the Indian subcontinent, the Indus watershed represents the only significant source of fresh water for agricultural and domestic use. At the present time, the Indus has a significant discharge only during the southwest monsoon; during the low water season, the fresh water is diverted by upstream dams and barrages. Water-resource development plans appear to be oriented toward the total capture and diversion of Indus River water for agricultural irrigation, suggesting, that at some point in the relatively near future (measured in decades), the Indus will no longer have a freshwater discharge into the delta area and the Arabian Sea.

During the 1982 overflight of the areas visited in 1977, it was apparent that the forested areas had further deteriorated and were wholly restricted to the well-flushed banks of the tidal channels; the interiors of the deltaic islands appeared bare. The author's visual comparison of the Indus delta with the Indian part of the Gangetic delta (i.e., that portion experiencing accelerated hypersalinity) suggested that the deterioration of the Indus delta mangrove ecosystem is much more extensive and severe. Based on the domestic and agricultural demand for fresh water in Pakistan, it appears that this process of deterioration is irreversible and will eventually lead to the loss of the remaining remnants of the mangrove ecosystem. It should be noted, however, that the annual harvests of shrimp/prawns remain high in apparent contrast with the conclusions of Martosubroto and Naamin (1977) and Turner (1977) who infer that a stock decrease should occur in proportion to progressive decreases in the area of mangrove forests.

RESEARCH RECOMMENDATIONS

General Recommendations

In consideration of the fact that: (1) the Indus River is the sixth largest river in the world, (2) the delta area is, or was, one of the world's largest mangrove areas in the world, (3) the Indus is to be totally diverted for upstream agricultural use, and (4) a significant regional change is occurring in the delta, it is recommended that an intensive study of this change be undertaken. This recommendation is based on two objectives. The first suggested objective is to study and document what happens to a deltaic estuary in an arid tropical environment when the freshwater input is progressively reduced over time. Snedaker et al. (1977) have outlined a suite of documented hypotheses concerning the effects of freshwater withdrawal, the majority of which may lead to profound ecological and economic impacts. The

second objective should focus on using the collected data and information to identify and evaluate alternative uses of the delta area for the economic benefit of the region and country. For example, large exposed barren areas might be suitable for the mariculture of shrimp or other marine species that have commercial importance. However, it would be necessary to evaluate the potential for acid-sulfate problems and to select species that can be cultivated in relative high salinity water. The high population density of Pakistan and the availability of a large exposed delta, no longer suitable for the economic utilization of mangroves, presents a persuasive argument that economic-utilization alternatives be sought for this area. It is noted, for example, that in many countries of the world, mangrove areas are being sacrificed for shrimp production; this is a problem that has minor significance due to the deteriorating status of the Indus delta.

Specific Recommendations

The following specific recommendations focus on three areas close to Karachi and in the Indus delta that could provide a base for significant research (as outlined above). In addition, some specific research topics unique to each area are proposed for consideration.

Karachi Harbour

The first and most easily worked, site, is situated around the eastern periphery of Karachi Harbour. This represents the best developed stand of mangroves between Karachi and the distant Indus spill river close to the Pakistan/Indian border. The forest has a well developed structure, evidenced by a closed canopy, vigorous appearance and a relatively high basal area. It is a remarkable site because it is exposed to nearly all forms of pollution that might be expected to occur in association with an urban metropolitan area and a major seaport. Flotsam and jetsam, and surface deposits of what appear to be crude oil and bilge wastes, are interspersed among the prop roots and pneumatophores of the mangroves. These materials seem to have little or no effect on the structure and functioning of the forest stand. Recently, Snedaker and Brown (1981a) have argued that mangroves might be tolerant to water-borne pollutants as a result of the same physiological processes that preclude excessive uptake of salt (see also, Walsh et al., 1973, 1979).

Conversely, the introduction of domestic wastes, rich in nutrients, may present a positive influence counterbalancing the bioeffects attributable to toxic materials. This site thus presents an opportunity to test a variety of hypotheses concerning the functioning of mangroves and their tolerance or intolerance to various forms of pollutants. In addition, this experimental area could also be used for a variety of other research projects involving faculty and students from the National Institute of Oceanography and the Institute of Marine Biology. This site is accessible on a daily basis to the principals living and working in the Karachi urban area.

Korangi Creek

Situated approximately halfway between Karachi and Keti Bunder is an area of mangrove forest dominated by the highly saline Korangi Creek. Apparently this area has remained hydrologically stable for the past 200 years when the Indus made its last major shift westward. The area, dominated by *Avicennia* spp., of stunted character with a patchy, low-density population, could be considered to exist in a steady-state condition relative to the salinity regime. The area is not populated although small fishing camps dot the seafront and forays are made into the mangrove areas to harvest *Avicennia* spp. foliage for cattle and camel fodder; camels are also herded into this area during the dry season as the mangroves represent the only green foliage present in the region during that part of the year. This site affords the opportunity to evaluate environmental-gradient control over the distribution of the mangroves and, make assessments of the influence of heavy grazing pressure. Of the three sites, this is the most difficult to reach on a frequent basis as there is no convenient overland access and thus, a boat trip of several hours is required to reach the site. However, a steel mill is presently under construction in the Korangi Creek drainage and the National Institute of Oceanography is to establish a oceanographic laboratory on site. The presence of the laboratory and port facilities would provide a logistic base making the research area conveniently accessible to researchers living in Karachi.

The shipping channel leading to the steel mill has been dredged and it has been observed that the channel banks are beginning to erode posing a significant potential cost in maintenance dredging

in the future. The mangroves in this area are dominated by the dwarf forms of *Avicennia*, which are most heavily concentrated along the channel banks, in spite of the incipient erosion. A research topic of very practical importance could be implemented to test the hypothesis that mangrove growth and development is nutrient limited, and, that a stimulation of growth would stabilize the banks and minimize erosion. The hypothesis could be simply tested by supplying nutrients of different types and in different forms followed by an evaluation of the response by the mangroves compared to controls. The practical objective of the testing would be to devise a maintenance schedule of fertilization that would minimize bank erosion and the subsequent requirement for maintenance dredging of the shipping channel.

Keti Bunder

Keti Bunder is a small village situated within the delta to the north of the major Indus spill river. It is claimed to have been a major economic center 15 to 20 years ago based on the cultivation and harvesting of rice and as a local fisheries center. Within recent years, the freshwater flow into this portion of the delta has all but completely stopped except for monsoonal flooding and, the village has declined to less than 50 families who subsist on limited fishing and the harvesting of *Avicennia* spp. for local sale as firewood. Keti Bunder is approximately four hours from Karachi by an all weather hardtop road; the last several kilometers, however, involve crossing mud and salt flats and can only be transversed with 4-wheel drive vehicles during low tide in the dry season. This area abounds in a variety of birdlife which appears to feed almost exclusively on mud skippers (*Periothalmus* sp. and *Boleophthalmus* sp.) and several species of mollusc and gastropods exposed on the banks during low tides. The Sind Department of Forests and Wildlife maintains a substation at Keti Bunder which is suitable for overnight accommodations and thus, as a field station for research. If the National Institute of Oceanography and/or the Institute of Marine Biology had a small boat on site, Keti Bunder would be the logical place to initiate long-term research to monitor changes in the delta.

The research interest in this part of the Indus delta focuses on the incipient and chronic changes that are presently taking place as a result of the reduced freshwater input and the ostensibly increasing salinity of both the surface water and the sediments. It thus represents an ideal situation for monitoring the changes that are occurring in the physical environment and, the attendant biological responses. Specifically, simple experiments could be designed to distinguish between increasing salinity and decreasing nutrient supply (formerly entrained in the terrestrial freshwater runoff) as the probable major causes of the decline in the structure of the mangrove forest.

The area around Keti Bunder also appears to be highly suited to pilot-scale experimental work in mariculture due to the presence of abandoned polders formerly used in rice cultivation. The initial emphasis would necessarily focus on species selection and the development of pond management protocols unique to the Indus delta environment. The basic techniques for the pond culture of the decapod crustaceans is now a reasonably well-developed science, suggesting that Pakistan principals would access the necessary background information prior to embarking on a major pond development program.

REFERENCES

Barth, H., 1982. The biogeography of mangroves, p. 35-60. *In:* D. N. Sen and K.S. Rajpurohit (eds.) *Contributions to the ecology of halophytes.* Dr. W. Junk Publishers, The Hague.

Carter, M.R., L. A. Burns, T.R. Cavinder, K.R. Dugger, P.L. Fore, D. B. Hicks, H.L. Revells and T.W. Schmidt, 1973. Ecosystems analysis of Big Cypress Swamp and estuaries. EPA 904/9-74-002, U.S. Environmental Protection Agency, Region, IV, Atlanta. 478 p.

Chapman, V.J., 1975. Mangrove biogeography, p. 3-22. *In:* G. Walsh, S. Snedaker and H. Teas (eds.) Proc. International symposium on biology and management of mangroves, Vols. I & II. Held 8-11 Oct. 1974 in Honolulu. Inst. Food Agric. Sci., Univ. Florida, Gainesville. 846 p.

Chapman, V.J., 1976. *Mangrove Vegetation.* J. Cramer, Germany. 499 p.

Chapman, V.J. (ed.), 1977. *Ecosystems of the World.* Vol. I. *Wet Coastal Ecosystems.* Elsevier Scientific Publ. Co., New York. 428 p.

Cintron, G., 1981. El manglar de la costa Ecuatoriana. Depto. de Recursos Naturales, Puerto Rico 37 p.

Gale, J., 1975. Water balance and gas exchange of plants under saline condition, p. 168-185. *In:* A. Poljakoff-Mayber and J. Gales (eds.) *Plants in saline environments.* Springer-Verlag, New York.

Heald, E.J., 1971 The production of organic detritus in a south Florida estuary. Sea Grant Tech. Bull. No. 6. Univ. Miami. 110 p.

Hicks, D.B. and L.A. Burns, 1975. Mangrove metabolic response to alterations of natural freshwater drainage to southwestern Florida estuaries, p. 238-255. *In:* G. Walsh, S. Snedaker and H. Teas (eds.) *Proc. International symposium on biology and management of mangroves,* Vols. I & II. Held 8-11 Oct. 1974 in Honolulu. Inst. Food Agric. Sci., Univ. Florida, Gainesville. 846 p.

Khan, S.A., 1966. Working Plan of the Coastal Zone Afforestation Division from 1963-64 to 1982-83. Government of West Pakistan, Agriculture Dept., Lahore.

Kunstadter, P. and S.C. Snedaker (eds.), (In press) Proc. Unesco Regional Seminar on Human Uses of the Mangrove Environment and Management Implications. Held in Dacca, Bangladesh, 4-8 December 1978.

Linden, Olof and Arne Jernelov, 1980. The mangrove swamp—an ecosystem in danger. Ambio 9(2): 81-88.

Lugo, A.E. and S.C. Snedaker, 1974. The ecology of mangroves. Annual Review of Ecology and Systematics 5:39-64.

Lugo, A.E., G. Evink, M. Brinson, A. Broce and S.C. Snedaker, 1975. Diurnal rates of photosynthesis, respiration and transpiration in mangrove forests of south Florida, p. 335-350. *In:* F.B. Golley and E. Medina (eds.) *Tropical Ecological Systems,* Ecological Studies, Vol. 11. Springer-Verlag, New York. 398 p.

Martosubroto, P. and N. Naamin, 1977. Relationship between tidal forests (mangroves) and commercial shrimp production in Indonesia. Mar. Res. in Indonesia No. 18:81-86.

Newbould, P.J., 1967. Methods for Estimating the Primary Productivity of Forests. IBP Handbook No. 2. Blackwell Scientific Publ., Oxford. 62 p.

Odum, W.E., 1969. The structure of detritus-based food chains in a south Florida mangrove system. Ph. D. diss. Univ. Miami, Coral Gables, Florida.

Odum, W.E., 1971. Pathways of energy flow in a south Florida estuary. Sea Grant Tech. Bull. No. 7. Univ. Miami. 162 p.

Pool, D.J., S.C. Snedaker and A.E. Lugo, 1977. Structure of mangrove forests in Florida, Puerto Rico, Mexico and Costa Rica. Biotropica 9(3): 195-212. Also published *in* Memorias del II Simposio Latino-americano sobre Oceanografia Biologica 2:137-150. Universidad de Oriente, Cumana, Venezuela. 261 p.

Queen, W.H., 1974. Physiology of coastal halophytes, p. 345-353. *In:* R. J. Reimold and W. H. Queen (eds.), *Ecology of Halophytes,* Academic Press, New York. 605 p.

Rodin, L.E., N. I. Bazilevich and N. N. Rozov, 1975. Productivity of the world's main ecosystems, p. 13-26. *In:* Productivity of World Ecosystems. Nat. Acad. Sci., Washington, D.C.

Rollet, B., 1981. Bibliography on mangrove research 1600-1975. Unesco, 7 place de Fontenoy, 75700 Paris. 479 p.

Saenger, E., J. Hegerl and J.D.S. Davie, (eds.) 1981. First Report on the Global Status of Mangrove Ecosystems. Prepared by the International Union for Conservation of Nature and Natural Resources, Commission on Ecology, Working Group on Mangrove Ecosystems. Toowong, Queensland 4066, Australia. 132 p.

Saifullah, S.M., 1982. Mangrove ecosystem of Pakistan, p. 69-80. *In:* The Third Research on Mangroves in Middle East, Japan Cooperative Center for the Middle East. Publ. No. 137. Tokyo, Japan.

Scholander, P.F., H.T. Hammel, E. Hemmingsen and W. Garey, 1962. Salt balance in mangroves. Plant Physiol. (Penn.) 37:722-729.

Snedaker, S.C., 1979. Mangroves: their value and perpetuation. Nature and Resources 14(3): 7-15.

Snedaker, S.C., 1982a. Mangrove species zonation: why? p. 111-125. *In:* D.N. Sen and K.S. Rajpurohit (eds.) Tasks for Vegetation Science, V. 2, *Contributions to the Ecology of Halophytes.* Dr. W. Junk Publishers, The Hague.

Snedaker, S.C., 1982b. A perspective on Asian mangroves, p. 65-74. *In:* C.H. Soysa, W.L. Collier and C.L. Sien (eds.) *Man, Land and Sea: Coastal Resource Use and Development in Asia.* University of Singapore Press, Singapore.

Snedaker, S.C. and M.S. Brown, 1981a. Water quality and mangrove ecosystem dynamics. EPA-600/4-81-022. Prepared for Office of Pesticides and Toxic Substances. U.S. Environmental Protection Agency. Environmental Research Laboratory, Gulf Breeze, Florida. 80 p.

Snedaker, S.C. and M.S. Brown, 1981b. Primary productivity of mangroves. *In:* C.C. Black and A. Mitsui (eds.) *CRC Handbook Series of Biosolar Resources.* Vol. I. *Basic principles.* CRC Press, West Palm Beach, Florida.

Snedaker, S.C. and M.S. Brown, (In prep.) Biosphere inventory of mangrove forest lands: total area, current status, managing institutions and research initiatives. A project of the SCOR/Unesco Working Group 60 on Mangrove Ecology supported by the USDA Forest Service as part of the U.S. Man and Biosphere Program. Miami, Florida.

Snedaker, S.C., D. de Sylva and D.J. Cottrell, 1977. A review of the role of freshwater in estuarine ecosystems. Vols. I & II. Final report submitted to the Southwest Florida Water Management District. Univ. Miami, UM-RSMAS 77001, Miami, Florida. 420 p.

Thom, B.G. 1967, Mangrove ecology and deltaic geomorphology: Tabasco, Mexico. J. Ecol. 55:301-343.

Thom, B.G., L.D. Wright and J.M. Coleman, 1975. Mangrove ecology and deltaic geomorphology: Cambridge Gulf-Ord River, Western Australia. J. Ecol. 63:203-232.

Turner, R.E., 1977. Intertidal vegetation and commercial yields of penaeid shrimp. Trans. Amer. Fisheries Soc. 106(5): 411-416.

Waisel, Y., 1972. *Biology of Halophytes*. Academic Press, New York.

395 p.

Walsh, G.E., 1977. Exploitation of mangal, p. 347-362. *In:* Chapman (ed.) (1977).

Walsh, G.E., R. Barrett, G.H. Cook and T.A. Hollister, 1973. Effects of herbicides on seedlings of the red mangrove, *Rhizophora mangle* L. BioScience 23(6): 361-364.

Walsh, G.E., K.A. Ainsworth and R. Rigby, 1979. Resistence of red mangrove *(Rhizophora mangle* L.) to lead, cadmium and mercury. BioTropica 11(1): 22-27.

West, R.C., 1956. Mangrove swamps of the Pacific coast of Colombia. Ann. Assoc. Am. Geogr. 46(1): 98-121.

16

Environmental Studies in Support of Fisheries Development and Management Programs in Pakistan

F. WILLIAMS
University of Miami,
Miami, Florida

ABSTRACT

The current status of fisheries development and production in Pakistan is briefly described. A short review is next given of the state of knowledge of the effects of ocean climate on fisheries at various time and space scales. The demersal and pelagic fisheries of Pakistan, actual and potential, are then considered in the context of ocean climate in the northern Arabian Sea. Strategies for basic oceanographic research in Pakistan are suggested which have practical value by contributing directly to programs for national fisheries development. A final comment is made on the need for cooperation of nations bordering the Arabian Sea, so as to better understand the oceanic climate in general in the area.

INTRODUCTION

The total catch of fish in Pakistan was 279,300 metric tons (mt) in 1980 (ranked 38th in the world), only slightly down from the record level of 300,400 mt in 1979 (FAO, 1982). However, these levels represent an increase of 60-70% over those recorded only a decade earlier. In 1980 the catch in the marine sector (including hilsa shads) was 232,900 (83.2% of total catch) and that in freshwater, exclusively fishes, was 46,300 mt (16.8%). The principal components of the marine catch (Table 1) were Indian oil sardine (*Sardinella longiceps*), rays, shrimps, sharks, croakers and sea catfishes in that order. The priority now assigned by the Government of Pakistan to the development of the fisheries sector suggests that it should help double current production to about 600,000 mt by 1990 (Anon., 1982). It was further reported that the present extension of Karachi Fish Harbor, training programs and joint ventures will assist in increasing the capacity of the industry and perhaps enable produciton very rapidly to reach 400,000 mt annually. The increase in production has also resulted in an increase in fish exports of 11% to 14,820 mt in 1980/81; for the first 6 months of 1981/82, exports were up 35% over the

TABLE 1. Principal components of the marine fisheries catch of Pakistan, 1980 (from FAO, 1982).

	Metric tons x 10^3	% of marine catch
Indian oil sardine	52.3	22.5
Rays	41.2	17.7
Shrimps	25.9	11.1
Sharks	23.7	10.2
Croakers	18.7	8.0
Sea catfishes	17.7	7.0
Other	53.4	22.9
Total	232.9	100.0

same period in the previous 12 months.

Compared with many countries in the world, even many developing ones, Pakistan is in a period of continuing growth in fisheries catches from what is recognized overall as one of the more biologically productive areas of the world ocean, that is the Arabian Sea. Given that the problem of overfishing and depletion of stocks appears as yet not to be the major preoccupation of local

Fisheries Departments and that the National Institute of Oceanography is soon to commence active research operations, it seems timely to consider programs which could lead to beneficial application of future research results to fisheries development and management problems, as well as satisfying purely scientific objectives. One of the most important of such programs could be concerned with the effects of the environment (as a predictive tool) on the living marine resources exploited as food. Thus, it seems appropriate to discuss the effects of ocean climate on the fisheries resources of Pakistan.

EFFECTS OF OCEAN CLIMATE ON FISHERIES

Cushing and Dickson (1976) presented a comprehensive review of the effects of climate on fisheries and concluded that "viewing the complex record of climate, hydrographic and biological variation as a whole, it is nevertheless possible to find signs of an underlying order in these events." General aspects of the effects of climate on fisheries have been commonly discussed in Japan over many years, [see for examples, symposia in Bulletin of Japanese Society of Fisheries Oceanography, 34(1979) and 39(1981)], especially from a practical fisheries standpoint, and workshops have been held recently in the USA (URI, 1979) and Europe (Parsons et al., 1978). Certainly it has been generally accepted that oceanic and atmospheric climate affects fish stocks in an overall sense though the nature and mechanisms are poorly understood (Cushing, 1978a,b).

Laevastu and Hayes (1981) emphasized the need to apply results of research in oceanography and marine biology to benefit fisheries in a practical sense through an understanding of fisheries ecology (interaction among fish, the biota and the environment). This input of environmental data was also stressed by Laevastu and Larkins (1981) for simulation models for an ecosystem (multispecies) approach to fisheries management.

Most fish stocks are highly variable in terms of catches or yields e.g. in the North Sea, off Japan and Peru, at the annual or longer term (decadal) level mainly, but not exclusively, because of fluctuations in strengths of year-classes (cohorts). These variations appear to be associated with the availability of a food supply for the various early life history stages of the fishes. In turn, the food supplies (plankton) vary with local ocean productivity and oceanic and atmospheric climate e.g. surface winds, upwelling, circulation patterns, solar radiation. In addition, local climate may be greatly influenced by events such as air-sea interactions which take place at a great distance. An example of this is the relationship of the phenomena of El Nino off Peru with previous physical processes in the Southwest Pacific involving changes in the Southern Oscillation and the "stacking-up" of warm water in the equatorial West Pacific Ocean.

In the past, most studies of the effects of climate on fisheries (e.g., fisheries oceanography) concentrated on attempting to derive empirical correlations between environmental factors and yield or year-class strength (Cushing, 1978a). Such studies have rarely been unqualified successes because of the limited number of data points which can be obtained (usually one per year) and the wide choice of environmental factors which can be utilized, to say nothing of our lack of understanding of species interactions. Skud (1982) has reported that observed changes in species dominance support the hypothesis that climatic factors can affect abundance of fishes but not absolute density which is controlled by interspecific competition.

In the last five years, therefore, much research in advanced fishing nations has concentrated on fundamental scientific studies which are designed to produce a detailed understanding of the factors involved in the linkages between physical and biological events controlling year-class strength. For example, one such factor, food, has been the subject of intensive investigations for the California anchovy by Lasker and his colleagues in California. In particular, experiments involving larval fish behavior and food requirements in relation to microscale environmental conditions were carried out in the laboratory and *in situ* in the ocean (Lasker, 1978; Owen, 1981).

The apparent control by environmental physical processes of the distribution and abundance of fish stocks was documented by Botsford and Wickham (1975) in relating upwelling indices and Dungeness crab catch off Oregon. Subsequently, environmental parameters have been included in stock-recruitment models, for example, for Atlantic menhaden off North

Carolina (Nelson et al., 1977) and for Pacific mackerel off California (Parrish and McCall, 1978). The types of environmental variables to be considered are also discussed in Owen (1981), Bakun and Parrish (1981) and Parrish et al. (1981). More recently, the 1980 FAO/IOC-sponsored workshop in Peru concentrated on examining in detail the effects of environmental fluctuations on the survival of pelagic fishes, specifically anchovies from the eastern boundary currents off Peru and California and for which there are strong data bases (Sharp, 1981). It was concluded that among the many important items to be derived from the analyses were (i) the need for attention to time-dependence of local events on physical processes which affect biological processes; (ii) whether there may be a series of "critical periods" in the early life history stages, not a single one; (iii) the lack of understanding of cause and effect in the ecosystem; (iv) the need for validation of models and concepts through better sampling systems (continued experimentation); (v) the need for increased laboratory work to complement studies at sea; (vi) the need for progressing from single to multi-species studies more in keeping with most fisheries, especially those in the tropics. Finally, a plea was made to advise the appropriate authorities "that the uncertainties of population abundance predictions can probably be reduced by intense, integrated efforts to sample, study and interpret results of research at scales, namely ten meters or less and at time periods of hours to days, for individual larval fish problems, up through longer term, broader scale events to help reduce uncertainties regarding population and ecosystem status."

In addition to the long-term effects through year-class strength, climate may also influence the interannual and interseasonal distribution and aggregation of fish e.g., albacore tuna off the U.S. west coast (see Laurs, 1977; Laurs and Lynn, 1977). ACMRR (1978) suggested that developing countries might be more concerned with this type of strategic effect rather than with absolute (gross) variations in abundance as the former could affect catch distribution, economic cost of fishing effort, etc. In addition, in most tropical areas fishes are usually short-lived and hence, the effect of dominant year-classes on stock size is much less than among long-lived temperate water species.

For many nations, as exemplified by Japan, the effect of climate on fisheries at the tactical, operational time scales of days and weeks may be considered more directly valuable, or at a minimum equally useful, as that at the strategic level. Information such as mixed layer depth and temperature, location and intensity of horizontal and vertical fronts, localized upwelling, mesoscale meanders and eddies in major currents e.g., in the Kuroshio Current off the east coast of Japan, and similar phenomena have been correlated with localized aggregation and availability of fish, leading to larger catches. Tomczak (1977) discussed such fishery environmental services, the products of which are transmitted by radio (voice and facsimile print) to the fishing fleets. Certainly the ability to remotely sense (aircraft, satellites) a limited number of physical parameters such as sea surface temperature and color (for chlorophyll) has improved time and space scales in monitoring of ocean parameters which, coupled with the advent of cheaper, better electronic reception equipment for fishing vessels, should increase the efficiency of environmental services. Many of the latter are related to well-developed, pelagic fisheries such as those for tunas; the extensive work on tuna and their environment (in the Pacific) is indicated in the review by Sund et al. (1981). Present work is aimed at refining the inputs to fisheries forecasting systems through more scientific information on aspects of physiology and behavior of the fish and better knowledge of physical processes involved in the vulnerability of fish to fishing gear under various conditions.

PAKISTAN FISHERIES IN THE CONTEXT OF OCEAN CLIMATE WITH SUGGESTED RESEARCH STRATEGIES

The overall comments in the previous section might lead to an accusation of an ivory tower approach to marine science and fisheries which has little relevance or application to the real world situation of the present state of rapid development of Pakistan's fisheries. However, formulation of even simple physical and biological relationships amenable to hypothesis testing might be useful for fisheries in the region and could result from careful focusing and orientation of marine research programs built on critical syntheses and reviews of existing information. Presently forecasting services at the strategic and tactical level of fisheries operations are mostly utilized by developed fishing nations in home areas. Such services, however rudimentary, might be useful at some levels of development such as for small-scale fleet operations (national or joint-ventures) or through extension-type educational activities using, in particular, the radio media.

Given that the exact proportioning of tonnage between the pelagic and demersal segments of the catch is difficult to ascertain for the two coastal areas of Pakistan, west and south (following Banse's definition, this volume), discussion of potential research programs cannot be presented in terms of priorities. However, the apparent rapid build-up in trawling operations (for fish and shrimp) suggests, perhaps, an initial emphasis on programs likely to benefit demersal fisheries.

Qasim (1982) has recently discussed the oceanography of the northern Arabian Sea whilst Banse (this volume) has reviewed the hydrography and biological phenomena in this region directed especially at the seas adjacent to the Pakistan coast. From these (and other) papers, it has been possible to identify those environmental parameters which appear to affect fish distribution, abundance and availability in time and space off Pakistan. It is necessary to reiterate, nevertheless, the uniqueness of the monsoonal reversal of the wind systems and subsequent effects on the overall circulation in the northwestern Indian Ocean, including the North Arabian Sea, and events in the ocean environment off the coast of Pakistan.

It now seems clear that upwelling *per se,* does not take place along the Pakistan coast but rather that there is upsloping of cool water along the continental shelf on both the west and south coasts of Pakistan during the southwest (SW) monsoon. This event always takes place by June, often by May, and persists, at least on the south coast, until well into the start of the northeast (NE) monsoon (late November-early December). Following the onset of the NE monsoon there is cooling of the inshore surface waters by several degrees C but this has been shown to be due to convective cooling caused by the cold winds of the NE monsoon. The cool upsloped water (for example see Banse this volume, Fig. 2 b and c) is low in dissolved oxygen because of its origin in the large upper oxygen-minimum layer of the North Arabian Sea. Concentrations at depths greater than 100 m are rarely above 2 ml/l in this zone and often well below 1 ml/l. However, Banse suggests that increased aeration due to the formation of a salinity maximum (at sigma-t=25g/l) and winter convection probably results in the upsloping

waters on the continental shelf of Pakistan being about 2 ml/l higher in oxygen content than those further south off the west coast of India. Nevertheless, the work of Doe (1965), Zupanovic and Mohiuddin (1973) and Ali Khan (1976) show that at least in some months and years dissolved oxygen levels at the bottom are extremely low on the south coast of Pakistan.

The appearance on the shelf of cool bottom water with low oxygen levels clearly affects the distribution of demersal fishes and shrimp as noted in Zupanovic and Mohiuddin (1973 and references therein). The movement of demersal fishes and shrimp further up the shelf, that is into shallower water, or even up into the surface water mass, causes increased aggregation and greater availability and is very important to tactical fisheries operations (though clearly fishermen are aware of such annual phenomena). The short-time scale (daily) fluctuations in the location (geographic, bathymetric) of cool, low oxygen water along the shelf would be useful to know from the scientific viewpoint, though the time of overall relaxation and retreat of the water off the shelf (at the end of year) is probably of greater fisheries value as it permits the reopening of trawling in deep water. In addition to the need to ascertain seasonal and short-term movements of the upsloped bottom water, Banse has correctly indicated the need to determine the factors affecting this already oxygen depleted water whilst on the shelf, i.e., contact time with the bottom and the benthos, the "rain" of organic material from the photic zone above and rates of vertical mixing with the surface water mass.

The distribution and availability of the demersal fisheries of the Pakistan continental shelf in a biological sense appear to be effectively controlled by the physical phenomena of upsloping of cool, low oxygenated water during the SW monsoon. Hence, it seems perfectly rational to design an integrated oceanographic program directed at giving both practical (applied) information to the fishery scientists as well as attempting to understand the basic physical processes at work. However, a first step (one of the aims of the 1982 Pakistan-USA workshop) is the need for syntheses of existing information from both the published and so-called grey literature, so as to optimize the design of the research programs and also to draw on experience gained in other regions with analogous situations.

From a fisheries point of view, the most important aspects relative to the upsloped cool, poorly oxygenated water are (i) the time of movement of this water on and off the continental shelf and the geographic extent of the phenomena; (ii) the intensity of the oxygen depletion and temperature drop in the upsloped water and the speed of it crossing the shelf (hence, increasing fish availability or causing death, if excessive); (iii) whether the spawning seasons of certain demersal fishes and shrimp are triggered by this annual event. Clearly, such practical fisheries requirements could be coupled with the basic science needs for (i) an understanding of the physical process of upsloping, and related chemical and biological oceanographic events, as it affects specific shelf areas including topographic features such as the Swatch (Indus Canyon); (ii) space scales involved at seasonal and interseasonal, and even daily,time scales; (iii) basic biological studies of the life history stages of those species considered dominant in the demersal fisheries (this presumes basic parameters for population dynamics will be determined by fisheries scientists) for determination of critical (sublethal, lethal) oxygen and temperature levels, spawning behavior, predator-prey relationships including comparison of food items in the different seasons and locations, etc.; (iv) more detailed information [following on work such as that of Muhammed and Arshad (1966) and Savich (1972)] on the benthos (macro and meiofauna) in relation not just to sediment but to the seasonal hydrographic cycles. Parulekar et al. (1982) have recently discussed benthic production of Indian seas in relation to demersal fisheries potential.

Selection of perhaps two areas, one on the narrow western shelf and one on the wider southern shelf, for long-term detailed scientific study could lead to better understanding of the basic oceanographic conditions prevailing off the Pakistan coast rather than a dispersed effort attempting to cover most of the coastline. Two areas are suggested for study as upsloping on the narrow western shelf, may be a more rapid event and hence, have a more rapid effect on the fisheries than on the wider shelf of the south coast.

Much less appears to be known about the pelagic fisheries of Pakistan coastal waters than the demersal ones; here concern is mainly with the small schooling species such as the Indian oil sardine rather than the more oceanic species such as tunas. Catches of the oil sardine in Pakistan were relatively large, 52,300 mt in 1980 or 22.5% of the marine catch (18.7% of total landings). It should be noted that until 1977 catches of the oil sardine were less than 14,000 mt per year but there was a sharp increase to a high of 71,400 mt in 1978 followed by a drop to 32,800 mt in 1979. However, the FAO Indian Ocean Programme (FAO, 1978) indicated the large potential of small pelagic fishes in the entire northwestern Indian Ocean based on acoustic surveys supported by some experimental fishing. Although small schooling pelagic species were found off both coasts of Pakistan, the biomass appeared greater (by almost an order of magnitude) off the south coast than off the west coast. Sardines, anchovies and small carangids were said to be most common (though anchovies do not appear in current FAO catch and landing statistics). A considerable fraction of the very large total pelagic biomass in the region is made up of mesopelagic fishes (Gjösaeter, 1978; Gjösaeter and Kawaguchi, 1980) especially the genus *Benthosema*. Ali Khan (1976) indicated in surveys of fish larvae off Pakistan in the NE monsoon (1964, 1967, 1968) that larvae of the oil sardine were dominant with those of anchovy and *Benthosema* next in importance.

The oceanographic data synthesized in Banse (this volume) and Qasim (1982) for the north Arabian Sea coupled with a knowledge of the catch distribution of small schooling pelagics in time and space may permit definition of those surface oceanographic features likely to be of importance in determining availability of such species off Pakistan. Most of these small schooling pelagics are probably feeding on phytoplankton (first trophic level) with some on zooplankton (second trophic level) or a mixture of both.

It should be stated at this point that during the SW monsoon the coastal current is anticyclonic (clockwise) to the east and south off Pakistan and during the NE monsoon cyclonic (anticlockwise) to the north and west. According to Banse the available nutrient and phytoplankton data for the Pakistan shelf area show some conflicting results. However, off the west coast of Pakistan during the SW monsoon there is the considerable effect of nutrients and phytoplankton advected from the upwelling areas off Oman. Off the south coast of Pakistan there is an apparent, but equivocal, inflow of nutrients from the Indus delta together with possible entrainment across the pycnocline of cool subsurface, nutrient-rich water (of low oxygen content) on the shelf. The apparent phytoplankton minimum off the south coast of Pakistan is in spring when the abundance of small schooling pelagic fish may be low. The peak phytoplankton season, according to Banse, is either; late in the SW or very early in the NE monsoon; small schooling pelagics are apparently most abundant in spring in inshore waters and this perhaps confirms the time of the phytoplankton maximum. Zooplankton data are basically lacking on a quantitative basis in time and space for Pakistan waters.

It seems for the immediate future that oceanography cannot contribute much to an increased utilization of the small schooling pelagics because of (i) lack of adequate nutrient, phytoplankton and zooplankton data on a seasonal basis for even representative areas of the west and south coasts; (ii) lack of knowledge for the west coast of the scales of advection from Oman waters; (iii) lack of data for the south coast on apparent seasonal input of nutrients from the Indus delta and from subpycnocline upsloped water on the shelf; (iv) lack of information on time and place of spawning of the dominant species of the small schooling pelagics, such as the oil sardine, and distribution of eggs and larvae other than the limited work of Ali Khan (1976). It is considered that a better overall knowledge of the seasonal cycle of phytoplankton in particular, and the causative physical and chemical processes, would help perhaps in determining likely location and availability of the desired species. Coupled with this is the need for information on trophic levels and food (species) of the life history stages of the dominant species of small schooling pelagics i.e., oil-sardine and anchovy, which would assist in determining apparent relationships between the fish and the environment. A considerable amount of work has been carried out on identical or related species, particularly in India. Thus, a first step would be to summarize existing literature on these species and determine the needed areas for future research, possibly with colleagues from neighbouring countries. It should be pointed out that little is known about the ecology of the diurnal vertically migrating species of mesopelagic fishes such as *Benthosema* which form such a very large fraction of the pelagic biomass off Pakistan. However, the information needed for a better understanding of the currently exploited small

schooling pelagics would also be useful in meeting this other objective related to mesopelagic fishes.

The oceanographic parameters needed for understanding of the stocks of large pelagic species, other than tunas, which occur in Pakistan shelf areas are unknown. In the case of tunas, existing seasonal data on the general oceanography of the North Arabian Sea, especially temperature and oxygen at surface and depth, would be sufficient to at least define preferred habitat boundaries for the various species (see Sund et al., 1981; Silas and Pillai, 1982) and hence, zones of availability to longline (subsurface) gear. Availability to surface gear (purse seine, live bait, trolling, beach seines) within preferred habitats would depend on appearance and sighting of schools at the surface in inshore areas.

Even if one excludes the potential (and actual) fisheries for large pelagic species, including tunas, and for mesopelagic species off the shelf, there is still a need for the oceanographic events and processes off Pakistan to be placed in the context of the overall need for increased knowledge of seasonal ocean circulation patterns in the northwestern Indian Ocean, particularly the role of large eddies in the extreme northern part of the Arabian Sea. Thus, there will be a need for marine scientists in Pakistan to take part in joint activities with neighbouring countries e.g., India, Oman, Yemen, and perhaps with other nations willing to work with them in the region. The role of satellite remote sensing in helping to obtain the large-scale synoptic view of events in the North Arabian Sea, including Pakistan coastal areas, should be carefully evaluated with due regard to the advantages and restraints imposed by such systems.

While completing this paper a copy was received of the recent puplication by Bakun et al. (1982) entitled "Ocean Sciences in Relation to Living Resources", resulting from initiatives arising from an Intergovernmental Oceanographic Commission resolution (11th Assembly) on Ocean Sciences in Support of Living Resources. The authors have prepared a detailed review and stimulating discussion of current thought on the topic. Many of the points made in this paper (and others that are not) are covered, though less eloquently than by Bakun and his colleagues.

REFERENCES

ACMRR, 1978. Report of the Ninth Session of the Advisory Committee of Experts on Marine Resources Research, 5-9 June 1978, Rome. FAO Fish. Rep. (206): 23 p.

Ali Khan, J., 1976. Distribution and abundance of fish larvae of the coast of West Pakistan. Mar. Biol. 37(4): 305-324.

Anonymous, 1982. Pakistan moves to protect her shrimp. Fish. News Internatl. 21(7):73 (July 1982).

Bakun, A. and R. H. Parrish, 1981. Environmental inputs to fishery population models for eastern boundary currents, p. 67-104. In: G. C. Sharp (ed.). IOC Workshop Rep. (28): UNESCO, Paris 323 p.

Bakun, A., J. Beyer, D. Pauly, J. G. Pope and G.C. Sharp, 1982. Ocean sciences in relation to living resources. Can. J. Fish. Aquat. Sci. 39:1059-1070.

Botsford, L. W. and D. E. Wickham, 1975. Correlation of upwelling index and Dungeness crab catch. Fish. Bull. U.S. 73(4):901-907

Cushing, D. H., 1978a. Climatic variation and fisheries, p. 18-19. In J. A. Gulland (ed.). Problems and progress in oceanography in relation to fisheries. FAO Fish. Rep. (206) Suppl. 1: 56p.

Cushing, D. H., 1978b. Biological effects of climatic change. Rapp. P.-v. Reun. Cons. Int. Explor. Mer 173: 107-116.

Cushing, D. H. and R. R. Dickson, 1976. The biological response in the sea to climatic changes. Adv. Mar. Biol. 14: 1-122.

Doe, L. A. E., 1965. Physical conditions on the shelf near Karachi during the premonsoonal calm 1961. Ocean Sci. Ocean Engineering 1.079 292.

FAO, 1978. Report of the FAO/Norway Workshop on the fishery resources of the north Arabian Sea, Karachi, Pakistan, 16-18 January 1978. Dev. Rep. Indian Ocean Programme (43) Vol. 2:vii+204 p.

FAO, 1982. Yearbook of fisheries statistics: catches and landings, 1980. Vol. 50:x+386 p.

Gjösaeter, J., Aspects of the distribution and ecology of the Myctophidae from the western and northern Arabian Sea, pp. 62-108. In: FAO (1978). Report of the FAO/Norway Workshop on the fishery resources of the North Arabian Sea, Karachi, Pakistan, 16-18 January 1978. Dev. Rep. Indian Ocean Programme (43) Vol. 2:vii + 204p.

Gjösaeter, J. and K. Kawaguchi,1980. A review of the world resources of mesopelagic fish. FAO Fish. Tech. Pap. (193): 151 p.

Laevastu, T., and M. L. Hayes, 1981. Fisheries Oceanography and Ecology, Fishing News Books Ltd., Farnham, Surrey, England: xiv + 199 p.

Laevastu, T., and H.A. Larkins, 1981. Marine fisheries ecosystem: its quantitative evaluation and management. Fishing News Books Ltd., Farnham, Surrey, England: xiv + 162 p.

Lasker, R., 1978. The relation between oceanographic conditions and larval anchovy food in the California Current; identification of factors contributing to recruitment failure. Rapp. P.-v. Reun. Cons. Int. Explor. Mer 173:212-230.

Laurs, R. M., 1977. Albacore advisory programme at the Southwest Fisheries Center, La Jolla, California, USA. pp. 110-119. In: G. H.

Tomczak (ed.) (1977). Environmental analysis in marine fisheries research: fisheries environmental services. FAO Fish. Tech. Pap. (170): 141 p.

Laurs, R. M. and R. J. Lynn, 1977. Seasonal migration of north Pacific albacore, Thunnus alalunga, into North American coastal waters: distribution, relative abundance, and association with transition zone waters. Fish. Bull. U.S. 75(4):795-822.

Muhammed, A. and M. Arshad, 1966. Benthos in relation to hydrology in offshore water of Karachi. Agriculture Pakistan 17:113-123.

Nelson, W. R., M. C. Ingham and W. E. Schaaf, 1977. Larvae transport and year-class strength of Atlantic menhaden, Brevoortia tyrannus. Fish. Bull. U.S. 75(1):23-41.

Owen, R. W., 1981. Patterning of flow and organisms in the larval anchovy environment, pp. 167-200. In: G. D. Sharp (ed.) (1981). IOC Workshop Rep. (28): UNESCO, Paris; 323 p.

Parrish, R. H. and A. D. MacCall. 1978. Climatic variation and exploitation in the Pacific mackerel fishery. California Dept. Fish, Game. Fish. Bull. (167): 110 p.

Parrish, R. H., C. S. Nelson and A. Bakun, 1981. Transport mechanisms and reproductive success of fishes in the California Current. Biol. Oceanogr. 1(2):175-203.

Parsons, T. R., B. O. Jansson, A. R. Longhurst and G. Saetersdal (eds.), 1978. Marine ecosystems and fisheries oceanography. Rapp. R.-v. Reun. Cons. Int. Explor. Mer 173.

Parulekar, A. H., S. N. Harkantra and Z. A. Ansari, 1982. Benthic production and assessment of demersal fishery resources of the Indian Seas. Indian J. Mar. Sci. 11:107-114.

Qasim, S. Z., 1982. Oceanography of the Northern Arabian Sea. Deep-Sea Res., 29(9A):1041-1068.

Savich, M. S., 1972. Quantitative distribution and food value of benthos from the west Pakistan shelf. Oceanology 12:113-119.

Sharp, G. D. (ed.)., 1981. Report of the workshop on the effects of environmental variation on the survival of larval pelagic fishes, Lima, 20 April-5 May 1980. IOC Workshop Report (28): UNESCO, Paris; 323 p.

Silas, E. G. and P. P. Pillai, 1982. Resources of tunas and related species and their fisheries in the Indian Ocean. Central Mar. Fish. Res. Inst. Bull. (India) (32):vi + 174 p.

Skud, B. E., 1982. Dominance in fishes: the relation between environment and abundance. Science 216: 144-149.

Sund, P. N., M. Blackburn and F. Williams, 1981. Tunas and their environment in the Pacific Ocean: a review. Oceanogr. Mar. Biol. Ann. Rev. 19:443-512.

Tomezak, G. H. (ed.), 1977. Environmental analyses in marine fisheries research; fisheries environmental services, FAO Fish. Tech. Pap. (170): 141 p.

URI, 1979. Climate and Fisheries: Proceedings from a workshop held March 29-31, 1978. Center for Ocean Management Studies, Univ. Rhode Island: 136 p.

Zupanovic, S. and S. Q. Mohiuddin, 1973. A survey of the fishery resources in the northwestern part of the Arabian Sea. J. Mar. Biol. Assoc. India 15(2):496-537.

17

Overview of the Hydrography and Associated Biological Phenomena in the Arabian Sea, off Pakistan

K. BANSE
University of Washington,
Seattle, Washington

ABSTRACT

Hydrography, oxygen, plant nutrients, and biological observations other than species records in the upper layers of the Arabian Sea north of 20°N and on the shelf of Pakistan, outside of estuaries and embayments, are reviewed against the backdrop of the conditions known from other subtropical, open ocean areas. The biological review is accompanied by remarks on methods used in the Arabian Sea and by references to literature on current work elsewhere. Seasonal features are emphasized. Newly-summarized bathythermograph data reaffirm the upsloping of cool water onto the shelf during the southwest (summer) monsoon and the absence of large-scale coastal upwelling during the northeast (winter) monsoon. The upsloping water normally seems to be better oxygenated than is the case off the Indian west coast owing to winter convection off Pakistan and the formation of a salinity maximum at sigma-t of about 25 g/l (rather than at about 24 g/l as elsewhere in the Arabian Sea). The effect of the upsloping water on demersal fishes, therefore, may be less drastic than off India. Some regional hydrographic differences, e.g., the influence of advection of nutrients from Arabia to the west (Makran) coast of Pakistan, the role of large eddies offshore, and (closer to shore) emerging differences among years are mentioned. The annual phytoplankton production in the open sea poleward of 20°N seems to be higher than that in the central Arabian Sea beyond the influence of the Arabian upwelling. The seasonal cycles of algal production rates in the open sea and on the shelf appear to include a minimum after the establishment of the shallow seasonal thermocline. Offshore, the daily production of the southwest monsoon season might be smaller than that of the winter season and actually be quite low; on the shelf and in contrast to the Indian west coast, the summer monsoon might not be everywhere the period of highest algal production, but this inference needs testing. Very little is known about seasonal cycles of zooplankton and benthos on the shelf of Pakistan. The biomass of the benthic meiofauna may be larger than that of the macrofauna.

The period of scientific exploration in the area is largely past even if there are open questions about the pattern of seasonal biological cycles. Therefore, the conventional seasonal collection of hydrographic as well as plankton data at fixed stations, followed by a few expedition-like surveys, is regarded as being of little promise. Instead, rapid progress could be achieved by focusing on specific questions. The subject of animal production is used as an example where results relevant to the tropics at large could be achieved with limited resources.

INTRODUCTION

The paper was written in order to introduce attendees of the U.S.-Pakistan Workshop on Marine Sciences in Pakistan, November 1982, to the local scene if they were newcomers to the subject or the area, and to provide a partially fresh look at phenomena basically familiar to others. It concerns the northernmost Arabian Sea, roughly north of 20°N, and between about 62 and 70°E and thus excludes the Gulf of Oman to the west but takes in the region off the Gulf of Kutch and the Indian peninsula of Saurashtra (Kathiawar) to the east (see Fig. 1a). After describing hydrographic features of the open sea, largely as they are relevant to biological phenomena of the epipelagic zone, it proceeds to the waters on the shelf but does not treat strictly-nearshore problems or estuarine questions. General text accompanying the regional treatments for phytoplankton and pelagic animals is to facilitate reference to modern work in other seas. Some suggestions for future studies are made in the body of the review as well as the concluding discussion. The discussion also remarks on problems in conventional plankton work which may not be amenable to rigorous investigation with limited resources on continental shelves or in the open sea anywhere.

MATERIAL AND METHODS

For the regional treatment, I relied in part on previous summaries which emphasized seasonal aspects (Banse, 1968; Wyrtki, 1973. Atlases by Krey and Babenerd, 1978; Wooster et al., 1967; and Wyrtki, 1971), but mainly on original literature. I also surveyed all bottle cast data for temperature and salinity when accompanied by oxygen observations, made since 1965 north of 20°E and between 60 and 70°E (from U.S. National Oceanographic Data Center), plus observations by "Anton Bruun" cruise 4 A,B in late 1963 (Woods Hole Oceanographic Institution, 1965) and "Meteor" in winter 1965 (Dietrich et al., 1966).

From all bathythermograph observations from the same area and stored in the U.S. Oceanographic Data Center, the data for the open sea centered around 21°N and 66°E and for two coastal areas (Fig. 1; caption to Table 1) were evaluated in detail. To make the seasonal picture more complete, six additional observations from between 24°24' and 24°29'N were included for Table 1b; there was only one value north of 25°00'N, on the shelf proper, in the longitudinal interval chosen which was also used. For the shelf south of Karachi (Table 1c), one of the two available quasi-synoptic surveys of broad strips across the shelf, each by one ship, showed the expected deepening of the pycnocline with increasing bottom depth or distance from shore. There, the data base was restricted to the area east of 66°10'E, and shoreward of a straight line from this longitude and latitude 24°20'N, to 22°30'N, 68°00'E. This line eliminates or includes the northeastern or southwestern halves of the northeastern or southwestern half-degree quadrants of the one-degree squares such dissected, and runs roughly parallel to the coast and very approximately along the 50-fathom (90 m) line. Included in its entirety is the one-degree square northeast of 22°N and 68°E, off the Gulf of Kutch.

The depth of the mixed layer was estimated in two ways (Wyrtki, 1971: p. 326): Either, the shallowest 5-m interval with a temperature difference of at least 0.5°C was found. The mixed layer depth then was calculated as the apparent depth where a straight line through the temperature-depths pairs of that depth interval intersected the surface temperature. Or, the mixed layer was assumed to be the depth at which the temperature was 1.0°C lower than at the surface. Means were calculated when both methods were applicable, as principally during summer stratification. In addition, the depths of selected isotherms were estimated. While all this was usually unambiguous with the mechanical bathythermograph data which had been digitized for 5-m intervals, the printouts for the expendable bathythermograph observations left more room for guessing of gradients since only the minimum data points necessary for reconstructing the original trace had been printed.

The coast to the west of approximately Karachi will be referred to as the west coast, and that to the southeast and southwest as the south coast of Pakistan, as in Haq et al. (1973). Other authors refer to these stretches as the Makran (Mekran) and Sindh coasts.

THE OPEN SEA

By way of further introduction the open Arabian Sea as a whole is in several respects similar to the other tropical and subtropical oceans in spite of the proximity of the continents and the closure in the north at about 25°N. The similarity is particularly apparent from November through February or March when the direction of the winds of the northeast (or winter) monsoon resembles that of the trade winds of the other oceans, and a similar pattern of open-sea, surface currents results. The deep surface layer is depleted of inorganic compounds of nitrogen, the principal limiting nutrient element; the upper part of the permanent thermocline is reinforced by a salinity maximum, and in the lower layers an oxygen minimum occurs which is especially pronounced in the eastern part of the northern Arabian Sea, again somewhat similar to other oceans.

The great difference from the other major oceans, of course, is the seasonal reversal of the atmospheric and oceanic surface circulations. The winds of the southwest (or summer) monsoon blow from about May through about September

TABLE 1. Temperature Structure in the Northern Arabian Sea

Means and (standard deviations): number of observations below months. Surf., surface temperature (°C); depth of mixed layer (M.L.) and of selected isotherms in meters.

1A. Open Sea

One-degree squares N and E of 20° and 65°E, 21°N and 65°E, and 20° and 66°E (see Fig. la)

Month	Surf.	M.L.	26°	23°	20°
J (2)	24.1	103 (10)	—	106 (2)	146 (1)
F (7)	23.7	125 (36)	—	98 (23)	173 (30)
M (23)	26.0	35 (14)	Variable	109 (2)	167 (32)
A* (1)	28.1	19 (—)	33 (—)	107 (—)	—
M (3)	29.6	23 (6)	61 (5)	(126) (—)	>>135
J (1-2)	29.1	36 (6)	45 (4)	(100) (—)	>>120
J (2)	29.2	30 (4)	38 (7)	101 (6)	>>100
A (10)	27.1	47 (9)	49 (10)	98 (14)	143 (13)
S (2)	26.7	51 (4)	53 (2)	73 (3)	131 (13)
O (16)	27.7	42 (5)	47 (5)	68 (6)	125 (9)
N (4)	26.9	64 (8)	63 (5)	81 (10)	127 (14)
Aug$_1$ # (5)	26.8	39 (2)	39 (2)	89 (13)	132 (8)
Aug$_2$ # # (5)	27.3	55 (2)	58 (3)	108 (7)	153 (8)

— +° 30'N, 66° 40'E
20°N, 65° 00' to 66° 02'E
20°N, 66° 14' to 66° 51'E

1B. Off Pakistan's Western Shelf
(24°30' to 25°00'N, 61° to 63°E, see Fig. 1a)

Month	Surf.	M.L.	26°	23°	20°
J (4-5)	24.3	91 (13)	—	102 (18)	185 (26)
F (1-2)	22.5	126 (9)	—	—	183 (—)
M* (8-9)	24.6	33 (13)	—	⎡ 37 (11) 123 (46) ⎤ *	159 (34)*
A (9-10)	24.9	25 (5)	—	⎡ 31 (8) 78 (14) ⎤ **	157 (24)
M (2)	29.9	8 (0)	14 (0)	52 (6)	129 (16)
J (4)	30.1	17 (5)	28 (6)	61 (8)	170 (2)
J (3)	29.4	16 (3)	19 (2)	28 (7)	97 (6)
A (6)	29.2	17 (3)	24 (6)	30 (6)	103 (23)
S (9)	28.2	20 (6)	23 (5)	30 (7)	85 (17)
O (3)	28.0	25 (6)	31 (7)	44 (3)	115 (25)
N_1 # (11)	27.3	31 (11)	29 (7)	48 (6)	116 (20)
N_2 # (5)	26.3	64 (12)	54 (6)	106 (19)	177 (7)
D (1)	25.6	50 (—)	—	57 (—)	132 (—)

* 23°, 4 shallow, 5 deep
** 23°, 5 shallow, 4 deep
 # early and late November. respectively

1C. Inshore on Pakistan's Southern Shelf
(See Fig. 1a)

Month	Surf.	M.L.	26°	23°	20°
J (3)	23.5	64* (50)	—	—	—
F (2,4)	23.1	>49;>76	—	10; 25 >94; >125	—
M (5)	24.7	41 (21)	—	71 (21)	138; 198; >210
A (1-2)	24.1	24 (19)		36; >76	168
M (1)	27.7	28 —	30 —	>57	—
J (1-2)	28.3	25 (5)	—	95 —	—
J (1)	26.9	13 —	15 —	34 —	88 —
A (1,3)	27.3	10 (5)	12 (1)	19 (3)	50 —
S (3-4)	27.7	20 (6)	16 (8)	35 (10)	75; >75; 104
O (7)	27.2	12 (4)	13 (7)	30 (9)	39,50,50; >55; >65; >85
N (11)	26.1	31** (16)	27° (15)	46# (18)	--
D (1-3)	24.8	100 —	—	100 # #	—

    *   also much deeper values
  **   also >35, >35, >45
    °   7 values only; also >20, >35' >35
    #   4 values only; also >60, >65, >80, >80
 # #   elsewhere, 39 and 90

and are during June to September appreciably stronger than those of the northeast monsoon. During the summer monsoon, the currents in the open sea and particularly along the coasts; the shoaling of the thermocline towards all coasts; the upwelling on both sides of the Arabian Sea; and partially the plankton concentrations, provide a picture quite different from the northeast monsoon season. The proximity of the land causes strong seasonal signals also in other repects, for instance as an appreciable annual range of surface temperatures even in the open sea, and as marked sea level changes nearshore. Finally, the coastal current tends to be anti-clockwise during much of the northeast monsoon, and clockwise prior to and during much of the southwest monsoon seasons; this is at least in part a response to the large-scale mass distribution and the resulting pressure gradients over the entire Arabian Sea rather than to the local winds. The following is to substantiate the overall picture but also will point out that many of these generalities do not hold north of 20° because of the proximity of the continent.

Hydrography

In the surveyed region poleward of 20°N, the offshore surface temperatures range from 23 to 24°C in February to about 29° in May and June (July), with a secondary minimum in August or September as a consequence of the southwest monsoon; a more or less marked secondary maximum is observed in October (means from monthly charts by Wooster et al., 1967). Above the continental slope off Pakistan's west coast, the means range from about 22 to above 30° (cf. Table 1b).

Wyrtki's (1971) bi-monthly charts show the mixed layer (see Methods, herein) as being deeper than 80 or even 100 (nearshore) m in January/February (limited data); it becomes shallower than 40 m in the western part of the area and somewhat deeper in the easternmost parts during March/April, and is above 40 m everywhere from

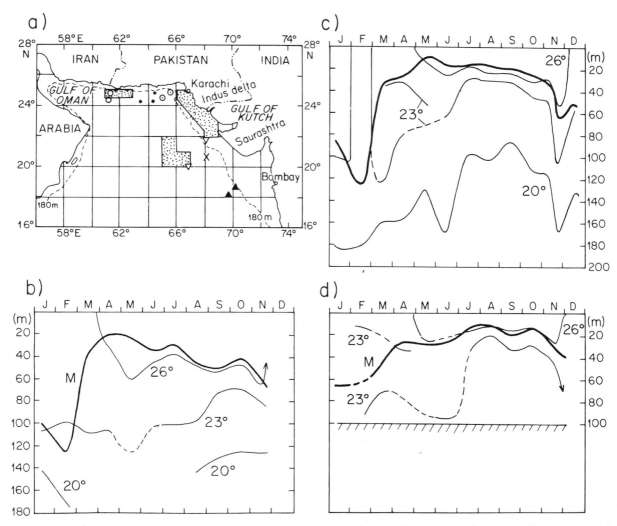

Figure 1. Sampling locations and seasonal trends of isotherms. 1A. Location of bathythermograph (BT) observations (shaded; see Table 1) and bottle casts (for ships' symbols, see Fig. 2; the southern ML 17 and LF 242, the northern ML 18 and LF 243; and AN 9 and 18 each were essentially on the same locations). 1B. Offshore BT observations (Table 1A). 1c. West coast BT observations (Table 1B). 1d. South coast BT observations (Table 1C; M in Figs. 1b-d, bottom of mixed layer).

July through November/December; mixed layer depths between 20 and 40 m are widely present from July through October. Accordingly, the maximal temperature gradient jumps from a depth of about 120 m in March/April to about 25 m in May to August, deepens to less than 50 (40?) m in September/October, and stays at about 40 m in November/December. This shallow gradient results in the open sea largely from the formation of the seasonal thermocline: The depth of 20°-isotherm, in the middle of the permanent thermocline, ranges between 120 and 160 m year-round; only toward the continental slopes are large seasonal changes in depth of this isotherm apparent, the shallowest positions being attained during July/August.

The bathythermograph observations in Fig. 1b,c suggest that the seasonal warming reduces the depth of the mixed layer drastically already in March, and that the reverse process of cooling becomes effective during November. Also (Fig. 1b), there is a seasonal rise of the 20° isotherm, as well as of the isotherms above (23°) which nearshore leads to upsloping of cold water (see

later). This upsloping persists after the southwest monsoon has ceased; off Arabia and the Gulf of Oman, it affects surface temperatures and primary production far into the open sea.

The surface salinity of the northernmost Arabian Sea ranges from about 36.2 to above 36.5 ‰ S. The usual salinity maximum of the tropical ocean, in the top of the thermocline, seems here to be formed in the middle of the northern Arabian Sea (Rochford 1964; his salinity maximum D, with values of 36.5 to 36.6 ‰ S, 26 to 27°C, and sigma-t of somewhat less than 24 g/l). Wooster et al. (1967) suggested tentatively that this water forms during spring and summer but the seasonal map of surface density in Wyrtki (1971:56) suggests otherwise. Additional sinking of water of about 36.6 ‰ S and 22 to 24°C (or slightly higher; sigma-t about 25 g/l) takes place on and beyond the northern shelf during late fall and early winter; the properties about fall into the envelope of Rochford's salinity minimum C but the present water is well-ventilated (Banse, 1968; see also below). This weak maximum, which replaces geographically the one at 24 g/l, is illustrated in

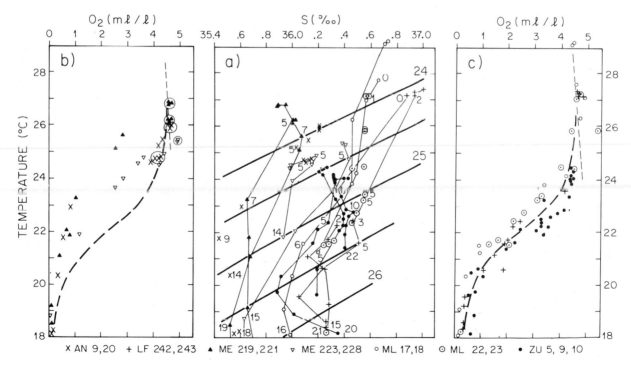

Figure 2. Bottle cast data (for locations, see Fig. 1a). 2a. Temperature and salinity (heavy lines, density; numbers in figures, depths in decameters). 2b,c. Temperature and oxygen of stations in Fig. 2a (heavy broken line, eye-fitted through points in Fig. 2c; thin broken line in upper corners of Figs. 2b, c, relation for saturated water). 2b. Eastern, low-salinity stations (upper samples for each station circled. 2C. Western, high-salinity stations. Ship names and dates: AN, "Atlantis II", Jan. 1977. LF, "Lesnoi", Sept. 1968. ME, "Meteor", March 1965. ML, "Mikhail Lomonosov", May 1966. ZU, "Zulfiquar", March 1967.

Fig. 2a (e.g., observations by PNS "Zulfiquar"; "Mikhail Lomonosov", stas. 22, 23). The water is absent off the Bombay shelf (cf. Fig. 2a,b, "Meteor" stas. 219, 221; "Atlantis II" stas. 9, 20), as well as off Arabia and the Gulf of Oman (Anton Bruun Cruise 4A, November; not depicted). Finally, the outflow from the Persian or Arabian Gulf (Rochford's salinity maximum B; with sigma-t values of about 26.3 to 26.8 g/1 according to Kuksa, 1972; see also Duing and Schwill, 1967; Wyrtki, 1971:261) is well below.

The seasonal change of dynamic heights penetrates north of 20° to less than 200 m, quite in contrast to the areas off Somalia and southwest India where depths of more than 350 m are being reached (Düing, 1970).

It is difficult to proceed beyond such broad statements because even the monthly charts of surface temperature and ship drift (e.g., Warners, 1952), based on a large number of years, show appreciable (i.e., persistent) regional differences in the area of concern; this is even more true for maps from quasi-synoptic surveys during a single month. For example, in 1968, I mentioned the complicated average current trends in the area during November, the month of transition from the clockwise (summer) to the anti-clockwise (winter) pattern. Further, for the shelf off Saurashtra and the Bombay area during spring, I had shown in 1973 that a long, narrow trough of the thermocline is a regular feature which cannot be reconciled with the notion of a simple clockwise current near and above the shelf during this season.

In addition, there is transient variability within season: For each of the five two (three)-months periods depicted, Düing's (1970) charts of dynamic heights show large cyclonic and anticyclonic eddies which presumably move horizontally; north of 20°, though, Düing's coverage is inadequate except perhaps for March/April (see Das et al., 1980 for February/April data from this area). Owing to these features, one may encounter great monotony, as in the "Anton Bruun" meridional section across two degrees of latitude on which the October means in Table la are based; at other times, fronts or eddies are very obvious, as in the "Atlantis II" section at 20°N which yielded the August values in the same table. Perhaps not incidentally, the chlorophyll concentrations on "Atlantis II" stas. 76 and 77 (sta. 76, 2 miles W of

the area covered and presumably still in the shallow-mixed layer regime designated as Aug1; sta. 77, in fact in the deep-mixed layer regime called Aug2) were 17 and 49 mg/m², in the shallow and deep-mixed regimes, respectively (Laird et al., 1964; cf. Yentsch, 1965: fig. 1); it will be argued later that the open Arabian Sea is nutrient limited during the southwest monsoon so deep mixing by an eddy should result in enrichment of the upper layers. Finally, for January/February 1974, meridional sections at about 61°E (Varkey and Kesavadas, 1976) and about 64°E (Varma et al., 1980) indicate a cyclonic and an anticyclonic eddy, respectively. The top of the thermocline was raised from 110 or 120 m to about 60 m, and descended from 40 or 60 m to 175-200 m, respectively. At 61°E, the accompanying uplift of water with high nutrients resulted in a marked reduction of transparency, presumably due to development of algae (cf. high rates of algal production in Radhakrishna et al., 1978: 271, on same cruise).

Düing (1971) suggested that the number of eddies may be higher in the transitional periods than in the summer or winter periods, i.e., that the hydrography should be quite complicated. This is also indicated by the dynamic heights for the Arabian Sea during November/December 1960 in Fomitchev (1964). Further, for much of the area off Pakistan, quasi-synoptic surface and 50-m maps as well as hydrographic sections are available for mid-November/first days of December of 1975 and 1976 (Yamanaka et al., 1978; Yamanaka et al., 1977). Both surveys show upward ridging of the thermocline between 23° and 24°N, extending from the west to about 64°E; the corresponding currents would set to the west near the shelf and the coast, but to the east in the open sea, south of the ridge. Moreover, upsloped water was present in 1975 on part of the shelf off the Indus mouth but downward-warping of the thermocline was marked along the shelf break off western Pakistan, with the mixed layer reaching 60 m, or even 80 m; again, different currents during this transitional period must have been present. As to be expected for such times, differences between years are present, e.g., the topography of the mixed-layer depth along the shelf during 1976 was more flat than in 1975. Nutrient distribution and supply to the upper layers, with all biological consequences, are bound to respond to such hydrographic differences.

Oxygen and Nutrient Salts

In the Arabian Sea anywhere, the concentrations of dissolved oxygen decline rapidly with depth; poleward of 20°N, the oxygen content of the thermocline tends to decrease at temperatures of 26 to 27° (cf. Banse, 1968: fig. 7), signifying that the oxygen demand outruns the supply. Values at 100 m almost nowhere seem to be above 2 ml/l, and commonly are below 1 ml/l (Wyrtki, 1971). These low concentrations form the top of the huge oxygen minimum layer at intermediate depths in which values below 0.2 ml/l occur north of 15°N throughout. In the northeastern part of the sea, concentrations even decline to below 0.05 ml/l, and reduction of nitrate takes place in midwater (cf. references in Qasim, 1982: p. 1057; also Anderson et al., 1982).

At the shallow depths of principal concern to this review, enhanced aeration owing to the salinity maximum at a sigma-t of about 25 g/l is apparent from Fig. 2c. The comparison with concentrations in water collected, for example, well south of Saurashtra, at the latitude of Bombay ("Meteor" stas. 219, 221), illustrates that there, water at comparable temperatures contains very approximately 2 ml/l less oxygen than off Pakistan. "Meteor" stas. 223 and 228 southwest of Saurashtra (cf. Fig. 2a) are intermediate in t-S characteristics and aeration. In consequence, if water of 23° upslopes onto the shelf, the effect on bottom animals and demersal fishes ought to be different off Bombay and off the central Pakistani coast (see later).

For the nutrient salts in the Arabian Sea, Wyrtki's atlas shows surface nitrate values as being below 0.5 μg-at/l N and often undetectable all year, the exception, during the summer, is the southwestern corner of the area of concern which is affected by the broad upwelling off Arabia. Only a few ammonium determinations seem to have been made; they show mostly the customary low values of the subtropics (e.g., Dietrich et al., 1965; Lukashev, 1980). Measurements of other nitrogenous algal nutrients appear to be almost absent (see, however Verlencar, 1980 for urea). The concentrations of reactive phosphate range between 0.4 and 0.6 μg-at/l P, and those of silicate from below 3 to above 5 μg-at/l Si, again with the exception of the southwestern corner during summer.

As a consequence of the general uplift of the pycnocline during the southwest monsoon along the northwestern, northern, and eastern coasts of the Arabian Sea, the pycnocline and the associated gradient of plant nutrients are quite shallow (25 to 35 m) and at well-lit depths in the open sea off the mouth of the Gulf of Oman and south of the Pakistani-Iranian border (for early September, see "Lesnoi" stations in Fig. 2a, c; for October/early November, Gilson, 1937; Ryther and Menzel, 1965; Woods Hole Oceanographic Institution 1965). Further in the northern Arabian Sea, horizontal advection must also play an appreciable role because of the relatively small distances involved; in fact, Ryther and Menzel (1965) demonstrated how far the chemical/biological effects of the upwelling off Arabia extend into the open sea.

The nutrient concentrations in the salinity maximum underlying the mixed layer are in part quite low (e.g., below 1 μg-at/l nitrate over much of the central Arabian Sea), in part they are very high (above 10 μg-at/l nitrate, off Arabia and toward Karachi, Wyrtki, 1971: p. 278/9) so that this upper salinity maximum might be an important source of nutrients to the mixed layer during the seasonal vertical convection. Since Wyrtki did not realize the occurrence of two geographically separate upper salinity maxima (one at a sigma-t of roughly 24 g/l and often only moderately-well aerated, the other at 25 g/l and well-aerated), his cited map possibly reflects partly the geographical distribution of these maxima. In any case, high rates of primary production in the open sea discussed below indicate an appreciable injection of nutrients into the mixed layer in broad areas during this time. A critical station-by-station review of the existing data on hydrography, nutrients, and primary productivity for the cool months should be worthwhile and be undertaken before any more efforts are spent at sea (also see General Discussion).

At greater depth, the nutrient concentrations increase rapidly in correspondence to the decline of dissolved oxygen. Poleward of 20° N at 100 m, according to Wyrtki (1971), nitrate ranges from approximately 15 (or slightly lower) to above 20 μg-at/l N, and reactive phosphate from about 1.0 to above 2.0 μg-at/l P. I presume that these very large concentrations are generally below the depth of seasonal mixing and thus not accessible to plant production except by eddy diffusion; the exception might be areas of seasonal sinking (convection) of surface water alluded to above. Certainly, the upper salinity maximum which overlies the oxygen minimum shows that vertical eddy diffusion cannot transport the enormous store of nutrient salts, equivalent to the consumed oxygen, upward into the photic zone; the only material injection from greater depth must take place by mixing at the edge of the sea.

Phytoplankton

The conventional view of the algal production process in the open tropical and subtropical oceans with a permanent deep thermocline comprises low production rates per unit of sea surface as well as low algal specific growth rates (carbon uptake per unit of chlorophyll or algal carbon, and time), equivalent to perhaps 0.1 to 0.3 doublings per day (cf. Eppley, 1981 for such rates). The algae are eaten about as fast as they are being produced, and periods of unbalance ("blooms") between their production and death by grazing normally do not occur. Most (often 90% or more) of the algal growth is based on ammonium (and some nitrogenous organic compounds) as well as phosphate excreted by these grazers and their predators (cf. Cushing, 1980, and measurements in the North Pacific by Eppley et al., 1973; Sharp et al., 1980).

The reason for the low algal production is thought to be the low nutrient concentration, principally of inorganic nitrogen compounds, which is ultimately caused by the continuous loss of organic matter from sinking of fecal pellets and other dead material and less so from vertical migrations of animals (diurnal or seasonal/ontogenetic). This loss is balanced by the upward-directed eddy diffusion of nitrate and phosphate which supports the rest (often 10% or less) of the daily algal production. Supply of nutrient salts by rain, and fixation of nitrogen gas by cyanobacteria (formerly called blue-green algae) is regarded as very minor (Capone and Carpenter, 1982; Harrison, 1980). In the absence of horizontal gradients of nutrient salts in the surface layer, as for example in the central North Pacific gyre, horizontal advection is also considered as negligible.

To be added to this picture is the occurrence of a deep chlorophyll maximum which is present in the open ocean where the photic zone extends beyond the bottom of the nutrient-depleted mixed layer and into the stratified water of the thermocline. Then, shade-adapted algal growth

may occur even at approximately 0.1% of the irradiance just below the surface (rather than being terminated at the too-often-cited 1% level); it draws on the nutrients diffusing upward from below (see Cullen, 1982 for a general review, and Jamart et al., 1977 for a model for the temperate zone that might apply also in the Arabian Sea). Photosynthetic carbon uptake in the maximum has been observed to contribute 15% (Anderson, 1969 for a temperate site) and up to 20% (Venrick et al., 1973 for tropical South Pacific stations) of the column production. The occurrence of the deep chlorophyll maximum in the Arabian Sea was briefly mentioned by Yentsch (1965) but otherwise largely overlooked, I believe, because of the too widely-spaced standard depths of water collection. Also previous determinations of carbon uptake rates in the Arabian Sea are apt to have terminated at too shallow depths (i.e., the 1% light depth) during periods without large algal concentrations near the surface, i.e., under well-stratified conditions. They thus underestimated the column production as demonstrated, for instance, by the high uptake rates at the lowest sampling depth at 5 out of 6 stations in Radhakrishna et al. (1978: 271, fig. 1).

The transparency in the Arabian Sea (Krey and Babenerd, 1976) should allow photosynthesis at 100 m or even deeper when the upper layers are poor in particulate matter but the mentioned open-sea, winter section at 61°E by Varkey and Kesavadas (1976) shows a 1% depth only about half as deep as the mixed layer, inside as well as outside the cyclonic gyre. Obviously, the absolute rather than the relative irradiance determines the bottom of the photic zone, as vividly illustrated by Steemann Nielsen and Hansen (1961) for Danish waters. In any case, these offshore data from the Arabian Sea may serve as a warning not to take irradiance for granted.

Phytoplankton field studies nowadays tend to be closely related to the results of laboratory studies: the connection between the environment and the physiology of cells that are bathed by the sea water itself and have rather high growth rates is, of course, close. The books edited by Falkowski (1980) and Platt (1981) comprehensively introduce phytoplankton physiology in ecological contexts as well as to some field studies (see also Holm-Hansen, this vol.) A very serious current question for field work is whether the ^{14}C method of measuring photosynthesis might grossly, by more than a factor of two, underestimate the actual rate of particulate algal net production (cf. Carpenter and Lively, 1980; Peterson, 1980), quite apart from the great variability among the numerous versions of the technique. Obviously, if the method is systematically far off the mark, all the maps of primary production rates to be discussed below will be erroneous.

Turning now to the phytoplankton of the open Arabian Sea as a whole, the hydrographic analogies between it and the other subtropical oceans during the northeast monsoon should make the above conventional view of primary production in the low-latitude, oligotrophic ocean applicable, with the qualification that horizontal advection should play a larger role than in, e.g., the North Pacific gyre. Indeed, the map of photosynthetic carbon uptake for December to May by Kabanova (1968; no data north of 20°N) supports the analogy for the offshore regions: Essentially, the rates are below 0.1 g C/m² d, a range regarded by Blackburn (1981: table I) as typical for the subtropical gyres. The average chlorophyll concentrations in the upper 50 m during November to April (Krey and Babenerd, 1976) are definitely higher, though, than in other subtropical gyres: South of about 20°N, the values range between 0.1 and 0.2 (0.3) μg/l. The concentrations depicted for, e.g., the lower latitudes of the South Indian Ocean, are below 0.05 μg/l throughout.

During the southwest monsoon season, the western third or more of the open Arabian Sea is greatly affected by the upwelling along the coast of Arabia. Carbon uptake rates above 0.5 or even 1.0 g C/m²d and chlorophyll concentrations (means for the upper 50 m) above 0.5 μg/l occur over wide areas (Krey and Babenerd, 1976). The eastern half of the open Arabian Sea, however, beyond the zone of coastal influence from India, exhibits rates below 0.1 g C/m²d and slightly lower pigment values. Note that Krey's map of carbon uptake for the eastern half is principally based on May observations (cf. Babenerd and Krey, 1974) and thus may not be entirely applicable to the southwest monsoon season.

A detailed look at these and subsequently published maps, however, yields a rather confusing picture, both for the Arabian Sea as a whole and for the area poleward of 20°N; the cause largely is inappropriate temporal lumping of data. For example, the map of carbon uptake for November to April by Krey and Babenerd (1976) depicts high rates (above 0.5 g C/m²d) over wide areas almost in the center of the Arabian Sea, in contrast to Kabanova's (1968) earlier map. Poleward of 20°N, during the same period, Krey and Babenerd (1976) show likewise rates above 0.5 g C/m²d almost all over. However, according to Babenerd and Krey (1974), the high rates in the western section, north of 20°, were almost entirely determined during early November when the summer pattern of upwelling or uplift of the thermocline still persisted (Ryther and Menzel, 1965); possibly the entire west coast of Pakistan may have been under this influence (cf. Ryther et al., 1966). The very few measurements available to Krey and Babenerd in the eastern section were apparently made during March. In contrast, offshore stations in the eastern area (north of 18°), occupied during November by Radhakrishna et al. (1978: 137), yielded rates of only 0.1 and 0.2 g C/m²d, which is not surprising in view of the summer thermocline presumably persisting there until then (cf. Jayaraman and Gogate, 1957). Yet, north of 20°, observation during December through February of another winter season show high values (generally above 0.5 or often 1.0 g C/m²d) west of 64°E as well as off Saurashtra and the Bombay area (Radhakrishna et al., 1978: 271, fig. 3; I had to infer the meaning of some of the isolines; dates in Varkey and Kesavadas, 1976; hydrographic sections in Varma et al., 1980). To conclude the review of papers for the winter season, Kuz'menko (1977) depicts high algal biomass from cell counts to the north and south of 20°N in the open sea (above 0.2 or even 1.0 mg/l wet mass; average for 100-m column) but suggests low values (generally below 0.1 mg/l) on the shelf. He states that the offshore values were accompanied by surface concentrations of about 1 μg-at/l phosphate-P caused by cyclonic water movements; I assume from Kuz'menko (1974) that the observations were made after November so that in addition to eddies, vertical overturn and convection from surface cooling might not be out of the question in some places (see the mixed layer of 70 to over 80 m depth, of 25 to 26°C and 36.4 to 36.5‰ S, during late November/early December 1976 in the area [Yamanaka et al., 1977: figs. 15-19; 27]). An east-west section across the middle Arabian Sea during winter by Savich (1968) likewise shows low plankton biomass near the shelf, relative to that in the open sea.

During March, poleward of 20°N, a few "Meteor" stations exhibited appreciable supersaturation of oxygen in the upper layers at temperatures of roughly 24 to 26°C, i.e., before the marked seasonal heating (Stas. 237, 238, to a lesser degree 233, 234; all southwest of Karachi; see also Fig. 2b). The oxygen must have been produced by vigorous photosynthesis. In contrast, Radhakrishna et al. (1978: 271) who had found high rates earlier in the year, measured carbon uptake values below 0.5 and even 0.3 gC/m² during March and April off Pakistan, east of 64°, including the area of the Indus delta. The hydrographic sections of the cruise, in Varma et al. (1980), show an already marked summer thermocline, with surface temperatures above 25.5°C. Possibly the thermocline had cut off the diffusive nutrient supply from below and reduced photosynthesis, but Radhakrishna et al. state that nitrate was undetectable in the surface water on only 5 out of their 22 stations. A different but not mutually exclusive explanation for the lowered primary production rates per unit of sea surface is the enhanced zooplankton mass observed on this cruise in the same area (see below): If grazing had lowered the algal biomass without affecting the specific production rates (e.g., the carbon uptake per unit of chlorophyll and light), reduced carbon uptake per unit volume of water or of sea surface would have resulted. I could not find the corresponding chlorophyll data to test this alternative (see the Discussion for the need of determining specific production rates and P versus I curves). As an aside, by its ammonium regeneration the zooplankton is the key to answering the thorny, currently much debated question of the nutrition of phytoplankton in oligotrophic waters, an issue that goes beyond the scope of this review (cf. Goldman, 1980; Goldman and Glibert, 1982; Lehman and Scavia, 1982; McCarthy and Goldman, 1979).

For the entire region north of about 21°N during the southwest monsoon (now, during May to October), Krey and Babenerd (1976) showed offshore production rates above 1.0 g C/m²d while Kuz'menko (1969) had reported rates of 0.3 to 0.5 g C/m²d for June 1966. Later, Kuz'menko (1974, not included in Krey and Babenerd's data base) observed rates below 0.1 g C/m²d for about the same area during September/November 1969. The apparent contradictions regarding the carbon uptake during the southwest monsoon season

seem to stem, again, from the temporal grouping of the data: The high southwest monsoon values depicted by Krey and Babenerd are in part based on four stations in June 1966 off the Gulf of Oman (upwelling observed) and near the Pakistani coast (cf. Kuz'menko, 1969) but essentially reflect observations of late October 1963 when the advective influence of the Arabian upwelling extended far across the Arabian Sea, with surface phosphate values above 0.5 μg at/l, and had appeared off Karachi (Ryther et al., 1966; Ryther and Menzel, 1965). The low rates of June 1966 and September/November 1969 are in line with the data of March/April 1974 (Radhakrishna et al., 1978: p. 271) and also with the mentioned May (1963 and 1964) data used by Krey and Babenerd (1976) for the eastern Arabian Sea south of 20°N, all measured above the seasonal thermocline. As inferred earlier, at least the eastern section of the area may be impoverished until the seasonal thermocline weakens or vanishes (November observation by Radhakrishna et al., 1978: p. 137). Thus, during the southwest monsoon season, Krey and Babenerd's picture may not be representative for the open sea, north of 20°N, beyond the area under the influence of the Arabian upwelling.

According to Krey and Babenerd (1976), the chlorophyll concentrations north of 20° are largely above 0.3 μg/l (average for 50-m column) for both halves of the year, and above 0.5 μg/l in the western part of the region. In contrast, the maps of offshore distributions of algal mass from cell counts in Kuz'menko (1977: fig. 2) during "summer", and during September/November in Kuz'menko (1974) illustrate wide areas of low biomass (below 0.1 mg/l wet mass).

In conclusion, the central Arabian Sea seems to be more productive than the subtropical gyres elsewhere (as stated by others before), and the area poleward of 20°N to be richer (in terms of carbon uptake rates and chlorophyll concentrations) than the central Arabian Sea. For the seasonal cycle of the open northern Arabian Sea, the following suggestion may be made: During the winter, vertical convection and moderate enrichment of the surface layer with nutrients at latitudes lower than in a truly oceanic climate lead to high rates of photosynthesis which might prevail until early March. The seasonal thermocline established during spring and persisting through late fall diminishes the nutrient supply so that the southwest monsoon season is a period of low primary production rates except in the broad western area influenced by the upwelling off Arabia and the Gulf of Oman. This view differs radically from that expressed by Qasim (1982: table 8) by calculating the same average column production rates for the southwest and northeast monsoon seasons.

In effect, I suggest that the continental effect might lead at low latitude to winter convection and a production maximum similar to that observed in the North Atlantic only at about 32°N, off Bermuda (Menzel and Ryther, 1961), but not at 2-3° of latitude to the south of the island (Ryther and Menzel, 1961). Deep mixing occurs in winter also in the North Pacific central gyre (Bathen, 1972) where the productivity has been well studied at 28°N (McGowan and Hayward, 1978; Sharp et al., 1980). Apparently, the mixing there does not reach nutrient-rich layers so that a clear seasonality of the production rates is absent. The connection between the aeration and nutrient content in the upper salinity maximum of the Arabian Sea and the enrichment by vertical convection during winter was raised already in the nutrient section. As indicated in that discussion, the ideas presented here about the annual productivity cycle of the northern Arabian Sea can probably be tested by a review of the existing biological measurements which makes critical use of the hydrographic and nutrient data of the same cruises and groups then the results in a convincing manner within the major seasons on the basis of hydrographic (i.e., independent) evidence as well as by geography. Satellite data can also be consulted (see General Discussion).

Turning now to the algal species involved, it appears by inference from adjoining regions that the species list of the area north of 20° will be long: It is in part open ocean but comprises also coastal waters of various kinds. Nearshore at Calicut on the Indian west coast, for example, Subrahmanyan (1958a) reported 226 diatom species collected by nets. On the basis of settled water samples, Subrahmanyan and Sarma (1967) could add only a few further species, presumably because every so often in the earlier study, even very small species became entangled in the meshes of the nets which then were woven from threads rather than today's smooth monofilaments (Wood, 1963 published a checklist for diatoms from the entire Indian Ocean). For dinoflagellates, the other large phytoplankton group of these seas (in terms of species, at least), Taylor (1976) listed 219 species from net collections in the Indian Ocean. Most of these will occur also in the Arabian Sea (F.J.R. Taylor, pers. comm.). Finally, Kuz'menko (1975) named 59 coccolithophorids for the Arabian Sea.

As in other areas with long species lists, also here only a restricted number is common. For instance, in the coastal data set of Subrahmanyan which extended over many years, only about 29 out of 226 diatom species, and 7 out of 120 dinoflagellates, dominated numerically (Subrahmanyan and Sarma, 1961). The largest number of species occurred at intermediate population levels of the net plankton (Subrahmanyan, 1959). Similarly, Noor-Uddin (1967a) noted for the inshore waters near Karachi that the heavy diatom blooms between December and May (net-collected plankton) were usually dominated by two or three species only. The dominance by a very few species is even more drastic when biomass instead of cell numbers is considered, as can be seen from the offshore data by Kuz'menko (1974), principally for bloom situations. I wonder, however, whether the distribution of mass over the species is offshore more even after long-lasting periods with little temporal change of biomass, i.e., when algal growth is nearly in balance with loss.

In remarks about more recent offshore samples, again collected by net, Subrahmanyan et al. (1982) stated that quite a few of the diatom species had not been encountered nearshore although the total number of species was less than inshore; usually, more than one or two species (as is the case inshore) dominated among the diatoms. Further, on the open shelf and offshore, dinoflagellates might dominate the net plankton to the almost total exclusion of the diatoms; this had never been observed by Subrahmanyan during his very extended nearshore collections.

Little attention seems to have been paid to species ecology in the offshore area of concern, beyond the work by Kuz'menko (1974, 1977), this author converted the cell numbers also into cell volumes and thus could report biomass data. As mentioned, a special but not well studied feature is the deep chlorophyll maximum. Some of the species occurring there are more or less restricted to this habitat and form a shade flora. The assemblage was reviewed by Taylor (1973) for the Indian Ocean and the species list discussed more generally by Sournia (1982). Venrick (1982) set species occurrence in the central North Pacific gyre into the context of our current theories of community regulation.

Animals.

Limited as our understanding of the conditions of growth and death of the phytoplankton may be, we know much less about zooplankton and nekton. First, sampling and study have been focused on organisms that can be caught by plankton or fishing nets. We have not collected or investigated most of the small forms like colorless flagellates or ciliates. We only are beginning to estimate the amount of fish in the sea by acoustical means; and we know that we do not capture at all the larger squids: the yield of these molluscs taken by sperm whales alone is estimated to be somewhat more than the yearly fish catch by the entire human species (Clarke, 1977; for the sampling issues see Angel, 1977 and Haury et al., 1978). Second, we do not have a method that measures the growth rate of animals, in analogy to the technique for photosynthetic uptake of ^{14}C. Third, for understanding zooplankton population dynamics, we need to know not only growth (development) rates and mortality as with the phytoplankton, but also fecundity (number of eggs per female). Fourth, marine zooplankton populations ordinarily cannot be revisited because of the ever-present advection so that the usual techniques of population dynamics and production measurements cannot be employed. The only areas where research on zooplankton may be more advanced than that on algae are natural history (largely, ontogenetic development and seasonal occurrence), biogeography, and the role of predation in influencing species composition of the prey community (cf. articles in Kerfoot 1980). Size-selective predation which has been mostly studied in fresh water, is now understood to structure plankton communities from the top downwards while hydrography and nutrient supply may do the same from the bottom up (cf. Landry, 1977; Steele and Frost, 1977).

In their world map of biomass of zooplankton collected in the upper 100 m by nets, Bogorov et al. (1969) depicted the open Arabian Sea, with 101-200 mg/m³ wet mass except in the very center (south of 20°N; 51-100 mg/m³), as being several times richer than much of the subtropical gyres of the other oceans on the same latitutdes: con-

centrations were drawn even higher (201-500 mg/m³) off the west coast of India including Saurashtra. Little detail, however, is known about the region poleward of 20°N. The seasonal coverage by the International Indian Ocean Expedition (IIOE) was inadequate as seen from the bi-(tri-)monthly charts for the material by Rao (1973); Prasad (1968, 1969) had therefore presented the IIOE catches grouped into two six-month periods. Except off Saurashtra, no hauls were taken in the area of concern during the summer monsoon; however, the collections off Arabia at and south of 20° suggest that the strong gradient of biomass from west to east observed in the entire area south of 20° and caused by the Arabian upwelling, will also be present north of 20° during the summer monsoon. A weak gradient, with concentrations decreasing from west to east by less than a factor of two, was depicted similarly by Prasad (1968) for the period from mid-October to mid-April. However, it is worth noting that the richer samples from the western side were taken during late October/early November when upwelling still prevailed off Arabia (Ryther and Menzel, 1965; Brinton and Tranter, 1969: Fig. 6) and may therefore not be representative for the northeast monsoon season. In fact, no east-west gradient is apparent for collections made from December 1973 to February 1974 and largely with the same equipment (Paulinose and Aravindakshan, 1977; dates in Varkey and Kesavadas, 1976). For March and April 1974, east of about 64°E and off the Indus Delta, and for May also off Bombay, Paulinose and Aravindakshan (1977) reported biomass values appreciably enhanced over those of the preceding period in the adjoining areas. On the same cruise, low algal production rates per unit of sea surface prevailed by and large; I suggested a possible connection in the discussion of phytoplankton.

Observations for four seasons by Tyuleva (1974), essentially from the Indian shelf and somewhat beyond, confirm the greater richness of the area north of 20°N as compared to the central Arabian Sea. The very few stations for each of the four periods from beyond the shelf off Saurashtra suggest a low zooplankton biomass (50-100 mg/m³) for the upper 100 m during January and June, and higher values (generally 200-300 mg) during April/May of the same year (the same trend as in Paulinose and Aravindakshan, 1977) and

September/October of another year. In the open sea off Saurashtra, March/April yielded higher zooplankton concentrations than did May/June also during the IIOE (Rao, 1973).

Day-night changes of biomass in the Arabian Sea by less than a factor of two are apparent in Prasad's (1968) maps for the net-collected zooplankton in the upper 200 m during the northeast monsoon season, and are negligible, if any, during the southwest monsoon. However, for fast swimming animals which can dodge the ordinary plankton net even at night, larger day-night changes can be expected. Drastic day-night changes in the northern Arabian Sea are exhibited by the mesopelagic fishes which live by day at 150-200 m depth at fairly low oxygen concentrations and occur at night in large concentrations in the upper 50 m (Gjösaeter, 1978). Other, invertebrate members of the deep scattering layer of the Arabian Sea spend the day at very low oxygen concentrations, often below 0.1 ml/l (Kinzer, 1969). A similar striking example for day-night changes was given by Boucher and Thiriot (1972: fig 3) for the oligotrophic western Mediterranean Sea: While the mesoplankton (in their paper, most copepods and similar animals collected by nets of 0.2 mm mesh size) did not exhibit significant changes, the biomass of the macroplankton (chaetognaths, medusae, salps) increased between two and threefold, and that of the micronekton (large copepods, hyperiid amphipods, euphausiid shrimps, fishes) by more than tenfold during the night in the upper 200 m.

Incidentally, as yet unexploited mesopelagic fishes appear to be the by far largest living resource of the area: Their biomass in the northern and western Arabian Sea was estimated to be between a staggering 56 and 148×10^6 t (Gjösaeter, 1978) compared with a few million tons for sardines and similar resources (Institute of Marine Research, Bergen, 1977b). For a strip along the Pakistani coast and 150 to 200 km wide, 5 acoustical estimates between spring 1975 and autumn 1976 ranged between 5 and 8×10^6 (average, 7) tons; a similar study covering approximately the same area 5 times from January to June 1977 yielded estimates between 3 and 13×10^6 t (Gjösaeter and Kawaguchi, 1980: pp. 71, 74). Abundant species among the mesopelagic fish species of the Arabian Sea might have a life cycle of one year or less (Gjösaeter, 1978) so that their annual production rate would be high; Gjösaeter (1978: p. 81)

believes that the annual production by these species would be as high as, or higher than their standing stock. By comparison, the annual marine fish catch of Pakistan in 1980 was 0.23×10^6 t (Food and Agriculture Organization, 1981). Rather large concentrations of dielly migrating, exploitable squids also occur in the region (Yamanaka et al., 1977).

The estimates of fish abundance may be compared with the concentrations of zooplankton collected by nets as reviewed earlier. The mentioned 7×10^6 of mesopelagic fish correspond roughly to 50-60 g/m^2. The plankton concentrations per m^2 in the upper 200 m seem to be less than that: The Soviet collections, ordinarily made by nets with 0.18 mm mesh size (Bogorov et al., 1969) and usually in the tropics determined as displacement volume of the entire catch minus the salps (Vinogradov, 1970: p. 33, and captions of tables cited below), correspond in the area of concern to 10 to 20 g/m^2 wet mass (occasionally 30 g) in the upper 100 m. On the average, the layer from 100 to 200 m may add perhaps 10 to 20% to these figures (cf. Vinogradov, 1970: tables 7, 71). The other data reviewed here were largely collected with a mesh size of 0.33 mm; the displacement volume per catch (ordinarily including salps) is numerically equivalent to grams wet mass below about 1 m^2, and refers in the open sea to the upper 200 m. The averaged IIOE data Prasad (1968: fig. 9) for the northernmost Arabian Sea range from about 10 to 40 g/m^2. Paulinose and Aravindakshan (1977) noted that the concentrations of zooplankton in their collection were several times those reported by Prasad (1968, 1969) for the material from the IIOE but did not attempt to explain the difference. For the waters near Pakistan, Haq et al., 1973 (full data in Ali Khan, 1976) likewise reported appreciably higher zooplankton biomass than shown in Prasad (1968).

Considering that different nets were used (even within expeditions, e.g., Paulinose and Aravindakshan 1977); that the determination of displacement volume is approximate; that the "true" displacement volume in a bottle is time-dependent (Ahlstrom and Thrailkill, 1963); that zooplankton volume in any case contains more water than does the wet mass of fish; and, finally, taking the fish estimates at face value; then it appears that on the grand average there is somewhat less (as little as one-half?) dry matter in the net-collected zooplankton than in the fish concentrations of the upper 200 m. The inclusion of the layers from 200 to 500 m will not make a great difference (cf. Vinogradov, 1970). The inclusion of larger zooplankton, likely to dodge the standard plankton nets, may not make much difference either if Vinogradov's observations (1970: table 24) are applicable in this region, or would make the zooplankton dry mass about equal to that of the mesopelagic fishes if the mentioned observations by Boucher and Thiriot (1972) pertained. In any event, the northern Arabian Sea seems in respect to the ratio of mesopelagic fish (micronekton)/zooplankton to be intermediate between other low-latitude open oceans and the western Mediterranean: Bigger animals avoiding conventional plankton nets, apart from large fishes and squids, may make up very roughly one-tenth of the biomass of the zooplankton (Blackburn et al., 1970 for the eastern tropical, moderately oligotrophic Pacific; Blackburn, 1981 for subtropical gyres in general), or amount to a few times the net-collected zooplankton, especialy at night (Boucher and Thiriot, 1972, for the oligotrophic western Mediterranean). In all these areas, of course, the production by the fish (the predators) is only a fraction of that by the zooplankton (their prey); the ratios of biomass maintained by either group, however, is a function of the production/biomass (or turnover) ratios which generally decrease with body size (cf. Banse and Mosher, 1980: fig. 6) and can well lead to more predator than prey biomass.

Regarding individual species of zooplankton as collected by nets, much more is known from the Arabian Sea than about the phytoplankton. The knowledge, however, is almost entirely restricted to morphological descriptions of development, biogeographical features, and seasonal occurrences on single stations. Some biogeographical principles emerge as shown for example by the paper by Rao (1979), and by articles in Zeitzschel (1973) for the Arabian Sea, as well as in van der Spoel and Pierrot-Bults (1979) in general. Few ecological studies of assemblages or communities exist for the Arabian Sea but meridional sections through the equatorial upwelling region have shown co-occurrences of species as well as horizontal zonation according to feeding types (Timonin, 1974; see also Vinogradov 1981). These rules for individual species should apply also to the upwelling areas in the northern Arabian Sea (see Rao and Nair, 1973). Other

examples for some regularities or rules are provided by Tsalkina (1973) regarding the diel vertical migration of cyclopoid copepod species in the eastern and southeastern Indian Ocean. For instance, by no means all species migrate; among those that do, the day depth and the range of vertical movement are greater in clear, oligotrophic water than below the less transparent equatorial zone (see also Naumov and Ponomareva, 1964). Finally, a connection between geographic and vertical distributions is found in the open Arabian Sea, as reviewed by Vinogradov (1970: p. 277). Numerous surface species which spend their entire life in the upper layers occur above areas with and without severe oxygen depletion below, while others which live also at depth are encountered at the surface only outside the area of greatest reduction of oxygen below.

Almost all the collection reviewed here were made with nets which by their nature do not retain small animals, destroy some gelatinous forms, and are avoided by large plankton as well as nekton. The ratio of loss of small species to retained mass of large animals is a function of the mesh size of the net, its size and towing speed, and the unknown size composition of the sampled assemblage and thus not constant or predictable. In the oligotrophic central North Pacific gyre, Beers et al. (1975: p. 628; see also Beers et al., 1982) estimated that the biomass of protozoans and metazoans lost through the meshes of a 0.2-mm mesh net approaches about half that of the animals caught by day, and about one third of those collected by night; these estimates do not include the very small, colorless flagellates which are now believed by many to play a large role in the food web (e.g., Fenchel, 1982; Sorokin, 1979; Williams, 1981a). Soviet investigators have likewise used ratios of about 1:2 for modelling purposes in oligotrophic waters (Vinogradov et al., 1972) and have reported similarly large though variable ratios for stations in equatorial upwelling (Vinogradov, 1977). The larger animals which avoid the conventional plankton nets, have been mentioned already.

The contribution by the various size classes of pelagic animals to the total consumption of food, regeneration of nutrients, and production of animal body mass will be slanted toward the small organisms because of their higher specific (per unit weight) rates of metabolism and growth. Note, however, that protozoans have very much lower specific rates of metabolism and growth than metazoans of the same body mass when both are compared under optimal conditions (cf. Banse, 1982a). In that paper as well as in Banse (1982b) I suggested that the metabolic role of protozoans may be less important in the pelagic zone relative to that of the metazoans than previously thought; however, this suggestion did not take the very small, usually neglected colorless flagellates into account. At any rate, Williams (1981b) measured in a temperate water body that by far the greatest oxygen consumption of the entire sample is due to organism smaller than net-collected animals, and largely smaller than the small metazoans. There is no reason to think that this would be different in the tropics.

While the large animals not collected by nets thus may not greatly matter in respect to their, say, excretion of nitrogenous materials, they have one important and in some respects overriding role: Their feeding influences or determines the size and species composition of the zooplankton and hence may influence the size composition of the phytoplankton (see introduction to the zooplankton section).

THE WATERS ON THE CONTINENTAL SHELF

General references for oceanographic studies of continental shelves (e.g., Emery, 1960, Mann, 1982; Pomeroy, 1979; and Walsh, 1981) focus on processes at cool- and warm-temperate latitudes. In contrast, the following is largely restricted to the northernmost Arabian Sea.

Hydrography

The differences between the Arabian Sea and the other oceans at the same latitude appear most clearly in the coastal waters: During the northeast monsoon season, the coastal currents tend to flow anti-clockwise rather than clockwise as elsewhere; also, a strong inshore-offshore temperature gradient appears off Saurashtra, with the cold water inshore, but the usual upwelling of the eastern boundary currents of the other oceans is absent. Conversely, during the southwest monsoon season, with the winds near Pakistan blowing essentialy onshore, the coastal current runs clockwise; deep water upslopes onto the shelf

everywhere and upwells to the surface off Africa and Arabia as well as southwest India, which again contrasts to the western and eastern sides of the other oceans. The temporal plankton distributions on the shelf and in the adjoining part of the open sea are affected accordingly, as are the distributions of demersal fishes.

The earlier hydrographic observations were reviewed by Banse (1968; see also Banse, 1973). For the Indo-Pakistani coast I showed (1968) that the maxima and minima of sea level (monthly means from tide gauges, corrected by monthly means of air pressure) occur in December/January and August/September, respectively. These sea level changes correspond roughly to the direction of the coastal currents, thus overriding the opposing effect of changes of water density. The lowered sea level of the summer also corresponds to the upsloping of cool water onto the shelf in the north and to the actual upwelling off southwest India. These conclusions about the causes of the sea level changes are corroborated by new data from Goa (Das, 1979).

From October onward (during the northeast monsoon but prior to and after the establishment of the anti-clockwise current pattern) the surface temperatures off Saurashtra and on the southern shelf of Pakistan decline nearshore and reach 20-22°C, while out at sea the temperatures are much higher; this distribution suggests broad-scale coastal upwelling. However, the mixed layer on the shelf during the period is deep and well-oxygenated, and the bottom water temperatures increase seaward; this would not be the case in an upwelling situation. Instead, the cause for the lowered nearshore temperatures is direct cooling by the cold and dry, offshore-directed wind (Banse, 1968). Accordingly, the water temperatures at a 12-m deep station in Kandla harbor, in the innermost Gulf of Kutch with its great tidal range, fall to 15 to 16°C at this time (Dhavan, 1972). Cool aerated water (24-26°C) generated during the season covers the open shelf and persists off Saurashtra under the subsequently established seasonal thermocline (Banse, 1968). I presume that similar water is present also on the adjoining Pakistani shelf during spring.

As in open sea, the effect of this convective mixing on nutrients at the surface is not known but perhaps the map for coastal water off India and Pakistan for February 1966 by Elizarov (1968) reflects such effects: North of 21°N, 0.7 to 1μg at/l

P-PO$_4$ were depicted (see also below); to judge from later observations during autumn, nitrate would be present above phosphate concentrations of very roughly 0.5 μg at/l (cf., for example, Woods Hole Oceanographic Institution, 1965).

As I showed previously, poorly-aerated water upslopes onto the shelf at about the onset of the southwest monsoon. The water may appear during late May ("Mikhail Lomonosov" sta. 18 in Fig. 1a, 27 May 1966, with 24.2° C and 4.50 ml/l O$_2$ at 25 m, and 21.3° and 0.93 ml/l O$_2$ at 50 m), but is always present after June; accordingly, the mixed layer is quite shallow (Fig. 1d). The uplift of the 23°-isotherm in this water, from its lowest position in about March, is by approximately 50 m and often amounts to 75 m, as also illustrated by Fig. 1d. With the data available in 1968, I suggested that the uplifted water reaches to 25 or 35 (40) m depth. However, Kuz'menko (1977) mentioned that the summer pycnocline on the Pakistani shelf can occur even at 12-17 m, with elevated phosphate values of about 0.7 to 1.7 μg-at/l P at the surface; he observed (1974) a pycnocline depth of 10-15 m for the September/November period. Similarly shallow thermoclines are seen in the data in Table 1c and Fig. 1d. Finally, a map for temperature at 10 m depth in the northern Arabian Sea for summer 1976 shows water below 25°C directly off the mouth of the Indus (August; with less than 3 ml/l O$_2$) and below 24°C (close inshore to the Pakistani west coast (September; Institute of Marine Research, Bergen, 1977a: figs. 38, 40). Therefore, the uplift of the 23°-isotherm may exceed 75 m.

The upsloped water persists on the shelf off Karachi and Pakistan's south coast well into November or even the first days of December, i.e., clearly into the period of the northeast monsoon (see also recent observations in mid-November/ early December 1975 for the area off Karachi, Yamanaka et al., 1978, and for late November/ early December 1976 for the region southeast of Karachi, Yamanaka et al., 1977). I had surmised in 1968 that the disappearance of the poorly-oxygenated water from the shelf in the northernmost Arabian Sea is connected with the reversal of the coastal current into the anti-clockwise pattern, as observed off southwest India. New data by Yamanaka et al. (1978) show that the reversal may not proceed in an anti-clockwise sequence. Instead, in 1975, the current on the western shelf set to the west, coupled with a deep downwarping of the thermocline, while a shallow pycnocline persisted on the southern shelf. The deep pycnocline on the western shelf was not caused by convective overturn which, judging from the stratification of the surface layer, had not taken place.

The large-scale (gross) upsloping or sinking on the shelf does not seem to be due to locally-caused divergence or convergence but to seasonal change of mass distribution and currents in the Arabian Sea at large (Banse, 1968). The local events may result in some fine-tuning, e.g., daily advances and retreats of the cool water on the shelf, but do not prevent the coastal current setting against the wind for about two months, as happens off the northern west coast of India around the turn of the year. Because of the regional cause, significant advances in understanding or predicting the hydrography off Pakistan is unlikely to come from local, limited efforts (see Discussion).

Phytoplankton

Waters on continental shelves generally exhibit higher primary production rates than those of the open sea because of better utilization of nutrients through more frequent recycling on the sea bed; increased vertical mixing from bottom-induced turbulence; coastal upwelling (if any); increased supply of nutrients through the estuarine mechanism (i.e. maintaining the salt balance on the shelf requires drawing-in of large volumes of deep water from offshore); and near estuaries, also river-borne nutrients. On the shelf in this arid region, the first three mechanisms presumably will be the most important, except perhaps off the Indus mouth, where the last two also merit consideration.

Maps for the southwest monsoon and the following period by Kuz'menko (1974, 1977) show wide distribution of high algal biomass and production rates on the shelf (but not in the open sea) as one may expect from the presence of the poorly-oxygenated and hence nutrient-rich, upsloped water (see also Kuz'menko, 1969; Radhakrishna, 1978: 137, November cruise). In contrast, Kuz'menko's (1977) map for the winter season shows appreciably lower algal biomass values on or near the shelf than in the open sea except in the rich area to the west of about 65°E. If these winter data are representative, the vertical convection in a deep layer of nutrient-poor water does not provide a great deal of nutrient salts because of the absence of high nutrients below. In the absence of low-salinity (low density) water, the high sea level near the coast would suggest a deep thermocline (cf. Figs. 1c, d) and sinking which are not conducive to supplying nutrients to the surface. On the other hand, the schematic plot of surface phosphate values off the southern tip of Saurashtra (Khimitsa, 1972; cf. Russian original for correct caption in fig. 1B, inset) indicate almost the opposite scenario, i.e., values up to about 1 μg-at/l P during the winter, low values during spring (as to be expected from the hydrography of that period) and about 0.7 μg-at/l P during the southwest monsoon. (Note that fig. 1A of Khimitsa, for the northeast monsoon season, refers to a period when the shallow pycnocline with poorly-oxygenated water beneath persists on the shelf). In line with the high winter phosphate data but in contrast to the mentioned map by Kuz'menko (1977), carbon uptake rates observed nearshore off Saurashtra between December through February by Radhakrishna et al. (1978: 271) were generally about as high as recorded in other areas during the southwest monsoon. The few shelf measurements by Radhakrishna et al. off the Gulf of Kutch, made in spring after establishment of the seasonal thermocline (cf. hydrographic sections in Varma et al., 1980), do show the expected lowering of production rates. As expected, a clear onshore-offshore gradient is absent.

In view of the scarcity of data, a regional treatment of algal production and distribution on the shelf will not be attempted. The available maps from regional surveys show along-shore differences, but it is not clear yet whether these are persistent regional features or reflect temporal changes within the periods of collecting. Kuz'menko (1977), though, observed high cell numbers in the vicinity of the Indus delta during three out of four seasons studied (not in spring). However, the implied effect of nutrient supply by the runoff is less obvious in terms of algal biomass; further, the lowering of the surface temperature on these stations during the summer by 3-4°C suggests upwelling to the surface, or entrainment of sub-pycnocline water into the river plume (estuarine mechanism) as important agents, in addition to any nutrients contributed by the river water itself. A map of temperature at 10 m depth for August 1976 (Institute of Marine Research, 1977), is strongly suggestive of entrainment of

deep water because of the pocket of low temperature just off the river mouth. (The time of peak discharge is June to August, Quraishee, 1975). For the west coast of Pakistan I expect that during and after the southwest monsoon it will always be under the influence of nutrient salts and plankton advected from the upwelling off Arabia.

Microscopic investigations of seasonal cycles of Pakistani nearshore phytoplankton have been performed but only in net-collected samples. The monthly means of weekly collections during three years at four sites near Karachi (Noor-Uddin, 1967a) indicate the highest concentrations to recur during the first five months of the year on three sites, and during the last four months on the fourth site. According to Noor-Uddin, the differences in species occurrence within months and between stations were remarkable; further, I note that the sites were heavily influenced by runoff since the monthly means of salinity fell to 27‰S. Centric diatoms and a few dinoflagellate taxa also were studied for about two years in Karachi Harbor a polluted embayment which I do not regard either as indicative for seasonal cycles of abundance in the open sea (for diatoms in the collections, Saifullah and Moazzam, 1978; for dinoflagellates, Saifullah, 1979). Therefore, the conditions on the open shelf off Pakistan remain to be explored in detail.

In the outer Gulf of Kutch, adjoining Pakistan's southern shelf, Gopalakrishnan (1972) investigated seasonal cycles of diatoms, again collected by nets, for two years in shallow (0-10 m) water off Port Okha. The highest cell numbers were apparently encountered during the southwest monsoon (see also Bhaskaran and Gopalakrishnan, 1972, on same collection). Mahayavanshi (1977; see Patel, 1976, for data on environment and zooplankton) added two further years of sampling by nets at Port Okha and at several other stations along the coast of Saurashtra, reporting likewise the highest diatom concentrations during the southwest monsoon but finding a secondary peak during the northeast monsoon season. The large salinity range off Port Okha, from slightly below 35 to about 39‰S (Bhaskaran and Gopalakrishnan, 1972), suggest marked coastal (i.e., local) features in the observed pattern (see also under zooplankton).

In conclusion, I am unsure about the nature of the seasonal cycle of the phytoplankton on the Pakistani shelf. I think it likely that the minimum of primary production occurs in spring, after the establishment of the seasonal thermocline and prior to the upsloping of the deep water at the onset of the southwest monsoon. However, with the exception of the Pakistani west coast which is affected by the upwelling off Arabia during the southwest monsoon, it is unclear from the phytoplankton data whether the period of maximal algal development will occur everywhere during the southwest monsoon, owing to the upsloping, or during the northeast monsoon, owing to convective mixing. It may also be inferred from nearshore zooplankton collections off Saurashtra (see later) that appreciable algal production might occur in both periods.

In the context of studies of algae (collected by nets) showing a maximum during the southwest monsoon, it is worth keeping in mind that the numbers of cells retained by nets is far from a constant proportion of the total population. Hence such time series do not necessarily demonstrate temporal trends in the total phytoplankton. Further, settling volumes (as used in these waters, for example, by Noor-Uddin, 1967a) reflect largely the abundance of spiny algae which neither correctly show the seasonal trend of the true algal biomass nor the total plankton biomass in the water (Lohmann, 1908: figs. 8, 9, for temperate coastal waters; see also Malone, 1980: tables 12.1, 12.2, for lower latitudes). In fact, most biomass is lost through the meshes of nearly all net collections (e.g., Subrahmanyan and Sarma, 1967; see also the discovery of photosynthetic cells smaller than 1 μm, Johnson and Sieburth, 1982, which may be the cause of losses of chlorophyll through glass fiber filters measured by Humphrey and Wootton, 1966). Finally, the maps by Kuz'menko (1974, 1977) illustrate that the trends of the cell numbers may be quite different from those of the biomass data based on the very same cell counts, depending on the unpredictable size composition of the assemblages. The same holds for the relation between animal specimen numbers and zooplankton biomass.

As implied, the entire plankton flora (including small species) has not been studied off Pakistan. The species list on the shelf might be somewhat shorter than in the open sea because of the more stressful environment. The list may be appreciably smaller for dinoflagellates in inshore water (F.J.R. Taylor, pers. comm.). Indeed, Subrahmanyan (1958) off Calicut recorded only

121 dinoflagellate species (the large majority being thecate) on inshore locations, to about 38 m depth, out of more than 200 expected for the Arabian Sea by Taylor (1976). Saifullah (1979) observed 44 thecate species in net samples from Karachi Harbor (in the wider sense). In the innermost Gulf of Kutch, very few dinoflagellate species (three!) were collected with a net of a mesh size somewhat larger than that employed earlier by Subrahmanyan (Dhawan, 1972). On the other hand, benthic algal species will often occur in inshore collections of plankton as demonstrated, for instance, by benthic diatoms listed by Noor-Uddin (1967a) and Saifullah and Moazzam (1978) from near Karachi, and "littoral" diatoms in Dhawan (1972) from the outer Gulf of Kutch.

Regarding seasonal occurrences of algal species off Pakistan, observations off southwest India suggest that the algal flora during blooms will be dominated by diatoms in regard to cell numbers (e.g., Subrahmanyan, 1959 for nearshore) and biomass (Banse, unpub. observations on the open shelf off Cochin, 1958/60). With the nutrient limitation temporarily removed, the diatoms can fully exploit their maximal growth rates which seem to be appreciably larger than those of most dinoflagellates and other algae of the same cell size (Banse, 1982b). Note that the inoculum size for the various species at the onset of conditions favoring a bloom will also heavily influence the species composition during the peak periods of biomass: Seed stocks may arise from resting spores (cf. Garrison, 1982), or may be advected from elsewhere. The lack of repetition of species occurrence from year to year, i.e., the near absence of regular "succession", during the upwelling off Calicut (Subrahmanyan, 1958b) and also near Karachi (Noor-Uddin, 1967a) may well have to do with the source of the upwelled water as quantified for other upwelling regions (Estrada and Blasco, 1979; Blasco et al., 1980; see also Torrington-Smith, 1976 for the western Indian Ocean).

Other variables leading to seasonal occurrence of species may be as follows: If the strength of the turbulence has a seasonal character, as is the case under the prevailing wind regime off Pakistan, the benthic species may appear seasonally for purely physical reasons. Turbulence also favors planktonic diatoms directly over other motile groups (cf. experiments by Eppley et al., 1978; Lund, 1978). Further, within each of the major taxonomic groups, nutrient salts will affect species competition through nutrient uptake kinetics (see also Harrison and Turpin, 1982). In addition, the nitrate cycle on the shelf may lead indirectly to seasonal cycles because depletion can initiate the formation of resting spores which are an important device for survival of diatoms through unfavorable seasons (Garrison, 1982). Finally, some species may condition the water by affecting the growth rates of co-occurring forms and thus influencing the subsequent composition of the assemblage; succession in this strict sense, thought to be rare in the sea, as well as the apparent sequence of species in seasonal cycles, has been reviewed by Smayda (1980) who emphasized the role of algal physiology.

Note that the entire above paragraph has implied that algal population changes are principally caused by algal specific growth rates when horizontal water movements (advection) are neglected; at best, however, this may be approximately so during the development of an algal bloom. Under most circumstances, i.e., when algal cell numbers change only slowly, the rate of change with time of the algal concentration N equals the product of N and the difference between the instantaneous rates of cell growth k and death d, $dN/dt = N (k-d)$. Hence, little or no temporal change of N in the presence of the usually appreciable carbon uptake rate means that the grazing mortality, the principal cause of algal death in the open sea, is almost as large as the algal growth rate (Frost, 1980). This fact has been neglected by almost all phytoplankton workers.

Zooplankton

Little is known about zooplankton distribution in nearshore waters of Pakistan, especially in regard to seasonal aspects, with the exception of Noor-Uddin (1967b, see below). To judge from observations off India, one can expect higher concentrations on the shelf than beyond (e.g., Grobov, 1968: table 6; Menon and George, 1977; Tyuleva, 1974) but in many sections across the shelf the trend is by no means obvious and the average relative increase cannot be estimated for the coasts of Pakistan. In any event, nearshore zooplankton concentration may be less than at some distance from it (Goswami et al., 1977 for sections to 40 m depth).

The zooplankton counts made on the samples from three years of weekly surface collections of phytoplankton from four stations near Karachi are summarized by Noor-Uddin (1967b; without data). According to the author, the specimen numbers showed a bimodal cycle, with high values from January through March or early April and from late June through October. On the Indian side of the Gulf of Kutch, seasonal collections were made off Port Okha by Bhaskaran and Gopalakrishnan (1972) and subsequently off Port Okha and off Rozi, in the middle reach of the Gulf, by Patel (1976). Both studies were based on two years of sampling with coarse, non-metered nets towed horizontally near the surface over 0-10 m depth (for Patel, see Mahyavanshi, 1977). The highest biomass (as displacement volume) occurred in Patel's two data sets during the southwest monsoon and the succeeding months of October and November as off southwest India (e.g., Menon and George, 1977); Bhaskaran and Gopalakrishnan's observations lacked these pronounced peaks. It is not clear, though, which contribution was made to the volume by the spiny diatoms which are principally retained by such nets. The total specimen numbers of copepods (without nauplii; to judge from the detailed identification by Patel who considered adults only) varied by five- to tenfold during the year. The seasonal peaks do not coincide except that maxima between January and March seem to be consistently present. Haq et al. (1973) found substantially more zooplankton volume inshore and offshore of Pakistan during November/December of one year than March of two much later years. They also remarked that the observed concentrations were high relative to those reported by Prasad (1968) from elsewhere in the Arabian Sea. Finally, Tyuleva (1974) concluded from collection in four seasons that June is the poorest period off Saurashtra and the Gulf of Kutch; this is borne out off Karachi by the scarce zooplankton from late April through June (Noor-Uddin, 1967b), and off Port Okha by the broad minima of copepod occurrences around this time of the year (Bhaskaran and Gopalakrishnan, 1972; Patel, 1976). The inshore cycle of Port Okha is different from that offshore where March/April yielded larger zooplankton volumes than January/February and June (Tyuleva 1974; Paulinose and Aravindakshan, 1977; see also the large salinity range off Port Okha).

As with the coastal phytoplankton, the list of dominant species inshore is manageably short. For example, Haq et at. (1973; table 2) named 16 species which contributed 63 to 90% of the average numbers of total copepods between October and March; 2 of these were considered as oceanic. Similarly, Patel (1976) listed 12 copepod species as dominant in his two-year collections in shallow waters of northern Saurashtra (see also remarks for Karachi waters in Noor-Uddin, 1967b). Patel's data for individual species also show clear seasonality (as also mentioned by Noor-Uddin, 1967b) but it is not indicated whether this is caused by advection by the seasonally reversing currents, or is of local origin

Low-Oxygen Bottom Water and Demersal Fishes

In 1968, I suggested that the oxygen content of upsloping water during the southwest monsoon season becomes so low that trawl catches of bottom fishes can be negatively affected. Bottom water temperature is not a predictor for catch per unit effort. I surmised that large parts of the outer (perhaps also middle) shelf, "approximately from Cochin to Karachi", might be devoid of commercially exploitable concentrations of demersal fish during that period. Doe (1965; observations also depicted in Ali Khan, 1976) had previously shown that water well below 1 ml/l O_2 covered the shelf between Karachi and the submarine canyon of the Indus in early November 1964 almost up to the 40-m line; he related this to reported unsuccessful fishing on the bank northwest to the canyon. I note that offshore during that year, the sub-thermocline water had an unusually low salinity (35.8‰S, see Doe, 1965) and hence was poorly oxygenated (Fig. 2b). Although there are other, scattered correlations between fish (and shrimp) catches and poorly oxygenated bottom water off Pakistan (Zupanovic and Mohiuddin, 1976), it appears that near-bottom water on the Pakistani shelf is less likely to become as deoxygenated as off the Indian west coast, including the shelf off Saurashtra. This is because of the higher initial oxygen content of the water upsloping off Pakistan (Fig. 2a,c; cf. Ramamirtham and Rao, 1974, for new data from the Indian west coast). Of course, the degree of depletion on the shelf will also depend on the residence time of the water in contact with the bottom, the rate of supply of organic matter from the photic zone above, and the rate of mixing, none of which are known.

Benthos

The bottom animal life on the shelf will only briefly be touched upon in spite of its large interaction with the overlying water. A cross-sectional review of current research is provided by Tenore and Coull (1980). Chapters in Fasham (in press) may also be consulted. Harkantra et al. (1980) list papers on quantitative studies of bottom fauna in Indian waters.

Quantitative work off Pakistan on animals collected by grabs and retained by fine screens was done by Muhammed and Arshad (1966a) near Karachi, and over the entire shelf by Savich (1972, with references to earlier Soviet collections; see also Grobov and Michaylov, 1968; Kondritskiy, 1975). Benthic forminifera were studied by Zobel (1973). For 10 out of 21 major animal groups listed, Savich (1972) observed an increase in biomass from the first survey in January to the second in March but it is not clear from the paper whether stations were revisited nor whether the increase was statistically significant. Savich (1972) further reported a very rapid decline of wet weight per unit area beyond about the 100-m isobath and tentatively related it to the low oxygen concentration of the overlying water. Likewise, Zobel (1973) regarded the low oxygen content of the bottom water at the depth of the oxygen minimum as the reason for a zone on the continental slope of Pakistan and India where specimens of benthic species made up less than 5% of all foraminifera on the sediment surface (see also meiofauna, below).

As with zooplankton, bottom animals sampled with one type of gear represent only a certain, and variable size fraction of the entire assemblage. In the papers reviewed above (except Zobel, 1973), sediment was collected with grabs and washed on shipboard through screens. Grabs will miss deep-digging animals as well as mobile or rare forms. Of the animals collected, the small ones will be lost in the washing process, as will be the bacteria. While the metabolic contribution of the unsampled, generally large forms is not well known, but is presumably small, it is certain that the major part of the food and oxygen consumed and the nutrients released are due to organisms not retained by the screens (e.g., Banse et al., 1971, with improvements by May, 1972; see also Kemp and Boyton, 1981; Pamatmat, in press). The principal role of the animals conventionally sampled is serving as the principal food for demersal fishes, and—together with unsampled large forms—to provide for mechanical disturbance of the sediment. This bioturbation is all-important for the chemistry of the pore water, exchanges with the overlying water column, and early diagenesis of the sediments.

The small animals lost during the sieving process on shipboard can be collected by other means. The metazoans (meiofauna) usually contribute a small fraction of biomass relative to that of the macrofauna (Fenchel, 1978). To judge from the means and ranges given by Qasim (1982), the mean weight per unit area of the combined meiofauna and foraminifera, between 0 and 200 m off the northern part of the west coast of India, including off Saurashtra, is about twice that of the larger animals although optimal weights for the latter exceed those of the former. Further, the "meiofauna" mass increases with depth, which does not seem to have been observed elsewhere; it would be useful to learn more details, in particular about how much of the weight is contributed by the foraminiferal tests.

The quantitative observations on macrobenthos off Pakistan were qualitatively related to depth and sediment type (Savich, 1972) but not subjected to statistical analysis. That approach would permit recognition of species co-occurrence and quantification with environmental parameters (e.g., Lie and Kelley, 1970). Subsequent studies, principally in temperate waters, have shown, however, that the causes underlying such co-occurrence of animals, as well as the absence of others, usually cannot be well understood (if at all) without experimental manipulation (cf. Paine, 1980; Peterson, 1978). The brunt of this currently active branch of research is directed at understanding which of the species potentially able to populate a habitat actually do so, and at explaining the reason for the number of species present. To my knowledge, however, previous research had done little to explain why the common species are as common as they are. No doubt, much more of this experimental work which tests specific hypotheses needs to be done.

GENERAL DISCUSSION

Introduction

This section will expand on a few previous suggestions about future work. Principally,

however, I will emphasize that in these waters the era of hydrographic "exploration" is largely past and may soon come to an end with regard to delineating plankton distributions in space and time. Keeping in mind the general understanding of pelagic ecosystems, I will suggest that Pakistani marine workers might wish to focus on a few answerable questions (cf. Platt, 1964) rather than repeat the traditional sequence of development of plankton research in less-known regions of the world.

The traditional approach starts out with extended seasonal studies on single stations or with broad surveys. Neither of these address specific issues but consume considerable time and effort, while usually yielding only general notions about possible causes. Because we have enough environmental and biological understanding in general as well as locally, it should be possible to plan future studies by anticipating their likely results. This would allow one to weigh beforehand the advisability of the investment in terms of manpower, time, and funds, rather than judge the matter with the benefit of hindsight (see also Düning, 1970, p.65 and Banse, 1973).

Such planning of specific research will be more successful if existing observations and theories can be evaluated beforehand. For example, before continuing future hydrographic investigations off Pakistan, it may be useful to go through all existing hydrographic data from the region. In addition, once these data are assembled, they can be used to reinvestigate older questions, e.g., the formation of the intermediate waters in the Arabian Sea. Rochford (1964) had only a limited number of stations, both geographically and temporally. It would seem promising now to go beyond his and Wyrtki's (1971) work by using more data for t-S diagrams and core layer studies, but also incorporating oxygen observations (cf. the two oxygen minima mentioned in Qasim, 1982: p. 1050). Another question might be the extent to which sea surface temperatures in the northern Arabian Sea reflect the global secular trends of warming as is true for the southern Arabian Sea (Fieux and Stommel, 1976), or whether the local continental regime overrides such small signals.

A data file would also be helpful for biological work. Earlier in this review, I proposed a hypothesis about the seasonal cycle of phytoplankton production. For testing, it might be sufficient to classify the measured rates of carbon uptake into Low, Medium, and High, in order to avoid the agonizing task of comparing various techniques, so evident from Kabanova (1968) (see also Doty et al., 1965) and Babenerd and Krey (1974) (see also Radhakrishna et al., 1978: p. 271, who suggest that their reported regional values are almost twice the actual rates). Photosynthesis per unit of sea surface could be compared station by station with hydrographic and nutrient data. Clarifying the relation between the rates and the environment, however, will require normalized rates, i.e. the quotient of carbon uptake per unit of chlorophyll. A new, critical review of the existing data is needed before discussing local or temporal differences in absolute terms (e.g., the area off the Gulf of Oman versus that off Pakistan's southern shelf).

The suggested re-evaluation of observations can help clarify the normal seasonal cycles, and show particularly for the open sea if in areas with deep convection the surface water is fertilized during the winter because of erosion of a poorly-ventilated, nutrient-rich salinity maximum below. Presumably, however, the areal coverage of the ship data will neither allow establishing regularity of processes nor guiding expeditions into rather localized areas. J.K.B. Bishop (pers. comm.) drew my attention to satellite data (Nimbus-7, since October 1978) providing observations of surface temperature and sea color. The latter can be converted into estimates of chlorophyll concentrations in the perhaps upper 20 to 30% of the photic zone (for methods, see Smith and Baker, 1981; Pelaez and Guan, 1982). Since during the winter, the northern Arabian Sea has very little cloud cover (Warners, 1952; U.S. Navy Hydrographic Office, 1960), satellite data thus may show whether cooling of the surface, leading to convection (not measurable from space) will result in locally elevated chlorophyll concentrations in surface layers which are visible from space. The regions of convection would also be of physical and chemical oceanographic interest.

The difficulties encountered by Winter et al. (1975) in modelling and thereby quantitatively interpreting observed temporal changes of chlorophyll concentration by means of simulated *in situ* ^{14}C-data, illustrate another important point. Quantitative interpretation of productivity data, as well as calculation of such rates from environmental information, require physiologically defined observations, i.e., the carbon uptake rate

per unit of chlorophyll under a range of irradiance including saturating intensities (so-called P versus I curves, cf. Steemann Nielsen and Hansen, 1961). In the future, therefore, efforts must be made to determine P versus I curves in preference to the conventional *in situ* measurements of ^{14}C uptake. Again, the specific question is more likely to lead to good answers than the unspecific question.

Plankton Investigations on Single Station

On the shelf the principal source of nutrients, besides grazing and regeneration by zooplankton, are the mineralization of organic matter on the sea bed and subsequent upward transport by physical processes (see Harrison, 1980; Rowe, 1981; Zeitzschel, 1980). Mineralization rates are temperature-dependent (see Nixon et al., 1980) and, thus, are presumably continually high in the Arabian Sea. Therefore, wherever a well-stirred, transparent water column touches the bottom, high primary production is assured (e.g., Prasad and Nair, 1963).

The ease with which remineralized nutrient salts are mixed in the water column will vary across an open shelf, resulting in a longshore zonation of primary production. This zonation should not remain geographically fixed, but rather be shifted across the shelf by physical processes (Fig. 3). On an open shelf, one can distinguish at least four zones: Nearshore, in the first zone, continual inter-action of waves with the sea bed results in nutrient supply to the entire column. Suspended sediment, however, may cause light limitation. Off Pakistan, the seaward limit of this zone must vary greatly with season, extending farthest offshore during the southwest monsoon.

A second zone lies offshore, where the bottom depth is shallower than the seasonal thermocline. Here, the water column may or may not be stirred, depending, to a considerable degree, on the vertical salinity gradients which can be quite marked in the tropics proper. For example, a haline-stratified water column on the middle of the shelf, as off southwest India, may be rather barren and oceanic in character. Near the sea bed there may be a slight decrease in dissolved oxygen and an appearance of nitrite, the result of reduced vertical mixing (Banse, 1968: fig. 2; unpubl. data on [14]C- uptake on same station. Off Pakistan, away from the influence by the Indus, I presume that little salinity stratification is present in the mixed layer. Certainly, the low salinity from the Gulf of Bengal that moves up the Indian west coast during the period of the anti-clockwise current does not reach these waters (Wyrtki, 1971).

In a third zone, a seasonal thermocline will temporarily isolate the upper layers from contact with the bottom, with the same effect as indicated for haline stratification (Zones 2 and 3 are identical during other seasons). With the increased transparency resulting from stratification in the upper layers, light in zones 2 and 3 may be sufficient for algal growth beneath the pycnocline, perhaps even on the bottom; this should be true off Cochin (Banse, unpubl. observations, 1959/60) but may not be universally applicable in the northern Arabian Sea (Qasim, 1982: fig. 2; cf. Krey and Babenerd, 1976:18). Finally, a fourth zone is the outer shelf where the permanent thermocline is shallower than the bottom depth. Here, for all practical purposes, offshore conditions prevail in the mixed layer, and the effect of the bottom is hardly felt except by a somewhat increased eddy diffusion through the pycnocline.

This picture can be more complicated where and when incoming freshwater forms a density front that separates the nearshore zone from the open shelf (e.g., Bigham, 1973). The restriction of river-borne sediment to a narrow nearshore band off the Indus delta during the southwest monsoon, is indicative of such a situation; during the northeast monsoon, the turbidity was observed to spread broadly over the shelf (Quraishee, 1975). Once established, the position of such fronts will vary somewhat with wind and tides, the density difference resulting in longshore transport. The cross-front mixing of freshwater is treated by Blanton (1980), and an overview of plankton production at shelf fronts can be found in Mann (1982, p. 225), and Bowman and Esaias, (1978).

The four zones depicted in Figure 3 presumably do not remain as fixed features throughout the year because of seasonal movement of the thermocline and temporal changes in wind, etc. Within any zone, the temporal changes of vertical stability, nutrient supply, etc., will be reflected in seasonal, recurrent changes of biota. Differences at any time between zones will lead to onshore-offshore gradients of concentration. Therefore, observations at a geographically fixed station could result in confusing within-zone changes with the seasonally-caused, between-zone differences brought by movements of the zones. Only concurrent hydrographic observations can define water mass changes. In addition, the station can be strongly affected by short-term water movements,

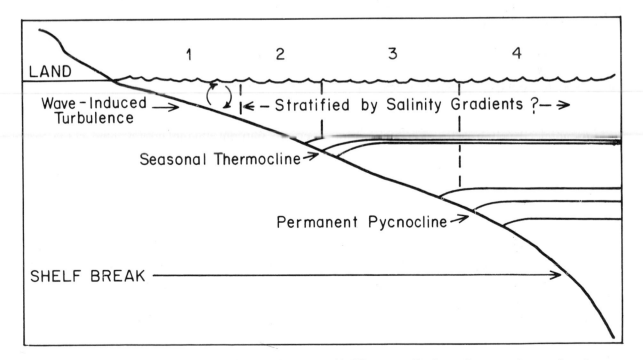

Figure 3. Sketch of a tropical continental shelf showing four zones with different stratification and hence nutrient supply to the upper water layers.

both tidal and wind-related. The importance of these effects, of course, will depend on the steepness of the onshore-offshore gradients and the extent of the currents. An example for such tidal effects on biological observations off Nova Scotia was provided by Sameoto (1978).

On the shelf off Pakistan, the strongest seasonal signal might be by the across-shelf advection of the permanent pycnocline. Because of the regularity and importance of this process, the observer on the fixed station would want to study its effects but small short-term (hours to days) changes can confuse the picture. Therefore, station location for any time series of plankton observations must be chosen carefully (see the quasi-synoptic quite irregular zooplankton data off Karachi in Muhammad and Arshad, 1966b). Finally, because of the critical importace of local hydrography to a fixed station, it is difficult to extrapolate the plankton observations from one station (e.g., off Saurashtra) to another area (e.g., off Karachi), even if the general hydrography of both localities is comparable.

The effect by along-shore advection further complicates the understanding of seasonal changes of organisms at a fixed station. The life expectancy of animals collected in conventional plankton nets may be 3-4 weeks, while those caught in larger nets or trawls may live for a year. If an animal does not migrate vertically but spends its entire life (e.g., one month) in a coastal current of 1/2 knot, a specimen could traverse the length of Pakistan's west coast (cf. Warners, 1952). How would one understand the population dynamics of species at a station off Karachi without knowing the upstream gradients?

Horizontal advection is the principal reason for the near-absence of population dynamics and production studies of marine zooplankton. This contrasts to the situation in lakes where populations can be more easily revisited. The issue of advection in a coastal sea was specifically addressed by Bernal and McGowan (1981) who found that in the upwelling system off California the monthly or quarterly means of zooplankton biomass collected over a span of 21 years were statistically better correlated with lateral advection than with a physical upwelling index (i.e., the local vertical, or onshore-offshore term). Since the coastal currents off Pakistan and India are appreciable, similar relations during upwelling periods may not be out of the question. Thus,

can the temporal distribution of zooplankton on a section (off e.g., Karachi) be explained from concurrent hydrographic and biological observations?

Finally, consider the so-called local time change of algal concentration (cf. Sverdrup et al., 1942: 160) in an imaginary large container with natural plankton. As before, the algal concentrations will change following $dN/dt = N\,(k\text{-}d)$. With a given N, a low or negligible change ($dN/dt \approx$ zero) may be caused by a negligible difference between large k and d, or between small k and d. Except in the rare situation of an algal bloom where $k \geqslant d$, it is therefore rather futile to try and quantify the spatial and temporal changes of phytoplankton merely from those physical and chemical variables that determine the algal instantaneous growth rate k. The same argument holds for zooplankton (see below). Finally, at the fixed station, one observes not only this local time change but the total, comprising advection and eddy diffusion; only in the case of small horizontal gradients can the two latter terms be safely neglected.

Quantitative explanations of plankton distributions are within reach if major efforts can be directed at specific problems in a well-studied area, e.g., the mechanisms of maintenance of a copepod population off a coast with upwelling, in the presence of marked vertical and horizontal advective term (Wroblewski, 1980; 1982). This type of directed effort has been accomplished only in a few places so far.

One can conclude from this discussion that major advances in understanding of the hydrography-nutrient-phytoplankton-zooplankton system is unlikely to come from addressing unspecific questions. (Perhaps, one might add this observation to the review by Prasad (1973) who deplored the lack of specialization of oceanography within Indian universities and the corresponding lack of co-ordination and co-operation among institutions). Even with limited resources, however, local studies can make major contributions to a more general understanding, by focusing on issues and effecting field collections accordingly. Bluntly stated, good science ("basic" or fundamental, as well as "applied") depends principally on asking the right questions (cf. Platt, 1964), and less so on having access to the latest instrumentation.

Animal Population Dynamics and Production

As stated earlier, understanding spatial and temporal distribution of animals lags that for phytoplankton, because of our inability to measure production directly (references for spatial problems of distributions may be found in Longhurst, 1981, and Steele, 1978). Furthermore, zooplankton (and fish) production in a particular area cannot yet be accurately calculated or predicted from algal production rates or concentrations.

Algae are eaten by suspension (filter) feeders which select their food principally by size (rather than kind) so that linear food chains do not exist. A ciliate living from algae may as easily be ingested as an algal cell of the same size, and supposedly herbivorous suspension feeders may even eat larvae of carnivores, their presumed predators. Because of these unpredictable "foodchains", laboratory-derived coefficients of food conversion (efficiencies) for individual species or food chains cannot be applied to the actual food web (i.e., a trophic level), and animal production cannot be accurately estimated from ^{14}C data.

Moreover, benthic marine ecologists, as well as freshwater plankton workers, have shown that the mechanisms involved in the maintenance of species composition of communities usually cannot be inferred from the observed distributions in space and time. Instead, manipulations (field experiments) are required (Peterson, 1979; Paine, 1980, for benthos; Kerfoot, 1980, for zooplankton). In my view, reasons for changes of abundance (with advection excluded), also cannot be easily understood from statistical analyses between abundance and environmental variables, as suggested by, e.g., Hazelwood and Parker (1961) or Angino et al. (1973) for plankton.

Environmental variables often do not change abundance directly (as will predation or, in the intertidal, excessive rain and heat), but they will affect the balance between the antagonistic processes of birth and death. Thus rates of fecundity, development (growth), and mortality (both usually stage-specific) are needed for understanding changes in abundance of a population not subject to vertical migration, advection, and eddy diffusion (plankton), or immigration or emigration (benthos). As indicated earlier, opportunities usually do not exist in the pelagic realm for revisiting populations so that the standard procedures of population dynamics developed for benthic organisms and the plankton of ponds or small lakes cannot be unambiguously applied. Even so, the important fecundity term can be obtained easily for pelagic species that carry their eggs or brood for some time (see Thompson and Easterson, 1977). Similar estimates have been made recently for copepods that broadcast their eggs (Runge, 1981).

The seasonal coastal studies of zooplankton from single stations in the Arabian Sea, so far, have provided astonishingly little on any of these aspects; laboratory work on fecundity, rate of development, specific growth rate (by weight), and generation time is apparently not actively pursued either. The papers by Checkley (1980), Hirota (1974), Landry (1978), and Reeve and Baker (1975) are examples (admittedly, ambitious ones) of what can and should be done with dominant species. Because of his critical approach to such experimental work on field problems, the article by Bartram (1981) is also cited here. In benthic animals with pelagic larvae, studies of population dynamics usually concern only the bottom-living stages and require assessment of spatfall, or an abundance measure at some arbitrary time shortly thereafter. Usually, production estimates for the entire population of a species are not markedly influenced by neglecting the very small individuals.

In determining production in a population, the classical (benthic) approach requires revisiting of the population and the enumerating measuring (weighing) of specimens. Optimally, animals should breed over short time intervals, interrupted by extended periods without new offspring. Also, emigration or immigration must be absent. Calculations yield the mean individual growth rate, the rate of tissue production including that by animals that died between sampling dates, the mortality rate, and tissue lost through death (cf. Crisp, 1971; Winberg et al., 1971). For accurate results, visits must be frequent relative to the life cycle of the species; often more than one collecting method has to be used (e.g., Nichols, 1975). Although warm-water bottom fauna often breed semi-continuously, a review by Ahmed (1980) shows that a large fraction of invertebrates and fishes off Pakistan are seasonal spawners (see also Kutty and Nair, 1973, for pelagic copepods). In plankton studies, advection of populations makes the calculation of population production more difficult (cf. Hirota, 1974, and reference to Fager, therein). A second choice is to combine abundance data as before with experimentally determined instantaneous growth rates (Crisp, 1971; Winberg 1971). The comparison of predicted numbers and mean weights with those observed on the subsequent visit yields information about mortality, which in plankton work is complicated by advection.

There are other techniques for estimating animal production, which, however, can show deviations from the calculated rates of more than a factor of two. One is dividing empirically determined ratios of production by biomass (P/B, usually extrapolated from one species to another), as for example in Banse and Mosher (1980; no equation yet for tropical organisms). The other combines estimates of respiration (from abundance data) with estimates of net growth efficiency (NGE, production/<production + respiration>; see Winberg, 1971). The problem of the definition of NGE (e.g., Dagg, 1976; Mann, 1982: p. 204), the variability of respiration values, and the dependency of NGE on age (size), food, and temperature result in uncertainties so great that good values can only be obtained if the NGE has been determined for a particular species. NGE of 0.20 and 0.40 (a modest range given the known variations among species; cf. Banse and Mosher, 1980: fig. 5; Winberg, 1971: fig. 5.9) translate to respiration/production ratios of 4.0 and 1.5 respectively. A refinement of this approach requiring additional measurements for the assemblage of concern has been advanced by Le Borgne (1982).

In models of plankton production, more involved, experimentally determined relations between temperature, food, ingestion, egestion, respiration, and excretion in dependence of animal size or kind have been employed (e.g. Steele and Frost, 1977; Steele and Mullin, 1977; Vinogradov et al., 1972). While the former authors did not compare model prediction of abundance with particular field situations, the latter found fair agreement between calculations and the field for animals grouped by size or feeding type.

All four calculations of production can be applied in principle to the melange of species as collected by a plankton net in order to attempt a total secondary (animal) production estimate. However, only mathematical models of production would permit semi-rigorous sensitivity analyses from which the error (confidence interval) of the estimate might be estimated. I am afraid that at present the error will be so large that decisions based on such estimates (e.g., the number of trawlers to be ordered, or answers to more esoteric questions like the transfer efficiency from one trophic level to the next) cannot be made soundly. The fishing yield of an area still is best assessed by experimental fishing interpreted by fisheries theory.

Considering the present uncertainties of the approaches mentioned above, and the general ignorance about basic parameters of life cycles even for common planktonic or benthic species, I believe that the most rapid and fundamental progress in understanding of animal production in the Arabian Sea might come from focusing on the study of some common species. Much of this can be based on field sampling and relatively simple laboratory set-ups. If the material is collected on fixed stations, and ancillary hydrographic observations are made concurrently, qualitative information on seasonal biological cycles will accrue as a fringe benefit. From such focused work, general principles or at least generalizations of broad interest, applicable to warm seas in general, will emerge once we have studied, for example, ten common pelagic copepods. Subsequent considerations of zooplankton production in the Arabian Sea could then be less daring than they would be at the present time. The same approach of specializing and of asking specific questions now rather than sometime in the furture, will work also for subject areas other than animal production.

ACKNOWLEDGEMENTS

It is a pleasure to thank the coordinators, Drs. B.U. Haq, J.D. Milliman, and G.S. Quraishee for giving me the opportunity to write this review and participate in the workshop. I am likewise grateful to Dr. S.I. Ahmed in Seattle, and to Dr. G.S. Quraishee and his colleagues in Karachi for their help in small and large matters during the workshop as well as for hints to some local literature. S. Fagerberg was a very patient typist.

Contribution No. 0000, School of Oceanography, University of Washington, 98195, Washington, Seattle, USA.

REFERENCES

Ahlstrom, E.H. and J.R. Thrailkill, 1963. Plankton volume loss with time of preservation. Calif. Coop. Oceanic Fish. Invest. Rept., 9:57-73.
Ahmed, M., 1980 The breeding and recruitment of marine animals of the coast of Pakistan bordering the Arabian Sea. Proc. Ist Pakistan

Congr. Zool., 55-96.

Ali Khan, J., 1976. Distribution and abundance of fish larvae off the coast of West Pakistan. Mar. Biol, 37:305-324.

Anderson, G.C., 1969. Subsurface chlorophyll maximum in the northeast Pacific Ocean. Limnol. Oceanogr., 14:386-391.

Anderson, J.J., A. Okubo, A.S. Robbins, and F.A. Richards, 1982. A model for nitrite distributions in oceanic oxygen minimum zones. Deep-Sea Res., 29A:1113-1140.

Angel, M.V., 1977. Windows into a sea of confusion: Sampling limitations to the measurement of ecological parameters in oceanic mid-water environments. Pp. 217-248, In: N.R. Anderson and B.J. Zahuranec (eds.), Oceanic Sound Scattering Prediction. Plenum Press, New York-London.

Angino, E.E., K.B. Armitage, and B. Saxena, 1973. Population dynamics of pond zooplankton. II. Daphnia ambigua Scourfield. Hydrobiologia, 42:491-507.

Babenerd, B. and J. Krey, 1974. Indian Ocean: Collected data on primary production, phytoplankton pigments, and some related factors. University of Kiel, Institut für Meereskunde, Kiel, 521 pp.

Banse, K., 1968. Hydrography of the Arabian Sea shelf of India and Pakistan and effects on demersal fishes. Deep-Sea Res., 15:45-79.

Banse, K., 1973. Upsloping of isotherms on the continental shelf off Goa and Bombay in June 1967.. J. Mar. biol. Ass. India, 14:344-356.

Banse, K., 1982a. Mass-scaled rates of respiration and intrinsic growth in very small invertebrates. Mar. Ecol. Progr. Ser., 9:281-297.

Banse, K., 1982b. Cell volumes, maximal growth rates of unicellular algae and ciliates, and the role of ciliates in the marine pelagial. Limnol. Oceanogr., 27:1059-1071.

Banse, K., D.R. May, and F.H. Nichols, 1971. Oxygen consumption by the sea bed. III. On the role of macrofauna at three stations. Vie et Milieu 22A, Suppl., 31-52.

Banse, K. and S. Mosher., 1980. Adult body mass and annual production/biomass relationships of field populations. Ecol. Monogr., 50:355-379.

Bartram, W.C., 1981. Experimental development of a model for the feeding of neritic copepods on phytoplankton. J. Plankton Res., 3:25-51.

Bathen, K.H., 1972. On the seasonal change in the depth of the mixed layer in the North Pacific Ocean. J. Geophys. Res., 77:7138-7150.

Beers, J.R., F.M. Reid, and G.L. Stewart, 1975. Microplankton of the North Pacific central gyre. Population structure and abundance, June 1973. Int. Revue ges. Hydrobiol., 60:607-638.

Beers, J.R., F.M. Reid. and G.L. Stewart, 1982. Seasonal abundance of the microplankton population in the North Pacific central gyre. Deep-Sea Res., 29A:227-245.

Bernal, P.A. and J.A. McGowan, 1981. Advection and upwelling in the California Current. Pp. 381-399, In: F.A. Richards (ed.), Coastal Upwelling. Am. Geophys. Union, Washington, D.C.

Bhaskaran, M. and P. Gopalakrishnan, 1972. Observations on the marine plankton in the Gulf of Kutch, off Port Okha. Indian J. Fish., 18:99-108.

Bigham, G.N., 1973. Zone of influence-inner continental shelf of Georgia. J. Sedim. Petrol., 43:207-214.

Blackburn, M., 1981. Low latitude gyral regions. Pp. 3-29. In: A.R. Longhurst (ed)., Analysis of Marine Ecosystems. Academic Press, London-New York.

Blackburn, M., R.M. Laurs, R.W. Owen, and B. Zeitzschel, 1970. Seasonal and areal changes in standing stock of phytoplankton, zooplankton and micronekton in the eastern tropical Pacific. Mar. Biol., 7:14-31.

Blanton, J.O., 1980. The transport of freshwater off a multi-inlet coastline. Pp. 49-64, In: P. Hamilton and K.B. Macdonald (eds.), Estuarine and Wetland Processes with Emphasis on Modeling. Plenum Press, New York-London.

Blasco, D., M. Estrada, and B. Jones, 1980. Relationship between phyto-plankton distribution and composition and the hydrography in the northwest African upwelling region near Cabo Corbeiro. Deep-Sea Res., 27 A:799-823.

Bogorov, V.G., M.E. Vinogradov, N.M. Voronina, I.P. Kanaeva, and I.A. Suetova, 1969. Zooplankton biomass in the ocean surface layer. Doklady Akad. Sci. USSR, Earth Sci. Section, 182:235-237 (transl. from Dokl. Akad. Nauk SSSR 182, 1968).

Boucher, J. and A. Thiriot, 1972. Zooplancton et micronecton estivaux des deux cents premiers metres en Mediterranee occidentale. Mar. Biol., 15:47-56.

Bowman, M.J. and W.E. Esaias (eds.), 1978. Oceanic Fronts in Coastal Processes. Springer-Verlag, New York-Heidelberg. 114 pp.

Brinton, E. and D.J. Tranter, 1969. Handbook to the International Zooplankton Collections. Vol. 1. Station List. Council of Scientific and Industrial Research, National Institute of Oceanography, Indian Ocean Biological Centre, Cochin. Unpaginated.

Capone, D.G. and E.J. Carpenter, 1982. Nitrogen fixation in the marine environment. Science (Washington, D.C.), 217:1140-1142.

Carpenter, E.J. and J.S. Lively, 1980. Review of estimates of algal growth using ^{14}C tracer techniques. Pp. 161-178, In: P.G. Falkowski (ed.), Primary Production in the Sea. Plenum Press, New York-London.

Checkley, D.M., Jr., 1980. Food limitation of egg production by a marine, planktonic copepod in the sea off southern California. Limnol. Oceanogr., 25:991-998.

Clarke, M.R., 1977. Beaks, nets and numbers. Symp. Zool. Soc. London, 38:89-126.

Crisp, D.J., 1971. Energy flow measurements. Pp. 197-297, In: N.A. Holme and A.D. McIntyre (eds.), Methods for the Study of Marine Benthos. International Biological Programme, Handbook, No. 16. Blackwell Scientific Publications, Oxford-Edinburgh. 334 pp.

Cullen, J.J., 1982. The deep chlorophyll maximum: Comparing vertical profiles of chlorophyll a. Can. J. Fish. Aquat. Sci., 39:791-803.

Cushing, D.H., 1980. Production in the central gyres of the Pacific (Bruun memorial lectures 1979). Unesco Intergovt. Oceanogr. Comm. Techn. Ser., 21:31-39.

Dagg, M.J., 1976. Complete carbon and nitrogen budgets for the carnivorous amphipod, Calliopius laeviusculus Kröyer. Int. Revue ges. Hydrobiol., 61:297-357.

Das, V.K., 1979. Seasonal variation in mean-sea level at Mormugao, west coast of India. Mahasagar (Bull. Nat. Inst. Oceanogr. India), 12(2):59-67.

Das, V.K., A.D. Gouveia, and K.K. Varma, 1980. Circulation and water characteristics on isanosteric surfaces in the northern Arabian Sea during February-April. Ind. J. Mar. Sci., 9:156-165.

Dhawan, R.M., 1972. Plankton and hydrological factors at Kandla in the Gulf of Kutch during 1960-1963. Indian J. Fish., 17:122-131.

Dietrich, G., W. Düing, K. Grashoff, und P.H. Koske, 1966. Physikalische and chemische Daten nach Beobachtungen des Forschungsschiffes "Meteor" im Indischen Ozean 1964/65. "Meteor" Forschungsergebn. A 2: 5 pp and tables. Bornträger Berlin.

Doe, L.A.E., 1965. Physical conditions on the shelf near Karachi during the postmonsoonal calm, 1964. Pp. 278-292 In: Ocean Science and Engineering, Vol. 1. Marine Technology Society/American Society of Limnology and Oceanography. Washington, D.C.

Doty, M.S., H.R. Jitts, O.J. Koblentz-Mishke, and Y. Saijo, 1965. Inter-calibration of marine plankton primary production techniques. Limnol. Oceanogr., 10:282-286.

Düing, W., 1970. The Monsoon Regime of the Currents in the Indian Ocean. East-West Center Press, University of Hawaii, Honolulu, Hawaii. 68 pp.

Düing, W. and W.-D. Schwill, 1967. Ausbreitung und Vermischung des salzreichen Wassers aus dem Roten Meer und aus dem Persischen Golf. "Meteor" Forschungsergebnisse, A, 3:44-66. Bornträger, Berlin.

Elizarov, A.A., 1968. Preliminary results of oneanographic investigations of the west coast of India. Proc. All-Union Inst. Mar. Fish. Oceanogr. (Trudy VNIRO), 64:94-101 (In Russian).

Emery, K.O., 1960. The Sea off Southern California. Wiley, New York-London. 366 pp.

Eppley, R.W., 1981. Relations between nutrient assimilation and growth in phytoplankton with a brief review of estimates of growth rate in the ocean. Can. Bull. Fish. Aquat. Sci., 210:251-263.

Eppley, R.W., P. Koeller, and G.T. Wallace, 1978. Strirring influences the phytoplankton species composition within enclosed columns of coastal water. J. Exp. mar. Biol. Ecol., 32:219-239.

Eppley, R.W., E.H. Renger, E.L. Venrick, and M.M. Mullin, 1973. A study of plankton dynamics and nutrient cycling in the central gyre of the North Pacific Ocean. Limnol. Oceanogr., 18:534-551.

Estrada, M. and D. Blasco, 1979. Two phases of the phytoplankton community in the Baja California upwelling. Limnol. Oceanogr., 24:1065-1080.

Falkowski, P.G. (ed.), 1980. *Primary Production in the Sea.* Plenum Press, New York-London. 531 pp.

Fasham, M.J. (ed.), (In press). *Flows of Energy and Material in Marine Ecosystems: Theory and Practice.* Plenum Press, New York-London.

Fenchel, T., 1978. The ecology of micro-and meiobenthos. Ann. Rev. Ecol. Syst., 9:99-121.

Fenchel, T., 1982. Ecology of heterotrophic microflagellates. I. Some important forms and their functional morphology. Mar. Ecol.-Progr. Ser., 8:211-223.

Fieux, M. and H. Stommel, 1976. Historical sea-surface temperatures in the Arabian Sea. Ann. Inst. Oceanogr. Paris, 52:5-15.

Fomitchev, A.V., 1964. The investigation of the North Indian Ocean currents. Trudy Inst. Okeanol. Akad. Nauk SSSR, 64:43-50 (In Russian).

Food and Agriculture Organization (UN), 1981. *Yearbook of Fisheries Statistics,* 50:386 pp.

Frost, B.W., 1980. Grazing. Pp. 465-491, *In:* I. Morris (ed.), *The Physiological Ecology of Phytoplankton.* Blackwell Scientific Publications, Oxford.

Garrison, D.L., 1981. Monterey Bay phytoplankton. II. Resting spore cycles in coastal diatom populations. J. Plankton Res., 3:137-156.

Gilson, H.C., 1937. The nitrogen cycle. Sci. Rept. John Murray Exped., 1933-34, 2(2):21-81.

Gjösaeter, J., 1978. Aspects of the distribution and ecology of the Myctophidae from the western and northern Arabian Sea. Food and Agriculture Organization, UNDP, Indian Ocean Fishery Commission, Indian Ocean Programme, Development Rept. 43, 2: (IOFC/DEV/78/43.2):62-108.

Gjösaeter, J., and K. Kawaguchi, 1980. A review of the world resources of mesopelagic fish. Food and Agriculture Organization, Fish. Techn., Paper 193 (FIRM/T193): 151 pp.

Goldman, J.C., 1980. Physiological processes, nutrient availability, and the concept of relative growth rate in marine phytoplankton ecology. Pp. 179-194, *In:* P.G. Falowski (ed.), *Primary Production in the Sea.* Plenum Press, New York-London.

Goldman, J.C. and P.M. Glibert, 1982. Comparative rapid ammonium uptake by four species of marine phytoplankton. Limnol. Oceanogr., 27:814-827.

Gopalakrishnan, P., 1972. Studies on the marine planktonic diatoms off Port Okha in the Gulf of Kutch. Phykos, 11:37-49.

Goswami, S.C., R.A. Selvakumar, and S.N. Dwivedi, 1977. Zooplankton production along central west coast of India. Pp. 337-353, *In: Proceedings of the Symposium on Warm Water Zooplankton.* National Institute of Oceanography, Goa.

Grobov, A.G., 1968. Quantitative distribution of zooplankton within the 0-100 m layer in the north-western part of the Indian Ocean. Proc. All-Union Inst. Mar. Fish. Oceanogr. (Trudy VNIRO), 64:260-270 (In Russian).

Grobov, A.G. and B.N. Michaylov, 1968. Quantitative distribution of benthos in shelf waters of the north-western part of the Indian Ocean. Proc. All-Union Inst. Mar. Fish. Oceanogr. (Trudy VNIRO), 64:196-203 (In Russian).

Haq, S.M., J. Ali Khan, and S. Chugtai, 1973. The distribution and abundance of zooplankton along the coast of Pakistan during postmonsoon and premonsoon periods. Pp. 257-272, *In:* B. Zeitzschel (ed.), *The Biology of the Indian Ocean.* Springer-Verlag, New York-Heidelberg.

Harkantra, S. N., A. Nair, Z.A. Ansari, and A.H. Parulekar, 1980. Benthos of the shelf region along the west coast of India. Indian J. Mar. Sci., 9:106-110.

Harrison, P.J. and D.H. Turpin, 1982. The manipulation of physical, chemical, and biological factors to select species from natural communities. Pp. 275-289, *In:* G.D. Grice and M.R. Reeve (eds.), *Marine Mesocosms.* Springer-Verlag, New York-Heidelberg.

Harrison, W.G., 1980. Nutrient regeneration and primary production in the sea. Pp. 433-460, *In:* P.G. Falkowski (ed.), *Primary Production in the Sea.* Plenum Press, New York-London.

Haury, L.R., J. A. McGowan, and P.H. Wiebe, 1978. Patterns and processes in the time-space scales of plankton distributions. Pp. 277-327. *In:* J.H. Steele (ed.), *Spatial Patterns in Plankton Communities.* Plenum Press, New York-London.

Hazelwood, D.H., and R.A. Parker, 1961. Population dynamics of some freshwater zooplankton. Ecology, 42:266-274.

Hirota, J., 1974. Quantitative natural history of *Pleurobrachia bachei* in La Jolla Bight. Fish. Bull. (U.S.), 72:295-335.

Humphrey, G.F., and M. Wootton, 1966. Comparison of the techniques used in the determination of phytoplankton pigments. Monogr. Oceanogr. Methodol. (Unesco), 1:37-63.

Institute for Marine Research, Bergen, 1977a. Indian Ocean Fishery and Development Programme, Pelagic Fish Assessment Survey North Arabian Sea. Report on Cruise No. 6 of R/V "Dr. Fridtjof Nansen" (Preliminary report, mimeo).

Institute for Marine Research, Bergen, 1977b. Final report survey results of Dr. Fridtjof Nansen (Pelagic Fish Assessment Survey North Arabian Sea). Food and Agriculture Organization, UNDP, Ocean Fishery Commission, Indian Ocean Programme, Development Rept. 43, vol. 2 (IOFC/DEV/78/43.2):13-61.

Jamart, B.M., D.F. Winter, K. Banse, G.C. Anderson, and R.K. Lam, 1977. A theoretical study of phytoplankton growth and nutrient distribution in the Pacific Ocean off the northwestern U.S. coast. Deep-Sea Res., 24:753-773.

Jayaraman, R., and S.S. Gogate, 1957. Salinity and temperature variations in the surface waters of the Arabian Sea off the Bombay and Saurashtra coasts. Proc. Indian Acad., Sci., B, 44:151-164.

Johnson, P.W., and J. McN. Sieburth, 1982. In-situ morphology and occurrence of eucaryotic phototrophs of bacterial size in the picoplankton of estuarine and oceanic waters. J. Phycol., 18:318-327.

Kabanova, Yu.G., 1968. Primary production of the northern part of the Indian Ocean. Oceanology, 8:214-225 (transl. from Okeanologiia 8, 1968).

Kemp, W.M. and W.R. Boyton, 1981. External and internal factors regulating metabolic rates of an estuarine benthic community. Oecologia (Berl.), 51:19-27.

Kerfoot, W.C. (ed.), 1980. *Evolution and Ecology of Zooplankton Communities.* Universities Press of New England, Hanover, N.H.-London. 793 pp.

Khimitsa, V.A., 1972. Phosphate distribution off the west coast of Hindustan. Oceanology, 11:632-635 (transl. from Okeanologiia 11, 1971).

Kinzer, J., 1969. On the quantitative distribution of zooplankton in deep scattering layers. Deep-Sea Res., 16:117-125.

Kondritskiy, A.V., 1975. Distribution of the zoobenthos biomass in the northwestern Indian Ocean. Oceanology, 15: 98-99 (transl, from Okeanologiia 15, 1975).

Krey, J., and B. Babenerd, 1976. *Phytoplankton Production. Atlas of the International Indian Ocean Expedition.* University of Kiel, Institut für Meereskunde. 70 pp.

Kuksa, V.I., 1972. Some peculiarities of the formation and distribution of intermediate layers in the Indian Ocean. Oceanology, 12: 21-30 (transl. from Okeanologiia 12, 1972).

Kutty, A.G.G., and N.B. Nair, 1973. Observations on the breeding periods of certain interstitial nematodes, gastrotrichs and copepods of the southwest coast of India. J. Mar. biol. Ass. India, 14: 402-406.

Kuz'menko, L.V., 1969. Primary production in the Arabian Sea in the summer monsoon period. Oceanology, 8:367-370 (transl. from Okeanologiia 8, 1968).

Kuz'menko, L.V., 1974. Primary production of the northern Arabian Sea. Oceanology, 13: 251-256 (transl. from Okeanologiia, 1973).

Kuz'menko, L.V., 1975. Size-weight structure of Arabian Sea

phytoplankton. Biologiia Moria (Akad. Nauk Ukrain. SSR, "Naukova Dumka" Kiev), 34: 26-38 (In Russian).

Kuz'menko, L.V., 1977. Distribution of phytoplankton in the Arabian Sea. Oceanology, 17: 70-74 (trans. from Okeanologiia 17, 1977).

Laird, J., B.B. Breivogel, and C.S. Yentsch, 1964. The distribution of chlorophyll in the western Indian Ocean during the southwest monsoon period, July 30-November 12, 1963. Woods Hole Oceanographic Institution, Ref. 64-33. 52 pp. (mimeo).

Landry, M.R., 1977. A review of important concepts in the trophic organization of pelagic ecosystems. Helgoländer wiss. Meeresunters, 30:8-17.

Landry, M.R., 1978. Population dynamics and production of a planktonic marine copepod, Acartia clausii, in a small temperate lagoon on San Juan Island, Washington. Int. Revue ges. Hydrobiol., 63: 77-119.

Le Borgne, R., 1982. Zooplankton production in the eastern tropical Atlantic Ocean: Net growth efficiency and P:B in terms of carbon, nitrogen, and phosphorus. Limnol. Oceanogr., 27:681-698.

Lehman, J.T. and D. Scavia, 1982. Microscale patches produced by zooplankton. Proc. Natl. Acad. Sci. USA, 79: 5001-5005.

Lie, U. and J.C. Kelley, 1970. Benthic infauna composition off the coast of Washington and in Puget Sound: Identification and distribution of the communities. J. Fish. Res. Bd. Canada, 27: 621-651.

Lohmann, H. 1908, Untersuchungen zur Feststellung des vollständigen Gehaltes des Meeres an Plankton. Wiss. Meeresunters. NF Kiel, 10: 129-370.

Longhurst, A.R., 1981. Significance of spatial variability. Pp. 415-441, In: A. R. Longhurst (ed.), Analysis of Marine Ecosystems. Academic Press, London-New York.

Lukashev, Yu. F., 1980. Deep-sea nitrite maximum and denitrification in the Arabian Sea. Okeanology, 20: 164-167 (transl. from Okeanologiia 20, 1980).

Lund, J.W.G., 1978. Experiments with lake phytoplankton in large enclosures. Repts. Freshw. biol. Ass., 46: 32-39.

McCarthy, J.J. and J.C. Goldman, 1979. Nitrogenous nutrition of marine phytoplankton in nutrient-depleted waters. Science (Washington, D.C.), 203: 670-672.

McGowan, J.A. and T.L. Hayward, 1978. Mixing and oceanic productivity. Deep-Sea Res., 25: 771-793.

Maliyavanshi, I.N., 1977. Further studies on the marine planktonic diatoms in the coastal waters of Saurashtra. Phykos, 14: 99-110.

Malone, T.C., 1980. Algal size. Pp. 433-463, In: I. Morris (ed.), The Physiological Ecology of Phytoplankton. Blackwell Scientific Publications, Oxford.

Mann, K.H., 1982. Ecology of Coastal Waters. A Systems Approach. Blackwell Scientific Publications, oxford. 322 pp.

May, D.R., 1972. The effects of oxygen concentration and anoxia on respiration of Abarenicola pacifica and Lumbrineris zonata (Polychaeta). Biol. Bull. (Woods Hole), 142: 71-83.

Menon, M.D. and K.C. George, 1977. On the abundance of zooplankton along the southwest coast of India during the years 1971-75. Pp. 205-213, In: Proceedings of the Symposium on Warm Water Zooplankton. National Institute of Oceanography, Goa.

Menzel, D.W. and J.H. Ryther, 1961. Annual variations in primary production of the Sargasso Sea off Bermuda. Deep-Sea Res., 7:282-288.

Muhammad, A. and M. Arshad. 1966a, Benthos in relation to hydrology in off-shore waters of Karachi. Agriculture Pakistan, 17: 113-123.

Muhammad, A., and M. Arshad, 1966b. Preliminary observation on the composition of zooplankton along the Karachi coast. Agriculture Pakistan, 17: 227-237.

Naumov, A.G. and L.A. Ponomareva, 1964. The vertical distribution and the diurnal migration of the ordinary species of the zooplankton from the northern part of the Indian Ocean. Trudy Inst. Okeanol., 64: 250-256 (In Russian).

Nichols, F.H., 1975. Dynamics and energetics of three deposit-feeding invertebrate populations in Puget Sound, Washington. Ecol. Monogr., 45: 57-82.

Nixon, S.W., J.R. Kelly, B.N. Furnas, C.A. Oviatt, and S.S. Hale, 1980. Phosphorus regeneration and the metabolism of coastal marine communities. Pp. 219-242, In: K.R. Tenore and B.C. Coull (eds.), Marine Benthic Dynamics. Univ. of S. Carolina Press, Columbia, S.C.

Noor-Uddin, 1967a. An account of the phytoplankton of Karachi with a note on their seasonal variations, fluctuations and distribution. Agriculture Pakistan, 18: 51-83.

Noor-Uddin, 1967b. An account of the zooplankton of Karachi coast with a note on their seasonal variations, fluctuations and distribution. Agriculture Pakistan, 18: 197-224.

Paine, R.T., 1980. Food webs: Linkage, interaction strength and community infrastructure. J. Anim. Ecol., 49: 667-685.

Pamatmat, M.M., (In press). Measuring the metabolism of the benthic ecosystem. In: M.J. Fasham (ed.), Flows of Energy and Material in Marine Ecosystems. Plenum Press, New York-London.

Patel, M.I., 1976. Seasonal variation in zooplankton from the coastal waters off Saurashtra. Indian J. Mar. Sci., 5: 140-144.

Paulinose, V.T. and P.N. Aravindakshan, 1977. Zooplankton biomass, abundance and distribution in the north and northeastern Arabian Sea. Pp. 132-136, In: Proceedings of the Symposium on Warm Water Zooplankton. National Institute of Oceanography, Goa.

Pelaez, J. and F. Guan. 1982. California Current chlorophyll measurements from satellite data. California Coop. Oceanic Fish. Invest. Rep. 23: 212-225.

Peterson, B.J., 1980. Aquatic productivity and the ^{14}C-CO$_2$ method: A history of the productivity problem. Ann. Rev. Ecol. Syst., 11: 359-385.

Peterson, H.P., 1979. Predation, competitive exclusion, and diversity in the soft-sediment communities of estuaries and lagoons. Pp. 233-264, In: R.J. Livingston (ed.), Ecological Processes in Coastal and Marine Ecosystems. Plenum Press, New York-London.

Platt, J.R., 1964. Strong inference. Science (Washington, D.C.), 146: 347-353.

Platt, T. (ed.), 1981. Physiological Bases of Phytoplankton Ecology. Can. Bull. Fish. Aquat. Sci., 210. 346 pp.

Pomeroy, L.R., 1979. Secondary production mechanisms of continental shelf communities. Pp. 163-186, In: R.J. Livingston (ed.), Ecological Processes in Coastal and Marine Ecosystems. Plenum Press, New York-London.

Prasad, R.R., 1968. Maps on the Total Zooplankton Biomass in the Arabian Sea and the Bay of Bengal. International Indian Ocean Expedition, Plankton Atlas, Vol. 1, Fasc. 1. Indian Ocean Biological Centre, National Institute of Oceanography, Council of Scientific and Industrial Research, New Delhi 18 pp.

Prasad, R.R., 1969. Zooplankton biomass in the Arabian Sea and the Bay of Bengal with a discussion on the fisheries of the regions. Proc. Nat. Inst. Sci. India, B35: 399-437.

Prasad, R.R., 1973. Indian oceanography in perspective. Pp. 40-45, In: Marine Biological Association of India. (ed.), Special Publication Dedicated to Dr. N.K. Panikkar. Cochin.

Prasad, R.R. and P.V.R. Nair, 1963. Studies on organic production. I. Gulf of Mannar. J. Mar. biol. Ass. India, 5:1-26.

Qasim, S.Z., 1982. Oceanography of the northern Arabian Sea. Deep-Sea Res., 29A:1041-1068.

Quraishee, G.S., 1975. Influence of the Indus River on marine environment. Pp. 111-118 In: Pakistan Acad. Sci., International Conference on the Management of Environment, February 17-19, 1975, Vol. II University of Islamabad, Islamabad.

Radhakrishna, K., V.P. Devassy, P.M.A. Bhattathiri, and R.S. Bhargava, 1978. Primary productivity in the northeastern Arabian Sea. Indian J. mar. Sci., 7:137-139.

Radhakrishna, K., V.P. Devassy, R.M.S. Bhargava, and P.M.A. Bhattathiri, 1978. Primary production in the northern Arabian Sea. Ind. J. mar. Sci., 7: 271-275.

Ramamirtham, C.P. and D.S. Rao, 1974. On upwelling along the west coast of India. J. Mar. biol.Ass. India, 15: 306-315.

Rao, T.S.S., 1973. Zooplankton studies in the Indian Ocean. Pp. 243-255, In: B. Zeitzschel (ed.), The Biology of the Indian Ocean.

Springer-Verlag, Heidelberg-New York.

Rao, T.S.S., 1979. Zoogeography of the Indian Ocean. Pp. 254-292, In: van der Spoel and Pierrot-Bults (eds.), Zoogeography and Diversity of Plankton. Arnold, London/Bunge Scientific Publishers, Utrecht.

Rao, T.S.S. and V.R. Nair, 1973. Chaetognaths in the upwelling areas of the Arabian Sea. Pp. 183-192, In: Marine Biological Association of India (ed.), Special Publication Dedicated to Dr. N.K. Panikkar. Cochin.

Reeve, M.R. and L.D. Baker, 1975. Production of two planktonic carnivores (chaetognath and ctenophore) in South Florida inshore waters. Fish. Bull. (U.S.), 73: 238-248.

Rochford, D.J., 1964. Salinity maxima in the upper 1000 metres of the North Indian Ocean. Aust. J. mar. freshw. Res., 15: 1-24.

Rowe, G.T., 1981. The benthic processes of coastal upwelling ecosystems. Pp. 464-471, In: F.A. Richards (ed.), Coastal Upwelling. Am. Geophys. Union, Washington, D.C.

Runge, J.R., 1981. Egg production of Calanus pacificus Brodsky and its relationship to seasonal changes in phytoplankton availability. Ph.D. thesis, Univ. of Washington, Seattle WA., 110 pp.

Ryther, J.H., J.R. Hall, A.K. Pease, A. Bakun, and M.M. Jones, 1966. Primary organic production in relation to the chemistry and hydrography of the western Indian Ocean. Limnol. Oceanogr., 11:371-380.

Ryther, J.H. and D.H. Menzel, 1961. Primary production in the southwest Sargasso Sea, January-February 1960. Bull. Mar. Sci. Gulf and Caribbean, 11:381-388.

Ryther, J.H. and D.W. Menzel, 1963. On production, composition and distribution of organic matter in the western Arabian Sea. Deep-Sea Res., 12:199-209.

Saifullah, S.M., 1979. Occurrence of dinoflagellates and distribution of chlorophyll a on Pakistan shelf. Pp. 203-208, In; D.L. Taylor and H.H. Seliger (eds.) Toxic Dinoflagellate Blooms. Elsevier/North Holland, New York-Amsterdam. 505 pp.

Saifullah, S.M. and M. Moazzam, 1978. Species composition and seasonal occurrence of centric diatoms in a polluted marine environment. Pak. J. Bot., 10: 53-64.

Sameoto, D.O., 1978. Zooplankton sample variation on the Scotian shelf. J. Fish. Res. Bd. Canada, 35: 1207-1222.

Savich, M.S., 1968. Phytoplankton of the Gulf of Aden and the Arabian Sea during winter monsoon. Proc. All-Union Res. Inst. Mar. Fish. (Trudy VNIRO), 64: 243-251 (In Russian).

Savich, M.S., 1972. Quantitative distribution and food value of benthos from the west Pakistan shelf. Oceanology, 12: 113-119 (transl. from Okeanologiia 12, 1972).

Sharp, J.H., M.J. Perry, E.H. Renger, and R.W. Eppley, 1980. Phytoplankton rate processes in the oligotrophic waters of the central North Pacific Ocean. J. Plankton Res., 2: 335-353.

Smayda, T.J., 1980. Phytoplankton species succession. Pp. 493-570, In: I. Morris (ed.), The Physiological Ecology of Phytoplankton. Blackwell Scientific Publications, Oxford.

Smith, R.C. and K.S. Baker. 1981. Oceanic chlorophyll concentrations as determined by satellite (Nimbus-7 Coastal Zone Color Scanner). Mar. Biol. 66: 269-279.

Sorokin, Yu. I., 1978. Decomposition of organic matter and nutrient regeneration. Pp. 501-616, In: O. Kinne (ed.), Marine Ecology. VI. Dynamics. Wiley/Interscience, Chichester-New York.

Sorokin, Yu. I., 1979. Zooflagellates as a component of the community of eutrophic and oligotrophic waters in the Pacific Ocean. Oceanology, 19: 316-319 (transl. from Okeanologiia 19, 1979).

Sournia, A., 1969. Cycle annuel du phytoplancton et de la production primaire dans les mers tropicales. Mar. Biol., 3: 287-303.

Sournia, A., 1982. Is there a shade flora in the marine plankton? J. Plankton Res., 4: 391-399.

Steele, J.H. (ed.), 1978. Spatial patterns in Plankton Communities. Plenum Press, New York-London. 460 pp.

Steele, J.H. and B.W. Frost, 1977. The structure of plankton communities. Proc. R. Soc. London, B280: 485-534.

Steele, J.H., and M. M. Mullin, 1977. Zooplankton dynamics. Pp. 857-890, In: E. D. Goldberg et al. (eds.), The Sea, Vol. 6. Wiley-Interscience, New York.

Steemann Nielsen, E. and V.K. Hansen, 1961. Influence of surface illumination on plankton photosynthesis in Danish waters (56°N) throughout the year. Physiol. Plant., 14: 595-613.

Subrahmanyan, R., 1958a. Phytoplankton organisms of the Arabian Sea off the west coast of India. J. Ind. Bot. Soc., 37: 435-441.

Subrahmanyan, R., 1958b. Ecological studies on the marine phytoplankton on the west coast of India. Mem. Indian Bot. Soc., 1: 145-151.

Subrahmanyan, R., 1959. Studies on the phytoplankton of the west coast of India. Part I. Quantitative and qualitative fluctuations of the total phytoplankton crop, the zooplankton crop and their interrelationships, with remarks on the magnitude of the standing crop and production of matter and their relationship to fish landings. Proc. Nat. Inst. Sci. India, 50: 113-187.

Subrahmanyan, R., C.P. Gopinathan, and C.T. Pillai, 1982. Phytoplankton of the Indian Ocean: Some ecological problems. J. Mar. biol. Ass. India, 17: 608-612.

Subrahmanyan, R. and A.H.V. Sarma, 1961. Studies on the phytoplankton of the west coast of India. Part III. Seasonal variation of the phytoplankters and environmental factors. Ind. J. Fish., 7: 307-336.

Subrahmanyan, R. and A.H.V. Sarma, 1967. Studies on the phytoplankton of the west coast of India. Part IV. Magnitude of the standing crop for 1955-1962, with observations on nanoplankton and its significance to fisheries. J. Mar. biol. Ass. India, 7: 406-419.

Sverdrup, H.U., M.W. Johnson, and R.H. Fleming, 1942. The Oceans— Their Physics, Chemistry, and General Biology. Prentice-Hall, Englewood Cliffs, NJ. 1087 pp.

Taylor, F.J.R., 1973. General features of dinoflagellate material collected by the "Anton Brunn" during the International Indian Ocean Expedition. Pp. 155-169, In: B. Zeitzschel (ed.), The Biology of the Indian Ocean. Springer-Verlag, Heidelberg-New York.

Taylor, F.J.R., 1976. Dinoflagellates from the International Indian Ocean Expedition. Bibliotheca Botanica (Schweizerbarth, Stuttgart) 132. 234 pp.

Tenore, K.R. and B.C. Coull (eds.), 1980. Marine Benthic Dynamics. University of South Carolina Press, Columbia, S.C. 451 pp.

Thompson, P.K.M. and D.C.V. Easterson, 1977. Dynamics of cyclopoid population in a tropical estuary. Pp. 486-496, In: Proceedings of the Symposium on Warm Water Zooplankton. National Institute of Oceanography, Goa.

Timonin, A.C., 1974. Structure of pelagic communities. Trophic structure of zooplankton communities in the northern part of the Indian Ocean. Oceanology, 13: 85-93 (transl. from Okeanologiia 13, 1973).

Torrington-Smith, M., 1976. The distribution of abundance, species diversity, and phytogeographic regions in west Indian Ocean phytoplankton. J. Mar. biol. Ass. India, 16: 371-380.

Tsalkina, A.V., 1973. Vertical distribution and diurnal migration of Cyclopoida (Copepoda) in the northeastern Indian Ocean. Oceanology, 12: 566-576 (transl. from Okeanologiia 12, 1972).

Tyuleva, L.S., 1974. Seasonal occurrence of productive zones off the west coast of India. Oceanology, 13: 268-273 (transl. from Okeanologiia 13, 1973).

U.S. Navy Hydrographic Office. 1960. Summary of Oceanographic Conditions in the Indian Ocean. U.S. Navy Hydrogr. Off. (Oceanogr. Analysis Divn., Mar. Sci. Dept.), Spec. Publ. 53., 142 pp.

van der Spoel, S. and A.C. Pierrot-Bults (eds.), 1979. Zoogeography and Diversity of Plankton. Arnold, London/Bunge Scientific Publishers, Utrecht. 410 pp.

Varkey, M.J. and V. Kesavadas, 1976. Light transmission characteristics of the northern Arabian Sea during December-May. Indian J. mar. Sci., 5: 147-151.

Varma, K.K., V.K. Das, and A.D. Gouveia, 1980. Thermohaline structure and watermasses in the northern Arabian Sea during February-April. Indian J. Mar. Sci., 9: 148-155.

Verlencar, X.N., 1980. Distribution of urea in the waters of the west coast of India. Indian J. Mar. Sci., 9: 230-233.

Venrick, E.L., 1982. Phytoplankton in an oligotrophic ocean:

observations and questions. Ecol. Monogr., 52: 129-154.

Venrick, R.L., J.A. McGowan, and A.W. Mantyla, 1973. Deep maxima of photosynthetic chlorophyll in the Pacific Ocean. Fish. Bull. (U.S.), 71: 41-52.

Vinogradov, M.E., 1970. *Vertical Distribution of the Oceanic Zooplankton*. Israel Prog. Sci. Transl., Jerusalem. 339 pp. (transl. of the Russian original of 1968).

Vinogradov, M.E., 1977. Pelagic ecosystem studies on the upwellings of the eastern Pacific ocean: Cruise 17 of the R/V "Akademic Kurchatov". Pol. Arch. Hydrobiol., 24 (Suppl.): 7-19.

Vinogradov, M.E., 1981. Ecosystems of equatorial upwellings. Pp. 69-93, *In:* A.R. Longhurst (ed.), *Analysis of marine Ecosystems*. Academic Press, London-New York.

Vinogradov, M.E., V.V. Menshutkin, and E.A. Shushkina, 1972. On mathematical simulation of a pelagic ecosystem in tropical waters of the ocean. Mar. Biol., 16: 261-268.

Walsh, J.J., 1981. Shelf-sea ecosystems. Pp. 159-198, *In:* A.R. Longhurst (ed.), *Analysis of marine Ecosystems*. Academic Press, London-New York.

Warners, C.J., 1952. *Indian Ocean Oceanographic and Meteorological Data*. Publ. Kon. Nederl. Meteorol. Inst. 135 (2nd ed.), 31 pp., plates.

Wiebe, P.H., S. Boyd, and J.L. Cox, 1975. Relationships between zooplankton, displacement volume, wet weight, dry weight, and carbon. Fish. Bull. (U.S.), 73: 777-786.

Williams, P.J. leB., 1981a. Incorporation of microheterotrophic processes into the classical paradigm of the planktonic food web. Kieler Meeresforsch. Suppl., 5: 1-28.

Williams, P.J. leB., 1981b. Microbial contribution to overall marine plankton metabolism: Direct measurements of respiration. Oceanol. Acta, 4: 359-364.

Winberg, G.G. (ed.), 1971. *Methods for the Estimation of Production of Aquatic Animals*. Academic Press, New York-London. 175 pp. (transl. of the 1968 Russian version).

Winberg, G.G., K. Patalas, J.C. Wright, A. Hillbricht-Ilkowska, W.E. Cooper, and K.H. Mann, 1971. Pp. 296-319, *In:* W.T. Edmondson and G.G. Winberg (eds.), *A Manual on Methods for the Assessment of Secondary Production in Fresh Waters*. International Biological Programme, Handbook No. 17. Blackwell Scientific Publications, Oxford-Edinburgh. 358 pp.

Winter, D.F., K. Banse, and G.C. Anderson, 1975. The dynamics of phytoplankton blooms in Puget Sound, a fjord in the northwestern United States. Mar. Biol., 29: 139-176.

Wood, E.E.J., 1963. Checklist of diatoms recorded from the Indian Ocean. Rep. Div. Fish. Commonw. Sci. Industr. Res. Org. Austr. 36. 311 pp.

Woods Hole Oceanographic Institution, 1965. U.S. Program in Biology. International Indian Ocean Expedition. Final Cruise Report, *Anton Bruun* Cruises 4A and 4B. Unpaginated.

Wooster, W.S., M.B. Schaefer, and M.K. Robinson, 1967. *Atlas of the Arabian Sea for Fishery Oceanography*. University of California, Institute of Marine Resources, Ref. IMR 67-12. 35 pp. Charts.

Wyrtki, K., 1971. *Oceanographic Atlas of the International Indian Ocean Expedition*. National Science Foundation, Washington, D.C., 531 pp.

Wyrtki, K., 1973. Physical Oceanography of the Indian Ocean. Pp. 18-36, *In:* B. Zeitzschel (ed.), *The Biology of the Indian Ocean*. Springer-Verlag, Heidelberg-New York.

Wroblewski, J.S., 1980. A simulation of the distribution of *Acartia clausi* during Oregon upwelling. J. Plankton Res., 2: 43-68.

Wroblewski, J.S., 1982. Interactions of currents and vertical migration in maintaining *Calanus marshallae* in the Oregon upwelling zone. Deep-Sea Res., 29 A: 665-686.

Yamanaka, H., M. Yukinawa, and I. Nakamura, 1977. Summary report on cruise of the R/V Shoyo Maru in the north Arabian Sea, 2 October 1976-13 January 1977. Food and Agriculture Organization, United Nations, UNDP, Indian Ocean Program, Tech. Repts., 77/14. 84 pp.

Yamanaka, H., T. Nishigawa, and J. Morita, 1978. Summary report on cruise of the R/V Shoyo Maru in the north Arabian Sea, 2 October 1975-14 January 1976. Food and Agriculture Organization, United Nations, UNDP, Indian Ocean Program, Tech. Repts., 76/11. 47 pp.

Yentsch, C.S., 1965. Distribution of chlorophyll and phaeophytin in the open ocean. Deep-Sea Res., 12: 653-666.

Zeitzschel, B., 1980. Sediment-water interactions in nutrient dynamics. Pp. 195-218, *In:* K.R. Tenore and B.C. Coull (eds.), *Marine Benthic Dynamics*. Univ. of S. Carolina Press, Columbia, S.C.

Zeitzschel, B. (ed.), 1973. *The Biology of the Indian Ocean*. Springer-Verlag, New York-Heidelberg. 549 pp.

Zupanovic, S. and S.Q. Mohiuddin, 1976. A survey of the fishery resources in the northeastern part of the Arabian Sea. J. Mar. biol. Ass. India, 15: 496-537.

18

Controls on Dissolved Oxygen Distribution and Organic Carbon Deposition in the Arabian Sea

R. D. SLATER
University of Chicago,
Chicago, Illinois

and

P. KROOPNICK
Exxon Production Research Company,
Houston, Texas

ABSTRACT

The mechanisms controlling the deposition of sediments rich in organic matter are of vital concern in our search for hydrocarbons. The Arabian Sea is unique among present day environments in that it exhibits an upwelling mechanism different from other previously studied areas of the world; the whole sea has a severe oxygen minimum about 1300 m thick; and the sediments are rich in organic carbon, even away from sources of high productivity.

The upwelling mechanisms operating off Somalia and Arabia are those of geostrophic adjustment and wind induced Ekman transport, respectively. The cause for upwelling off India is not yet understood, but it may also be due to geostrophic adjustment.

The oxygen distribution is controlled by the following factors:

1. High primary production, averaging twice that of the world ocean, supplies a large amount of organic matter to the water column. The subsequent oxidation of this organic matter causes an intense mid-depth oxygen minimum.
2. Strong stratification in the upper 200 m reduces the exchange of oxygen between the atmosphere and subsurface layers.
3. Monsoonal circulation increases the residence time of the intermediate waters, thus allowing for more oxygen depletion.
4. Inflow of high salinity, organic-rich water from the Persian Gulf and Red Sea increases the stratification and intensifies the oxygen minimum zone.

The TOC values for Arabian Sea sediments are higher off the Indian coast than near Arabia. This preferential preservation of organic carbon is primarily controlled by the concentration of dissolved oxygen in the waters overlying the sediments.

INTRODUCTION

To understand the factors that were important in the production of organic rich source beds in the past we must understand the mechanisms operating today. One important mechanism is upwelling. Smith (1968) and Summerhayes (1981) have examined many upwelling areas for the present day oceans. The locations of the major upwelling areas along with the production mechanism are shown in Figure 1.

Upwelling brings nutrient-rich water to the surface, stimulating high biological productivity. This in turn produces large amounts of organic matter which will remove oxygen from the water as it decays. If the oxygen depletion is large enough, substantial amounts of organic matter can be preserved in the sediments.

There are three things that make the Arabian Sea unique. First, it is an area exhibiting an upwelling mechanism different from other areas of the world (Smith, 1968). Second, the whole Arabian Sea has a severe oxygen minimum (<0.5 ml/l), not just areas near the upwelling centers.

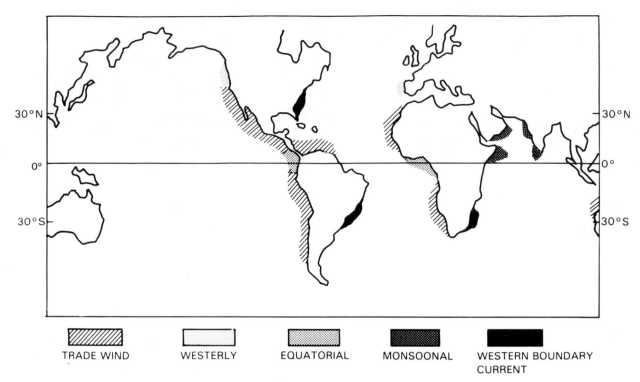

Figure 1. Major locations of different types of upwelling.

Third, the sediments are rich in organic carbon, even away from the sources of high productivity.

THE ARABIAN SEA

The Arabian Sea (Fig. 2) is bounded by India to the east, the Arabian peninsula to the west, and Iran and Pakistan to the north; the south is open to the Indian Ocean. Both the Red Sea and the Persian Gulf border on the Arabian Sea. The Indus river is the major river of the area, transporting over 400 million tons of sediment a year. The Narmada and Tapti rivers in India are the only other rivers of significance.

Two items should be noted about the bathymetry of the Arabian Sea: the continental slope and shelf are wider on the Indian side than the Arabian side, therefore exposing a larger area of the sea floor to the oxygen-poor water; and the Arabian Basin is separated from the Somali Basin by the Carlsberg Ridge, reducing the flow of bottom water into the Arabian Basin.

Hydrography

The surface currents in the northern Indian Ocean are controlled by the seasonally reversing monsoonal winds (see Fig. 3). During the winter, the winds are predominantly from the northeast,

and the water circulation is cyclonic (counterclockwise). In the summer, the winds change to the southwest and the current pattern is anticyclonic. Both the winds and the currents are stronger during the southwest monsoon. The flow in the intermediate layers is not well known. It appears, however, that it may also be affected by the monsoonal wind changes (Wunsch, 1977; Warren et al., 1966; Wyrtki, 1971). The abyssal circulation in the western part of the Indian Ocean is restricted by a series of submarine ridges; water

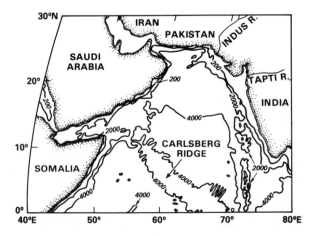

Figure 2. Map of the Arabian Sea showing 200, 2000, and 4000 m depth contours and major rivers (after Wyrtki, 1971).

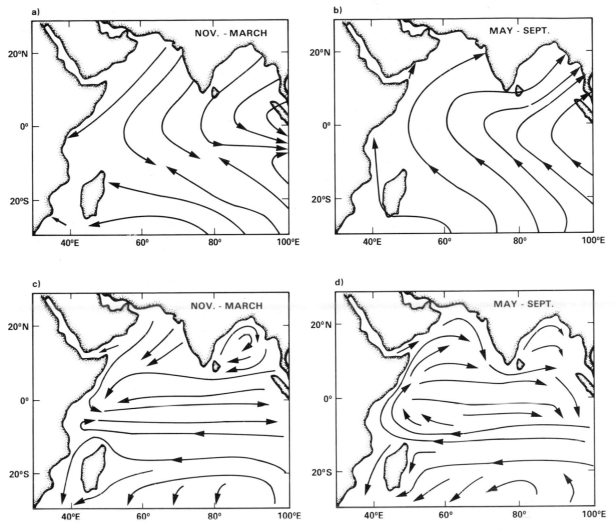

Figure 3. Seasonal circulation patterns for the northern Indian Ocean; a and b surface winds (after Hastenrath and Lamb, 1980); c and d surface currents (after Pickard, 1963).

can only flow through relatively shallow gaps in these ridges, thus the northward flow of abyssal water is severely limited (Warren, 1974, 1978; Kolla et al., 1976; Johnson and Damuth, 1979).

Upwelling occurs along the coasts of Somalia and the Arabian peninsula (Bruce, 1974), and India (Sharma, 1978; Hinton and Wylie, 1982) during the summer months. Off Somalia, the upwelling is controlled by the Somali current which accelerates during the southwest monsoon and causes the interior density surfaces to tilt upwards near the coast to establish a geostrophic balance (Smith, 1968). Off the Arabian peninsula, wind induced Ekman-type upwelling occurs during the southwest monsoon (Prell and Streeter, 1982: Smith, 1968).

Along the west coast of India, upwelling begins in the south around the end of March and propagates northwards until about the end of May. Downwelling then commences in August (Sharma, 1978). The cause of this upwelling along the coast of India is not yet understood. The local winds blow in the wrong direction to produce upwelling during the southwest monsoon when they are strongest, and the winds are too weak during the northeast monsoon when they blow in the proper direction (Sharma, 1978). Hinton and Wylie (1982) show that the upwelling may be caused by non-local winds, but the resulting vertical velocities from their calculations are too small. Sharma (1978) argued that the upwelling is due to geostrophic adjustment (similar to the Somalia upwelling), but more evidence is needed before this hypothesis can be accepted.

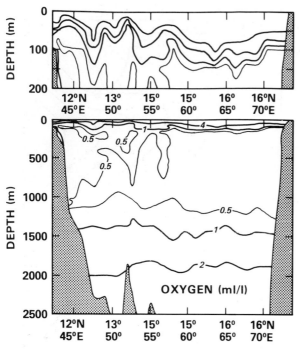

Figure 4. East-west vertical profiles of dissolved oxygen in the Arabian Sea and Red Sea (after Wyrtki, 1971). The oxygen minimum ranges from about 150-1500 m.

Vertical sections of the dissolved oxygen distribution in the Arabian Sea are shown in Figures 4 and 5. We can see that a thick layer of oxygen-poor water (<0.5 ml/l) extends from 200-1500 m everywhere in the Arabian Sea. Figure 6 gives a typical plan view of the oxygen distribution showing its zonal nature. The oxygen concentrations are lowest in the northeast corner of the sea, and are lower off India than off Arabia at the same latitude. The surface waters are generally saturated with oxygen, but the deep water is relatively low in oxygen because of the restricted flow with concentrations usually less than 4.0 ml/l (Wyrtki, 1971).

The nutrient distribution is generally the inverse of dissolved oxygen. Phosphate concentrations are greater to the north and west. Nitrate, however, has greater concentrations to the north and east (Wyrtki, 1971). The largest amounts of nutrients in the surface waters occur in the upwelling regions off Arabia and Somalia. The nitrate concentrations in the near surface and intermediate water are lower near India than near Arabia (Wyrtki, 1971) due to more vigorous upwelling off Arabia and to denitrification in the northeastern Arabian Sea.

Denitrification, an indication of low oxygen concentration, has been observed by Sen Gupta et al. (1975, 1976), and Deuser et al. (1978). The denitrification takes place in the outflowing waters of the Persian Gulf at oxygen concentrations less than 0.16 ml/l (Deuser et al., 1978). Sen Gupta et al. (1976) found denitrification to also take place in the outflow from the Red Sea. They give a maximum oxygen concentration of 0.3 ml/l below which denitrification takes place.

Productivity

As yet, no comprehensive study of the productivity in the Arabian Sea has been attempted. Previous studies have almost always been separated in time (both season and year) and space. This is a serious problem when one considers the extreme variability in the Arabian Sea due to its monsoonal circulation. In addition to this seasonal variation, there are also yearly differences as in other areas of the world. The following conclusions about the productivity take the above considerations into account.

The Arabian Sea is an area of high productivity, averaging about twice that of the world ocean (Ryther et al., 1966). The highest values occur in the north, being typically greater than 1 g Cm^{-2} day^{-1} (Kuz'menko, 1973; Ryther et al., 1966). Productivity along the coast of India is about the same as that off Arabia (see Fig. 7). The phytoplankton productivity is very patchy throughout the Arabian Sea (Ryther and Menzel, 1965; Radhakrishna et al., 1978). Ryther and Menzel (1965), trying to explain this patchiness, noted that since the subsurface waters were so rich in nutrients, anywhere some mechanism could bring this water to the surface high productivity would result. Some authors have used nutrient concentrations as a relative measure of primary productivity. However, neither Radhakrishna et al. (1978) (for February, March, and November, 1977) nor Bhargava et al. (1978) (for October and December, 1976) found a significant correlation between nutrients and primary productivity, indicating that nutrients are not a limiting factor.

Sediments

The marginal areas of the northern and western Arabian Sea exhibit coarse grained sediment that is low in carbonate content with large amounts of quartz and feldspar indicating significant terrigenous sedimentation (Kolla et al., 1981). The percentage of $CaCO_3$ in the sediments

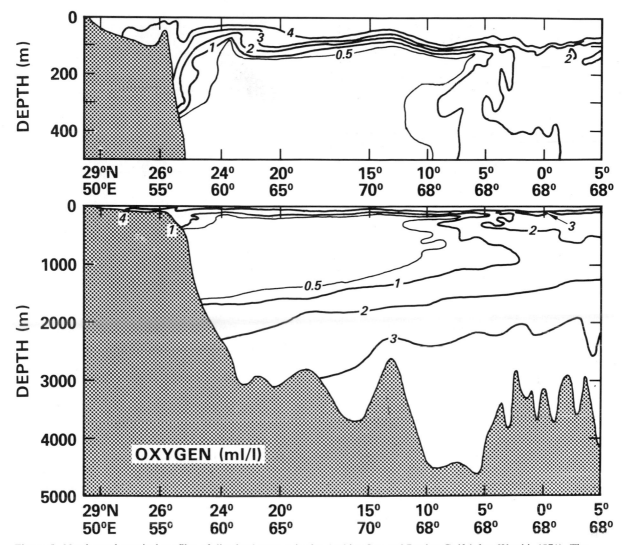

Figure 5. North-south vertical profiles of dissolved oxygen in the Arabian Sea and Persian Gulf (after Wyrtki, 1971). The oxygen minimum ranges from about 150-1500 m.

is low on the continental shelf off India and near the Indus river, indicating dilution of marine sediment by terrigenous input. However, the continental slope is very high in carbonate, indicating mostly marine sedimentation. The major input of terrestrial matter is from the Indus river. River input from the Arabian peninsula and northern Africa is negligible, making aeolian transport significant (Kolla et al., 1981).

The distribution of organic carbon in the sediments is shown in Figure 8. Large values (greater than 1.0%) are seen all along the continental margins. Even in the deep ocean, up to 500 km off the Arabian peninsula, values can exceed 1.0%.

Little data is given for the sedimentation rates

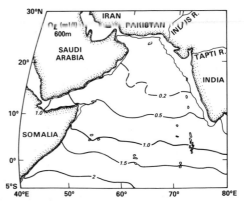

Figure 6. Dissolved oxygen (ml/l) distribution at 600 m for the northwest Indian Ocean (after Wyrtki, 1971). This depth falls in the oxygen minimum layer. Note the zonal structure of the isopleths; also note the lowest concentrations occur in the northeast Arabian Sea.

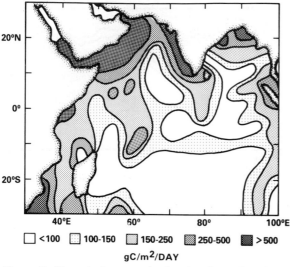

Figure 7. Phytoplankton production for the Indian Ocean (after FAO Department of Fisheries, 1972).

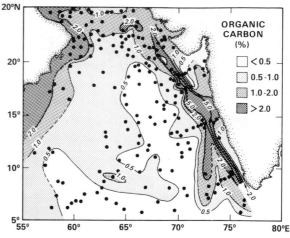

Figure 8. Total organic carbon in the surface sediments of the Arabian Sea (after Kolla et al., 1981).

in the Arabian Sea. Zobel (1973), in her study of the sediments of the Indian and Pakistan margins, was not able to determine accurate sedimentation rates. However, she did give rough estimates of "1-2 cm/1000 years in the deep basin and up to 40 cm/1000 years on the upper continental slope." Sedimentation rates are lower off southern India than at the mouth of the Indus river (Zobel, 1973). Kolla et al. (1981) cite a rate of 4 cm/1000 years in water depths of 2000-3500 m off the Arabian peninsula.

CONTROLS ON DISSOLVED OXYGEN DISTRIBUTION AND ORGANIC MATTER DEPOSITION

Dissolved Oxygen

There are five main factors that affect the distribution of dissolved oxygen in the Arabian Sea. First, the high productivity averaging about twice that of the world ocean (Ryther et al., 1966), creates a large oxygen demand on the water column.

Second, the Arabian Sea is strongly stratified in the upper 200 m (Ryther and Menzel, 1965; Gangadhara Rao et al., 1974). This reduces the exchange of oxygen between the atmosphere and the subsurface layers.

Third, the circulation in the northern Indian Ocean is heavily influenced by the monsoonal nature of the winds. It appears from maps of the dynamic topography presented in Wyrtki (1971),

that not only the surface currents reverse but so do the intermediate layers. This will cause the residence time of the water to increase, thus allowing for more organic matter to be oxidized and oxygen consumed. This, coupled with the lack of transequatorial water transport (Sharma, 1976) tends to isolate the northern Indian Ocean.

Fourth, there is an inflow of water rich in organic matter from the Red Sea and the Persian Gulf (Ryther and Menzel, 1965; Szekielda, 1970). These waters not only increase the stratification because of their high salinities but also help in the depletion of oxygen. In fact, the oxygen depletion in these waters is great enough to cause denitrification (Sen Gupta et al., 1976; Deuser et al., 1978).

Fifth, the abyssal water in the Arabian Sea is low in oxygen because of the restricted flow path between the polar source regions of the deep water and the Arabian Sea (Warren, 1974, 1978). This will affect the positioning of the lower boundary of the oxygen minimum layer, lowering it because of the smaller oxygen gradient.

Organic Matter

Kolla et al. (1981) briefly examined the influence of marine versus terrestrial organic matter type for the Arabian Sea. Their results show three broad areas. In two areas terrestrial organic matter predominated over marine. The first area starts 500 km off the southern tip of India and extends seaward about the same distance. The second area is located on the Carlsberg Ridge in the middle of the Arabian Sea. Predominantly marine organic matter is found along the coasts, with mixed types occurring in the central areas. None of the areas exhibited any significant correlations between grain size and organic carbon or nitrogen.

The percent total organic carbon (TOC) in the sediments of the Arabian Sea is typically greater than 2.0% in the coastal areas. These high values can even extend 500 km offshore (Fig. 8). Highest TOCs are found along the shelf break near India. TOCs in excess of 4.0% have been reported in the Gulf of Aden (Udintsev, 1975; Wiseman and Bennett, 1940).

Three oceanographic factors control the percentage of TOC in the sediments: primary productivity, rate of sedimentation, and the concentration of dissolved oxygen in the water overlying the sediments. Care must be taken in relating these three factors with TOC, since in the marine environment (with little terrestrial input) the sedimentation rate is strongly related to primary production, and bottom water dissolved oxygen (BWDO) can either control TOC or be controlled by TOC.

Müller and Suess (1979) determined a relationship between TOC and sedimentation rate and primary production. There is some degree of uncertainty whether their relation is valid for the Arabian Sea since they had only 4 of 26 samples from depths of less than 2000 m, and much of the area of interest is in the shallower depths. This, along with the lack of both good sedimentation rate data and reliable measurements of the primary productivity in the Arabian Sea, make it impossible to test Muller and Suess's relation.

Since the productivity appears to be about equal off both Arabia and India, and since there is little data regarding sedimentation rates, that leaves us with BWDO to be considered. Figure 9 is a plot of BWDO versus TOC for the Arabian Sea. The data points that fall far above the curve are all from relatively deep water and their high TOCs are probably caused by turbidity currents, while those below the curve are from shallow depths and are most likely the result of scouring, given their sandy texture. Other authors (Correns, 1939; Richards and Redfield, 1954; Calvert, 1964; Gross, 1967; Reimers, 1981) have noted relations between TOC and BWDO for different locations of the world. The relations for the Oregon coast (Gross, 1967) and Peru coast (Reimers, 1981) are very different from that for the Arabian Sea.

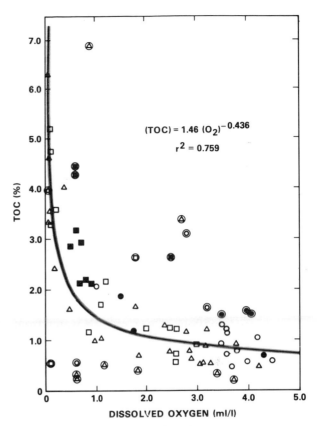

Figure 9. Total organic carbon (TOC) in the surface sediments versus the concentration of dissolved oxygen in the water overlying the bottom. Data from Wiseman and Bennett (1940) and Stackelberg (1972). The calculation of the regression curve did not utilize the circled points. Circled points above the curve are from the lower continental slope and are probably enriched by turbidity currents. The circled points below the line are from above the shelf break and may be influenced by erosion.
Key to symbols: open circle, open ocean; closed circle, equatorial ocean; open square, Gulf of Oman; closed square, Gulf of Aden; triangle, Indian coast.

However, the data presented by Richards and Redfield (1954), for the Gulf of Mexico fits the Arabian Sea curve very well. This may be due to the fact that in the Arabian Sea and Gulf of Mexico the oxygen distributions are of non-local origin (Richard and Redfield, 1954), whereas the oxygen distributions off Peru and Oregon are probably of local origin.

The differences in TOC between Arabia and India, seen in Figure 8, can thus be explained by the higher dissolved oxygen concentrations off the Arabian peninsula (Fig. 4, 5, and 6).

DISCUSSION AND CONCLUSIONS

The Arabian Sea upwelling, different from other areas of the world (Fig. 1), is strongly influenced by the monsoonal circulation. Global climate model results indicate that the strength of the monsoons, and therefore strength of upwelling, has fluctuated over the past 20,000 years. The monsoons were less intense than at present during the glacial maximum (about 18,000 years before present), and more intense during the transition from glacial to interglacial (about 9000 years before present; Prell and Streeter, 1982). This change in upwelling will affect the productivity of the Arabian Sea, and possibly the distribution of dissolved oxygen and the preservation of organic carbon in the sediments.

The distribution of dissolved oxygen is controlled by five factors: the high productivity of the coastal and even open ocean areas; strong stratification of the water column in the upper 200 m; the seasonal reversal of the intermediate depth currrents, and lack of transequatorial water transport; the inflow of organic-rich water from the Red Sea and Persian Gulf; and the low oxygen content of the abyssal waters.

The preservation of organic carbon is influenced by primary productivity and rate of sedimentation, but primarily controlled by the concentration of dissolved oxygen in the waters overlying the sediments. While the relation between TOC and bottom water dissolved oxygen (BWDO) (Fig. 9) may indicate either a control of TOC by BWDO or control of BWDO by TOC, we believe the former to be the case. This is because the overall dissolved oxygen distribution is controlled by remote, not local, factors.

ACKNOWLEDGEMENTS

We thank the Exxon Production Research Company for permission to publish this work. Colin Summerhayes provided the initial stimulus to investigate the Arabian Sea. E-Chien Foo helped us to understand the dynamics of the wind and ocean current patterns for the Indian Ocean.

REFERENCES

Bhargava, R.M.S., P.M.A. Bhattathiri, V.P. Devassy, and K. Radhakrishna, 1978. Productivity studies in the southeastern Arabian Sea. Indian Journal of Marine Science, 7:267-270.

Bruce, J.G., 1974. Some details of upwelling off the Somali and Arabian coasts. J. Mar. Res., 32:419-423.

Calvert, S.E. 1964. Factors affecting distribution of laminated diatomaceous sediments in Gulf of California. In T.H. van Andel and G.G. Shor, Jr., (eds), Marine Geology of the Gulf of California, American Association of Petroleum Geologists Memoir, 3, Tulsa, Oklahoma, pp. 311-330.

Correns, C. W., 1939. Pelagic sediments of the North Atlantic Ocean. In P.D. Trask (ed.) Recent Marine Sediments, A Symposium. American Association of Petroleum Geologists, Tulsa, Oklahoma, pp. 373-395.

Deuser, W.G., E.H. Ross, and Z.J. Mlodzinska, 1978. Evidence for the rate of denitrification in the Arabian Sea. Deep-Sea Research, 25:431-445.

Food and Agriculture Organization Department of Fisheries, 1972. Atlas of the Living Resources of the Sea. Food and Agriculture Organization of the United Nations, Rome.

Gangadhara Rao, L.V., T. Cherian, K.K. Varma, and V.V.R. Varadachari, 1974. Hydrographical features of the inner shelf waters along the central west coast of India during winter, spring, and summer. Mahasagar: 7, 15-26.

Gross, M.G., 1967. Organic carbon in surface sediment from the Northeast Pacific Ocean. International Journal of Oceanology and Limnology, 1:46-54.

Hastenrath, S. and P. J. Lamb, 1980. On the heat budget of hydrosphere and atmosphere in the Indian Ocean. Journal of Physical Oceanography, 10:694-708.

Hinten B.B. and D.P. Wylie, 1982. The wind stress patterns over the Indian Ocean During the summer Monsoon of 1979. Journal of Physical Oceanography, 12: 186-199.

Johnson, D.A. and J.E. Damuth, 1979. Deep thermohaline flow and current-controlled sedimentation in the Amirante Passage: Western Indian Ocean. Marine Geology, 33:1-44.

Kolla, V., L. Sullivan, S.S. Streeter and M.G. Langseth, 1976. Spreading of antarctic bottom water and its effects on the floor of the Indian Ocean inferred from bottom-water potential temperature, turbidity, and sea-floor photography. Marine Geology, 21:171-189.

Kolla, V., P.K. Ray, and J.A. Kostecki, 1981. Surficial sediments of the Arabian Sea. Marine Geology, 41:183-204.

Kuz'menko, L.V., 1973. Primary production of the northern Arabian Sea. Oceanology, 13:251-256.

Muller, P.J. and E. Suess, 1979. Productivity, sedimentation rate, and sedimentary organic matter in the oceans—I. organic carbon preservation. Deep-Sea Research, 26:1347-1362.

Pickard, G.L., 1963. Descriptive Physical Oceanography. Pergamon Press, Oxford, 200 pp.

Prell, W.L. and H.F. Streeter, 1981: Temporal and spatial patterns of monsoonal upwelling along Arabia: A modern analogue for the interpretation of Quaternary SST anomalies. J. Mar. Res. 40:143-155.

Radhakrishna, K., V.P. Devassy, R.M.S. Bhargava, and P.M.A. Bhattathiri, 1978. Primary production in the northern Arabian Sea. Indian Journal of Marine Science, 7:271-275.

Reimers, C.E., 1981. Sedimentary organic matter: distribution and alteration processes in the coastal upwelling region off Peru. Ph. D. dissertation, Oregon State University, 219 pp.

Richards, F.A. and A.C. Redfield, 1954. A correlation between the oxygen content of sea water and the organic content of marine sediments. Deep-Sea Research, 1:279-281.

Ryther, J.H. and D.W. Menzel, 1965. On the production, composition, and distribution of organic matter in the western Arabian Sea. Deep Sea-Research, 12: 199-209.

Ryther, J.H., J.R. Hall, A.K. Pease, A. Bakun, and M.M. Jones, 1966. Primary organic production in relation to the chemistry and hydrography of the western Indian Ocean. Limnology and Oceanography, 11:371-380.

Sen Gupta, R., S.P. Fondekar, V. N. Sankaranarayanan, and S. N. De Sousa, 1975. Chemical oceanography of the Arabian Sea: Part I—Hydrochemical and hydrographical features of the northern basin. Indian Journal of Marine Science, 4:136-140.

Sen Gupta, R., M.D. Rajagopal, and S. Z. Qasim, 1976. Relationship between dissolved oxygen and nutrients in the north-western Indian Ocean. Indian Journal of Marine Science 5:201-211.

Sen Gupta, R., V.N. Sankaranarayanan, S.N. De Sousa, and S.P. Fondekar, 1976. Chemical oceanography of the Arabian Sea. III Studies on the nutrient fraction and stoichiometric relationships in the northern and eastern basins. Indian Journal of Marine Science, 5:58-71.

Sharma, G.S., 1976. Transequatorial movement of water masses in the Indian Ocean. J. Mar. Res., 34:143-154.

Sharma, G.S., 1978. Upwelling off the southwest coast of India. Indian

Journal of Marine Science, 7:209-218.

Smith, Robert L., 1968. Upwelling. *In* H. Barnes (ed.), *Oceanography and Marine Biology Annual Review,* George Allen and Unwin Ltd., London, pp. 11-46.

Stackelberg, U.V., 1972. Faziesverteilung in Sedimenten des indisch-pakistanischen Kontinentalrandes. "Meteor" Forschung-sergebnisse, C, 9:1-73.

Summerhayes, C.P., 1981. Sedimentation of organic matter in upwelling regimes. *In* E. Suess and J. Thiede (eds), *Coastal Upwelling: Its Sediment Record.* Advanced Research Institute, NAT.

Szekielda, K-H., 1970. The liberated energy potential available from oxidation processes in the Arabian Sea. Deep-Sea Research, 17:641-646.

Udintsev, G.B., 1975: *Geological-Geophysical Atlas of the Indian Ocean.* UNESCO, 151 pp.

Warren, B.A., 1974. Deep flow in the Madagascar and Mascerene basins. Deep-Sea Research, 21:1-21.

Warren, B.A., 1978. Bottom water transport through the Southwest Indian Ridge. Deep-Sea Research, 25:315-321.

Warren, B.A., H. Stommel, and J.C. Swallow, 1966. Water masses and patterns of flow in the Somali Basin during the southwest monsoon of 1964. Deep-Sea Research, 13:825-860.

Wiseman, J.D.H. and H. Bennett, 1940. The distribution of organic carbon and nitrogen in sediments from the Arabian Sea. John Murray Expedition 1933-1934, Scientific Reports, 3:193-221.

Wunsch, C., 1977. Response of an equatorial ocean to a periodic monsoon. Journal of Physical Oceanography, 7:497-511.

Wyrtki, K., 1971. *Oceanographic Atlas of the International Indian Ocean Expedition.* National Science Foundation, Washington, D.C., 531 pp.

Zobel, B., 1973. Biostratigraphische Untersuchungen an Sedimenten des indischpakistanischen Kontinentalrandes (Arabisches Meer). "Meteor" Forschungsergebnisse, C, 12:9-73.

G. MARINE NUTRIENT CYCLE

19

The Ecological and Biogeochemical Roles of the Bacterioplankton in Coastal Marine Ecosystems

F. AZAM
Scripps Institution of Oceanography,
La Jolla, California

ABSTRACT

The recent application of new techniques has shown that the bacterioplankton play a much more active role in coastal marine ecosystems than previously believed. Fluorescence microscopy shows that bacteria are a significant fraction of the total living biomass, comparable to the zooplankton. Several different methods indicate that natural bacterial assemblages double once a day, similar to the phytoplankton growth rate. Bacterial abundance and growth are strongly correlated with phytoplankton biomass, suggesting a tight coupling between the production of organic matter by algae and its utilization by bacteria. The bacterioplankton are likely to be a major route for energy and material fluxes in marine foodwebs, consuming one-third or more of the primary production. Recent evidence suggests that heterotrophic flagellates consume a significant fraction of the bacterial production, though many other organisms also eat bacteria. This predation probably aids in nutrient recycling by releasing the nitrogen and phosphorus sequestered by the bacteria. Since bacteria account for about eighty percent of the living surface in seawater, they are also likely to be important in the absorption and transformation of dissolved pollutants. Therefore, it is clear that any study of coastal marine ecosystems should consider the role of the bacterioplankton.

INTRODUCTION

In recent years, there has been a fundamental change in our view of the role of bacteria in marine ecosystems. Historically, the models of marine foodwebs (e.g. Steele, 1974) have been based on the notion that primary production is consumed exclusively by the herbivores, which in turn are eaten by the carnivores. It now appears that a large fraction of the primary production is, in fact, consumed by bacteria (Hagström et al., 1979; Fuhrman and Azam, 1980, 1982), and that it supports a dynamic microheterotrophic foodweb. These new findings and the methods which have led up to them are still being scrutinized. However, there is an atmosphere of general acceptance among the biological oceanographers that the existing conceptual models of the foodweb are in need of revision to accommodate the bacteria as significant secondary producers and agents of organic matter transformation.

In this paper, I will describe some recent observations on the patterns of distribution and activity of bacteria in the sea, and remark on the new methods employed in these studies. I will then address the following questions: (1) What is the magnitude of bacterial biomass and production in relation to other components of the foodweb? Does bacterial production and its consumption by animals constitute a major pathway for material and energy transfer in the foodweb? (2) How important are bacteria in converting organic to inorganic matter and recycling nutrients to primary producers? How does this process occur? This discussion will make clear why it is important to incorporate bacterial studies into the research goals of oceanography in Pakistan. I am not aware of any pertinent studies in the coastal waters off Pakistan, but generalizations arrived at in this paper are expected to apply to the Pakistani waters as well.

BACTERIAL ABUNDANCE AND BIOMASS

A basic goal of marine biology is to describe the distributional patterns of organisms, and to find out what causes these patterns. This knowledge is needed for developing a model of the organization of the foodweb, and to answer questions about how the foodweb functions, and what controls the productivity of the sea. If bacteria are indeed a significant part of the marine biota then it is necessary to consider them in such an ecosystem context.

In the past, bacterial abundance in seawater had been greatly underestimated, because most bacteria were too small to be seen by phase-contrast microscopy, the technique used. Most bacteria as they occur in the sea are only 0.2-0.6 μm in diameter (Fuhrman, 1981), much smaller than the laboratory cultures of terrestrial and other bacteria with which microbiologists were familiar. Culture techniques used to determine the abundance of viable bacteria yielded even lower estimates (10^3 ml^{-1} : compared with 10^4-10^5 ml^{-1} by phase-contrast microscopy). As a result, it was thought that the bacterial biomass was only a trivial fraction of the total biomass in seawater and that most bacteria were probably dormant. It was, therefore, felt that bacteria could not play a significant role in marine ecosystems.

The application of fluorescent stains to the enumeration of bacteria in seawater (Francisco et al., 1973; Hobbie et al., 1977) has led to reliable, and much higher, estimates (approximately 10^6 cells ml^{-1}). The method of Hobbie et al., (1977) (acridine orange direct counts, AODC) has been generally accepted as standard. The cells are stained with acridine orange (which fluoresces when bound to DNA and RNA) and counted in an epifluorescence microscope. The method is simple, straight-forward and precise. Some other fluorescent stains, for example DAPI (Porter and Feig, 1980) and Hoechst 33285 (Paul, 1982), may be preferable if particle-bound bacteria are to be counted.

The size of bacteria may vary considerably within and between samples, therefore bacterial numbers are inadequate for comparing standing stocks. The cell volume must also be known. This tedious task can be performed either by epifluorescence microscopy (sizing directly or from photomicrographs) or by scanning electron microscopy. Because of the difficulty of the procedure, very few biomass estimates have been published so far; most authors report only bacterial abundances. In the next few years, automated cell counting and sizing with computer-assisted image analyzers should become possible

(Hagström and Larsson, in press; Sieburth, personal communication).

BACTERIAL SECONDARY PRODUCTION

There have also been rapid developments in measuring bacterial secondary production. At least six independent methods have been proposed since 1977 (for a review see Azam and Fuhrman, in press). The most promising methods are the frequency of dividing cells (FDC) (Hagström et al., 1979), and tritiated thymidine incorporation (TTI) (Fuhrman and Azam, 1980, 1982). The FDC method is conceptually elegant. The percentage of cells in the division phase is uniquely related to an assemblage growth rate. Microscopic examination of the sample is the only manipulation required to determine growth rate, and this can be done days to weeks after sample preservation in the field. However, the FDC method is difficult to calibrate and is somewhat tedious in requiring extensive microscopy. The TTI method is based on measuring the rate of bacterial DNA synthesis in natural samples and, with the help of a set of assumptions, converting it into the rate of bacterial production (Fuhrman and Azam, 1982). The method is rapid, convenient, and highly sensitive but requires the use of radioactively labeled (tritiated) thymidine as the metabolic probe for measuring the rate of DNA synthesis. Moreover, TTI yields minimum estimates because of the conservative nature of the assumptions made.

The above methods have permitted, and prompted, quantitative studies of bacterioplankton distribution and production. Although published data are still scant, some generalizations have already emerged. Bacterial abundance in the coastal euphotic zone is on the order of 1×10^6 ml^{-1}, which roughly translates into 0.1 mg wet weight (or 0.01 mg cell carbon) per liter of seawater. There is a decrease in bacterial abundance and size as one goes from coastal to off-shore areas.

Since the methods for measuring bacterial production are so recent, only a few reliable estimates have been made so far. These suggest that the bacterial production rate in coastal waters varies over a broad range, between 2-250 ug C l^{-1} d^{-1}. Table 1, taken from a review by Ducklow (in press), summarizes the recent measurements of bacterial biomass and production at different sites and by a variety of techniques. It should be pointed out that not all methods are reliable, and this may account for some variability. Despite considerable

TABLE 1. Estimates of bacterial abundance, biomass, production and specific growth rates for various ocean regions, classified by analytical techniques used to measure production. Bacterial production is also expressed as a percentage of primary production by phytoplankton, from Ducklow (in press).

| Region | Method | PRODUCTION RATE | | | BIOMASS | | References |
		μg C l^{-1}d^{-1}	μ(d^{-1})	% of Primary Production	μg C l^{-1}	10^9 cells l^{-1}	
Coastal waters and Shelves							
English Channel	[14]C flux from algae to bacteria	15-60	ND[1]	2-33	ND	ND	Derenbach and Williams, 1974
Baltic Sea	FDC[2]	4-9[3]	0.5	18-24	6-18	ND	Larsson and Hagström, 1982
S. California Coast	change in cell no. in culture	10-35	0.4-1.3	9-17	6-9	1-1.5	Fuhrman and Azam, 1980
S. California Coast	[3H]thymidine incorporation	0.7-50	0.01-2.0	9-17	8.13	1.0-1.5	Fuhrman and Azam, 1980
Georgia Coast	FDC, [3H] thymidine incorporation	40-130	0.9-1.0	5-25	45-150	3-10	Newell and Fallon, 1982
New York Bight	change in cell no. in culture	24[4]	1.4	ND	12[5]	1.2	Kirchman et al., 1982
New York Bight	[3H] thymidine incorporation	16	1.2	25	13[5]	1.3	Ducklow and Kirchman, 1983
Kiel Bight	change in cell no. in diffusion culture	10-500	0.3-1.0	15-30	40-500	0.8-5.7	Meyer-Reil, 1977
Open Ocean							
N. Atlantic-Azores	change in ATP in diffusion culture	122	4	ND	30	ND	Sieburth et al., 1977
N. Atlantic	monosaccharide disappearance in open water	30-50	2.6-20	ND	2-15	ND	Burney et al., 1979
Caribbean[6]	[3H] adenine incorporation	2-6	4-11	ND	0.5	ND	Karl, 1979
Gulf Stream Warm Core Ring	[3H] thymidine incorporation	0.5- 3[7]	0.2-0.5	6-25	2-6[7]	0.3-1	Ducklow, unpublished
Peruvian upwelling area	[14]C dark uptake	4-30	0.3-0.4	3-75	15-85	0.8-4	Sorokin, 1978
S. California Bight[8]	[3H] thymidine incorporation	0.6	0.2	10-25	3[7]	0.5	Fuhrman et al., 1980
Caribbean	diel carbohydrate change in open water	15-40	5-15	ND	1-8	0.2-1.4	Burney et al., 1982

[1]No data.

[2]Frequency of dividing cells.

[3]Yearly average.

[4]Estimated from exponential growth in a 1:10 dilution in culture; abundance and biomass values are for open water.

[5]Estimated from cell biomass of 10^{-14} g C cell^{-1}

[6]450 m deep.

[7]Estimated from cell biomass of 6×10^{-15} g C cell^{-1} and the factor 2.1×10^5 cells produced per nmol [3H] thymidine incorporated (Fuhrman and Azam, 1982).

[8]Stations more than 10 km offshore only.

variability, most measurements fall around a growth rate of one doubling per day, comparable to phytoplankton growth rates. However, some studies yield much faster growth rates. For example, Sieburth et al. (1977) claim that bacteria in the oligotrophic Caribbean Sea double 5-15 times per day. These and other findings have brought into question the validity of primary production measurements. To satisfy the energy requirement for such high bacterial production the primary production rate would have to be about 10 times greater than presently believed. The estimates of bacterial production by other methods, such as TTI, require 10-50% of the presently measured primary production, but these may be underestimates (Fuhrman and Azam, 1980).

Carbon Conversion Efficiency of Marine Bacteria

A central issue concerning the role of bacteria in marine ecosystem is: what fraction of organic matter used by bacteria is assimilated and what fraction is mineralized? In the past, it had been assumed that bacteria in the sea grew inefficiently, respiring away much of the carbon in their food. They were, therefore, believed to be mineralizers. It now appears that a wide range of conversion efficiencies may obtain under different nutritional conditions. Pure culture studies give values in the range 40-80% (Townsend and Calow, 1981; Payne and Wiebe, 1978), while measurements for natural bacterial assemblages are technically difficult. Williams (1973) found 71% and 77% efficiency of ^{14}C assimilation for seawater samples incubated with tracer levels of ^{14}C-labeled glucose and amino acids, respectively. These values may be overestimates if some of the label was excreted during incubation (Itturiaga and Zsolnay, 1981). Newell and co-workers (Newell et al., 1981; Stuart et al., 1982) report much lower efficiencies (6 to 33%) for carbon conversion of a variety of plant debris, but high values were obtained in nitrogen-rich conditions (Newell, personal communication, quoted in Azam et al., 1983).

It is clear that no single figure for conversion efficiency is likely to be valid for all nutritional conditions. For bacteria growing in the coastal euphotic zone, a carbon conversion efficiency of 50% may be reasonable in view of the discussion above while recognizing that this value may vary somewhat (Fuhrman and Azam, 1980, 1982; Ducklow, in press). This permits one to convert

bacterial production rate (in terms of biomass carbon) into a rate of utilization of organic matter, or flux of carbon through the bacterioplankton. If the rate of primary production is known (and assuming steady-state) one can compute what fraction of primary production would be channeled through bacterioplankton. This approach has provided the most direct argument, so far, that bacterioplankton is a major route for material and energy flux in pelagic marine foodweb, accounting for one-third to one-half of the primary production (Hagström et al., 1979; Fuhrman and Azam, 1980, 1982; Table 1).

The Fate of Bacterial Secondary Production

The above discussion has shown that bacterial biomass is comparable to the zooplankton biomass, and that bacteria double roughly once each day in the coastal euphotic zone. The bacterial abundance does not change much from day to day; in the coastal surface waters it is generally between 0.5 to 2.0 million bacteria per ml. This means that bacteria do not accumulate in the seawater to a significant extent. The production and removal of bacteria must therefore be tightly coupled processes. These observations raise the question: What are the routes and mechanisms of removal of bacterial production?

There are several possible routes which bacterial secondary production may follow; each has implications for the flow of matter and energy in marine foodweb. Heterotrophic flagellates and zooplankton may prey on bacteria; some predaceous bacteria may prey on other bacteria; bacteria may lyse spontaneously due to nutrient stress or other metabolic stresses. There is very little information at present on the last two processes, but considerable research has been done in the last few years on the role of zooplankton as predators of bacteria.

A variety of marine zooplankton have been shown to prey on bacteria. These include heterotrophic flagellates (Haas and Webb, 1979; Fenchel, 1982a, b, c; Kirk and Azam, unpublished), ciliates (Hollibaugh et al., 1980; Fenchel, 1980), larvaceans (Hollibaugh et al., 1980; King et al., 1980), and salps (Harbison and McAlister, 1979; Mullin, 1983). While some organisms obtained significant parts of their food from bacteria, most removed only a small fraction of the measured bacterial productivity (King et al., 1980). Heterotrophic flagellates appear to be the main consumers of the bacterial secondary production. Fenchel (1982b, c) has shown that microflagellates in the size range of 3-10 μm are effective bacteriovores in the sea, capable of filtering 12-67% of the water column per day. These are mainly choanoflagellates and colorless chrysomonads which are common in seawater and reach densities on the order of 10^3 ml^{-1} (Sieburth, 1979; Fenchel, 1982b, c). Kirk and Azam (unpublished), measured bacterial production in parallel with predation on natural assemblages of bacteria by natural assemblages of flagellates.

They found that about 25% of the bacterial productivity could be accounted for by flagellate grazing. It seems probable that many other zooplankton also obtain part of their nutrition from bacteria (King et al., 1980). It appears, therefore, that there is a strong predation pressure on bacteria.

FACTORS AFFECTING BACTERIAL GROWTH

It is now possible to map regions of the sea with respect to the distributional patterns of bacteri-

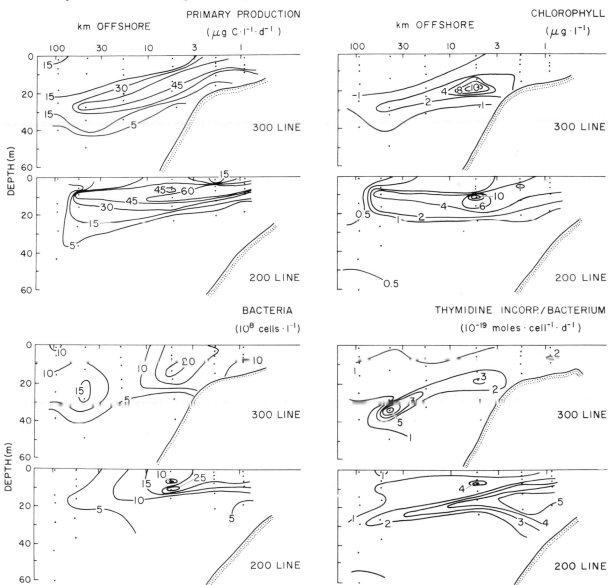

Figure 1. Onshore-offshore vertical sections of primary productivity, chlorophyll, bacteria, and thymidine incorporation per bacterium. The 200 and 300 lines are Southern California Bight Stations 201 to 206 and 301 to 306, respectively; each dot represents one water sample. The relationship between thymidine incorporation per cell and specific growth rate is that 5×10^{-19} mol cell^{-1} d^{-1} is roughly equivalent to one doubling per day, from Fuhrman and Azam (1982).

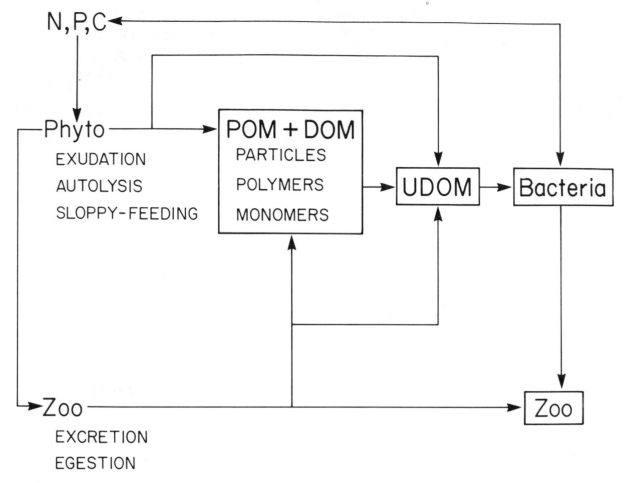

Figure 2. Sources and pathways of directly utilizable dissolved organic matter (UDOM) in seawater. DOM=dissolved organic matter; POM=particulate organic matter; sloppy-feeding=spillage of algal cell contents during feeding by zooplankton; phyto=phytoplankton; zoo=zooplankton; N, P, C=inorganic nitrogen, phosphorus, and carbon, respectively, from Azam and Ammerman (in press).

oplankton abundance, biomass and production. Such maps can be compared with distributional patterns of phytoplankton, nutrients, and other environmental factors, to try to identify factors affecting bacteria. Results of detailed studies in the Southern California Bight (Fuhrman et al., 1980; Fuhrman and Azam, 1982; Fig. 1) show that bacterial abundance and growth rates are strongly correlated with chlorophyll concentration (an indicator of phytoplankton abundance) and follow sharp onshore (high) offshore (low) gradients as does chlorophyll. Further evidence of the relationship between algae and the bacterial growth is that the specific growth rate of heterotrophic bacteria in seawater follows a diel pattern increasing during the daylight and decreasing at night (Sieburth et al., 1977; Azam and Fuhrman, in press; Fuhrman et al., unpublished). It appears that there is a tight coupling between the production of organic matter by algae and its utilization by bacteria.

Sources and Modes of Production of Bacterial Nutrients

The discovery that a significant fraction of daily primary production is utilized by the heterotrophic bacterioplankton is in apparent contradiction with the conventional wisdom that the dissolved organic matter (DOM) is too dilute to support rapid bacterial growth (Stevenson, 1978). Concentration of most components of DOM are extremely low, in the range 10^{-12} M to 10^{-8} M (Mopper and Lindroth, 1982; Ammerman and Azam, 1981). Figure 2 illustrates the main mechanisms of DOM production in seawater. Phytoplankton cells are the major source of bacterial nutrients into the environment, via exudation (Larsson and Hagström, 1982; Williams, 1981), autolysis, or loss of cellular pools during predation by herbivores (Lampert, 1978; Copping and Lorenzen, 1980). Organic detritus undergoing hydrolysis by colonized bacteria is also a source of DOM for other bacteria.

Azam and Ammerman (in press) argue that since DOM is produced by discrete particulate sources (Fig. 2) there would be high nutrient concentrations near a source. Sustained sources of DOM (exuding phytoplankton, decomposing detritus, etc.) would create persistent nutrient gradients in their vicinity. Bacteria, by chemotaxis and motility, may respond to the presence of the microenvironmental nutrient gradients and position themselves in space and time to maximize nutrient acquisition. The result, they hypothesize, is a clustering of bacteria around the sustained sources of DOM (cluster hypothesis; Azam and Ammerman, in press).

The cluster hypothesis is a useful conceptual framework for organizing the recent findings concerning the role of bacteria in the pelagic marine foodweb and in the cycling of organic matter. The bacteria-algae association functions as a coordinated metabolic unit wherein algae provide the energy and carbon for the heterotrophic bacteria, and bacterial mineralization of nitrogen and phosphorus containing DOM components helps maintain high concentrations of plant nutrients in the microenvironment. The result is a rapid growth of both the algae and bacteria. This scenario has important implications for models of microbial foodweb.

THE ROLE OF BACTERIA IN NITROGEN AND PHOSPHORUS MINERALIZATION

Since bacteria are so small, they have very large surface: volume ratios. It has been calculated that in coastal surface waters about 80% of "living surface" is associated with bacteria (Azam et al., in press). This means, from first principles, that bacteria would be very effective in uptake of soluble substances, both organic nutrients and inorganic minerals.

Historically, it has been thought that bacteria mineralize N and P for plant growth. Although bacteria play a central role in this process, it is becoming increasingly clear that bacteria alone are not sufficient. Bacteria have lower C/N and C/P ratios than their food, therefore, they in fact sequester N and P. They are also effective

competitors with algae for the uptake of inorganic N and P (Harrison et al., 1977). Therefore, the release of mineral nutrients accumulated by bacteria occurs when animals eat and metabolize bacteria.

The above argument applies only to the N and P taken up by bacteria. It may still be true that the extracellular phosphatases of bacteria produce more mineral P than bacteria need, thus they may be net mineralizers (Ammerman and Azam, unpublished). Moreover, the nitrogen conversion efficiency may not be high under all nutritional conditions, thus bacteria at times may liberate mineralized ammonia into the environment. It is important, therefore, to make a distinction between the mineralization of sequestered N, P and the adventitious mineralization especially of P.

THE MICROBIAL FOODWEB

It is now clear that most of the biomass and metabolism in the seawater is due to bacteria, phytoplankton and microzooplankton. The metabolic and trophic interactions of these organisms can now be organized into models of a microbial foodweb ("Microbial loop"; Azam et al., 1983). Figure 3, taken from a paper by Azam et al. (1983), is used here to illustrate the interactions within the microbial foodweb. The main feature of the model (based on Sheldon et al., 1972) is that organisms effectively utilize particles one order of magnitude smaller than their own size. DOM, mainly of phytoplankton origin, is utilized by heterotrophic bacteria, which supports their growth. Heterotrophic microflagellates keep bacterial populations within narrow limits (0.5-2 x 10^6 ml^{-1}) by predation. Cyanobacteria, which are in the same size range as bacteria, are also eaten by the microflagellates. Flagellates are preyed upon by ciliates and other microzooplankton which are in the same size range as the large phytoplankton (10 to 80 μm). The organic matter derived from phytoplankton is thus channeled into the herbivore food chain by the microbial foodweb, but very inefficiently (since it involves multiple trophic transfer *en route*). The microbial foodweb would, therefore, be important for mineralization of organic matter.

Several types of ecological relationships are involved in the functioning of the microbial foodweb. Azam et al. (1983) have suggested that these are: commensalism, competition and preda-

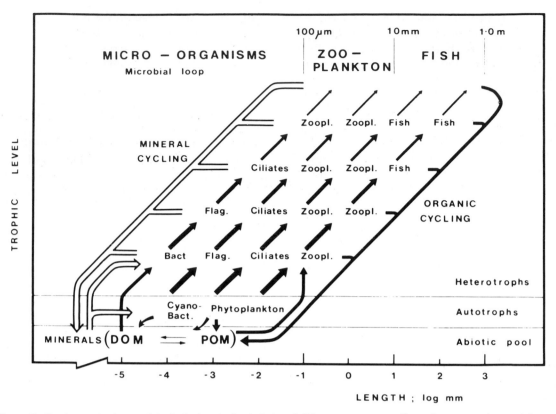

Figure 3. Semi-quantitative model of planktonic food chains. Solid arrows represent flow of energy and materials; open arrows, flow of materials alone. It is assumed that 25% of the net primary production is channelled through DOM and the "microbial loop", bacteria (Bact.), flagellates (Flag.) and other microzooplankton (e.g. ciliates). It is further assumed that the most efficient predator: prey size ratio is 10:1, hence the slope of the lines relating trophic status to log body length is 1:1. The food chain base represents a size range 3 orders of magnitude (smallest bacteria 0.2 μm, largest diatoms 200 μm); therefore, a trophic level will have a size-range factor of 10^3 and conversely each size class of organisms (100 μm) will represent at least 3 trophic levels. The thickness of open arrows (left) represents the approximate relative magnitude of minerals released in excretion at each trophic level; corresponding organic losses (faeces, mucus, etc.) are shown on the right hand side, from Azam et al. (1983).

tion. Commensalism may be important in algae-bacteria interaction (discussed under "cluster hypothesis"). However, there is competition between algae and bacteria for mineral nutrients. This competition is influenced by predation by microflagellates on bacteria. Predation releases inorganic nutrients which constitute a feedback into the nutrient pools for algae and bacteria.

The above discussion reinforces the notion that components of the microbial foodweb are tightly linked in a metabolic and trophodynamic sense. In order to understand the regulation of material and energy fluxes through the microbial foodweb, it is necessary to know the nature of the couplings in greater detail.

ROLE OF BACTERIA IN POLLUTED MARINE ECOSYSTEMS

The methodological and conceptual advances

in studying the role of bacteria in marine ecosystems have also created renewed interest in the role of bacteria in pollution ecology. The existence of a large bacterial biomass and rapid rates of bacterial production would suggest that bacteria could be a significant link in pollutant transfer to bacteriovores. The efficiency of such transfer will be a function of the interactions within the microbial foodweb (above). It is therefore important to study these interactions.

As stated above, bacteria represent about 80% of the biosurface in seawater (Azam et al., in press). It is likely therefore that bacteria would dominate the biotic adsorption and absorption of dissolved pollutants. It is important to study these properties of the natural population of marine bacteria to quantify the role bacteria play in the ecology of coastal marine ecosystems.

CONCLUSIONS

It should be apparent from the arguments advanced here that bacterioplankton play significant and varied roles in the structure, functioning, and metabolism of coastal marine foodwebs. Bacteria almost exclusively utilize DOM, and this apparently represents a major link in material and energy transfer. They are significant secondary producers, and this supplies the particulate base for a microbial foodweb, which is important for mineral nutrient regeneration.

The ability of bacteria to accumulate dissolved substances from low concentrations in seawater may mean that they can also accumulate trace pollutants, organic and inorganic. This point needs attention since bacteria could introduce the accumulated pollutants into the microbial foodweb.

It is suggested that comprehensive research programs in the future should incorporate studies of bacterioplankton in an ecosystem context.

REFERENCES

Ammerman, J.W. and F. Azam, 1981. Dissolved Cyclic adenosine monophosphate (cAMP) in the sea and uptake of cAMP by marine bacteria. Mar. Ecol. Prog. Ser., 5: 85-89.

Azam, F., T. Fenchel, J. G. Field, J.S. Gray, L.-A. Meyer-Reil and F. Thingstad, 1983. The ecological role of water-column microbes in the sea. Mar. Ecol. Prog. Ser., 10: 257-263.

Azam, F. and J.W. Ammerman, 1984 (in press). Cycling of organic matter by bacterioplankton in pelagic marine ecosystems: Micro-environmental considerations. In: M.J. Fasham (ed.), Flows of energy and material in marine ecosystems: Theory and practice. NATO-ARI. May 1982.

Azam, F., J.W. Ammerman, J.A. Fuhrman and A. Hagstrom, 1984 (in press). Role of bacteria in polluted marine ecosystems. In: H. White (ed.), Proceedings of the workshop on meaningful measures of marine pollution effects, NOAA, April, 1982.

Azam, F. and J.A. Fuhrman, 1984 (in press). Measurement of growth of bacteria in the sea and the regulation of growth by environmental conditions. In: J. E. Hobbie and P. J. leB. Williams (eds.), Heterotrophic activity in the sea. Plenum Press.

Burney, C.M., P.G. Davis, K.M. Johnson and J. McN. Sieburth, 1982. Diel relationships of microbial trophic groups and in situ dissolved carbohydrate dynamics in the Caribbean Sea. Mar. Biol., 67: 311-322.

Burney, C. M., K. M. Johnson, D. M. Lavoie and J. McN. Sieburth, 1979. Dissolved carbohydrate and microbial ATP in the North Atlantic: Concentrations and interactions. Deep-Sea Res., 26A: 1267-1290.

Copping, A. E. and C. J. Lorenzen, 1980. Carbon budget of a marine phytoplankton-herbivore system with carbon-14 as a tracer. Limnol. Oceanogr., 25: 873-882.

Derenbach, J.B. and P.J. leB. Williams, 1974. Autotrophic and bacterial production: Fractionation of plankton populations by differential filtration of samples from the English Channel. Mar. Biol., 25: 263-269.

Ducklow, H. W., 1983 (in press). The production and fate of bacteria in the oceans. Bioscience.

Ducklow, H.D. and D. L. Kirchman, 1983. Bacterial dynamics and distribution during a spring diatom bloom in the Hudson River plume, USA. J. Plankton Res., 5: 333-355.

Fenchel, T., 1980. Suspension feeding in ciliated protozoa: Feeding rates and their ecological significance. Microbial Ecol., 6: 13-25.

Fenchel, T., 1982a. Ecology of heterotrophic microflagellates. I. Some important forms and their functional morphology. Mar. Ecol. Prog. Ser., 8: 211-223.

Fenchel, T., 1982b. Ecology of heterotrophic microflagellates. II. Bioenergetics and growth. Mar. Ecol. Prog. Ser., 8: 225-231.

Fenchel, T., 1982c. Ecology of heterotrophic microflagellates. IV. Quantitative occurrence and importance as bacterial consumers. Mar. Ecol. Prog. Ser., 9: 35-42.

Francisco, D.E., R. A. Mah and A. C. Rabin, 1973. Acridine orange epifluorescence technique for counting bacteria. Trans. Am. Micros. Soc., 92: 416-421.

Fuhrman, J. A., 1981. Influence of method on the apparent size distribution of bacterioplankton cells: epifluorescence microscopy compared to scanning electron microscopy. Mar. Ecol. Prog. Ser., 5: 103-106.

Fuhrman, J.A., J.W. Ammerman and F. Azam, 1980. Bacterioplankton in the coastal euphotic zone: distribution, activity, and possible relationships with phytoplankton. Mar. Biol., 60: 201-207.

Fuhrman, J.A. and F. Azam, 1980. Bacterioplankton secondary production estimates for coastal waters of British Columbia, Antarctica, and California, Appl. Environ. Microbiol., 39: 1085-1095.

Fuhrman, J.A. and F. Azam, 1982. Thymidine incorporation as a measure of heterotrophic bacterioplankton production in marine surface waters: Evaluation and field results. Mar. Biol., 66: 109-120.

Haas, L.W. and K.L. Webb, 1979. Nutritional mode of several non-pigmented microflagellates from the York River estuary, Virginia. J. exp. mar. Biol. Ecol., 39: 125-134.

Hagstrom, A. and U. Larsson, 1984 (in press). Diel and seasonal variation in growth rates of pelagic bacteria. In: J.E. Hobbie and P.J. leB. Williams (eds.), Heterotrophic activity in the sea. Plenum Press.

Hagstrom, A., U. Larsson, P. Horstedt, and S. Normark, 1979. Frequency of dividing cells, a new approach to the determination of bacterial growth rates in aquatic environments. Appl. Environ. Microbiol., 37: 805-812.

Harbison, G.R. and V.L. McAlister, 1979. The filter-feeding rates and particle retention efficiencies of three species of Cyclosalpa (Tunicata, Thaliacea). Limnol. Oceanogr., 24: 875-892.

Harrison, W.G., F. Azam, E. H. Renger and R. W. Eppley, 1977. Some experiments on phosphate assimilation by coastal marine phytoplankton. Mar. Biol. 40: 9-18.

Hobbie, J.E., R.J. Daley and S. Jasper, 1977. Use of Nuclepore filters for counting bacteria by fluorescence microscopy. Appl. Environ. Microbiol., 33: 1225-1228.

Hollibaugh, J.T., J.A. Fuhrman and F. Azam, 1980. Radioactively labeling of natural assemblages of bacterioplankton for use in trophic studies. Limnol. Oceanogr., 25: 172-181.

Itturiaga, R. and A. Zsolnay, 1981. Transformation of some dissolved organic components by a natural heterotrophic population. Mar. Biol., 62: 125-129.

Karl, D.M., 1979. Measurement of microbial activity and growth in the ocean by rates of stable ribonucleic acid synthesis. Appl. Environ. Microbiol., 38: 850-860.

King, K.R., J.T. Hollibaugh and F. Azam, 1980. Predator-prey interactions between the larvacean Oikopleura dioca and bacterioplankton in enclosed water columns. Mar. Biol., 56: 49-57.

Kirchman, D., H. Ducklow and R. Mitchell, 1982. Estimates of bacterial growth from changes in uptake rates and biomass. Appl. Environ. Microbiol., 44: 1296-1307.

Lampert, W., 1978. Release of dissolved organic carbon by grazing zooplankton. Limnol. Oceanogr., 23: 831-834.

Larsson, U. and A. Hagstrom, 1982. Fractionated phytoplankton primary production, exudate release, and bacterial production in a Baltic eutrophication gradient. Mar. Biol., 67: 57-70.

Meyer-Reil, L.-A., 1977. Bacterial growth rates and biomass production, p. 223-235. In: G. Rheinheimer (ed.), Microbial ecology of a brackish water environment. Springer-Verlag, Berlin.

Mopper, K. and P Lindroth, 1982. Diel and depth variations in dissolved free amino acids and ammonium in the Baltic Sea determined by shipboard HPLC analysis. Limnol. Oceanogr., 27: 336-347.

Mullin, M.M., 1983. *In situ* measurement of filtering rates of the salp, *Thalia democratia,* on phytoplankton and bacteria. J. Plankton Res., 5: 279-288.

Newell, R.C., M.I. Lucas and E.A.S. Linley, 1981. Rate of degradation and efficiency of conversion of phytoplankton debris by marine microorganisms. Mar. Ecol. Prog. Ser., 6: 123-136.

Newell, S.Y. and R.D. Fallon, 1982. Bacterial productivity in the water column and sediments of the Georgia (USA) coastal zone: Estimates via direct counting and parallel measurement of thymidine incorporation. Microbial Ecol., 8: 33-46.

Paul, J.H., 1982. Use of Hoechst dyes 33258 and 33342 for enumeration of attached and planktonic bacteria. Appl. Environ. Microbiol., 43: 939-944.

Payne, W. T. and W. J. Wiebe, 1978. Growth yield and efficiency in chemosynthetic microorganisms. Ann. Rev. Microbiol., 32: 115-183.

Porter, K.G. and Y. S. Feig, 1980. The use of DAPI for identifying and counting aquatic microflora. Limnol. Oceanogr., 25: 943-948.

Sheldon, R.W., A. Prakash and W.H. Sutcliffe, 1972. The size distribution of particles in the ocean. Limnol. Oceanogr., 27: 327-340.

Sieburth, J. McN., 1979. *Sea Microbes.* Oxford Univ. Press, New York. 491 p.

Sieburth, J. McN., K.M. Johnson, C.M. Burney and D.M. Lavoie, 1977. Estimation of *in situ* rates of heterotrophy using diurnal changes in organic matter and growth rates of picoplankton in diffusion culture. Helgolander Wiss. Meeresunters., 30: 565-574.

Sorokin, Y.I., 1978. Decomposition of organic matter and nutrient regeneration, p. 501-616. *In:* O. Kinne (ed.), *Marine Ecology,* Vol. 4. Wiley Interscience, Chichester.

Steele, J.H., 1974. *The structure of marine ecosystems.* Harvard Univ. Press, Cambridge, Mass., 128 p.

Stevenson, L.H., 1978. A case for bacterial dormancy in aquatic systems. Microbial Ecol., 4: 127-133.

Stuart, V., J.G. Field and R.C. Newell, 1982. Evidence for absorption of kelp detritus by the ribbed mussel *Aulacomya ater* using a new ^{51}Cr-labelled microsphere technique. Mar. Ecol. prog. Ser., 9: 263-271.

Townsend, C.R. and P. Calow (eds.), 1981. *Physiological ecology: an evolutionary approach to resource use.* Sinauer Associates, Inc., Sunderland, Mass., 393 p.

Williams, P.J. leB., 1973. On the question of growth yields of natural heterotrophic populations. *In:* T. Rosswall (ed.) *Modern Methods in the Study of Microbial Ecology.* Bull. Ecol. Res. Comm. (Stockholm) 197. Swedish Natural Science Research Council.

Williams, P.J. leB., 1981. Incorporation of microheterotrophic processes into the classical paradigm of the planktonic foodweb. 15th European Symposium on Marine Biology, Kiel F.R.G. Kieler Meeresforsch. Sonderh., 5: 1-28.

20

Phytoplankton Dynamics in the Marine Food-Web

O. HOLM-HANSEN

Scripps Institution of Oceanography,
La Jolla, California

ABSTRACT

Measurement of primary production is an important component of any ecological study involving biological resources in the marine environment. It is also important, however, to consider all other sources of reduced carbon that can serve as a food base for organisms. Primary production is a function of phytoplankton biomass times growth rates. This paper discusses factors which control plant biomass (grazing and losses to deep water), as well as those which control cellular growth rates (light, temperature, and nutrients). Our present concepts of the marine food-web are discussed, specifically in regard to the importance of the heterotrophic microbial populations (bacterioplankton and protozoans), which may metabolize up to 50% of the total primary production. Recent studies utilizing ^{15}N-labelled substrates point out the importance of these microbial populations in mineralization processes. Some examples are discussed to illustrate the practical importance of phytoplankton studies when combined with data from other disciplines. These include (a) application of remote sensing for locating and predicting behavior of economically important fish stocks, and (b) the importance of the characteristics of the phytoplankton crop (cell size, species and chemical composition) in regard to recruitment success of annual year classes of fish or invertebrate larvae.

INTRODUCTION

If one scans the world's phytoplankton literature of the past 20 years and then compares such work to present day studies in the marine environment, one discerns a fairly recent change in the questions being asked by biological oceanographers. For a long time oceanographers were concerned with describing phytoplankton distribution, species composition, and primary production with minimal amounts of ancillary data. Often such observations were limited to surface waters, so that no data were obtained regarding biological conditions throughout the euphotic zone (commonly defined as the depth at which ambient light intensity is 1% of that incident upon the sea surface). Although such studies still have much merit in regard to specific problems, much phytoplankton research has evolved whereby dynamics of phytoplankton populations are studied relative to physical mixing phenomena and to all chemical and biological factors which affect microbial cells. These studies, which may be termed "process-oriented," are concerned with increasing our understanding of the routes and fluxes of organic carbon (or energy) flow within the entire water column throughout the entire year.

Interdisciplinary group efforts concerned with process-oriented studies unfortunately require much time, equipment, and money, and hence are often difficult to undertake for small groups or laboratories. Individual researchers, or small groups, have to focus on a limited number of specific problems relating to the overall food-web in the marine environment. It is important, however, that the broad, dynamic objectives of phytoplankton research be recognized as they can help to define and to direct individual research programs. In this paper I would like to discuss briefly selected topics illustrating the interaction between physical, chemical, and biological factors in the oceans as they affect primary production, and also to convey some understanding and appreciation of the dynamics of the food-web involving microbial cells. Hopefully, these comments will be of interest to those students and researchers who (a) have basic interests in the functioning of phytoplankton in the sea, (b) are concerned with the marine carbon cycle in regard to either utilization of natural marine resources or to effects of pollutants on marine life, and (c) wish to increase food produciton by mariculture methods.

INPUTS OF ORGANIC CARBON

In open ocean studies of the marine food-web we commonly ignore all inputs other than that of primary production by phytoplankton. In coastal areas, however, one should consider all sources which can have an impact on the carbon budget. These other sources include (i) dissolved and detrital material originating from benthic algae, marsh grasses, mangroves, etc., (ii) terrestrial inputs (from both fresh water and atmospheric routes), which would include all natural organic materials in addition to anthropogenic substances, and (iii) products formed by chemoautotrophic organisms which reduce CO_2 with the energy obtained from oxidation of reduced substrates such as ammonia in the water column or sulphides in sediments. Recent studies on deep-sea thermal vents (Karl et al., 1980) have demonstrated that large and unique communities of organisms may subsist on a food chain based on chemosynthesis in the absence of solar radiation. In coastal areas some of these additional sources of organic carbon (particular mangrove and sea grass communities) are of prime importance in regard to support of invertebrate or fish resources (Kumari et al., 1978). Qasim et al. (1979) have discussed the importance of detrital organic compounds in the Laccadive Sea.

CONCEPT OF THE FOOD CHAIN

For many years the prevailing thoughts regarding the cycle of organic carbon in the sea were that (i) there was a fairly simple "pyramid" of producers, with a large base formed by the phytoplankton and progressively smaller populations of consumers, starting with the herbivorous zooplankton and culminating with the carnivorous vertebrates, and (ii) the marine food chain was more or less limited to the upper 200 m of the ocean, with very little biological activity occurring beneath that depth. Little attention was paid to the smaller cells, and the bacterial trophic level was thought of mostly in regard to mineralization processes. Some of the inadequacies of such a classical food chain were pointed out by Pomeroy (1974) and recently

discussed by Williams (1981). The phytoplankton trophic level was classically conceived of in terms of those cells which could be concentrated with a mesh with openings of 35 μm. Today our concepts of life in the sea are very different. It is now recognized that the picoplankton (less than 2.0 μm) often account for a significant proportion of total primary production. These small autotrophic cells consist mostly of procaryotic cyanobacteria (Johnson and Sieburth, 1979), as well as eucaryotic cells (Johnson and Sieburth, 1982). Nannoplankton (between 2.0 and 20 μm) are now known often to account for over 50% of the total phytoplankton biomass (Malone, 1980) and up to 90% of the primary production (Bröckel, 1981). The nannoplankton fraction includes many heterotrophic cells which apparently subsist largely on bacterial cells (Fenchel, 1982). One of the reasons for our relative lack of knowledge concerning the composition and functioning of the nannoplankton fraction has been the inadequacy of our analytical methods. The method most commonly used for such studies relies on settling techniques followed by inverted microscope analysis (Utermöhl, 1958). There are several limitations to this method, especially in regard to fragile nannoplankton cells that lack thick cell walls. Some of these limitations have been eliminated recently by development of a Filter-Transfer-Freeze technique (Hewes and Holm-Hansen, 1983) which allows all such cells to be transferred quantitatively onto an optically clear glass slide, which then permits a variety of observations utilizing transmission and epifluorescence microscopy. The microplankton (> 20 μm), which often contains less biomass than the smaller cells, are of great importance as food reserves for many invertebrate and vertebrate species of economic importance. The complexity in cellular organization, as well as in nutritional mode, of all these microbial assemblages has recently been described by Sieburth (1979). Thus, instead of a simple progression from large phytoplankton to zooplankton to fish, etc., in most waters there is a large number of complex routes describing the flow of organic carbon between organisms of varying size and nutritional mode. A relatively large percentage of primary production, for instance, may be assimilated by bacterial cells, which serve as the food base for many heterotrophic protozoan organisms (see Azam, this volume). A carbon atom may thus be "recycled" many times by microbial cells before it is liberated as CO_2 or incorporated into macroscopic organisms. The expression "unstructured food-web" conveys a more accurate description of the actual carbon flow than the older term "food chain" (Isaacs, 1973). The size distribution, as well as chemical composition, of the autotrophic cells is very important in regard to the nature of the higher trophic levels. As many grazing organisms show selectivity in food consumption and efficient grazing only above a threshold value in regard to particle size, survival or rate of growth of certain species is primarily dependent on the characteristics of the food base and less so on the total biomass of food particles.

Just as our concepts of the food-web in the euphotic zone have changed considerably in recent years, our views of the other components of the carbon cycle in the marine environment have also changed significantly. The entire water column must be viewed as a variable, changing, environment characterized by a heterogeneous assemblage of microbial and macroscopic organisms which are subsisting on and altering the concentration, composition, and form of organic matter as it descends toward the sediments. The sediment surface (or the "mixed layer," which may be 5-10 cm in depth) is an area of greatly enhanced biological activity that is important not only to the sediment-dwelling organisms but also to organisms living within the nepheloid layer where sedimented materials are maintained in suspension. Although such environments are usually far below the depth at which net photosynthesis may occur, chlorophyll-containing cells are found in small concentrations throughout the entire water column and in the sediments. We know relatively little about the viability and metabolic activity of these potentially autotrophic cells.

PRIMARY PRODUCTION

If one studies phytoplankton growth in a batch culture in the laboratory over a period of a few weeks, one can get good agreement of the rate of organic carbon production (primary production) as measured with radiocarbon assimilation with the concentrations of total particulate organic carbon (POC), cellular content of adenosine triphosphate (ATP), or chlorophyll. If you attempt to do the same measurements on a natural water sample taken from the euphotic zone (prefiltered through 200 μm mesh to remove macro-zooplankton), the correlation between primary production data and total POC, ATP, or chlorophyll-a in the sample after incubation is very poor. The chlorophyll content will often remain the same or actually decrease slightly for the first 1-2 days, in spite of measurable rates of light-activated radiocarbon assimilation. The reason for such "imbalance" between measured chemical entities is most likely related to the complexity of nutritional types in any natural water sample. Just as the phytoplankton concentration remains fairly uniform in the euphotic zone from day to day as the result of grazing zooplankton populations, the same processes occur in the water samples incubated for primary production measurement. The fact that one generally screens off the macroscopic zooplankton (over 200 μm) in primary production studies suggests that the grazing pressure exerted by heterotrophic microplankton is very significant as compared to that resulting from the net plankton.

Primary production (expressed in any biomass units such as POC, chlorophyll, ATP, etc.) will be a function of the autotrophic biomass times the cellular growth rates. It is generally assumed that phytoplankton are growing exponentially, but grazing pressure as mentioned above may invalidate this assumption. In natural waters the plant biomass is controlled to a large extent by grazing pressure and by losses to deep water through settling or physical mixing processes (Margalef, 1978), whereas growth rates will be controlled largely by nutrient availability, light conditions, and temperature as outlined below. When primary productivity measurements are made at sea, it gives relatively little information other than the rate of accumulated radiocarbon during the experimental period. In the absence of other data, no conclusions can be made regarding potential productivity, growth rates, or factors which are limiting the magnitude of primary production. Seasonal synoptic surveys over a period of some years can, however, give one reasonable estimates of primary production in the experimental area. Interpretation of primary productivity data is greatly enhanced when correlated with simultaneously obtained data on some of the factors mentioned below. Geographical differences in rates of primary production have recently been discussed in several books (Falkowski, 1980; Raymont, 1980; Longhurst, 1981) and will not be repeated here. Some discussion of primary productivity, including reference to waters bordering Pakistan, is given in the paper by Banse (this volume).

Grazing Pressure

The importance of grazing organisms in regard to measurement of primary production is that (i) the phytoplankton biomass is decreased, thus lowering photosynthetic rates, and (ii) a large fraction of the phytoplankton biomass that is consumed is oxidized to liberate fixed organic carbon as CO_2. Much organic carbon can also be expected to be liberated, either as soluble organic compounds or as detrital material, through "sloppy feeding" by grazing organisms. These materials (which are measured during primary productivity studies utilizing proper techniques) furnish much of the energy substrates for bacterial populations. As much of the grazing pressure comes from micro-zooplankton cells (< 200 μm) which completely overlap the size spectrum of phytoplankton cells, it is not possibly to achieve a good separation of autotrophic and heterotrophic cells. Assuming that one is interested in determining true *in situ* production rates, one should not alter the distribution of organisms in any way. The major reason that the larger zooplankton are commonly removed is that they exist in relatively low concentrations and are patchy in distribution. Unequal distribution of grazers in the productivity sample bottles would result in large variations in experimental data. This is not a serious problem when considering microplankton.

Losses to Deep Water

Phytoplankton cells are often transported out of the euphotic zone by (i) settling due to the specific density of cells being slightly higher than that of sea water (Walsby and Reynolds, 1980), (ii) convective, downward mass transport of water as at a convergence zone, and (iii) through circulation in a deep wind-mixed layer. When photoautotrophic cells sink below the light-compensation depth for photosynthesis (where rate of respiration equals that of photosynthesis) they either perish or adjust their metabolism to compensate for the loss of photosynthetically derived energy. Cells in the wind-mixed upper layer of the oceans will "circulate" up and down at speeds which are not reliably estimated by any direct measurement. As any vertical movement by cells in the upper water column may result in dramatic changes in light, temperature, and

nutrient conditions, such physical mixing forces may have profound effects on viability and metabolic functioning of phytoplankton cells. For further discussion on this critical depth consideration, see Sverdrup (1953) and Raymont (1980).

In recent years there has been much emphasis placed on deploying Particle Interceptor Traps at various depths in the water column to collect all particulate materials which are settling out (Fellows et al., 1981; Eppley et al., 1983). Although there are serious limitations to the use of these moored sediment traps, they do allow one to get some idea of the flux and chemical composition of organic matter as a function of depth, which is informative in regard to the chemical and biological processes occurring throughout the water column.

Factors Affecting Growth Rates

The major factors to be considered here are light, nutrients, and temperature. It should be pointed out, however, that much of our information on growth rates has been obtained with laboratory cultures which may respond quite differently than most species in a natural water sample. It is very difficult to determine growth rates of natural phytoplankton populations because of various "bottle effects," grazing effects during incubation, possibly artifacts associated with radiocarbon procedures, and great difficulty in estimating initial phytoplankton biomass. The most reliable methods for determination of specific growth rates would appear to be those microscopic methods (e.g., direct counting of one or more species, auto-microradiography, counting of binucleated cells) which permit one to differentiate between the many species in any natural water sample. Such methods though are very time-consuming and are often restricted to cells which are most amenable to such microscopic enumeration. At present there is considerable disagreement regarding determination of algal growth rates, particularly in oligotrophic waters (Goldman, 1980). Resolution of this problem will necessitate the development and application of other techniques to support the traditional radiocarbon measurements.

Light

Solar radiation is the energy source for all photoautotrophic production of reduced organic carbon in the sea. The magnitude of primary production will be limited by the quantum flux which is absorbed by photosynthetic pigments. A large percentage of solar radiation will be unavailable for photoplankton growth due to reflectance from the sea surface, absorption by water molecules, and absorption by all dissolved and particulate materials which do not actively transfer energy to chlorophyll-a (Morel, 1976). Optical characteristics of marine waters have been documented by Tyler and Smith (1970) and Yentsch (1980). There has been much research on the relationship between light intensity and the rate of photosynthesis (Parsons et al., 1977; Fogg, 1975; Yentsch, 1980; Platt and Gallegos, 1980). These "P vs I" curves, as they are commonly called, are generally done either with laboratory cultures (which may be very different physiologically from phytoplankton in natural water samples) or with natural samples containing a heterogeneous assemblage of cells, all of which may have different photo-responses. These data suggest that: (a) the light compensation intensity is actually closer to 0.1% of the intensity of sunlight incident upon the sea surface instead of the 1% figure which is usually mentioned; (b) the range where light is limiting the rate of photosynthesis (between the compensation point and the saturating light intensity, which is often approximately 20% of incident sunlight) is commonly described as a linear relationship, but is actually closer to a hyperbolic function; (c) above the light saturation intensity the photosynthetic rate either remains the same or it declines. This photo-inhibition of the photosynthetic rate is commonly ascribed to photorespiratory reactions (Burris, 1980). There has been relatively few investigations where wavelength of light is controlled to simulate that at various depths in the water column. Recent work has also shown that middle ultraviolet radiation (280-340 μm) can cause marked reduction in primary production in the upper portion of the euphotic zone (Smith et al., 1980).

Temperature

Eppley (1972) has assembled much of the available data on growth rates of phytoplankton as influenced by temperature and has suggested that the upper limits for maximal growth rates (μ_{max}), are described by the equation:

$$\mu_{max} = 0.851 \, (1.066)^T$$

where T is degrees centigrade. A slightly different temperature response equation has been described in Goldman and Carpenter (1974), which predicts a fairly similar Q_{10} response (2.09) as compared to 1.88 from Eppley's data, but is based on a lower photosynthetic rate per unit biomass. Recent field data from the Antarctic has been discussed in regard to the above temperature equations (Neori and Holm-Hansen, 1982).

Nutrients

It can be safely assumed that marine phytoplankton requires the same essential macro-nutrients and micro-nutrients (Mo, Mn, Co, B, Zn, Cu, Si) required for most higher plants, and possibly others such as Se and some of the halogens. Twenty years ago the prevailing view of most biological oceanographers was that marine phytoplankton was never limited by any nutrient deficiency. There are now good data in the literature showing that nitrogen (McCarthy, 1980), phosphorus (Nalewajko and Lean, 1980), and silicon (Paasche, 1980) often do limit the rate of primary production. Barber and Ryther (1969) have presented data which suggest that iron availability may limit photosynthetic rates, particularly in upwelling areas. At the present time the pendulum has swung to the extent that many investigators now think it likely that marine phytoplankton may be limited by all the nutrient deficiencies described for fresh waters, which basically implies all the known essential mineral elements.

There has been much effort to study the assimilation of nutrients by phytoplankton, especially in regard to determining K_S values (the nutrient concentration which results in half-maximal growth rate) for nitrogen, phosphorus, and silicic acid. These K_S values are generally in the range of 0.1 to 2.0 micromolar, which causes a serious problem when trying to relate these values to observed phytoplankton growth rates in natural samples. In oligotrophic areas of the oceans nitrate and phosphate are barely detectable (about 0.02 μM or less) in the upper 80 m using standard analytical methods, but yet the photosynthetic assimilation numbers (rate of photosynthesis per unit chlorophyll per unit time) are not far different from those found in coastal waters. It is obvious that we have a relatively poor understanding of the dynamics of nutrient assimilation by marine phytoplankton, and the extent to which nutrient deficiencies limit primary production. Many of the above questions are discussed in the recent book by Platt (1981).

PHYSIOLOGICAL ADAPTATION

Some appreciation of the diversity of environmental stresses imposed upon marine phytoplankton may be had by comparing the marine environment to terrestrial habitats. There is great diversity in size, morphology, and physiology of land plants in response to varying conditions of temperature, light, nutrients, moisture, predators, etc. Most plants are highly adapted to survival under specific environmental conditions. The same holds true for phytoplankton in the oceans, but their problems are compounded by the fact that they must accommodate themselves to a three-dimensional existence, where settling in the water column will dramatically alter nutrient, light, and temperature conditions. Phytoplankton cannot maintain their position in one spot as can land plants (or benthic algae), and hence they must be capable of chemical and physiological adaptations to meet the ever-changing environmental conditions.

Conditions in the Euphotic Zone

If one looks only at light and nutrient factors, it is evident that all depths in the euphotic zone present special problems in regard to phytoplankton growth. Light conditions are optimum in the upper portion of the water column, but it is at these depths that nutrients are often undetectable and that conditions favor photorespiratory reactions. Deep in the euphotic zone nutrients are in much higher concentration, but here light intensity will limit productivity. There is thus a continuum of varying environmental stresses imposed on cells throughout the depth of the euphotic zone. Within the wind-mixed upper layer cells are being circulated up and down and thus continually being exposed to variable conditions of light, nutrients, and temperature. The oligotrophic waters of the central oceans offer a good example of the interaction of these variables with phytoplankton biomass and productivity (Kiefer et al., 1976; Eppley, 1981). There are many studies demonstrating that species composition of the phytoplankton often change with depth (Sournia, 1982) and that cells show chemical adaptations (e.g., higher ratios of chlorophyll-a per unit organic carbon) in response to the light and nutrient fields. One problem which is receiving much attention at the present time is whether or not cells can show physiological adaptations in relatively short time periods (hours) that would enable them to cope better with changing conditions as they circulate within the upper mixed layer of the oceans.

NITROGEN METABOLISM AND MINERALIZATION

It has long been known that ammonia is generally preferentially assimilated by plant cells as compared to nitrate. The importance of ammonia in marine waters was not generally recognized, however, until development and application of ^{15}N isotope methodology (Dugdale and Goering, 1967; Eppley et al., 1983). By use of ^{15}N-labelled nitrogenous substrates, it has been demonstrated that ammonia usually accounts for over half the total nitrogen assimilated by marine phytoplankton, with the balance being furnished mostly by nitrate, and lesser amounts by urea and other organic substrates. As ammonia is usually undetectable in deep waters, the ammonia that is serving as the substrate for phytoplankton assimilation must arise from mineralization processes occurring within the euphotic zone. There are three major applications of such ^{15}N uptake studies: (i) one can determine the rate of assimilation of individual nitrogenous substrates (and hence the rate of supply, assuming a quasi-steady state); (ii) as nitrate-based productivity represents "new" production, it should be indicative of the amount of primary production which can be removed from the euphotic zone to deeper waters (ammonia-based production is termed "regenerated" production); (iii) ammonia formed in the euphotic zone must result primarily from grazing organisms and the bacterioplankton. The magnitude of this production indicates the rate of mineralization processes occurring with the euphotic zone and hence should be able to be correlated with studies on the biomass and nutritional mode of microbial cells.

One potential difficulty with the ^{15}N method is that one generally has to add fairly high levels of enriched substrates (about 0.5 μM for ammonia, and about 5.0 μM for nitrate) in order to obtain reliable uptake data. These concentrations of added substrates may be so large relative to the ambient nutrient concentrations (especially in oligotrophic waters) that uptake data represent artificially high rates. This potential limitation of the method is not significant when working with nutrient rich waters, such as in upwelling areas or in the seas surrounding Antarctica. Recent studies in the Antarctic (Olson, 1980; Glibert et al., 1982; Rönner et al., 1983) have shown that between 50 to 90% of all the nitrogen taken up by phytoplankton is in the form of ammonia, in spite of the high nitrate levels (20-30 μM) which are found in all waters south of the Antarctic Convergence. Concomitant studies have shown that the net zooplankton can furnish only a few percent of the ammonia required to maintain this flux into the phytoplankton cells (Biggs, 1982). This suggests that heterotrophic microbial cells constitute a very dynamic and important component of the food-web in Antarctic waters.

MISCELLANEOUS RESEARCH PROJECTS

Banse (this volume) has suggested that priority be given to trying to get answers to a limited number of specific problems, in contrast to large surveys or baseline studies. I think this is a good suggestion, and below I have listed a few problems which I think could be profitably pursued in waters close to Pakistan.

Applications of Remote Sensing

In an earlier section I mentioned the necessity to approach many biological questions by an interdisciplinary approach which also involves geographical coverage to some extent and temporal studies so that one can distinguish between normal seasonal events and the occasional episodic event. Such studies are, however, very expensive in regard to ship costs. The development of remote sensors mounted on satellites, however, gives one the opportunity to follow physical and biological events (thermal and color imagery) in the upper water column over extensive areas (Smith and Baker, 1982; Smith et al., 1982). Such data are very useful in directing the time and place where shipboard studies and observations would be most profitable. Research scientists at the National Oceanographic and Atmospheric Administration (NOAA) laboratories in La Jolla have been using this approach in conjunction with studies of the tuna and anchovy fisheries. Thermal images extending out a few hundred miles from the California coast

nicely delineate the thermal fronts between the colder coastal water and the warmer waters off shore. The cold waters are rich in phytoplankton, but yet the albacore tuna are found to be concentrated in the warm water very close to the cold water fronts. Such information can be used not only to "guide" the fishing fleet, but can also be used to help formulate research problems in regard to the interaction between physical and biological characteristics of oceanic areas as related to the distribution of specific fish species. A slightly different situation prevails concerning the distribution of spawning northern anchovy (*Engraulis mordax*) in southern California waters. They were found in high concentrations in warm water, but without any correlation to the thermal fronts. Color imagery of the warm water areas showed that the spawning areas of *E. mordax* coincided with areas of high chlorophyll content (Lasker et al., 1981).

Food Reserves for Economically Important Species

In regard to studies on the annual recruitment success of economically important fish or invertebrate species, it is important to have some understanding of the specific food requirements throughout the period of development, growth, and reproduction (Lasker, 1981). It is not sufficient merely to document the total food supply (e.g., total phytoplankton biomass as estimated by chlorophyll-a measurement). This is illustrated by two examples. The success of the anchovy year classes off California were found to be correlated with (i) a minimum density of food particles, and (ii) the species composition of the phytoplankton. The anchovy thrived on dinoflagellates (with the exception of *Gonyaulax*), but did not thrive when diatoms were abundant. This selective feeding requirement is also illustrated by the food particles which are ingested by Antarctic krill (*Euphausia superba*). Laboratory studies (Meyer and El-Sayed, 1983) have indicated that krill prefer particles larger than 20 μm in size and are relatively poor grazers on particles 5-10 μm in size. This conclusion is also supported by our field data in the Scotia Sea, which showed that diatoms and dinoflagellates were the preferred food items as contrast to the large biomass of nannoplankton (mostly flagellates and monads). Our krill studies were one component of a large interdisciplinary program which was concerned with the distribution, behavior, and feeding requirements of all developmental stages of *Euphausia superba* relative to the distribution and species composition of the phytoplankton populations (Holm-Hansen and Huntley, 1983). I would suggest that the nutritional requirements of your very important shrimp fishery would be an interesting and hopefully profitable enterprise in regard to long-term management of the resource.

Mariculture Projects

Many of the points I have discussed are "observational" whereby the environmental conditions cannot be controlled. Such ecological studies hopefully will enable us to better our understanding of the growth and physiology of both plant and animal cells in the oceans and how they are affected by changes in environmental conditions. I would expect that such knowledge will be of much importance when shallow coastal regions are used in semi-controlled mariculture programs for rearing of desired fish or crustacean species. Extensive mariculture programs have been initiated in recent years by various countries, and I should think that the coastal areas of Pakistan would offer many opportunities for increasing the food productivity from marine waters.

Individual Projects

Many of the problems and questions I have discussed today involve a considerable amount of equipment and often require coordination with expertise from other fields. I would like to point out to you (and to potential readers) that there are still many important problems which can be approached by a single, interested and imaginative researcher. I will cite only two such examples. Within the last ten years there has arisen a dogma in the phytoplankton literature that approximately 50% of most phytoplankton cells in the euphotic zone are dead. This conclusion is based largely on micro-autoradiographic studies with radiocarbon incorporation. Although I am not convinced that this conclusion is valid, I do not know of any data to disprove it, or even any alternate method to test the hypothesis. Hopefully, someone with a little imagination and patience in the laboratory will find some way to determine the viability of individual marine phytoplankton cells.

The second problem concerns the distribution

and significance of blue-green algae in the marine environment. The Arabian Sea is well-known for its blooms of the nitrogen fixing *Trichodesmium*. There are many interesting biochemical, physiological, and ecological questions concerning the growth, death, and fate of these blue-green cells. Although they are widespread in tropical waters, I do not think anyone has succeeded in maintaining *Trichodesmium* in a healthy state for a long period of time in the laboratory. For such studies one, therefore, must have ready access to natural populations. We do not have such filamentous blue-green algae in the plankton off the coast of California, but they could conveniently be studied in Pakistan.

REFERENCES

Barber, R.T. and J.H. Ryther, 1969. Organic chelators; Factors affecting primary production in the Cromwell Current upwell. J. Exp. Mar. Biol. Ecol., 3:191-199

Biggs, D.C., 1982. Zooplankton excretion and NH_4^+ cycling in near-surface waters of the Southern Ocean. I. Ross Sea, Austral Summer 1977-1978. Polar Biol., 1:55-67.

Bröckel, K. von, 1981. The importance of nanoplankton within the pelagic Antarctic ecosystem. Kieler Meeresforsch., 5:61-67.

Burris, J.E., 1980. Respiration and photorespiration in marine algae. *In:* P.G. Falkowski (ed.), *Primary Productivity in the Sea,* Plenum Press, N.Y., 411-432.

Dugdale, R.C. and J.J. Goering, 1967. Uptake of new and regenerated forms of nitrogen in primary productivity. Limnol. Oceanogr., 12:196-206.

Eppley, R.W., 1972. Temperature and phytoplankton growth in the sea. Fishery Bulletin, 70:1063-1085.

Eppley, R.W., 1981. Autotrophic production of particulate matter. *In:* A.R. Longhurst (ed.), *Analysis of Marine Ecosystems,* Academic Press, N.Y., 343-361.

Eppley, R.W., E.H. Renger, and P.R. Betzer, 1983. The residence time of particulate organic carbon in the surface layer of the ocean. Deep-Sea Res., 30:311-323.

Falkowski, P.G., 1980. Primary Productivity in the Sea. Plenum Press, N.Y., 531 pp.

Fellows, D.A., D.M. Karl, and G.A. Knauer, 1981. Large particle fluxes and the vertical transport of living carbon in the upper 1550 meters of the northeast Pacific Ocean. Deep-Sea Res., 28: 921-936.

Fenchel, T., 1982. Ecology of heterotrophic microflagellates, III. Quantitative occurrence and importance. Mar. Ecol. Prog. Ser., 9:25-42.

Fogg, G.E., 1975. Algal Cultures and Phytoplankton Ecology, Second Ed., The Univ. of Wisconsin Press, 175 pp.

Glibert, P.M., D.C. Biggs, and J.J. McCarthy, 1982. Utilization of ammonium and nitrate during austral summer in the Scotia Sea Deep-Sea Res., 29:837-850.

Goldman, J.C., 1980. Physiological processes, nutrient availability, and the concept of relative growth rate in marine phytoplankton ecology. *In:* P.G. Falkowski (ed.), *Primary Productivity in the sea,* Plenum Press, N.Y., 179-194.

Goldman, J.C. and E.J. Carpenter, 1974. A kinetic approach to the effect of temperature on algal growth. Limnol. Oceanogr., 19:756-766.

Hewes, C.D. and O. Holm-Hansen, 1983. A method for recovering nanoplankton from filters for identification with the microscope: The filter-transfer-freeze (FTF) technique. Limnol. Oceanogr., 28:389-394.

Holm-Hansen, O. and M. Huntley, 1983. Feeding requirements of krill in relation to food sources. J. Crustacean Biol., (In Press).

Isaacs, J.D., 1973. Potential trophic biomasses and trace substance concentrations in unstructured marine food-webs. Mar. Biol. 22:97-104.

Johnson, P.W. and J. McN. Sieburth, 1979. Chroococcoid cyanobacteria in the sea: A ubiquitous and diverse phototrophic biomass. Limnol. Occanogr., 24:928-935.

Johnson, P.W. and J. McN. Sieburth, 1982. *In situ* morphology and occurrence of eucaryotic phototrophs of bacterial size in the picoplankton of estuarine and oceanic waters. J. Phycol., 18:318-327.

Karl, D.M., C.O. Wirsen, and H.W. Jannasch, 1980. Deep-sea primary production at the Galapagos hydrothermal vents. Science, 207:1345-1347.

Kiefer, D.A., R.J. Olson, and O. Holm-Hansen, 1976. Another look at the nitrite and chlorophyll maxima in the central North Pacific. Deep-Sea Res., 23:1199-1208.

Kumari, L.K., S. Vijayaraghavan, M.V.M. Wafar, J.P. Royan, and A. Rajendran, 1978. Studies on detritus in a tropical estuary. Indian Jour. Mar. Science, 7:263-266.

Lasker, R., 1981. Factors contributing to variable recruitment of the northern anchovy (*Engraulis mordax*) in the California current: Contrasting years, 1975 through 1978. Rapp. P.-v. Reun. Cons. Int. Explor. Mer., 178:375-388.

Lasker, R., J. Pelaez, and R.M. Laurs, 1981. The use of satellite infrared imagery for describing ocean processes in relation to spawning of the northern anchovy (*Engraulis mordax*). Remote Sensing of Environment, 11:439-453.

Longhurst, A.R., 1981. *Analysis of Marine Ecosystems.* Academic Press, N.Y., 741 pp.

Malone, T.C., 1980. Algal Size. *In:* I. Morris (ed.), *The Physiological Ecology of Phytoplankton,* Univ. of California Press, Berkeley, 433-463.

Margalef, R., 1978. Life-forms of phytoplankton as survival alternatives in an unstable environment. Oceanol. Acta, 1:493-509.

McCarthy, J.J., 1980. Nitrogen *In:* I. Morris (ed.), *The Physiological Ecology of Phytoplankton,* University of California Press, Berkeley, 191-233.

Meyer, M.A. and S.Z. El-Sayed, 1983. Grazing of *Euphausia superba Dana* on natural phytoplankton populations. Polar Biology, 1:193-197.

Morel, A., 1976. Available, usable, and stored radiant energy in relation to marine photosynthesis. Deep-Sea Res., 26:673-688.

Morris, I. (ed.), 1980. *The Physiological Ecology of Phytoplankton.* University of California Press, Berkeley, 625 pp.

Nalewajko, C. and D.R.S. Lean, 1980. Phosphorus *In:* I. Morris (ed.), *The Physiological Ecology of Phytoplankton,* University of California Press, Berkeley, 235-258.

Neori, A. and O. Holm-Hansen 1982. Effect of temperature on rate of photosynthesis in Antarctic phytoplankton. Polar Biology, 1:33-38.

Olson, R.J., 1980. Nitrate and ammonium uptake in Antarctic waters.

Limnol. Oceanogr., 25:1064-1074.

Paasche, E., 1980. Silicon. *In:* I. Morris (ed.), *The Physiological Ecology of Phytoplankton,* University of California Press, Berkeley, 259-284.

Parsons, P.R., M. Takahashi, and B. Hargrave, 1977. *Biological Oceanographic processes,* 2nd ed., Pergamon Press, N.Y., 332 pp.

Platt, T. (ed.), 1981. *Physiological Bases of Phytoplankton Ecology.* Can. Bull. Fish. Aquat. Sci. 210:346 p.

Platt, T. and C.L. Gallegos, 1980. Modeling primary production. *In:* P.G. Falkowski (ed.), *Primary Productivity in the Sea,* Plenum Press, N.Y. 339-362.

Pomeroy, L.R., 1974. The ocean's food-web, a changing paradigm. Bioscience, 24:499-504.

Qasim, S.Z., M.V.M. Wafar, S. Vijayaraghavan, J.P. Roya, and L.K. Kumari, 1979. Energy pathways in the Laccadive Sea (Lakshadweep). Indian Journal Mar. Sciences, 8:242-246.

Raymont, J.E.G., 1980. *Plankton and Productivity in the Oceans,* 2nd ed., Pergamon Press, N.Y. 489 pp.

Rönner, U., F. Sörensson, and O. Holm-Hansen, 1983. Nitrogen assimilation by phytoplankton in the Scotia Sea. Polar Biology 2:137-147.

Sieburth, J. McN., 1979. *Sea Microbes.* Oxford University Press, N.Y., 941 pp.

Smith, R.C. and K.S. Baker, 1982. Oceanic chlorophyll concentrations as determined by satellite (Nimbus-7 coastal zone color scanner). Mar. Biol., 66:269-279.

Smith, R.C., K.S. Baker, O. Holm-Hansen, and R.J. Olson, 1980. Photoinhibition of photosynthesis in natural waters. Photochem. Photobiol., 31:585-592.

Smith, R.C., R.W. Eppley, and K.S. Baker, 1982. Correlation of primary production as measured aboard ship in southern California coastal waters and as estimated from satellite chlorophyll images. Mar. Biol., 66:281-288.

Sournia, A., 1982. Is there a shade flora in the marine plankton? Jour. Plankton Res., 4:391-399.

Sverdrup, H.U., 1953. On conditions for the vernal blooming of phytoplankton. J. Cons. Int. Explor. Mer., 18:287-295.

Tyler, J.E. and R.C. Smith, 1970. Measurement of spectral irradiance underwater. Gordon and Breach, N.Y., 103 pp.

Utermöhl, H., 1958. Zur vervollkommung der quantitativen phytoplankton-methodik. Mitt. Int. Ver. Theor. Angew. Limnol., 9:1-38.

Walsby, A.E. and C.S. Reynolds, 1980. Sinking and Floating. *In:* I. Morris (ed.), *The Physiological Ecology of Phytoplankton,* University of California Press, Berkeley, 371-412.

Williams, P.J. leB., 1981. Incorporation of microheterotrophic processes into the classical paradigm of the plankton food-web. Kieler Meeresforsch., Sonderh., 5:1-28.

Yentsch, C.S., 1980. Light attenuation and phytoplankton photosynthesis. *In:* I. Morris (ed.), *The Physiological Ecology of Phytoplankton,* University of California Press, Berkeley, 95-127.

H. PHYSICAL AND CHEMICAL OCEANOGRAPHY

21

Coastal Upwelling: A Synopsis of it's Physical, Chemical and Biological Characteristics

T. T. PACKARD
D. BLASCO
Bigelow Laboratory for Ocean Sciences,
West Boothbay Harbor, Maine,
and
R. C. DUGDALE
University of Southern California,
Los Angeles, California

ABSTRACT

Upwelling ecosystems are characterized by complex features such as the upwelling cell, upwelling plumes, relaxation, coastally trapped waves, El Niño, aguajes, nutrient enrichment, oxygen minimum zones, nutrient traps, short food chains. This paper examines the potential explanations of the richness of upwelling areas and argues that the unique counter-current system, the surface-offshore flow opposed to the subsurface onshore flow and the surface-equatorward current opposed to the poleward subsurface current, provides not only a nutrient trapping mechanism but also a plankton trapping mechanism. In addition, the short food chain, the large organisms at its base (diatoms and dinoflagellates) and the temporal intermittency of upwelling events all contribute to making upwelling ecosystems extraordinarily productive.

INTRODUCTION

Throughout most of the world's oceans there is little exchange of surface and subsurface seawater; rarer still is the systematic injection of the fertile subsurfaces waters into the surface layer. Because of this, the surface layers are infertile and only weakly support marine life. However, along the west coasts of the continents, an area comprising only 0.1% of the ocean, the fertile subsurface waters are injected into the surface waters and stimulate the growth of dense populations of marine organisms. These upwelling zones are characteristically rich fishing grounds from which 50% of the world's fish harvest is derived (Ryther, 1969). In addition, they have a major effect on the climate of the adjacent land mass. Thus, from an economic point of view upwelling zones are interesting. Oceanographically they are characterized by large gradients in the physical,

Fig. 1. The maxima and minima of upwelling intensity within a coastal upwelling system. This entire coast of southern Peru is characterized by upwelling, but during this period (28-30 May 1974) centers of intense upwelling occurred at Callao, San Juan, Atico and Mollendo. This figure is redrawn from Zuta et al. (1978).

chemical and biological properties of the ocean. In these zones, one can observe large volumes of water rising at one location and sinking at another, currents moving in opposite directions, oceanic fronts, oxygen minimum and denitrification zones, blooms of phytoplankton that may exceed 200 mg chlorophyll per liter, patches of zooplankton so dense that they clog an engine's cooling system within minutes (Bass and Packard, 1976; Packard et al., 1979).

Just outside these zones, sometimes less than 200 km from the coast, one can observe a stratified, nutrient-poor, oligotrophic ocean. Transects from this static, oligotrophic zone to the dynamic upwelling zone teeming with life, allow one to investigate easily many oceanographic processes. Because of their economic importance and because they serve as gigantic oceanographic laboratories, the upwelling areas of California, N.W. Africa, Peru and S.W. Africa have been intensely studied over the last decade. The Indian Ocean upwelling systems were largely ignored during this same period. As a result, the brief review of the major physical, chemical, and biological properties of coastal upwelling systems presented in this paper is largely based on the studies from the Atlantic and Pacific Oceans, but may have application in understanding the Indian ocean upwelling system.

PHYSICAL STRUCTURE AND DYNAMICS

The Upwelling Cell

Recent studies have shown that upwelling along a coast is characterized by centers of maximal and minimal activity (Fig. 1) that vary in response to the local wind (Halpern, 1974) as well as to the large scale wind conditions, that vary seasonally (Banse, 1968) and that vary in response to global meteorological changes (Quinn, 1974). Nevertheless, it is useful to outline a hypothetical "upwelling cell" to serve as a conceptual model for further discussion (Fig. 2). The most outstanding feature of this upwelling cell is the ascendance of the isopycnals in the upper 100 m near the coast. At the surface this is detected as a drop in temperature in the near-shore region (Region I, Fig. 2). At the offshore edge of the upwelling cell is a frontal region characterized by a large temperature gradient and convergence on the inshore side of the front (between regions III and IV in Fig. 2; Cushing, 1971; Boje and Tomczak,

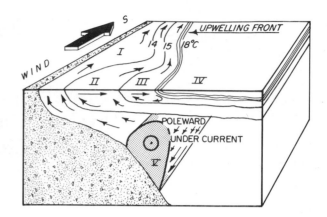

Fig. 2. A hypothetical upwelling cell in the northern hemisphere. The main features are the rising isotherms near the coast, the subsurface onshore flow, the surface offshore-equatorward flow, the poleward undercurrent and the upwelling front that separates the warm offshore oligotrophic water (Region IV) from the cold inshore eutrophic upwelled water (Regions I, II, and III).

1978; Andrews and Hutchings, 1980). Below the offshore edge of the upwelling system lies a poleward undercurrent (Fig. 2) which flows opposite to the surface flow (Smith, 1968; Barber and Smith, 1981). The thickness of this flow and its extent over the shelf varies with the upwelling area. Off N.W. Africa, it is largely a slope phenomenon while off Peru it extends over much of the shelf. The poleward undercurrent in the Arabian Sea upwellings (off Arabia and southwest India) has not been studied.

The surface current in upwelling systems is characteristic of eastern boundary currents; it flows equatorward in the upper 70 m off N.W. Africa and Oregon, USA and in the upper 30 m off Peru (Barber and Smith, 1981). Part of this surface current also flows offshore as would be expected from a consideration of continuity since new water is upwelling near the coast. In the three most intensely studied regions, the offshore flow was very shallow; only 19 m off Oregon, USA, 24 m off Peru and 33 m off N.W. Africa (Barber and Smith, 1981). Below this surface offshore flow there is an onshore flow to provide a source for the rising waters inshore. These source waters range from 20-80 m off Oregon, USA, 20 to 200 m off Peru and 30 m to the bottom off N.W. Africa.

The forces that drive the upwelling cell originate in the large-scale wind patterns. The high pressure systems that sit at midlatitudes over both the northern and southern parts of the Atlantic and Pacific oceans set up anticyclonic atmospheric systems that drive winds equatorward along the coasts of North and South America and Northwest and Southwest Africa. These winds, when blowing along western coastlines, drive the surface waters to the west by a combination of frictional and Coriolis forces known as the Ekman transport mechanism. The offshore flow along large sections of coast forces subsurface waters to rise from below. The strength of this upwelling will be greater if the topography of the local terrain intensifies the local winds (Barber and Smith, 1981). It will also be strengthened if the topography of the coastal sea floor funnels or constricts the longshore current (Mittelstaedt, 1974; Cruzado and Salat, 1981). Neither of these factors will destroy the basic configuration of the upwelling cell, they will simply distort its dimensions and extend its influence farther into the offshore waters.

Plumes

Small scale upwelling structures that appear to have an inshore "center" of cold water which spreads offshore in the shape of a ellipse, are often called plumes. The most well studied plume occurs off the Peruvian coast at 15°S (Brink et al., 1981), where it has been observed in the same place since 1966 (Ryther et al., 1971). It occupies the upper 25 m of a 30 km strip of coastal water and extends for 40 km offshore (Fig. 3). At its upwelling center the temperature often falls below 16°C and at its offshore edge the temperature may reach 20°C. The offshore flow ranges from 0.2 to 0.5 km/hr in the upper 20 m when upwelling-favorable winds are blowing. Below that depth the flow is reversed. The vertical upwelling velocity at the center was calculated to be as high as 36 m per day (Brink et al., 1981). Structures such as this one can be observed along the Peruvian coast as Zuta et al. (1978) have shown in their long-term hydrographic study of the Peruvian Upwelling System. A similar structure has been found off the Monterey Peninsula in California (Traganza et al., 1981) and a similar one apparently occurs off Point Conception, California. Off Cape Town, South Africa, there is another plume that extends into the South Atlantic (Andrews and Cram, 1969; Andrews and Hutchings, 1980). The upwelling velocity in this plume has been shown to be closely

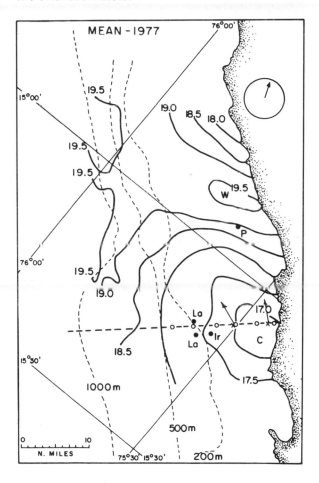

Fig. 3. The plume at 15°S in the Peruvian upwelling system. The contours depict the mean sea surface temperatures as measured from aircraft flying at 150 m above the ocean (redrawn from Brink et al., 1981).

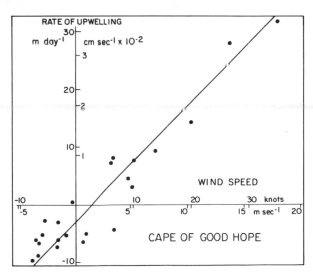

Fig. 4. Rate of upwelling as a function of local wind speed in the plume off the Atlantic side of the Cape of Good Hope. The upwelling rate was measured by the ascendence of isotherms in a 24 hr. period. The wind velocity represents the average for the comparable time period as measured at the Cape of Good Hope lighthouse (redrawn from Andrews and Hutchings, 1981)

coupled to the local wind speed (Fig. 4). The existence of plumes elsewhere is not documented, although they are commonly seen in satellite photographs. Brink et al. (1981) conclude that they are likely to be wide spread in the world's upwelling systems.

Relaxation

Just as upwelling is rarely continuous along a coast it is also rarely continuous with time. More commonly, upwelling occurs as events that coincide with variations in the wind field. It follows that when the wind field is unfavorable for upwelling, the upwelling system will tend to relax. Freshly upwelled water that has not been heated sufficiently to remain at the sea surface will subside and warm offshore waters will flow inshore to take its place. Such a relaxation can be detected by a vertical array of current meters as a reversal in the normal offshore flow at the surface and onshore flow at depth. Brink et al. (1981) recorded such a reversal and the resulting relaxation for two days in the plume at 15°S off Peru. During this period the surface water flowed onshore and "old" water relatively rich in phytoplankton, nitrite and ammonium accumulated and subsided inshore. Water at depth flowed offshore. Similar observations have been made in the northwest African (Barton et al., 1977) and Oregon, USA (Halpern 1974) upwelling systems using only hydrographic data. Halpern (1974) found that during a period of weak winds, the static stability in the surface waters increased, and decreased when the winds strengthened. He ascribed these differences to alternating periods of relaxation and upwelling where first the offshore waters converge inshore during relaxation from the first upwelling event and then are pushed offshore again when the winds strengthen during the second upwelling event. From the biological point of view, these relaxation events are important because they provide a physical mechanism for pumping particulate organic matter into the subsurface waters (Packard et al., 1977 and 1982).

Coastally Trapped Waves

Another source of variability in an upwelling system is from coastally trapped waves. The origin of these long period internal Kelvin waves is not well known, but their velocity fluctuations affect the density distribution of the water they pass

through. Off Peru, their wavelength is about 2000 km, their period is about 8-10 days, their speed is about 200 km/day in a poleward direction (Smith, 1978; Brink et al., 1980). Coastally trapped waves are not simple sinusoids. They display a complex pattern that reflects the irregular forcing of the large scale meteorological disturbances. They are known to have important effects on the inshore-offshore flow regime in an upwelling region and consequently, and they have a strong influence on planktonic organisms. In our study of the Peru Current, the passage of a coastally trapped wave caused the benthic boundary layer to thicken and caused plankton populations to move laterally rather than offshore. This increased their residence time in the upwelling zone, and decreased the effect of the counter current "seeding" mechanism that is common to upwelling systems (Brink et al., 1981).

El Niño

In the Peruvian upwelling system, the El Niño occurs when warm nutrient-poor equatorial waters flow south along the Peruvian coast and bring torrential rains and floods to the land and starvation to the fish at sea (See Fleming & Ostenso, this volume). The combination of destruction in the countryside and the collapse of the anchovy fishery spells economic disaster for Peru (Enfield, 1980). In the sea, the catastrophe starts when the nutrient-poor equatorial waters replace the nutrient-rich Peru undercurrent waters as upwelling source waters. When this nutrient-poor water is upwelled to the surface it fails to stimulate phytoplankton growth, forcing the anchovies to seek food elsewhere or starve. Many anchovies do starve but even if some survive, they become too dispersed to be fished, so the commercial fishery collapses.

The control over the timing of El Niño is beginning to be understood. Recent research has revealed a connection between onset of El Niño and changes in atmospheric pressure, trade winds and oceanic currents along the equator. The sequence of events starts when a low pressure center over Indonesia shifts eastward (Wyrtki, 1977; Quinn, 1974; Bjerknes, 1961). This results in stronger southeast trade winds over the equatorial Pacific, which strengthens the South Equatorial Current and causes a thick warm-water lens to develop in the western Pacific. At the same time, while the eastern Pacific is stripped of its warm surface water, cold nutrient-rich water is upwelled and the biological productivity is enhanced along the American coasts. This is considered the "normal" upwelling situation. Then the Indonesian low re-centers over Indonesia, the trade winds relax, and the excess warm water flow back toward the Americas. This transport is accomplished by a strengthened equatorial counter-current system. Tide gauges on Pacific Islands show clearly this intensified flow soon after the southeast trades relax. Upon reaching the Galapagos this water is diverted north and south and can be detected as far apart as Sitka, Alaska and Valparaiso, Chile (Enfield, 1980). Along the Peruvian coast this warm water is seen as El Niño. The El Niño phenomenon was previously thought

to be unique to the Peruvian upwelling system but similar events may occur elsewhere throughout the world (Walsh, 1978; Tont, 1981).

Often co-occurring with El Niño is a subsurface condition called aguaje. The aguaje probably results as organic-rich high-density water subsides under the low-density advancing El Niño waters, stimulates metabolism in the oxygen minimum zone, and forces the microbial populations to utilize sulfate to meet their respiratory requirement. Then when an upwelling event occurs, the product of sulfate respiration, hydrogen sulfide, is upwelled and fish kills occur along the coast. It is when the fish begin to die and the sea smells of sulfide that the Peruvians speak of aguaje.

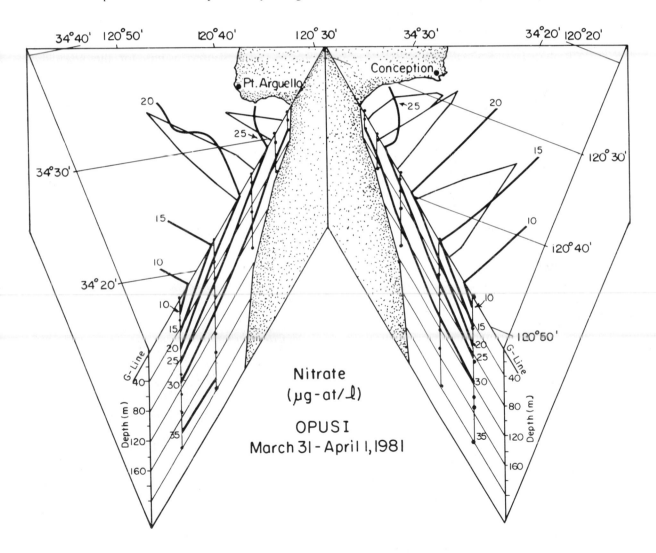

Fig. 5. The rising isopleths of nitrate in the California upwelling system off Point Conception. These are preliminary results of the OPUS program (Dugdale et al., 1982a).

THE CHEMICAL CHARACTERISTICS OF UPWELLING SYSTEMS

Nutrient Enrichment of the Surface Waters

The most obvious chemical characteristic of an upwelling area is the elevated level of inorganic nutrient salts at the sea surface. Figure 5 shows, as an example, the enhanced nitrate levels in the California upwelling off Cape Conception. Whenever one crosses into an upwelling area from the oligotrophic ocean, a sharp increase in the phosphate, nitrate and silicate content increase noticeably (Dugdale, 1972). Many times nitrite and ammonium are also abundant, whereas outside the upwelling area they are rarely detected. A more subtle change occurs in the pH of the seawater. It decreases several tenths of a pH unit in the upwelling area, which is a large change considering the strong buffering capacity of seawater (Simpson and Zirino, 1980). An additional chemical index of upwelling that can often be detected in the surface waters are low oxygen concentrations. Banse (1968) reports that during the southwest monsoon the oxygen levels off Calcutta (India) fall to 60-70% of saturation, once they were observed to fall as low as 30% (Subrahmanyan, 1959). Minas et al. (1982) have made a detailed study of the changes in oxygen in the upwelled waters off N.W. Africa and show that photosynthesis leads to a steady increase in surface-water oxygen as upwelled water flows away from the upwelling source. These three factors, low pH, high nutrients and low oxygen all indicate that the upwelled water has risen from depths at which the biological processes of remineralization and respiration have dominated photosynthesis. The great abundance of nitrate in the rising waters indicate the previous occurrence of extensive nitrification. The measurable levels of nitrite and ammonium indicate continual nitrification activity in the rising waters.

These characteristics of the upwelled waters make it easy to map the limits of an upwelling area by a ship steaming a "zig-zag" course over the suspected area (Walsh et al., 1977). The upwelled waters will easily be detected as the areas of elevated nitrate and depressed pH and temperature. This approach was used extensively in studies of the Californian, northwest African, Peruvian and Somalian upwelling systems (Dugdale, 1972; Walsh et al., 1977; Jones and Halpern, 1981; Smith and Codispoti, 1980). The

magnitude of the nutrients, oxygen, and the pH in the upwelled waters will depend on the strength of the physical forces driving the upwelling and the chemical characteristics of the subsurface water (Codispoti and Friederich, 1978; Smith and Codispoti, 1980). Since nutrients characteristically increase and the oxygen and pH decrease with depth, the stronger the physical driving forces the higher the nutrients and the lower the pH will be in the surface waters. However, if the subsurface waters are nutrient-poor, as off Cabo Frio, Brazil, the upwelling source water can come from as deep as 150 m and still contribute little to the surface nutrient content (Fahrbach and Meincke, 1979). The elevated nutrient levels in the surface waters of the southern end of the northwest African upwelling system have been traced to the nutrient differences in the North Atlantic Central Water (NACW) and the South Atlantic Central Water (SACW), the source waters for the upwelling systems. North of Cape Blanc the nutrient-poor NACW, is the source of the upwelled water. South of the Cape the upwelling source-water rises from the nutrient-rich SACW (Fraga, 1974). Neither water mass is as rich as the source water that feeds the Peruvian upwelling system even though they are much deeper layers. Off Peru, the waters rise from 50 to 75 m but contain nitrate levels of 20-25 μM. Off northwest Africa, they rise from more than 100 m but only contain 5 to 10 μM nitrate (Codispoti et al., 1982; Barber and Smith, 1981).

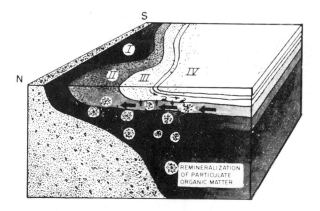

Fig. 6. The nutrient trapping mechanism in an upwelling system. Particulate organic matter sinks into the inflowing upwelling source waters. The remineralization of this organic matter enriches the already rich upwelling source water and serves to amplify the inflowing nutrient levels. The horizontal arrows reflect increasing nutrient concentrations as the upwelling source water moves onshore along isopycnal surfaces. The stippling reflects plankton density.

Nutrient Traps

A more subtle but universal characteristic of upwelling systems is their ability to trap nutrient salts (Redfield et al., 1963). This mechanism occurs because the offshore flowing upwelled water is raining particulate organic matter down on the on-shore flowing, upwelling source water (Fig. 6). Vertically migrating plankton can augment the transport of organic matter from the upper layer to the lower layer. The effect of this can be demonstrated by comparing the isopleths of nitrate, phosphate or silicate with the isohalines or isopycnals in a transect across an upwelling area (Friederich and Codispoti, 1979). Off Peru at 15°S, the nitrate in the water between the 25.7 and 25.8 isopycnal surfaces increases from 10 μM to 20 μM due to the remineralization of organic matter that occurs as the water between these surfaces moves onshore. In addition to the on-shore off-shore nutrient trap mechanism, the Peruvian upwelling system should have an "along-shore" nutrient trap mechanism that arises because the poleward undercurrent flows so close to the equatorial surface flow. The undercurrent serves as a source of upwelled water and as a remineralization zone

(Friederich and Codispoti, 1981). This may be an additional reason why the Peruvian upwelling is so productive.

Oxygen Minimum Zones

An additional chemical characteristic of upwelling areas is the oxygen minimum zone that lies below the source waters of the upwelling system (Margalef and Estrada, 1981). Figure 7 shows the very large oxygen minimum zone below the southwest Indian upwelling system (Sen Gupta, 1980). Oxygen utilization by the microbes that feed on the sinking particulate organic matter occurs throughout the water column in an upwelling area. But, above the euphotic zone photosynthesis masks the respiratory processes that utilize the oxygen, and advection in the upwelling source waters is sufficiently swift that the respiratory processes do not have enough time to deplete their oxygen supply. Consequently, the oxygen minimum zone is best developed below the upwelling source water where particulate organic matter has a longer residence time. In certain locations, such as off southwest Africa, Peru and the Arabian Sea, excessive amounts of organic matter sink into the subsurface waters, and during its decay the microbes consume all the oxygen and are forced to couple their respiration to alternative electron acceptors. Their first choice after oxygen is nitrate and successively as each nitrogenous electron acceptor becomes scarce they turn to nitrite (NO_2), nitrous oxide (N_2O) and finally when all the soluble combined nitrogen has been reduced to nitrogen gas, the microbes start reducing sulfate (SO_4). At this point, the end product, sulfide ($S^=$), starts to accumulate in the seawater column. This is disastrous for oceanic ecosystems because $S^=$ is toxic to fish, invertebrates and other higher organisms. As a result these organisms are killed or otherwise excluded from the seawater column as the zone of $S^=$ bearing waters expands. Since the mid water regions are relatively poorly inhabited the effect of $S^=$ in the seawater is rarely noticed, however, if the $S^=$ is upwelled into the surface, its presence is indicated by massive fish kills, the aguaje mentioned above. During a major aguaje in March 1976 the oxygen minimum zone off Peru exhausted its supply of nitrate, nitrite and nitrous oxide and began to reduce sulfate to sulfide (Dugdale et al., 1977; Packard et al., 1978). The $S^=$ bearing water upwelled and fish kills occurred along the coast.

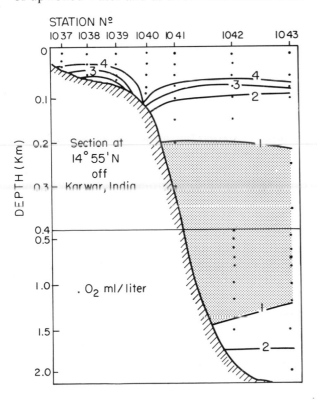

Fig. 7. The oxygen minimum zone off the southeast Indian upwelling system (redrawn from Sen Gupta et al., 1980).

Records of similar aguajes show that deep-water metabolic anomalies occur roughly every eight years, which is similar to the frequency of the other marine disaster off Peru, El Niño

THE BIOLOGICAL CHARACTERISTICS OF UPWELLING AREAS

The Biological Structure of the Upwelling Cell

Analysis of the three dimensional structure of an upwelling cell shows the subsurface onshore flow providing the water to be upwelled near the coast (Figs. 2 and 6). The inshore region of maximum upwelling and highest nutrient concentrations has the lowest phytoplankton biomass (Region I in Fig. 6 and Dugdale et al., 1982b). The region of highest phytoplankton abundances and greatest nutrient gradients (Region III in Fig. 6) occurs at the upwelling frontal zone (Blasco, 1971). Between these two zones the turbulence subsides sufficiently to permit phytoplankton populations to develop (Region II in Fig. 6). Beyond the frontal zone are regions of low nutrient concentrations and low phytoplankton biomass (Region IV in Fig. 6). A closer look at the zonation reveals that the phytoplankton species assemblages are grouped in specific bands in the upwelling cell according to the circulation and hydrography (Blasco et al., 1981). Using principal component analysis, Blasco et al., (1981) identified six phytoplankton assemblages that are defined by their relative affinity for upwelling, shelf influence, and alongshore flow. One group prefers the inner shelf, nearshore region, and two groups prefer the offshore region and only move in over the shelf during relaxation periods. The fourth group occurs during periods of strong upwelling and the fifth and sixth groups prefer weak upwelling.

Less is known about the distribution of zooplankton in an upwelling cell, because they are more difficult to sample than are phytoplankton. The observed distribution pattern suggests that high zooplankton populations may be found where abundant and appropriate food is available. In the Oregon upwelling, the various stages of the copepod, *Calanus marshallae* have different cross-shelf distributions (Peterson et al., 1981). Off northwest Africa, small copepods dominate the continental shelf populations and large zooplankton dominate the slope populations (Smith and Whitledge, 1977). Euphausids, which

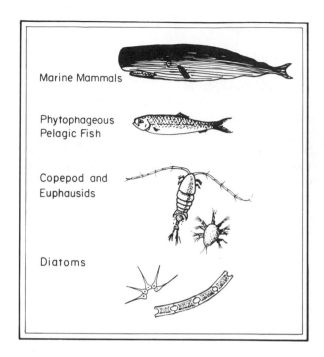

Fig. 8. Idealized food chain in an upwelling ecosystem. the high yield at the higher trophic levels is partially due to the autotrophic base being composed of large phytoplankters, principally diatoms. If the food chain were based on smaller autotrophic organisms, there would be more links in the food chain and a smaller amount of organic matter transferred to the top levels (adapted from Parsons, 1979).

are one of the largest zooplankters, are characteristically found at the shelf edge in all upwelling areas (Thiriot, 1978). The major fishes in upwelling areas are anchovies, sardines, jack mackerel and hake. The adults of these fish array themselves so that the anchovies and sardines can feed on the phytoplankton, the jack mackerel can feed on the copepods and euphausids and the hake can feed on the sardines, anchovies and euphausids (Cushing, 1978). In the upwelling cell, the hake will be offshore over the slope and around the frontal zone, the jack mackerel live at the shelf edge or beyond and the anchovies and sardines will live over the shelf. and inshore of the upwelling front (Fig. 8).

Biomass and Productivity

The most striking biological characteristic of upwelling areas is the magnitude of the biomass at all sizes of the biological size spectrum. From whales to phytoplankton, the concentration of plants and animals is usually many fold larger than it is in the rest of the ocean. The whaling ship captains of the last two centuries learned that the waters "downstream" from upwelling centers served as good hunting zones (Cushing, 1971). The large international fishing fleets of this century have all learned to locate in the Peruvian, southwest African, northwest African or Californian upwelling areas because of their abundance of fish (Walsh, 1977).

The abundance of life at the lower end of the size spectrum is usually only detectable by research vessels, and since the international focus on upwelling in the last two decades, we now have an abundance of studies of the phytoplankton, bacterioplankton and zooplankton levels (Boje and Tomczak, 1978; Richards, 1981). All the studies have found large numbers of organisms and high levels of productivity as expected, but they have also revealed a surprising degree of organization at the population level and a surprising level of time-dependent variability. As we have seen in the previous section, the phytoplankton species do not just co-occur randomly as in a large pot of pea soup; they group in specific assemblages according to the circulation (Blasco et al., 1981). This means that distinctly different assemblages will be found in the different provinces of an upwelling cell. The zooplankton also display an unexpected degree of organization within the system. Large organisms such as euphausids, large copepods and chaetognaths dominate the zooplankton from the shelf edge outward and small zooplankters (100-500 μ) dominate the inshore region (Smith and Whitledge, 1977; Thiriot, 1977). The euphausids and large copepods seem especially well-suited and may have evolved to exploit the large phytoplankton species characteristic of upwelling systems (Parsons, 1969).

To learn why upwelling areas are so biologically rich, one must look at what controls the first link in the food chain, the phytoplankton. First of all, extraordinary levels of both carbon and nitrogen uptake are reported. For two upwelling areas, northwest African and Peruvian, the mean carbon uptake rate per square meter is 2 g per day (Barber and Smith, 1981). This is 50 to 70 times the productivity in most other parts of the ocean. What characteristics of upwelling systems are responsible for stimulating this enhanced productivity? It is likely that there are important factors of which we are not yet aware, nevertheless, we do know that light, nutrients, circulation, water mass structure, time variability

and species composition are of major importance. Without light and nutrients phytoplankton cannot grow. Upwelling areas offer adequate but not exceptional light conditions, but as we have seen, they do offer exceptional nutrient conditions. Nevertheless, nutrients alone cannot explain the degree to which total biological productivity is enhanced in upwelling areas. (Codispoti, in press). The circulation is important in adding and distributing the nutrients throughout the upwelling area. Vertical motion is responsible for introducing and removing nutrients from the surface waters. Horizontal motion is responsible for distributing both nutrients and plankton throughout the upwelling system. The combination of both types of circulation provide the basis of the nutrient trap and helps to maintain the plankton in the upwelling system (Peterson et al., 1979 and Smith and Barber, 1981). So circulation is a key factor in conjunction with nutrients in augmenting the productivity in upwelling areas.

The water mass structure is another factor to be considered because if the upwelling is located in a zone of nutrient-poor subsurface waters then the upwelled waters will not stimulate plankton growth. This is why the productivity of the Peruvian upwelling is greatly weakened during an El Nino. The light is adequate, the upwelling still brings subsurface waters into the euphotic zone, and the horizontal circulation still provides a nutrient and plankton trapping mechanism; but the plankton do not bloom because the upwelled water is not sufficiently enriched with nutrients.

The time-variability factor is subtle. One might think that if upwelling areas were constantly injecting nutrients into the euphotic zone at a high rate, the upwelling areas would be even more productive than they are now. This might be true and Hutchings (1981) has documented such a case for a bay in the southwest African upwelling system, but most of the world's upwelling systems are episodic. Furthermore, the clupeoid fish that inhabit upwelling areas are adapted to this episodic behavior. The clupeoid larvae require certain phytoplankton species in high concentrations. Such concentrations would not be allowed to develop if the upwelling system supported a large steady population of microheterotrophs. Their grazing would prevent it. However, nutrient injection alternates with relaxation in todays upwelling systems and the microheterotrophs, with their relatively short generation time, cannot develop into large populations. As a result the large phytoplankton blooms that are vital to the survival of larval fish, can develop (Lasker, 1975, 1978; Mathisen et al., 1978). Once past this initial critical stage, the fish are not food-limited and develop into the large fish stocks characteristic of upwelling regions.

The species-composition factor is also subtle. Upwelling areas stimulate the growth of large-sized phytoplankters (Fig. 9). The turbulence of the ecosystem sustains the dominance of diatoms which have high growth rates and because of their large mass can in turn sustain large zooplankters (i.e. euphausids) or phytophageous fish. Thus the upwelling system tends to encourage high rates of carbon fixation into the phytoplankton and a rapid transfer of this carbon to higher trophic levels that can be exploited by man (Ryther, 1969, and Parsons, 1979). Because of the time variability factor, high growth rates and rapid transfer up the food chain are not always prevalent in upwelling areas, but they are prevalent with a greater frequency there than in other oceanic ecosystems. However, even when turbulent conditions in them do not occur and relaxation conditions encourage the development of dinoflagellate or mesodinium blooms, the ecosystem still provides a favorable nursery for fish larvae because both types of

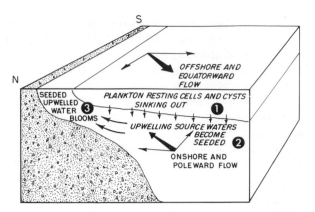

Fig. 9. The plankton trapping mechanism in upwelling areas. In addition to nutrient trapping, plankton trapping helps upwelling areas produce larger standing stocks of organisms than would be calculated from purely nutrient considerations.

phytoplankton are large and are preferred by the larvae. (Lasker, 1975, 1978).

Upwelling Areas as Plankton Traps and Larval Fish Nurseries

The increase in biological productivity in upwelling areas is not a linear extension of the productivity-nutrient relation found in other oceanic areas. The increase is much greater than expected (Codispoti, in press). A shorter, more

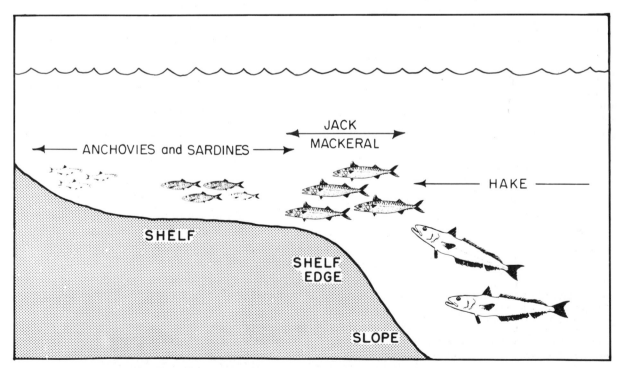

Fig. 10. The location in an upwelling area of the four characteristic fish found in these areas.

efficient food chain is one explanation for this increase as we have briefly discussed (Ryther, 1969; Parsons, 1979; Barber and Smith, 1980; Codispoti, in press), but an examination of the food chain in upwelling areas does not always support this argument (Cushing, 1971, 1978, and 1981). Another explanation can be derived by extending the nutrient-trap concept to plankton. The countercurrent circulation system in upwelling areas permits plankton reseeding and restocking (Peterson et al., 1979; Smith and Barber, 1981) and serves as a plankton trap (Fig. 10). Fish will not be concentrated by this mechanism, but if the mechanism helps to maintain high concentrations of phytoplankton and zooplankton in the upwelling area, then larval fish have a much higher probability of survival (Mathisen et al., 1978). This success will lead to higher adult food populations. The rationale for this statement is based on recent findings that: (1) adult fish are not food-limited in upwelling areas, and (2) the survival of larval fish depends on their encountering a dense plankton patch within a short time after hatching (Lasker, 1975; Mathisen et al., 1978). The probability of encountering such a patch in an upwelling area is much greater than elsewhere and for this reason large fish populations are found there. Thus upwelling areas contain high concentrations of plankton and fish because the counter current system traps the plankton and maintains it at such high concentrations that the survival rate of larvae fish is greatly increased. Since larval survival is the rate limiting step in fish production, such a larval nursery insures large adult fish populations.

ACKNOWLEDGEMENTS

We thank L. Codispoti and B. Jones for fruitful discussions that aided the development of this paper, and M. Colby, A. Levin, P. Oathout, and V. Wotton for their help in preparing it.
This is contribution No. 83002 from Bigelow Laboratory for Ocean Sciences. Our efforts were supported by ONR contract N00014-76-C-0271 and NSF grants OCE 8011187, and OCE 80-22631.

REFERENCES

Andrews, W.R.H. and D.L. Cram, 1969. Combined serial and shipboard upwelling study in the Benguela Current. Nature London, 24: 902-904.

Andrews, W.R.H. and L. Hutchings, 1980. Upwelling in the Southern Benguela Current. Prog. Oceanogr., 9(1): 1-81.

Banse, K., 1968. Hydrography of the Arabian Sea shelf of India and Pakistan and effects on demersal fishes. Deep-Sea Res., 15: 45-79.

Barber, R.T. and R.L. Smith, 1981. Coastal Upwelling Ecosystems. In: A. A. Longhurst, (ed.), Analysis of Marine Ecosystems. Acad. Press. pp. 31-68.

Barton, E.D., A. Huyer and R.L. Smith, 1977. Temporal variation

observed in the hydrographic regime near Cabo Correiro in the Northwest African upwelling region. February to April 1974. Deep-Sea Res. 24: 7-23.

Bass, A.E. and T.T. Packard, 1976. Physical, chemical and biological observations from JOINT II, R/V ALPHA HELIX Leg 0, 5-26 March 1976. CUEA Data Report 41.

Bjerknes, J., 1961. "El Nino" study based on analysis of ocean surface temperatures 1935-57. Inter-Amer. Trop. tuna Comm. Bull., 5: 219-307.

Blasco, D., 1971. Composition and distribution of phytoplankton in the region of upwelling off the coast of Peru. Inv. Pesq., 35: 61-112.

Blasco, D., M. Estrada and B.H. Jones, 1980. Relationship between the phytoplankton distribution and composition and the hydrography in the northwest African upwelling region near Cabo Corbeiro. Deep-Sea Res., 27A: 799-821.

Blasco, D., M. Estrada, B.H. Jones, 1981. Short term variability of phytoplankton populations in upwelling regions—The example of Northwest Africa. pp. 339-347. In: F.A. Richards (ed.), Coastal Upwelling. AGU publication 529 pp.

Boje, R. and M. Tomczak, 1978. Ecosystem Analysis and the definition of boundaries in upwelling region. pp. 3-11. In: R. Boje and M. Tomczak (eds.), Upwelling Ecosystems. Springer-Verlag New York, 303 pp.

Brink, K.H., D. Halpern and R.L. Smith, 1980. Circulation in the Peruvian upwelling system near 15°S. J. Geophys. Res., 85(C7): 4036-4048.

Brink, K.H., B.H. Jones, J.C. Van Leer, C.N.K. Mooers, D.W. Stuart, M.R. Stevenson, R.C. Dugdale and G.W. Heburn, 1981. Physical and biological structure and variability in an upwelling center off Peru near 15°S during March, 1977. In: F.A. Richards (ed.), Coastal Upwelling, Amer. Geophys. Union, Washington, D.C., pp. 473-495.

Brockmann, C., E. Fahrbach, A. Huyer and R. L. Smith, 1980. The poleward undercurrent along the Peru coast: 5 to 15°S. Deep-Sea Res., 27A: 847-856.

Codispoti, L.A., In press. Nitrogen in upwelling systems. In: E. J. Carpenter and D.G. Capone (eds.) Nitrogen in The Maine Environment. Elsevier.

Codispoti, L.A., R.C. Dugdale and H.J. Minas, 1982. A comparison of the nutrient regimes off Northwest Africa, Peru and Baja California. Rapp. P.-v. Réun. Cons. int. Explor. Mer, 1980: 184-201.

Cruzado, A. and J. Salat, 1981. Interaction between the Canary Current and the bottom topography. 167-175. In: F. A. Richards (ed.), Coastal Upwelling, Amer. Geophys. Union, Washington, D.C., 529 pp.

Cushing, D.H., 1971. Upwelling and the production of fish. Adv. Mar. Biol., 9: 255-334.

Cushing, D.H., 1978. Upper tropic levels in upwelling areas. pp. 101-110. In: R. Boje and M. Tomczak (eds.), Upwelling Ecosystems Springer-Verlag, New York, 303pp.

Cushing, D.H., 1981. The effect of El Nino upon the Peruvian anchoveta stock. pp. 449-457. In: F.A. Richards (ed.), Coastal Upwelling Amer. Geophys. Union. Washington, D.C., 529pp.

DeVries, T.J. and W.G. Peavey, 1982. Fish debris in sediments of the upwelling zone off central Peru: a late quaternary record. Deep-Sea Res., 28(1A): 87-109.

Dugdale, R.C., 1972. Chemical oceanography and primary productivity in upwelling regions. Geoforum, 11: 47-61.

Dugdale, R.C., J.J. Goering, R.T. Barber, R.L. Smith and T.T. Packard, 1978. Denitrification and hydrogen sulfide in the Peru upwelling region during 1976. Deep-Sea Research, 24: 601-608.

Dugdale, R.C., K. Brink and B. Jones, 1982a. OPUS. EOS, 63(3): 83.

Dugdale, R.C., J. MacIsaac and R. Barber. 1982b. Primary production in upwelling centers. EOS. 63(3): 62.

Enfield, D.B., 1980. El Nino Pacific eastern boundary response to interannual forcing. In: M.H. Glantz (ed.). Resource Management and Environmental Uncertainty. John Wiley and Sons., Inc. pp. 213-254.

Fahrbach, E. and J. Meincke, 1979. Some observations on the

variability of the Cabo Frio upwelling. CUEA Newsletter 8, 13-18.

Fraga, F., 1974. Distribution des masses d'eau dans l'upwelling de mauritanie. Tethys, 6(1-2): 5-10.

Friederich, G.E. and L.A. Codispoti, 1979. On some factors influencing dissolved silicon distribution over the northwest African shelf. J. Mar. Res., 37: 337-353.

Friederich, G.E. and L.A. Codispoti, 1981. The effects of mixing and regeneration the nutrient content of upwelling waters off Peru. pp. 221-227. In: F.A. Richards (ed.), Coastal Upwelling. AGU publication 529 pp.

Guillén, O. and R. Calienes, 1981. Upwelling off Chimbote. pp. 312-326. In: F.A. Richards (ed.), Coastal Upwelling. AGU publication 529 pp.

Halpern, D., 1974. Variations in the density field during coastal upwelling. Tethys, 6(1-2): 363-374.

Hart, T. and R. Currie, 1960. The Benguela current. Discovery Rep., 31: 123-298.

Hutchings, L., 1981. The formation of plankton patches in the southern Benguela current. In: F.A. Richards (ed.) Coastal Upwelling. AGU publication, pp. 496-506.

Jones, B.H. and D. Halpern, 1981. Biological and physical aspects of a coastal upwelling event observed during March 1974, off northwest Africa. Deep-Sea Research, 28A: 71-81.

Lasker, R., 1975. Field criteria for survival of anchovy larvae; the relation between inshore chlorophyll maximum layers and successful first feeding. Fish. Bull, U.S., 73: 453-462.

Lasker, R., 1978. The relation between oceanographic conditions and larval anchovy food in the California Current: Identification of factors contributing to recruitment failure. Rapp. P.-V. Reun. Cons. Int. Explor. Mer. 173: 212-230.

Margalef, R., 1978. What is an upwelling ecosystem. pp. 12-14. In: R. Boje and M. Tomczak (eds.), Upwelling Ecosystems, Springer-Verlag, New York. 303 pp.

Margalef, R., 1979. The organization of space. Oikos. 33: 152-159.

Margalef, R. and M. Estrada, 1981. On upwelling, eutrophic lakes, the primitive biosphere and biological membranes. pp. 522-529. In: F. A. Richards (ed.), Coastal Upwelling. AGU publication 529 pp.

Mathisen, O.A., R.E. Thorne, R.J. Trumble and M. Blackburn, 1978. Food consumption of pelagic fish in an upwelling area. In: R. Boje and M. Tomczak (eds.). Upwelling Ecosystems. Springer-Verlag, New York, 303 pp.

Minas, H.J., L.A. Codispoti and R.D. Dugdale, 1982. Nutrients and primary production in the upwelling region off Northwest Africa. Rapp. P.-v Réun. Cons. int. Explor. Mer, 180: 148-183.

Minas, H.J., T.T. Packard, M. Minas and B. Coste, 1982. Role of oxygen in the production-regeneration system of the coastal upwelling area off NW-Africa. J. Mar. Res., 40(3): 615-641.

Mittelstaedt, E., 1974. Some aspects of the circulation in the N.W. African upwelling area off Cape Blanc. Tethys, 6: 89-92.

Packard, T.T., H.J. Minas, T. Owens and A. Devol, 1977. Deep-sea metabolism in the eastern tropical North Pacific Ocean. In: N.R. Andersen and B.J. Zahuronec (eds.) Oceanic Sound Scattering Prediction. Plenum Press, New York, 859 pp.

Packard, T.T., R.C. Dugdale, J.J. Goering and R.T. Barber, 1978. Nitrate reductase activity in the subsurface waters of the Peru Current. J. Mar. Res., 36: 59-76.

Packard, T.T., D. Blasco and V. Jones, 1979. Special experiments with nitrate reductase and ETS in plankton from the Peru upwelling system. CUEA Tech. Rept. No. 54.

Packard, T.T., P.C. Garfield and L.A. Codispoti, 1982. Oxygen consumption and denitrification in the deep water of the Peru Current system. In: E. Suess and J. Thiede (eds.) Coastal Upwelling: Its Sediment Record. Proc. NATO Conf., Vilamoura/Algarve Portugal, Sept. 1-4, 1981.

Parsons, T.R., 1979. Some ecological experimental and evolutionary aspects of the upwelling ecosystem. South African Journal of Science. 75: 536-540.

Peterson, W.T., C.B. Miller and A. Hutchinson, 1979. Zonation and maintenance of copepod populations in the Oregon upwelling zone. Deep-Sea Research, 26: 467-494.

Quinn, W.H., 1974. Monitoring and predicting El Nino invasions. J. Appl. Meteor., 13: 825-830.

Redfield, A.C., B.H. Ketchum and F.A. Richards, 1963. The influence of organisms on the composition of sea-water. pp. 26-77. In; M.N. Hill (ed.), The Sea. Interscience Publishers, 1963, Volume 2.

Richards, F.A., 1981. Coastal Upwelling. Am. Geophys. Union. Washington, D.C. 529 pp.

Rojas de Mendiola, B., 1981. Seasonal phytoplankton distribution along the Peruvian coast. 348-356. In: F.A. Richards (ed.), Coastal Upwelling. AGU publication 529 pp.

Ryther, J.H., 1969. Photosynthesis and fish production in the sea. The production of organic matter and its conversion to higher forms of life vary throughout the world ocean. Science, 166: 72-76.

Ryther, J.H., D. Menzel, E. Hulburt, C. Lorenzen and N. Corwin, 1971. Production and utilization of organic matter in the Peru coastal current. Anton Bruun Rept. No. 4, Texas A&M Press.

Sen Gupta, R., A. Braganca, R.J. Noronha and S.Y.S. Singbal, 1980. Chemical oceanography of the Arabian Sea: Part V— Hydrochemical characteristics of the central west coast of India. Indian J. Mar. Sci., 9: 240-245.

Simpson, J.J. and A. Zirino, 1980. Biological control of pH in the Peruvian coastal upwelling area. Deep Sea Res., 27: 733-744.

Smith, R.L., 1968. Upwelling. Oceanogr. Mar. Biol. Ann. Rev., 6: 11-46.

Smith, R.L., 1978. Poleward propagating perturbations in currents and sea levels along the Peru coast. J. Geophys. Res., 83: 6083-6092.

Smith, R.L., 1981. A comparison of the structure and variability of the flow field in coastal upwelling region: Oregon, Northwest Africa and Peru. 107-118. In: F. A. Richards (ed.), Coastal Upwelling. Amer. Geophys. Union, Washington, D.C. 529 pp.

Smith, S.L. and T.E. Whitledge, 1977. The role of zooplankton in the regeneration of nitrogen in a coastal upwelling system off northwest Africa. Deep Sea Res., 24: 49-56.

Smith, S.L. and L.A. Codispoti, 1980. Southwest monsoon of 1979: Chemical and biological response of Somali coastal waters. Science, 209: 597-599.

Smith, W.O., Jr. and R.T. Barber, 1981. The role of circulation, sinking and vertical migration in physical sorting of phytoplankton in the upwelling center at 15°S. 366-371. In: F.A. Richards (ed.), Coastal Upwelling. Amer. Geophys. Union, Washington, D.C. 529 pp.

Subrahmanyan, R., 1959. Studies on the phytoplankton of the west coast of India. II. Physical and chemical factors influencing the production of phytoplankton, with remarks on the cycle of nutrients and on the relationship of the phosphate-content to fish landings. Proc. Indian Acad. Sci., B50(4): 189-252.

Thiriot, A., 1978. Zooplankton communities in the West African upwelling area. pp. 32-61. In: R. Boje and M. Tomczak (eds.). Upwelling Ecosystems. Springer-Verlag, New York. 303 pp.

Tont, S.A., 1981. Temporal variations in diatom abundance off southern California in relation to sea surface temperature, air temperature and sea level. J. Mar. Res., 39: 191-201.

Traganza, E.D., J.C. Conrad and L.C. Breaker, 1981. Satellite observations of a cyclonic upwelling system and giant plume in the California Current. 228-241. In: F. A. Richards (ed.), Coastal Upwelling. Amer. Geophys. Union, Washington, D.C. 529 pp.

Trumble, R.J., O.A. Mathisen, and D.W. Stuart, 1981. Seasonal food production and consumption by nekton in the northwest African upwelling system. pp. 458-463. In: F.A. Richards (ed.), Coastal Upwelling. AGU publication. 529 pp.

Walsh, J.J., 1977. A biological sketchbook for an eastern boundary current. pp. 923-968. In: E.D. Goldberg, I.N. McCave, J.J. O'Brien and J.H. Steele (eds.), The Sea. Volume 6. 1048 pp.

Walsh, J.J., T.E. Whitledge, J.C. Kelley, S.A. Huntsman and R.D. Pillsbury, 1977. Further transition states of the Baja California upwelling ecosystem. Limnol. Oceanogr. 22: 264-280.

Wyrtki, K., 1977. Sea level during the 1972 El Niño. J. Phys. Oceanogr., 7: 779-787.

Zuta, S., T. Rivera, A. Bustamante, 1978. Hydrolic aspects of the main upwelling areas off Peru. pp. 235-257. In: R. Boje and M. Tomczak (eds.), Upwelling Ecosystems. Springer-Verlag, New York, 303 pp.

22

El Niño and other Large-Scale Air-Sea Interactions

R. J. FLEMING and N. A. OSTENSO
National Oceanic and Atmospheric Administration,
Rockville, Maryland

ABSTRACT

The latest results from data analysis, field experiments and theoretical work have come together to create a clearer perception of the role of large scale oscillations in the atmosphere and ocean in influencing climate variability. The major oscillations discussed briefly in this paper are confined to Pacific and Indian Ocean Oscillations and include the Southern Oscillation, El Niño, and the Walker Circulation. A brief summary of how these "Oscillations" are related to each other and to other important oceanic and atmospheric changes is provided. An international observational and research program is being designed to advance our knowledge of these Oscillations. The observation phase of this project has several elements which require a composite observing system to measure important variables in the atmosphere and ocean. These variables include four major groups of parameters essential to the large-scale climate interactions: the wind field, surface heat and moisture fluxes, sea level and the thermal structure in the upper ocean. Specific observations of these parameters will build upon existing observing programs. However, much of the Pacific and Indian Ocean regions are not adequately observed either in the ocean or in the atmosphere. New observational techniques will have to be used to achieve the research goals of the program, and to help advance our current understanding to the point of practical applications.

INTRODUCTION

An international observational and research program is being designed to advance a theory for short-range climate variations which take place over major portions of the earth. Results from various national research programs, early results from the Global Weather Experiment, and recent theoretical work have come together to create a clearer perception of the role of large scale oscillations in the atmosphere and oceans in influencing climate variability.

This paper summarizes a portion of the current planning for this project within the United States. The following section briefly reviews the scientific basis for this air-sea interaction research program. The next section summarizes the expected elements of the project with emphasis on the observational requirements in the Pacific and Indian Oceans.

THE SCIENTIFIC PROBLEM

The major oscillations briefly discussed here are the Southern Oscillation, the Walker Oscillation and El Niño. (These will be collectively referred to as Oscillations in the remainder of this paper.) The Southern Oscillation, a large scale swaying of atmospheric mass between the Pacific and Indian Oceans, was first described in the literature as early as 1897 and clarified by Walker in 1923. This swing is seen in sea level pressure records as a seesaw of high pressure in the South Pacific Ocean and low pressure in the Indian Ocean at one end of the cycle, and opposite conditions at the other end of the cycle. The Walker Circulation is a zonal wind circulation system associated with the Southern Oscillation and is discussed later. The El Niño generally refers to an anomalous warming of the ocean off South America. During anti-El Niño conditions, the thermocline is shallow in the eastern tropical Pacific and the upwelling caused by the prevailing winds bring cold nutrient-rich waters near the surface. Phytoplankton (miniscule floating plants) flourish in their new sunlit environment and ignite a food chain which supports zooplankton, fish,

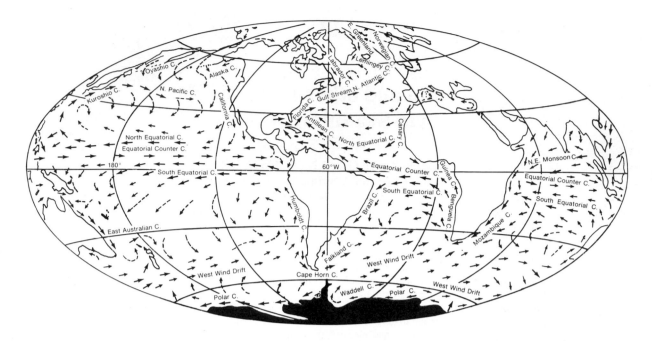

Figure 1. Surface currents of the World Ocean in the northern winter.

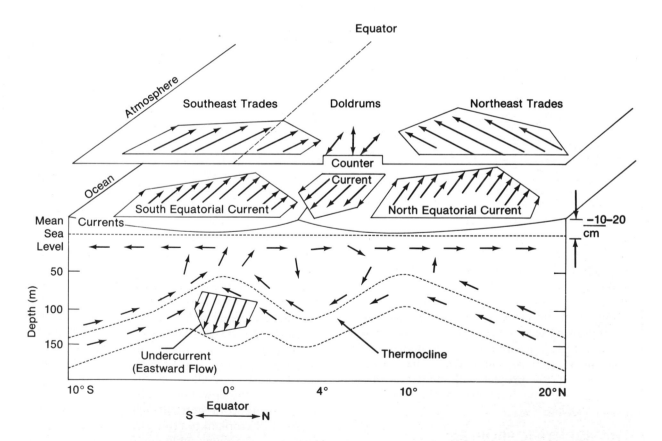

Figure 2. Schematic representation of surface winds and equatorial current systems in a meriodional vertical cross-section in the tropics (modified from Wyrtki 1978).

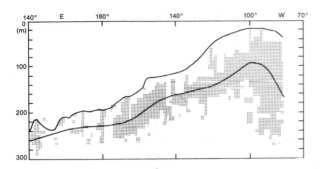

Figure 3. Depth (m) of the 14°C isotherm along the equator. The scatter diagram is from bathythermographic observations between 1°N and 1°S (after Meyers, 1979).

America which cause severe flooding and extensive crop damage. Australia gets more rain than normal just prior to El Nino and drought conditions after its onset. Studies have linked variations in the Indonesian winter monsoon region and the summer monsoon region of Pakistan, India and China to the Southern Oscillation and Walker Circulation. Over half the world's population is affected by these phenomena.

These Oscillations are physically connected. A simple chronology of their main features which link together as part of a large-scale air-sea interaction process is provided below.

Avoiding an extensive discussion of the general circulation of the atmosphere, suffice it to say that mid-latitude westerlies and low latitude easterlies drive large-scale ocean gyres as pictured in Fig. 1. Near the equator, a complex current system exists between the gyres. Fig. 2 shows a representation of these currents. The arrows above indicate the forcing southeast and northeast trade winds, the arrows below indicate the westward flowing ocean currents (the South Equatorial Current and the North Equatorial Current) and the eastward flowing currents (the Equatorial Countercurrent and the Equatorial Undercurrent).

birds and man. (Such coastal upwelling occurs over only 1% of the world's oceans, but supports more than half of the world's commercial fish stocks.)

When the warm waters of El Nino come, there is widespread destruction. Plankton, fish and birds die – man only loses economically. In Peru, the anchovy fishing is drastically reduced and the death of millions of sea-birds (which feed on the anchoveta) limits the annual take of nitrate rich guano. This event causes major fluctuations in the world fish, poultry and fertilizer markets. El Nino usually brings heavy rains along the coast of South

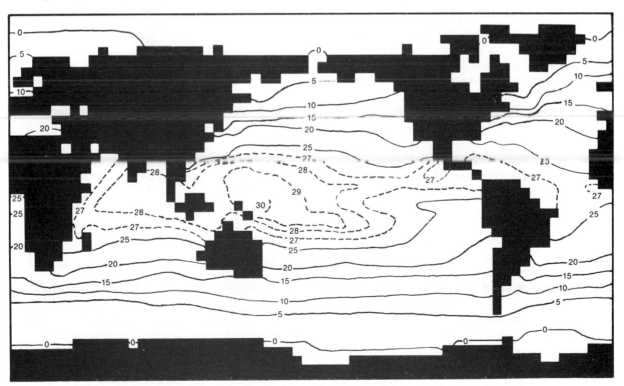

Figure 4. Observed annual mean surface temperature of the oceans (from Bryan, 1978).

The persistent trade winds drive these current systems and also systematically "pile up" a mass of warm water in the western Pacific; the latter is readily discernible from sea level records from coastal and island stations. The ocean adjusts to this excessive mass build-up in the surface layers of the western Pacific with the thermocline being deep in the west and extending upward toward the surface in the east (see Fig. 3). For reasons only partially understood, the world's warmest ocean temperatures appear in this warm pool of the western Pacific (an example is indicated in Fig. 4).

Let us now return to the atmosphere and see what role this persistent "hot spot" in the western Pacific plays with the overlying atmospheric circulation. Above this warm pool of water we have the energy source for a circulation made up of the zonal components of low and high level tropospheric equatorial winds. One cell extends westward over the Indian Ocean and another extends eastward over the Pacific Ocean (see Fig. 5). The top of Figure 5 shows the low level convergence over the warm pool (west of the dateline), warm moist air rising above it, the air moving eastward in the upper branch, descending as dry air over the convection-free eastern Pacific (where much cooler sea surface temperatures prevail), and finally returning westward at low levels as part of the trade wind system. This circulation was termed the Walker Circulation by Bjerknes (1969).

The normal position of the Walker Circulation is as shown at the top of Fig. 5. A shift or weakening of this circulation is statistically linked to the El Niño. When an El Niño situation occurs, it is as if the thermal low shifts east of Canton Island (170°W) and the anomalous circulation is as shown at the bottom of Fig. 5. Canton Island's winds may actually reverse at low and high levels.

The Walker Circulation is also strongly connected to the Southern Oscillation. The Southern Oscillation is a swing of the mass of the atmosphere as seen in the sea level pressure records as a seesaw of high pressure in the South Pacific and low pressure in the Indian Ocean at one end of the cycle, and opposite conditions at the other end of the cycle (see Fig. 6). The "period" of the Southern Oscillation varies from 2-10 years, but Trenberth (1976) finds the most coherent signal in fluctuations with periods of 3-6 years.

A brief summary of the main features in the chronology of a typical Oscillation follows. This has been drawn from the work of Wyrtki (1975,

1979), Philander (1981), Rasmusson and Carpenter (1982), and Busalacchi and O'Brien (1981). Six phases can be identified and these are (using the nomenclature of Rasmusson and Carpenter in describing the composite of several events): Build-up (-3), Onset (0) Peak (+4), Transition (+9), Mature (+13) and Termination (+18). The numbers in parentheses are the number of months relative to the Onset of El Niño, which usually occurs in December.

We caution that, as a composite, the indicated months of each phase are only representative. The precise timing of a particular phase of a particular event might differ from the months given above. Moreover the changes associated with the annual cycle will not be emphasized here in this chronology but are an important aspect to be studied (e.g., see Ramage, 1981). With these caveats, the following key equatorial atmospheric and oceanic features are associated with each phase of the Oscillation.

Build-up Phase: September (-3 months)

This first phase develops over a long period with the Southern Oscillation Index (e.g., the surface pressure at Tahiti minus the surface

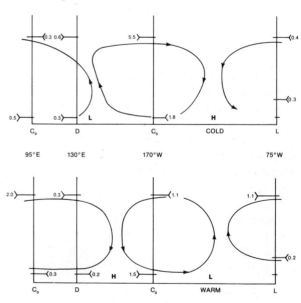

Figure 5. Schematic representation of the Walker Circulation. The top panel applies to the normal state with cold water along the South America Coast. The wind arrows denote actual average departures from normal of zonal winds (in m/s) at 200 and 850 mb. The cross section along the equator shows Cocos Island, Darwin, Canton Island and Lima. The bottom panel indicates the opposite anomalies during El Niño warm water months (from Julian and Chervin, 1978).

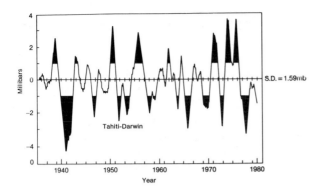

Figure 6. Time series of the Southern Oscillation Index given by the difference between the normalized sea level pressure anomalies at Tahiti and Darwin. The series has been smoothed with a low pass filter which effectively removes fluctuations of less than about 1 year (Trenberth, personal comm., 1981).

pressure at Darwin) slowly building to a positive maximum. (The indicated date of September (—3 months) is only a representative snapshot in time.) The following conditions have evolved and prevail: stronger than normal (anomalous) sustained easterly surface winds, strong upwelling along South America and the equator, intensified western equatorial currents (especially the South Equatorial Current), an accumulation of warm water and a rise of sea level in the western Pacific (a downwelled pycnocline), and a depressed sea level in the eastern Pacific (an upwelled pycnocline).

In the atmosphere, the so-called South Pacific Convergence Zone (SPCZ), an area of persistent cloudiness from the equator near New Guinea stretching diagonally southeast across the Solomon Islands toward higher latitudes, has been displaced southwest of its normal position. This location of the SPCZ brings considerably more rain to eastern Australia. At this point, for reasons open to conjecture, the seeds of change have already been sown. Anomalous westerly winds begin to provide an eastward wind stress on the ocean just west of the dateline—apparently initiating the next phase.

Onset Phase: December (0 months)

Anomalous atmospheric conditions now exist west of the dateline: surface easterly winds have been weakened or, in some cases, replaced by westerlies, and positive SST anomalies extend eastward from the usual warm pool. The Southern Oscillation Index is now falling rapidly as pressure is falling in the eastern Pacific and rising in the

west. From the pressure changes and anomalous conditions, west of the dateline, it is evident that the Walker Circulation has considerably weakened.

The collapse of the southeast trades, west of the dateline, destroys the force that has maintained the eastward pressure gradient in the ocean in that region. The Coriolis force being small near the equator, the water sloshes toward the east, down the pressure gradient toward South America. (In the mathematical terms of equatorial linear dynamics, meteorologists and oceanographers would agree that such an abrupt change in the zonal winds (or wind stress) would initiate equatorially trapped eastward propagating Kelvin waves.) At this time warm water begins to appear off the coast of South America.

Peak Phase: April (+4 months)

At this stage we have: a continuous draining of warm water from the west, intensified eastward flow in the ocean currents, and a poleward spreading of the thermocline disturbance along the South American coast. The warm SST near the dateline spreads eastward merging with the warm SST anomaly moving westward from the coast of Peru and Ecuador (the latter takes five months to reach the Line Islands). This westward movement is close to what one would expect from Rossby waves excited at the eastern coast—either from the impinging Kelvin waves or due to local changes in the wind stress.

Returning to the atmosphere, we find that the anomalous surface westerly winds have spread eastward to 160°W (past Canton Island at 170°W) and are consistent with a weakening of the Walker Circulation (as in bottom of Figure 5). The Intertropical Convergence Zone (ITCZ), a persistent east-west band of clouds normally located at about 7-8°N moves south of its usual position toward the equator. The SPCZ moves northeast of its normal location (see Fig. 7).

Transition Phase: September (+9 months)

Precipitation, wind and SST anomaly fields reveal a further displacement of the SPCZ to the northeast. An area of large positive SST anomalies now covers a vast area of the Pacific (see Fig. 8). This period is called the Transition Phase because presumably a weak El Niño would end at this time.

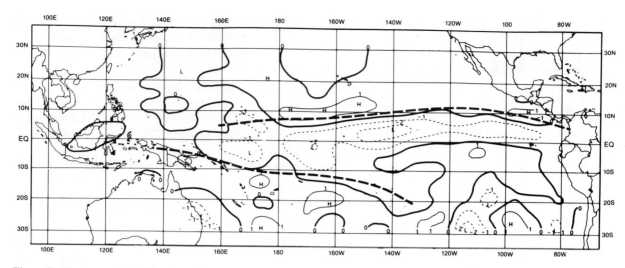

Figure 7. The heavy dashed line indicate the normal position of the Intertropical Convergence Zone (ITCZ) and the South Pacific Convergence Zone (SPCZ).

Mature Phase: January (+13 Months)

In a major El Niño, there is a second peak in SST, sea level and pycnocline depression in the eastern Pacific. The El Niño survives the semi-annual rise in the pycnocline that occurs in the Transition Phase and goes on to deepen in the Mature Phase. In the atmosphere, the shifting of the two major convergence zones (the ITCZ moving closer to the equator and the SPCZ moving to the northeast) leads to a much smaller wedge-shaped dry zone in the eastern tropical Pacific and anomalous precipitation in the central equatorial Pacific.

The results of Rasmusson and Carpenter show

that anomalous flow in the western Pacific favors a weakened northeast (winter) monsoon circulation. This agrees with other scientific studies that have correlated the Southern Oscillation to both the winter and summer monsoons. The anomalous circulation in the western Pacific and the significant northeast shift of the SPCZ set the stage for drought conditions over Australia.

Termination Phase: June (+18 months)

A rapid return to normal conditions occurs within a two-month period. Apparently several factors lead to this rapid return and produce the

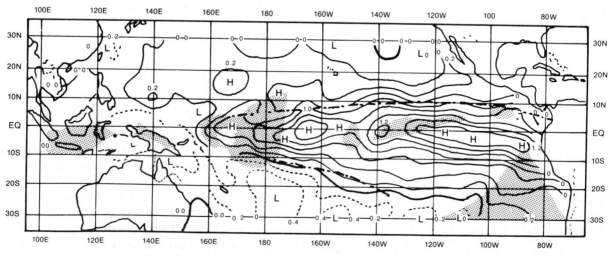

Figure 8. Transition phase SST anomaly (°C) composite pattern averages for August-October of the El Nino year (from Rasmusson and Carpenter, 1982).

following results: the trade winds increase, the South Equatorial Current increases, the Equatorial Countercurrent decreases, sea level drops along the coast of South America and rises even more rapidly in the western Pacific. Thus the stage is set for the gradual build-up for the next cycle.

The large pool of warm water which covers a vast part of the eastern and central Pacific from the Peak Phase through the Mature Phase represents a significantly different kind of forcing for the atmosphere. A number of recent scientific articles have related this equatorial forcing to weather and climate conditions over North America. Horel and Wallace (1981) have correlated SST time series to Northern Hemisphere geopotential height fields thus showing well defined teleconnection patterns. Opsteegh and Van den Dool (1980) have modeled such teleconnections, Hoskins and Karoly (1981) and Webster (1981) have explained these patterns based upon Rossby wave propagation on a sphere. Webster (1982) has shown that in the case of weak advection and westerly winds over the equatorial warm water anomaly, the maximum response is downstream at higher latitudes. These conditions

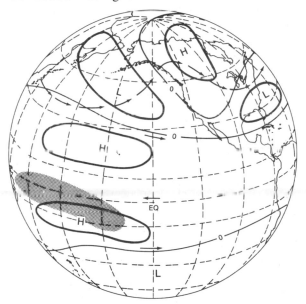

Figure 9. Schematic illustration of hypothetical global pattern of middle and upper tropospheric geopotential height anomalies during a Northern Hemisphere winter during an episode of warm SST in the equatorial Pacific. The heavy arrows reflect the strengthening of the subtropic jets in both hemispheres and stronger easterlies near the equator. The light arrows depict an actual streamline as distorted by the anomaly pattern, with pronounced troughing over the central North Pacific and ridging over western Canada. Shading indicates enhanced cloud and rain (from Horel and Wallace, 1981).

of weak advection and westerlies are present in the central Pacific sequence we have described above (recall the westerly wind intrusion past 160°W in the Peak Phase). Thus, we have a theory for tropical to high latitude teleconnections. Horel and Wallace have composited winter records associated with these warm pool situations and confirm that the observations match the theory (see Fig. 9).

When the atmospheric conditions are as indicated in Figure 9, the subtropical jet strengthens, the Aleutian Low deepens, the Polar jet sweeps much farther north over western Canada than normal and then swings back much deeper than normal to the south over the eastern United States. These conditions describe the situation which occurred during the 1976-1977 winter (one year after the onset of the 1976 warming). There were abnormally warm conditions in Alaska, the western part of the United States experienced a very severe drought (with water rationing in California), and the eastern half of the country had extremely severe winter weather. The adverse economic impact of that particular winter was the worst on record and has not been matched since.

Several studies have related these oscillations to the summer monsoon. The latest work by Rasmusson and Carpenter (private communication) indicates that over the 105 year period from 1875 to 1979 there have been 25 moderate/strong warming events in the central and eastern Pacific. In 19 of the 25 events the Indian monsoon rainfall was below normal. They also found that the five largest drought conditions occurred during El Niño events.

OBSERVATIONAL REQUIREMENTS

While these Oscillations described above always alter the climatic conditions, the details differ from event to event. For example, the response over the United States after the 1972 event was different from the 1976 warm water event. Thus, the large-scale air-sea interaction program (which is still in the planning stages and which is currently called the Tropical Ocean and Global Atmosphere (TOGA) program of the World Climate Research Program) must determine the cause and effect relationships associated with these Oscillations and determine the factors that make their influence different from event to event.

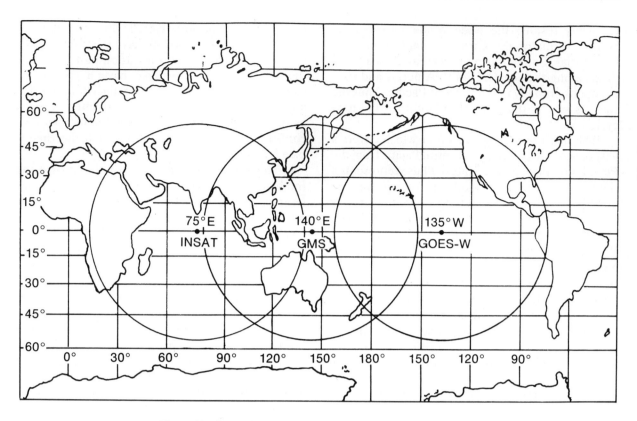

Figure 10. Geostationary Satellite Coverage expected during TOGA.

The major program elements to meet the scientific challenge outlined above are still being formulated. They will include: specific ocean and atmospheric monitoring, equatorial Pacific, Indian (and eventually, Atlantic) field studies, analysis and diagnostics, and modeling and prediction research.

It has taken many years to assemble our present knowledge of the chronology of these Oscillations. Many details are still not clear. This is so for several reasons: (1) the Pacific and Indian Oceans are vast areas, (2) the data are sparse in many large regions of the ocean and atmosphere (in some areas we do not even know the long-term mean values of key variables), and (3) some of the indices, which must be measured, suffer from a poor signal-to-noise ratio.

Specific long term (10 year) observations must be made over the entire Pacific basin and Indian Ocean region in order to provide the needed information to study the interannual variability of the Oscillations. A two-year intensive monitoring of a complete El Niño must be made. This will be done in different regions of the atmosphere and ocean in different phases of the event.

The monitoring must include measuring four major groups of parameters essential to the large-scale climate interactions: the wind field, surface heat and moisture fluxes, sea level and the thermal structure in the upper ocean. Specific observations of these parameters will build upon existing observing programs.

The World Weather Watch (WWW) network of the WMO is adequate for many aspects of the program, but a modest enhancement of upper air observations in the low latitudes is critical to achieving the program goals. The requirement for monitoring the atmospheric circulation near the equator can partially be met by establishing or continuing to maintain routine surface and upper air wind observation stations on a small number of islands which are located near the equator at strategic longitudes. These include the Galapagos, Christmas, Canton, Nauru and Tarawa Islands.

Satellite-derived winds (obtained from cloud motions) are extremely important and three geostationary satellites (from the United States, Japan and India) will be available (see Fig. 10). Winds from ships-of-opportunity and from the mobile ship program of the WWW can supplement

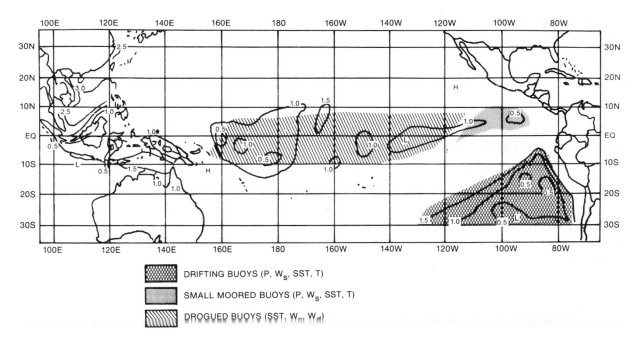

DRIFTING BUOYS (P, W_s, SST, T)

SMALL MOORED BUOYS (P, W_s, SST, T)

DROGUED BUOYS (SST, W_m, W_d)

Figure 11. Strategic areas for buoy deployment in the TOGA field program.

the satellite winds to obtain the surface wind stress. Surface wind stress measured directly from a satellite would be extremely beneficial, and efforts to have a scatterometer on a polar orbiting satellite are underway in several countries.

SST is a crucial parameter and efforts to combine SST's from every source (surface to satellite) into a quality data set will be an important part of the program. In addition to improved SST analyses, changes in the thermal structure of the upper ocean can be observed by XDT profiles taken from ships of opportunity. Basin-wide ship-of-opportunity programs are currently in operation but improvements can be made in the tropical regions. Additional ships will be recruited in the eastern Pacific. In the central Pacific, cooperative efforts between American and French scientists collect XBT sections between New Caledonia, Tahiti, Japan, California and Panama. In the western Pacific, Japanese fishing vessels, research vessels, and other voluntary observing ships are participating in an organized activity as an essential part of WESTPAC. An expanded XBT ship-of-opportunity program in the Indian Ocean would be extremely valuable.

Changes in the topography of the sea surface, as measured by sea level gauges on islands, also provide valuable information on these Oscillations. Sea level measurements provide a measure of the geostrophic flow in the major currents, and also indirectly provide information on the thermocline topography. The sea level network in the Pacific is now quite good and must be maintained. Altimeter measurements from satellites, determining the changes with time of the slope of the sea surface relative to the geoid, can also provide information on geostrophic flow. Satellites with this capability are being planned, but may not be available for the beginning of this program.

Various current meter moorings will be set. A current meter mooring at 110°W has been in place since 1979 as part of the United States program called Equatorial Pacific Ocean Climate Studies (EPOCS). This location is considered strategic and scientists have recommended that this be maintained for the full 10-year period. Other moorings at a few strategic locations in the Pacific and Indian Oceans are needed.

Certain geochemical measurements (e.g., freon, tritium) could provide additional evidence regarding oceanographic and biological processes. Recent progress in freon measurements and analyses indicates that freon tracers have the potential of providing information on: heat transport within the thermocline, vertical transfer processes, and water mass movements over interdecadal time scales.

Last, but far from least, an array of meteorological and oceanic buoys will be an absolute necessity for the success of the program. Buoys of the type used in the Global Weather Experiment and a new generation of drifting platforms will be deployed to help measure the atmospheric pressure and wind fields, ocean currents, surface heat and moisture fluxes, and the thermal structure in the upper ocean. These buoys will be deployed in at least the following three strategic areas: (1) the southeast tropical Pacific, in a region of almost non-existent surface observations, (2) the near-equatorial latitude band, across the entire breadth of the Pacific, and (3) in the poorly sampled region of the ITCZ. Plans for buoys in the Indian Ocean and at higher latitudes of the Southern Hemisphere are being formulated.

The buoy network in the southeast tropical Pacific must be adequate to resolve the major variations in the southeast trade winds and the large scale variations in the southeast Pacific high. Buoys with thermistor chains measuring ocean temperature as a function of depth are also required in this region. The combined meteorological/oceanographic buoy network in the near-equatorial belt must be adequate to resolve the wind fluctuations of importance to the large-scale equatorial ocean dynamics. Ideally, we would also like to measure the east-west gradient of the moisture flux in this region. A new generation of surface flux drifters is in the development stage now. It is hoped that platforms with the ability to measure parameters in addition to those mentioned above (e.g., air-sea temperature difference, humidity, solar radiation, and possibly wind stress and precipitation from subsurface acoustic data) will be available. The strategic region near the ITCZ (see Fig. 11) is particularly void of ship traffic and is an area of few observations of any kind. Here we may have to use small, inexpensive moored buoys in order to keep them in the region.

SUMMARY

It was indicated earlier that over half of the world's population is affected by the climatic variations induced by these Oscillations. Having recognized their importance, the challenge now is to understand the cause and effect relationships affecting the various phases of these Oscillations, and how and why they differ from event to event.

This remains a formidable task. Many factors can affect climate over interannual time scales. Nevertheless, the largest climatic signal (other than the annual cycle) in short-term climate change has been identified and a consistent theory is emerging. There is more than a ray of hope that seasonal forecasts might be provided several seasons in advance.

ACKNOWLEDGEMENTS

This paper considered only a part of a Program Development Plan prepared for the National Oceanic and Atmospheric Administration (NOAA). A great many meteorologists and oceanographers in the United States (both in NOAA and in other federal agencies and universities) contributed to that plan. Their efforts are hereby acknowledged.

REFERENCES

Bjerknes, J., 1969. Atmospheric Teleconnections from the Equatorial Pacific. Mon. Wea. Rev., 97: 163-172.

Bryan, K., 1978. The Ocean Heat Balance. Oceanus, 21(4): 19-26.

Busalacchi, A. J. and J. J. O'Brien, 1981. Interannual Variability of the Equatorial Pacific in the 1960's. J. Geophys. Res. 86(C11): 10901-10907.

Horel, J. D. and J. M. Wallace, 1981. Planetary Scale atmospheric phenomena associated with the Southern Oscillation. Mon. Wea. Rev., 109: 813-829.

Hoskins, B. J. and D. J. Karoly, 1981. The steady linear response of a spherical atmosphere in thermal and orographic forcing. J. Atmos. Sci., 38: (in press)

Julian, P. R. and R. M. Chervin, 1978. A study of the Southern Oscillation and Walker Circulation phenomenon. Mon. Wea. Rev., 106: 1433-1451.

Meyers, G., 1979. Annual variation in the slope of the 14°C isotherm along the equator in the Pacific Ocean. J. Phys. Oceanog. 9: 885-891.

Opsteegh, J. D. and H. M. Van den Dool, 1980. Seasonal differences in the stationary response of a linearized primitive equation model: prospects for long range forecasting. J. Atmos. Sci., 37: 2169-2185.

Philander, S. G. H., 1981. The response of equatorial oceans to a relaxation of the trade winds. J. Phys. Oceanogr., 11: 176-189.

Ramage, C. S. and A. M. Hori, 1981. Meteorological aspects of El Niño Mon. Wea. Rev., 1827-1835.

Rasmusson, E. M. and T. H. Carpenter, 1982. Variations in tropical sea surface temperature and surface wind fields associated with the Southern Oscillation/El Niño. Mon. Wea. Rev., (In press)

Trenberth, K. E., 1976. Spatial and Temporal Variations in the Southern Oscillation. Quart. J. Roy. Meteor. Soc., 102: 639-653.

Walker, G. T., 1923. Correlation in seasonal variations of weather, VIII: A preliminary study of world weather. Mem. of the Indian Meteor. Depart., 24(4): 75-131.

Webster, P.S., 1981. Mechanism determining the atmospheric response to sea surface temperature anomalies. J. Atmos. Sci., 38: 554-571.

Webster, P. S., 1982. Seasonality in the local and remote atmospheric response to sea surface temperature anomalies. (submitted to JAS)

Wyrtki, K., 1975. El Niño-The dynamic response of the equatorial Pacific Ocean to atmosphere forcing. J. Phys. Oceanogr., 5: 572-584.

Wyrtki, K., 1978. Lateral Oscillations of the Pacific Equatorial Counter-current. J. Phys. Oceanogr., 8(5): 530-532.

Wyrtki, K., 1979. El Niño. La Recherche, 10: 1212-1220.

23

Inorganic Interactions of Silica-containing Minerals with Seawater

D. C. HURD
Shell Development Company,
Houston, Texas

ABSTRACT

Equations and rate constants for silica-containing minerals dissolving in 25° C, pH 8 seawater suggested by previous workers must be corrected for the differences in suspended matter concentration (surface area of solid per unit volume of solution) and temperature between their work and that of average seawater before these equations can be applied to the problem of silica cycling in the oceans. Thus, correction relative to suspended matter concentration in average seawater reduces their rate constants by 6-7 orders of magnitude; correction for temperature, another order of magnitude; and another correction for that portion of the oceans which is undersaturated with respect to clays gives their corrected range of maximum dissolution rates as $7\text{-}700\text{x}10^{-6}$ micro-moles silica/liter/year. From this calculation it can be shown that $1.5\text{-}150\text{x}10^6$ years will be required for clay minerals dissolving to replace the present amount of dissolved silica in the oceans; since the residence time for dissolved silica in the oceans is approximately 1 to $2\text{x}10^4$ years, the above rate is considered to be somewhat sluggish. Possible flaws in the above reasoning and estimates of the heterogeneous solution rate constants for several clay minerals in cm/sec are discussed, as well as the role of clays as sinks for dissolved silica. The dissolution rates of decomposing phytoplankton and estimates for the heterogeneous solution rate constant for biogenic silica are 10 to 150 times smaller than acid-cleaned biogenic silica under the same conditions. Using these figures, the time required for the dissolution of various thickness tests can be estimated. Finally, the annual rates of silica production by organisms are considered and the dissolution of the bulk of this material compared with several insitu dissolution rates' estimates. The author concludes from the above lines of reasoning (rates of dissolution and adsorption; solubilities; production rates) that biogenic silica is responsible for the control of dissolved silica in the world's oceans.

INTRODUCTION

Calculations involving the cycling of silica in the oceans have been reviewed by numerous investigators within the past few years (Calvert, 1974; Heath, 1974; Wollast, 1974; Burton and Liss, 1973; Lisitzen et al., 1967); investigators have also studied the changes occurring in dissolved silica concentration as rivers mix with seawater (Fanning and Pilson, 1973; and others listed in Burton and Liss, 1973); finally, several researchers have begun to describe the actual dissolution rates (Lerman et al., 1975; Kido and Nishimura, 1975; Edmond, 1974; Tsunogai et al., 1974; Hurd, 1973; Lal and Lerman, 1973; Matsumoto, 1973; Wollast and Debroeu, 1971; van Lier, deBruyn and Overbeek, 1960; Hurd et al., 1979; Hurd and

Birdwhistell, 1983) and the solubilities (Maynard, 1975; Fournier, 1973; Hurd, 1973; Siever and Woodford, 1973; Siever, 1968; Mackenzie et al., 1967; Hurd and Theyer, 1975; Hurd et al., 1979) of alumino-silicate and silica minerals in seawater. The author has a number of basic disagreements with several of the above interpretations of experimental data and oceanographic observations; the following article does not attempt to re-do the entire silica cycle but rather to consider several sections in detail: the in situ dissolution rates of alumino-silicates versus biogenic silica—the latter appearing more likely to be controlling the dissolved silica concentrations in the deep oceans. It should also be noted that the author regards the following arguments as a useful exercise rather than the final word on the subject;

those interested in this subject should not hesitate to question any or all of the arguments presented here.

Dissolution Rates and Rate Constants of Alumino-silicate Minerals in the Oceans

Recently, Lal and Lerman (1973) and Lerman et al. (1975) have presented several mathematical models describing the dissolution kinetics of a variety of silicates in 20° to 25° C, approximately pH 8 seawater. The models were applied to a set of experiments begun some eight years earlier and described a number of times (Mackenzie et al., 1967; Mackenzie and Garrels, 1965). The authors suggest that alumino-silicates may be important in controlling the concentration of dissolved silica in seawater by both dissolution and precipitation. This concept has been contested by a number of

workers (Maynard, 1975; Schink et al., 1974; Siever and Woodford, 1973; Broecker, 1971; Grill, 1970; Fanning and Schink, 1969; Amit, 1969). I wish to extend the arguments of these latter authors and further to point out a number of possible inconsistencies in the reasoning behind the application of Lerman et al.'s results (1975) directly to the world's oceans. That these authors have chosen to ignore the effects of particle concentration per unit volume (i.e. surface area of solid per unit volume of solution) and temperature is central to my criticism; certain qualifying assumptions regarding these parameters as applied to the oceans do in fact greatly alter the authors' conclusions regarding the reaction times obtained from their solution rate constants.

Lerman et al. (1975) use a first-order dissolution equation to describe approximately the dissolution behaviour of their alumino-silicate

TABLE 1. Specific surface area of various clays and sediments

Clay mineral and description	Nitrogen adsorption m²/gm	Glycerol adsorption m²/gm
Kaolinite[1]	7-30	14, 23
Montmorillonite[1]	15-100 external	500-750 total
		15-35 external
"Illite"[1]	67-100	76, 91
Bentonite[1]		300-575 total
		35-116 external
Chlorite[1]		
Vermiculite[1]	.5	275
Halloysite[1]	ca. 45	76, 102
Endellite[1]		430 total 27 external
Allophane[1]	ca. 300	ca. 260
Red clay[2]	30-70	
Terrigenous[2]	11-17	
shelf sediment	11-17	
shallow, 50 m[4]	2.7	
deeper, 200 m[4]	45.7	
Palygorskite[3]		450-550
Diatoms[5]	89-123	
Diatoms[6]	80-150	
Sponge spicules[7]	0.1-1	
Radiolarians		
0-10 mybp[8]	55-270	
10-40 mybp[8]	2-50	
Silicoflagellates[6]	40-60	

[1]Diamond and Kinter (1956)
[2]Weiler and Mills (1965)
[3]Slabough and Stump (1964)
[4]Banin et al. (1975)
[5]Lewin (1961)
[6]Lawson et al. (1978)
[7]Lawson and Hurd (personal data)
[8]Hurd and Theyer (1975)

minerals:

$$dCsol/dt = K1 (Csat-Csol) \quad (1)$$

where Csol is the dissolved silica concentration at time t in moles/cm^3, Csat is the equilibrium concentration of dissolved silica under the conditions of the experiment, and K1 is a rate constant in sec.$^{-1}$. Others (see Hurd, 1973 and Hurd and Birdwhistell, 1983 and references therein for a more complete discussion of the problem) have preferred to use the following equation for relatively low specific surface area samples:

$$dCsol/dt = K2 (Csat-Csol)S \quad (2)$$

where S is the surface area of the solid per unit volume of solution cm^2/cm^3=cm^{-1}) and K2 is the heterogeneous rate constant in cm/sec. The main difference between the two equations is the separation of the rate constant K1 into another rate constant K2 and a surface area per unit volume term. Examples of the range in specific surface of common silicate minerals are given in Table 1. If S and K2 are combined to produce K1 because of the absence of knowledge about S, then it becomes difficult to compare one set of investigators' experiments with those of another. The following discussion points out several of these difficulties and makes reasonable estimates so that comparisons are possible.

Let us first consider the effect of altering their suspended particulate concentrations to fit deep ocean concentrations. In the experiments of Lerman et al. (1975), a suspension concentration of 5 gm/l was used. In the deep oceans, however, a concentration range of 1 to 10x10^{-7} gm/l is more commonly encountered or calculated (Kido, 1974; Chester and Stoner, 1972; Manheim et al., 1972; Copin-Montegut and Copin-Montegut, 1971; Feely et al., 1971; Lisitzin, 1970; Manheim et al., 1970; Toyota and Okabe, 1967; Armstrong, 1958). These findings suggest that all of the rate constants given as K1, in sec.$^{-1}$ (see Table 2), should be reduced by 6 to 7 orders of magnitude, assuming that the authors' assertion that the physical properties of the minerals they studied are in fact representative of suspended oceanic sediments is correct.

The effects of temperature are no less important to the rate of dissolution of alumino-silicates. The average temperature of the world's oceans is not 25° C as implied by the conclusions of Lerman et al. (1975) but rather 4° C. If the theoretical activity diagrams given by Helgeson and Mackenzie (1970) are correct, then the solubility of alumino-silicates decreases by at least a factor of 2 as the temperature of seawater decreases from 25° C to 0° C. Based on these diagrams as well as work by Hurd et al. (1979) for the pH range 7.5 to 8.3, at 4° C in seawater, I suggest the alumino-silicate minerals studied by Lerman et al. (1975) are associated with the following silica concentration ranges: kaolinite, 80±25 uM; montmorillonite, 200±50 uM; illite, analcite 60±25 uM; and chlorite, 40±35 uM.

If we then consider the distribution of both

TABLE 2. Estimated heterogeneous solution rate constants from data of Lerman et al. (1975)

Mineral and sample description	Rate constants of Lerman et al., seconds^{-1}	Estimated ranges of specific surface area, m^2/gm	Estimated value of S, cm^2/cm^3	Estimated value of K2 at 25°C, pH 8, cm/second
1. Quartz Lerman et al. (1975)	1 x 10^{-7}	0.4-0.7	230-350	2-3 x 10^{-8}
2. Kaolinite a. Lerman et al. (1975)	1-7 x 10^{-7}	7-30	350-1500	0.7-20 x 10^{-8}
b. Siever (1968)	*7 x 10^{-7}	30	3000	*2±.75 x 10^{-10}
3. Montmorillonites and bentonite Lerman et al. (1975)	.8-4 x 10^{-7}	15-115 (300-800)	750-5800 (15,000-40,000)	0.5-5 x 10^{-10}
4. Illite, muscovite Lerman et al. (1975)	.9-4 x 10^{-7}	30-100	1500-5000	0.2-2.5 x 10^{-10}

*pH 6, distilled water

temperature and dissolved silica as a function of depth and latitude in the Atlantic, Pacific and Indian Oceans (Fiaderro, 1975; Mann et al., 1973; Stefansson and Atkinson, 1971; Rakestraw, 1964; Rosanov, 1964), an interesting fact emerges: 75 to 85 percent of the Atlantic is noticeably undersaturated or near saturation with respect to most of the clay minerals, whereas the remainder of the Atlantic and 80 to 90 percent of the Pacific and Indian Oceans are near equilibrium or supersaturated.

Thus, I estimate that about 30 percent (10% Pacific, 80% Atlantic, and 25% Indian) of the world's oceans are probably undersaturated with respect to the solubility of the above clays at 4° C or above; the rest of the world's oceans are then near equilibrium or slightly supersaturated. But relative to this volume, only 1 to 3 percent of the world's open oceans are above 20° C. Therefore, we must either ignore that volume of the oceans whose temperature is less than the approximate temperature of the Lerman et al. (1975) experiments, or reduce their value of K1 again to allow for the lower temperatures involved (but having a larger volume of solution available for dissolution). I suggest that, considering the accuracy of the arguments involved, that the effect is approximately the same. The values of K1 should be reduced by a factor of 10 (analogous to those for biogenic silica, Hurd, 1973) for a 20° C decrease in temperature and allow an order of magnitude greater volume of water to be included in the reactions.

If we assume that all of the clays in the 30 percent of the oceans are now dissolving at their maximum rates (now reduced by 7 to 8 orders of magnitude because of the above reasoning), that the new range of rate K1 values is approximately 7 to 700×10^{-16} Sec^{-1}, that the average saturation value of the clays is 100 uM, and that the average dissolved silica in the water column considered is approximately 10 to 20 micromolar, then using the altered rate K1 of Lerman et al. dissolved silica is injected into the water column at a rate of 5 to 700×10^{-23} moles/cm^3/sec or 2 to 200×10^{-6} micromoles/1/yr.

If we further assume that the average dissolved silica concentration for all of the world's oceans is approximately 100 micromolar, and remember that this dissolution is only occurring in one-third of the oceans then 1.5 to 150×10^6 years would be required to replenish the present amounts of

dissolved silica in the oceans. Since the mixing time of the oceans is approximately 1500 to 2000 years (Broecker, 1974) and the residence time of dissolved silica in the oceans (corrected for diffusive flux from the sediment-type recycling) is approximately 8,000 to 10,000 years, the above-calculated silica replacement rate by clay mineral dissolution is somewhat sluggish.

The major flaw possible in the above reasoning is that neither I nor Lerman et al. (1975) know the true extent to which the minerals used in their experiments are representative of those found in the open oceans. However, Siever (1968) shows data for kaolinite dissolving in pH 6 distilled water (his Figure 1, the ground-only sample). To arrive at an estimate for Lerman et al.'s (1975) K1 and my K2, I assume the following: Csat is approximately 600 micromolar, and the surface area per unit volume term is 3×10^3 cm^2/cm^3 (there appears to be a typographical error concerning the specific surface area of the sample which should be 30 rather than 3 m^2/gm); this yields values of $7 \pm 1 \times 10^{-10}$ cm/sec for K2. Since Siever (1968) used twice the suspended solid of Lerman et al. (1975), the above value for K1 should be halved for possible direct comparison. The resulting value of $3.5 \pm 1.5 \times 10^{-7}$ Sec^{-1} is well within the range of 1 to 5×10^{-7} Sec^{-1} estimated by Lerman et al. (1975) for the two kaolinites used in seawater. It is entirely possible that the ionic strength difference between distilled water and seawater will alter the true rate K2, but considering the other possible errors involved in the calculations, the agreement is satisfactory.

The above calculations have led me to estimate ranges for the heterogeneous solution rate constant K2, in cm/sec, of a number of clays in Table 2. These estimates are still quite tentative and the order of magnitude range in values is a reasonable estimate.

The point of the calculations summarized in Table 2 is to discourage further use of the combined rate K2 and surface area per unit volume of solution term as was done by Lerman et al. (1975) and others. It should be apparent that there are a countably infinite number of such combinations possible depending on the mineral phase, specific surface area of the solid and the amount of suspended matter per unit volume of water involved, and that direct comparisons of such combined constant is at best, difficult.

If we next consider the possible role of clay

minerals as sinks of silica while suspended in the oceans, the arguments of Fanning and Schink (1969) and Mackenzie et al. (1967) must be extended.

Comparison of dissolved silica release and uptake profiles of several alumino-silicate minerals (Mackenzie et al., 1967) with the re-calculated dissolution rates and rate constants above suggest that if the process of dissolved silica uptake can be represented even crudely by a first-order equation similar to that describing dissolution, then the values of the rate constants will be less than or equal to those for dissolution. That is, time periods even longer than 1.5 to 150×10^6 years would be required for the removal of all of the dissolved silica now present in the oceans.

I conclude from the above rate calculations that clay minerals cannot control the concentration of dissolved silica in the present world oceans, a conclusion also reached by a number of previous investigators but for other reasons (Siever, 1968; Siever and Woodford, 1973; Fanning and Schink, 1969; Schink et al., 1974; Harris, 1966; Calvert, 1968, 1974).

Estimates of Biogenic Silica Dissolution Rates in the Oceans

Contrasting with the slow dissolution rates of clays are those of biogenically precipitated silica. At 3° C the solubility of acid-cleaned biogenic silica is 5 to 10 times greater than that estimated for clays, the specific surface area of partially dissolved recent radiolaria and diatoms is equal to or larger than clays (see Table 1), and the value of K2 for acid-cleaned material is approximately 100 times higher than those estimated for clays (Lewin, 1961; Hurd, 1973; Hurd and Theyer, 1975; Lawson et al., 1978; Hurd and Birdwhistell, 1983). Yet it would be unreasonable to assume that the mean value of K2 for decomposing phytoplankton or radiolarians falling through the water column would be the same as the acid-cleaned value.

As summarized and extensively described in Lawson et al. (1978) it is difficult to avoid the conclusion that the value of K2 for suspended biogenic silica is on the order of 10 to 150 times smaller for natural material decomposing than for acid-cleaned material in the laboratory. This conclusion is hardly surprising, but it is necessary to have an estimate for K2 for natural material.

The following example suggests a possible application:

How long will it take a diatom frustule of approximately 0.1 micron thickness to dissolve in 3° C, pH 8 seawater in the Antarctic?

1. Flux from the surface of the diatom in moles/cm²/sec=K2 (C sat-C sol) where K2=.06 to 1×10^{-9} cm/sec, Csat is approximately 900 micromolar, and Csol is approximately 80 to 100 micromolar. So, flux=4.8 to 80×10^{-17} moles/cm²/sec=2.9 to 48×10^{15} gm SiO_2/cm²/sec.

2. The density of biogenic silica is approximately 2.0 gm/cm³ (Hurd and Theyer, 1977)

3. Flux/Density=dissolution rate in units of length/time
   =1.5 to 42×10^{-15} cm/sec.
   =1.5 to 42×10^{-11} micron/sec

4. Frustule thickness/Linear dissolution rate = 0.1 micron/1.5 to 42×10^{-11} micron/sec=7.6 to 210 years.

If dissolution occurs only on one side and the surface remains flat. Further discussion of this general problem can be found in Hurd and Birdwhistell (1983).

Lal and Lerman (1973) who suggest that the same particle will dissolve in about 2 days, using their value of $G=2 \times 10^{-3}$ cm/yr for a linear dissolution rate. Part of the difficulty arises from their model of what a diatom or radiolarian looks like as well as their possible misinterpretation of Berger's (1968) field experiment and one of the author's earlier papers (Hurd, 1972).

Because of the relatively large mesh used in Berger's experiment (50 to 60 microns) it is not really possible to tell how much actual dissolution occurred as compared with loss by mechanical breakage and subsequent washing out; Berger realized this and only made depth-relative calculations which partially corrected for this loss. If we use my above estimate for a dissolution rate of approximately 3 to 50×10^{-15} gm SiO2/cm²/sec, and assume that the specific surface area of Berger's 15 mg sample is 50 to 100 m²/gm, giving a total of 7.5 to 15×10^{-3} cm², then the weight loss for this sample is 2.3 to 75×10^{-11} gm/sec. A five percent weight loss would require 1 to 33×10^6 seconds or 12 to 385 days. Since the duration of the experiment was approximately 120 days, and also since Berger pretreated the samples with acid and peroxide (which would initially allow order of magnitude higher dissolution rates, although see Hurd (1972) for further comments on this experiment) at least five percent of the observed weight loss could well have occurred by dissolution. Lal and Lerman appear to have used only the gross dimensions of diatoms to arrive at their 20 micron figure; since diatoms are not solid silica but more like hollow, thin-walled discs, cylinders and plates, to maintain their model we are forced to imagine that these hollow objects somehow shrink when dissolving. In addition, their in situ flux of 4×10^{-7} gm $SiO_2/cm^3/yr$ (approximately 7 microgm-atoms/l/yr) is fully two orders of magnitude higher than previously mentioned estimates (Edmond, 1974; Kido and Nishimura, 1973, 1975; Matsumoto, 1973; Tsunogai, 1972).

Lal and Lerman refer to Hurd's (1972) dissolution rates as being 100 to 1000 times higher, but this occurred because of their confusion

There are two basic approaches to estimating the amount of particulate matter produced. One is the geochemical budget type which has been thoroughly described by Broecker (1974). Although I differ with the values for average deep water, surface water and river water given by Broecker, the general conclusion still remains the same: 10 to 15 times the annual river input of about 4×10^{14} gm SiO_2 is precipitated by organisms in surface waters each year, and of this amount 90 to 95 percent redissolves in the water column and in the upper few centimeters of the sediments.

If all of the dissolution occurred only in the water column, the in situ production of dissolved silica would be 0.05 micromoles/l/yr, which agrees remarkably well with estimated rates in the North Pacific. Since production of biogenic silica is not everywhere uniform in the oceans, it should not be expected that dissolution would be uniform everywhere in deep waters. As suggested by Hurd

TABLE 3. Calculations relating to Si (OH)₄ uptake by diatoms from data of Goering et al. (1973)

Depth m	Dissolved Si (OH)₄ μM	Particulate Si (OH)₄ μM	In situ Vsi day^{-1}	Replacement time days	dC/dt, Si (OH)₄ μgmSiO₂/l/day	¹⁴C uptake μgmC/l/day	Weight ratio uptake gmSiO₂/l/day / gmC/l/day
0	1.7	1.8	0.16	6.3	16.2	880	0.018
3	1.7	1.9	0.35	2.9	36	600	0.06
6	2.4	1.9	0.2	5	30	670	0.04[5]
10	2.5	1.6	0.07	14	12	530	0.02[2]
21	7.5	2.2	0.02	50	9	(20)	0.4[5]

between the rate constant K2 and the actual linear dissolution rate. Thus Hurd's acid-cleaned value for K2 at 3° C of 1×10^{-8} cm/sec yields a linear dissolution rate of about 1×10^{-5} cm/yr, about 200 times less than their value. I leave the adjustment of their subsequent calculations to the interested reader.

Finally, several questions must be asked about the possibility of biogenic silica dissolution controlling the dissolved silica concentration in the oceans.

Is there enough biogenic silica produced by organisms to explain both the in situ dissolution rates and the net accumulation rates in the sediments?

(1973) the relative solution potential of the upper 1000 m of the water column varies by a factor of 20 when waters of various latitudes in the Pacific are considered.

A second method of estimating particulate SiO_2 has been advocated by Lisitzin et al. (1967). This method uses the weight ratio of particulate SiO_2 to organic carbon in suspended matter from the sea surface times the annual production of organic carbon as estimated from primary productivity measurements. Using Koblentz-Mishke's (1965) estimates of 0.9 to 1.7×10^{16} gm C/yr and the Lisitzin et al. factor of 2.3 weight ratio (or 0.46 mole ratio) one arrives at the figure 2.1 to 3.9×10^{16} gm SiO_2/yr or about 50 to 100 times higher than the annual river input. However, as mentioned in Hurd (1973), such estimates ignore production by non-diatom algae, possible differences in re-cycling rates of silica versus carbon, non-differentiation of live and decayed material, and incomplete removal of suspended matter from seawater samples by the separator. Further difficulties are reviewed in Dugdale (1972). If it is at all possible to estimate silica production as a function of productivity, a more useful measurement would be the simultaneous determination of both net carbon fixed and silica precipitated. This has recently been done by Goering et al. (1973). The author has rearranged a portion of their data to make Table 3. The particular system studied does not appear to be silicate-limited (during the time of analysis) and Blasco (1971) has suggested that diatoms are by far the most dominant group of phytoplankton present; that the SiO_2/carbon uptake ratio is on the order of 60 times smaller than the Lisitzin et al. ratio casts doubts on the universal use of their ratio. If experiments such as those of Goering et al. (1973) could be done on higher-latitude regions of high diatom productivity, perhaps this controversy would be solved.

.ACKNOWLEDGEMENTS

This paper was researched and written when the author was a visiting researcher at the Department of Analytical Chemistry, Faculty of Fisheries, Hokkaido University, Hakodate, Japan on a fellowship from the Japan Society for the Promotion of Science for Foreign Researchers. In particular the following individuals were most helpful: Drs. K. Kido, S. Noriki, and M. Nishimura, my host scientist; Messrs. S. Nakaya and M. Uematsu. This research was supported by grants from the JSPS and the Office of Naval Research. Hawai Institute of Geophysics Contribution.

REFERENCES

Amit, O., 1969. Silica content and silica-clay interactions in the water of the Dead Sea (Israel). Chem. Geol., 5: 121-129.
Armstrong, F.A.J., 1958. Inorganic suspended matter in seawater. J. Mar. Res., 17: 23-34.
Banin, A., M. Gal, Y. Zohard, and A. Singer, 1975. The specific surface area of clays in lake sediments—measurement and analysis of contributors in Lake Kinneret, Israel. Limnol. And Oceanogr., 20(2): 278-282.
Berger, W.H., 1968. Radiolarian skeletons. Solution at depths. Science, 159: 1237-1239.
Blasco, D., 1971. Composicion y distribucion del fito plancton en la region del aflorimento de las costas peruanas. Investigacion Pesq., 35: 61-112.
Broecker, W.S., 1971. A kinetic model for the composition of seawater. Quat. Res., 1: 188-207.
Broecker, W.S., 1974. Chemical Oceanography. (K. S. Deffreyes, General Editor), Harcourt Brace Jovanovich, Inc., 214 pp.

Burton, J.D., 1970. The behavior of dissolved silicon during estuarine mixing. II. Preliminary investigations in the Vellar Estuary, southern India. J. Cons. Perm. Int. Explor. Mer, 33; 141-148.
Burton, J.D. and P.S. Liss. 1973. Processes of supply and removal of dissolved silicon in the oceans. Geochim. Cosmochim. Acta, 37: 1761-1773.
Calvert, S.E., 1968. Silica balance in the ocean and diagenesis. Nature 219: 919-920.
Calvert, S.E., 1974. Deposition and diagenesis of silica in marine sediments. Spec. Pubs. Int. Assoc. Sediment, 1: 273-299.
Chester, R. and J. Stoner, 1972. Concentration of suspended particulate matter in surface sea water. Nature, 240: 552-553.
Copin-Montegut, C. and G. Copin-Montegut, 1972. Chemical analyses of suspended particulate matter collected in the Northeast Atlantic. Deep-sea Res. Oceanogr. Abs., 19(6): 445-452.
Diamond, S. and E.B. Kinter, 1956. Surface areas of clay minerals as derived from measurements of glycerol retention. Fifth National Conference on Clays and Clay Minerals, 334-347.
Dugdale, R.C., 1972. Chemical oceanography and primary productivity in upwelling regions. In: W. Tietze (ed.), Upwelling in the Oceans. Journ. Phys., Human and Reg. Geosci. Geoforum, 11: 47-61.
Edmond, J.M., 1974. On the dissolution of carbonate and silicate in the deep ocean. Deep-Sea Res., 21: 455-580.
Fanning, K.A. and M.E.Q. Pilson, 1973. The lack of inorganic removal of dissolved silica during river-ocean mixing. Geochim. Cosmochim. Acta, 37: 2405-2415.
Fanning, K.A. and D.R. Schink, 1969. Interaction of marine sediments with dissolved silica. Limnol. oceanogr., 14: 59-68.
Feely, R.A., W.M. Sackett, and J.E. Harris, 1971. Distribution of particulate aluminum in the Gulf of Mexico. J. Geophys. Res., 76(24): 5893-5902.
Fiaderro, M.E., 1975. Numerical modeling of tracer distributions in the Deep Pacific Ocean. Unpublished Ph.D. Dissertation, University of California, San Diego, 226 pp.
Fournier, R.O., 1973. Silica in thermal waters: Laboratory and field investigations. In: Proc. of Symp. on hydrogeochemistry and Biochemistry, Vol. I—Hydrochemistry, pp. 122-139. U.S. Geological Survey, Clarke Co., Wash. D.C.
Goering, J.J., D.M. Nelson, and J.A. Carter, 1973. Silicic acid uptake by natural populations of marine phytoplankton Deep-Sea Res., 20: 777-789.
Grill, E.V., 1970. A mathematical model for the marine dissolved-silicate cycle. Deep-Sea Res. 17(2): 245-266.
Harris, R.C., 1966. Biological buffering of oceanic silica. Nature, 212 (5059): 275-276.
Heath, G.R., 1974. Dissolved silica and deep-sea sediments. In: W. W. Hay (ed.) Studies in Paleoceanography. Soc. Econ. Paleo. Min. Spec. Publ., 20: 77-93.
Helgeson, H. C. and F. T. Mackenzie, 1970. Silicate-seawater equilibrium in the ocean system. Deep-Sea Res., 17: 877-892.
Hurd, D. C., 1972. Factors solution rate of biogenic opal in seawater. Earth Planet. Sci. Lett., 15(4): 411-417.
Hurd, D. C., 1973. Interactions of biogenic opal, sediment and seawater in the Central Equatorial Pacific. Geochim. Cosmochim. Acta, 37: 2257-2282.
Hurd, D. C. and F. Theyer, 1975. Changes in the physical and chemical properties of biogenic silica from the Central Equatorial Pacific. I. Solubility, specific surface area, and solution rate constants of acid-cleaned samples. In: T.R.P. Gibb, Jr., (ed.), Analytical Methods in Oceanography. Adv. in Chem. Ser., 147: 211-230.
Hurd, D. C. and F. Theyer, 1977. Changes in the physical and chemical properties of biogenic silica from the central equatorial Pacific: Part II. Refractive index, density, and water content of acid-cleaned samples, Am. J. of Sci., 277: 1168-1202.
Hurd, D.C., C. Fraley, and J. K. Fugate, 1979. Silica apparent solubilities and rates of dissolution and precipitation. In: ACS Symposium Series, No. 93. Chemical Modeling in Aqueous Systems Everett A. Jenne (ed.) 413-445.
Hurd, D.C. and S. Birdwhistell, 1983. On producing a more general model for biogenic silica dissolution, Am. J. of Sci., 283: 1-28.

Kido, K., 1974. Latitudinal distribution and origin of particulate silica in the surface water of the North Pacific. Marine Chem., 2: 277-285.

Kido, K. and M. Nishimura, 1973. Regeneration of silicate in the ocean. 1. The Japan Sea as a model of closed system. J. Oceanogr. Soc. Japan, 29(5): 185-192.

Kido, K. and M. Nishimura, 1975. Silica in the sea—its forms and dissolution rate. Deep-Sea Res., 22: 323-338.

Koblentz-Mishke, O. I., 1965. Primary productivity in the Pacific. Oceanol., 5(2): 104-116.

Lal, D. and A. Lerman, 1973. Dissolution and behavior of particulate biogenic matter in the ocean: Some theoretical considerations. J. Geophys. Res., 78(30): 7100-7111.

Lawson, D. S., D. C. Hurd, and H. S. Pankratz 1978. Silica dissolution rates of decomposing phytoplankton assemblages at various temperatures. Amer. J. of Sci., 278: 1373-1303.

Lerman, A., F. T. Mackenzie, and O. P. Bricker, 1975. Rates of dissolution of aluminosilicates in seawater. Earth and Planet. Sci. Lett., 25: 82-88.

Lewin, J. C., 1961. The dissolution of silica from diatom walls. Geochim. Cosmochim. Acta, 21: 182-198.

Lisitzin, A. P. 1970. Sedimentation and geochemical considerations. *In: Scientific Exploration of the South Pacific*, pp. 89-131. Nat'l Acad. of Sci., Wash. D. C. Stand. Book No. 309-01755-6.

Lisitzin, A. P., Y. I. Belyayev, and A. Bogoyavlenskiy, 1967a. Distribution relationships and forms of silicon suspended in waters of the World Ocean. Internat. Geol. Rev., 9(3): 253-274.

Lisitzin, A. P., Y. I. Belyayev, Y. A. Bogdanov, and A. Bogoyavlenskiy, 1967b. Distribution relationships and forms of silicon suspended in waters of the World Ocean. Internat. Geol. Rev., 9(4): 604-623.

Mackenzie, F. T. and R. M. Garrels, 1965. Silicates: Reactivity with seawater. Science, 150: 57-58.

Mackenzie, F. T., R. M. Garrels, O. P. Bricker, and F. Bickley 1967. Silica in seawater: Control by silica minerals. Science, 155: 1404-1405

Manheim, F. T., R. H. Meade, and G. C. Bond, 1970. Suspended matter in surface waters of the Atlantic continental margin from Cape Cod to the Florida Keys. Science, 167: 371-376.

Manheim, F. T., J. C. Hathaway, and E. Uchupi, 1972. Suspended matter in surface waters of the northern Gulf of Mexico. Limnol. and Oceanogr., 17(1): 17-27.

Mann, C. R., A. R. Coote, and D. M. Garner, 1973. The meridional distribution of silicate in the Western Atlantic. Deep-Sea Res., 20(9): 791-802.

Matsumoto, E., 1973. Geochemical study on the abyssal water in the Pacific. Bull. Geol. Surv. Japan, 24: 21 (79)-32(90).

Maynard, J. B., 1975. Kinetics of silica sorption by kaolinite with application to seawater chemistry. Amer, J. Sci., 275: 1028-1048. Agronomy, 15 (602): 339-396.

Rakestraw, N. W., 1964. Some observations on silicate and oxygen in the Indian Ocean. Rec. researches in the fields of hydrosphere, atmosphere, and nuclear-geochemistry. Sugaware Festival Vol: 243-255.

Rozanov, A. G., 1964. Distribution of phosphates and silicates in the water of the northern part of the Indian Ocean. Trud. Inst. Okeanol. LXIV: 102-114.

Schink, D. R., 1967. Budget for dissolved silica in the Mediterranean Sea. Geochim. Cosmochim. Acta, 31: 987-999.

Schink, D. R., K. A. Fanning, and M. E. Q. Pilson, 1974. Dissolved silica in the upper pore waters of the Atlantic Ocean Floor. J. Geophys. Res., 79(15): 2243-2250.

Siever, R., 1968. Establishment of equilibrium between clays and sea water. Earth and Planet. Sci. Lett., 5: 106-110.

Siever, R. and N. Woodford, 1973. Sorption of silica by clay minerals. Geochim. Cosmochim Acta 37: 1851-1880.

Slabough, W. H. and A. D. Stump, 1964. Surface areas and porosity of marine sediments. J. Geophys. Res., 69: 4773-4778.

Stefansson, U. and L. P. Atkinson, 1971. Relationship of potential temperature and silicate in the deep waters between Cape Lookout, North Carolina and Bermuda. J. Marine Res., 29(3): 306-318.

Toyota, Y. and S. Okabe, 1967. Vertical distribution of iron, aluminum, silicon and phosphorus in particulate matter collected in the Western North Pacific, Indian and Antarctic Oceans. J. Oceanogr. Soc. Japan., 23(17): 1-9.

Tsunogai, S., 1972. An estimate of the rate of decomposition of organic matter in the deep water of the Pacific Ocean. *In:* Y. Takenouchi et al., (ed.), *Biological Oceanography of the Northern North Pacific Ocean.* Idemitsu Shote, Tokyo, Japan, pp. 517-533.

Tsunogai, S., Y. Nozaki, and M. Minagawa, 1974. Behavior of heavy metals and particulate matters in seawater expected from that of radioactive nuclides. J. Oceanogr. Soc. Japan, 30(6): 251-259.

van Lier, J. A., P. L. deBruyn, and J. T. Overbeek, 1960. The solubility of quartz. J. Phys. Chem., 64: 1675-1682.

Weiler, R. R. and A. A. Mills, 1965. Surface properties and pore structure of marine sediments. Deep-Sea Res. 12: 511-529.

Wollast, R., 1974. The silica problem. *In:* E. D. Goldberg, (ed.), *The Sea,* Vol. 5, *Marine Chemistry.* Wiley Inter-Sceince Co., New York, pp. 359-392.

Wollast, R. and F. De Broeu, 1971. Study of the behavior of dissolved silica in the estuary of the Scheldt. Geochim. Cosmochim. Acta, 35: 613-620.

24

Integrated Approach to an Anaerobic Marine Ecosystems Study: An Example From Saanich Inlet, B.C., Canada

S. I. AHMED
S. L. KING
V. JOHNSON
and
B. L. WILLIAMS
University of Washington,
Seattle, Washington

ABSTRACT

Previous studies to understand the processes responsible for the diagenesis of organic matter in anoxic marine environments have largely relied on the measurements of "end-products" in such reactions, viz.: sulfide (sulfate), phosphate ammonia and silicate as well as pH, chlorinity and alkalinity. In some studies direct measurement of the process of sulfate reduction has also been carried out by the radioisotopic technique. In the expectation that a much more rigorous understanding of the overall processes in the early diagenesis of organic matter (OM) would accrue as a result of an integrated approach, we undertook to measure simultaneously a variety of the essential biological as well as chemical components of the anoxic system. In terms of the decomposition of OM in anoxic, sulfide-bearing environments, the fermentative group of microorganisms should represent the first line of attack on the OM and hence should be considered of prime ecological importance. The isolation, enumeration and classification of this group of microorganisms from the sediments of Saanich Inlet, an anoxic fjord off Vancouver Island, B.C., Canada is reported. A good degree of correlation was found between sulfate reduction rates and the respiratory enzymatic activities which reconfirms the validity of the subsurface reduction rates reported earlier. The results of experiments with laboratory cultures of a Saanich Inlet sulfate reducer as well as the results of the bottom water enrichment experiment elucidate the rates and pathways of decomposition of selected organic compounds.

INTRODUCTION

Recent observations have shown that early diagenesis of organic matter can occur in marine anoxic sediments (Berner, 1976; Nissenbaum et al., 1972). The conclusions with regard to the mineralization of organic matter are largely based on the concentration changes of interstitial waters of such biologically important elements as nitrogen, sulfur, phosphorus and silicon. While the concentration changes can occur in all sediments, they are greatest in anoxic sediments, rich in organic matter such as in Saanich Inlet. Since the alterations in the concentrations of these elements are time dependent processes, kinetic models have been developed to calculate the

various chemical reactions. Calculations and comparisons of dissolved ammonia, phosphate, and sulfide concentrations lead one to the conclusion that the organic matter diagenesis must be mediated by microorganisms (Berner, 1976). It should be pointed out that while determination of the concentrations of the reactants and products is useful to build an initial model for diagenesis, a model is only as good as the assumptions it is based upon. To minimize assumptive errors we clearly need direct knowledge and understanding of the rates and processes concerned with the transformation of organic matter as well as the nature, abundance and distribution of the participant microorganisms.

We have selected Saanich Inlet, an anoxic fjord

located on the southeast side of Vancouver Island, British Columbia, Canada as the study site. The inlet has a shallow sill (approx. 70 m) which restricts circulation of the deeper water (maximum depth 226 m) inside the basin so that for about eight months of the year the bottom waters contain no oxygen, nitrate or nitrite and significant concentrations of free H_2S are present (Anderson and Devol, 1973). Since the deep waters contain H_2S for most of the year, the inlet is ideally suited to studying anaerobic microbial processes free from competition with benthic meio- and macro-fauna. The unique nature of our study stems from our simultaneous consideration of microbial processes such as sulfate reduction rates, rates of degradation of selected organic compounds as well as the organisms involved in such processes. The latter, especially the fermentative groups of microorganisms have been largely ignored in most studies.

MATERIAL AND METHODS

Sediment Sampling

Sediment was collected with a geological, gravity corer fitted with a 5 cm diameter polyethylene sleeve. The core liner was pre-cut to 5 cm sections and taped together to enable the desired sampling depths to be separated from the rest of the core without disturbing the samples. Chemical analysis of interstitial water was carried out as described by Johnson (1981).

Bacteriological Sampling, Media for Growth, Procedure for Isolation of the Microbial Populations in Saanich Inlet Sediments

Bacteria were isolated from anoxic sediments in the central basin in October 1978 and in February and March 1979. Our isolation procedures were designed to detect the general type of bacteria present in low and high nutrient medium and incubation was conducted under aerobic as well as anaerobic conditions. The isolation procedure and growth conditions have been described elsewhere (Johnson, 1981). For the detection and enumeration of sulfate reducers, a separate set of isolation and enumeration procedures was used. For further details see Johnson (1981).

Characterization of the Isolate

A minimal set of taxonomic tests were used to achieve a preliminary characterization of the bacterial isolates from Saanich Inlet sediments. After the initial study, a program for a more thorough characterization of the isolates was developed including determination of the fermentation capacity on 27 different carbohydrates. After all the tests had been performed, a computer analysis program was instigated to organize various groups of organisms into clusters, on the basis of their Jacaard Similarity Coefficients (Sneath and Sokal, 1973).

Enzyme Assays

The measurement of 1) Cytochrome C reductase, and 2) Electron Transport System (ETS) activity was conducted in a *Desulfovibrio desulfuricans* strain $\neq$17990 obtained from ATCC (American Type Culture Collection, Maryland) and also in a Saanich Inlet isolate characterized as *Desulfovibrio salexigens*. Cell-free extractions and preparations of sediment samples for the enzymatic analyses were carried out according to Ahmed et al. (1984).

The cytochrome C reduction assay as described (Ahmed et al., 1984) is a modification of the assay of Bramlett and Peck (1975):

SO $+AMP+2$ cyto (oxid.) $->APS+2$cyto (red.)

Cytochrome C reducing activity

ETS activity was measured by a modification of the procedure developed by Kenner and Ahmed (1975). All assays were fully optimized before application to sediment samples.

Laboratory Culture Studies of *Desulfovibrio salexigens* Isolated from Saanich Inlet

A marine sulfate reducing bacterium characterized as *Desulfovibrio salexigens* isolated from Saanich Inlet was grown in 1.5 l batch cultures under anaerobic conditions in a basal salts medium supplemented with pyruvate or lactate, as described by King (Ph.D. Thesis, University of Washington, in prep.).

Bottom Water Enrichment Experiments

Anoxic, sulfide-bearing bottom water from Saanich Inlet was obtained in thoroughly cleaned

5-1 Scott-Richards bottles and transferred into thoroughly N_2-purged sterile flasks without exposure to the atmosphere. Each flask contained 1.5 g/l (final concentration of the sodium salt) of either acetate, lactate, glycerol or pyruvate. A 1-2 lb over pressure of N_2 was maintained to insure anaerobiosis and facilitate sampling. The cultures were incubated at 15°C and monitored over a 26-day period for cell number, sulfide, pH, volatile and nonvolatile organic acids as described by King (Ph.D. Thesis, University of Washington, in prep.).

RESULTS

Results of Isolation and Enumeration of the Microbial Populations in Saanich Inlet Sediments

Plate counts ranged from 10^3 to 10^5 colony forming units per gram of dry sediment. The general number and type of bacteria recovered did not appear to vary with the medium, the station location or the month in which the sampling occurred. The predominant groups of bacteria were characterized as Bacillaceae belonging to the genus *Bacillus* or *Clostridium*.

Although a high degree of diversity was noted among the isolated strains, certain significant differences were found in the variety and type of features observed according to the depth in the anoxic sediment from which the strains were recovered. The strains from 10-15 cm displayed the largest number of different biotypes and the greatest diversity in the features examined. A high percentage of the strains recovered between 0-5 cm produced acetate, lactate and succinate

compared to strains recovered below 5 cm. However, a significantly lower percentage of strains recovered from below 15 cm produced lactate compared to the strains from above 15 cm. A much larger percentage of the carbohydrates examined were acidified by the bacteria recovered between 5 and 15 cm than elsewhere in the sediment column. A significantly lower percentage and variety of carbohydrates were acidified by the bacteria from 50-55 cm depth.

Vertical Profiles of Enzymatic Activities in the Sediments of Saanich Inlet

a) *Cytochrome C reducing activity*

The vertical profiles of enzymatic activities as detected by Cytochrome C reduction are given in Table 1. From this table it appears that the maximum Cytochrome C reducing activity is located at 12-13 cm depth and not at the surface. However, when adjustment of the surface activity is made on a dry weight basis, the surface activity should increase by approximately two-fold (Murray et al., 1978; Johnson, 1981).

b) *ETS activity*

We have previously conducted measurements of ETS activity in selected sulfate-reducing bacteria isolated from Saanich Inlet. In these studies we have shown that when sulfate-reducing bacteria are grown in the presence of sulfate, ETS activity is proportional to the sulfide produced as well as other biomass parameters and substrates utilized (King and Ahmed, in prep.). On the basis of these results we decided to measure ETS

TABLE 1. Cytochrome C reducing activity in Saanich sediments.

Sediment sample depth (cm)	Cytochrome C red. act'y ($\triangle$D550 nm/cc sediment/min)
0	0.218
6-7	0.505
12-13	0.612
18-19	0.510
24-25	0.487
36-37	0.400
48-49	0.340

TABLE 2. ETS activities in various depths of Saanich sediments. $\triangle$D 490 nm/10 min/ml volume of sediment at 10°C.

Core #1		Core #2	Core #3	Core #4
0.81 (3-6 cm)		0.555 (0 cm)	*1.2 (0 cm)	0.87 (2.5-3.5 cm)
1.54 (6-9 cm)		0.30 (5 cm)		1.71 (8.5-9.5 cm)
3.065 (9-12 cm)		1.85 (10 cm)	0.8 (10 cm)	1.73 (14.5-15.5 cm)
1.59 (15-18 cm)		0.70 (15 cm)	0.4 (15 cm)	
1.66 (21-24 cm)		0.725 (25 cm)	0.4 (25 cm)	1.21 (26.5-27.5 cm)
0.83 (27-30 cm)				0.70 (38.5-39.5 cm)
0.288 (39-42 cm)				0.40 (45.5-46.5 cm)
0.10 (51-54 cm)		0.150 (50 cm)	0.10 (50 cm)	0.10 (51.5-52.5 cm)

* Surface O.D.'s are sometimes unreliable because they may be biased by a greenish-brown extract color.

activities in the extracts of Saanich Inlet sediments as presented in Table 2. In three out of four cores we observed an apparent subsurface maximum in the ETS activity. Even when the surface or near surface activity is expressed on a dry weight basis instead of volume basis as presented in Table 2, the subsurface maximum in ETS activity is clearly discerned. The ETS activity decreases somewhat more rapidly with depth of sediment than the Cytochrome C reducing activity, although it is still observable at ca. 50-55 cm depth, but in much reduced amounts, i.e., >10% of the surface activity, when calculated on a dry weight basis.

Culture Studies of *Desulfovibrio salexigens* Isolated from Saanich Inlet

Some of the results of the six experiments which were run with varying concentrations of lactate, pyruvate and sulfate are reported in Table 3. The data show that growth under these conditions can be monitored equally well by cell number, sulfide production, or ETS activity. Growth rate appears to be independent of sulfate concentration over the range of 1-25 mM and pyruvate appears to support a slightly higher growth rate than lactate. Table 4 shows the final

TABLE 3. Growth rate constant, u, according to the equation $X = X°Exp [\mu t]$ where t is in days and x is either cell number, sulfide concentration, acetate concentration, or ETS activity/unit volume of culture. Estimates are reported with one standard deviation; n is the number of replicate cultures.

Expt.	Electron Donor Conc. (mM/l) Lactate	Pyruvate	Sulfate Conc.	Cell No.	Mean Growth Rate (day^{-1}) H$_2$S	Acetate	ETS Act'y	n
1.	53.5	—	24.0	2.58±0.17	2.77±0.10	2.51±0.23	2.73±0.05	5
2.	13.4	—	11.6	2.22±0.17	2.48±0.37	2.61±0.41	2.13±0.24	4
3.	13.4	—	3.5	2.36±0.16	2.47±0.25	2.71±0.35	3.22±0.53	5
4.	—	26.8	11.6	2.58±0.27	2.72±0.07	2.55±0.31	2.60±0.28	4
5.	—	26.8	3.5	2.67±0.17	2.80±0.14	2.70±0.09	2.42±0.14	4
6.	—	26.8	1.0	2.55±0.12	2.74±0.20	2.65±0.15	—	5

TABLE 4. Initial substrate and final product concentrations for growth of *Desulfovibrio salexigens* isolated from Saanich Inlet. Growth at 20°C in basal salts supplemented with lactate or pyruvate.

Experiment	Substrates			Products		
	Lactate (mM)	Pyruvate (mM)	Sulfate (mM)	Cell Number (x 10^7/ml)	Sulfide (mM)	Acetate (mM)
1	53.5	—	24.0	30.4	18.6	35.9
2	13.4	—	11.6	6.41	5.07	13.4
3	13.4	—	3.5	3.07	3.05	7.8
4	—	26.8	11.6	29.2	2.99	27.1
5	—	26.8	3.5	20.7	2.33	21.8
6	—	26.8	1.0	9.08	0.92	10.0

yields in the six experiments. A most interesting result is that pyruvate supports a much greater growth with much less sulfide produced. This indicates that lactate and pyruvate are metabolized by different pathways for the following reasons: 1) Lactate is more reduced than pyruvate and, therefore, could be expected to provide greater growth when oxidized to acetate; 2) the acetate produced from lactate (Experiments 1,2,3) can almost be completely accounted for based on sulfate reduction whereas in the pyruvate cultures, more than twice as much acetate is produced than can be accounted for by sulfate reduction. It appears, therefore, that pyruvate can be metabolized to acetate by a pathway independent of sulfate reduction, and that energy is produced as a result.

Bottom Water Enrichment Experiments

The results of these experiments are presented in Table 5. The results can be summarized as follows: pyruvate supports the best overall growth in terms of both growth rate and total cell number. Lactate was the best substrate for the sulfate-reducing bacteria in terms of total sulfide produced, but the growth rate of the sulfate reducers was slightly higher on pyruvate. (We have shown in pure culture experiments with a Saanich isolate that the rate of sulfide production corresponds directly with the growth rate in exponentially growing cultures.) We, therefore, conclude that sulfate reducers have less competition for lactate than for pyruvate. The volatile fatty acid end-products from lactate,

TABLE 5. Enrichment of anoxic bottom water from Saanich Inlet. Td refers to doubling time taken from a semi-log plot of concentration vs. time. Y refers to final yield. Cell number concentrations per ml. Initial substrate concentration was 1.5 g/*l* as the sodium salt (except glycerol). T=15°C.

Substrate		Cell Number	Sulfide	Acetate	Proprionate	Butyrate	Isobutyrate
Acetate	Td	230 hr	180 hr	—·	—	—.	—
	Y	5×10^5	0.3 mM	(15mM)	—	—	—
Lactate	Td	112 hr	20 hr	40 hr	110 hr	—	—
	Y	1×10^7	3 mM	11 mM	2 mM	—	trace
Glycerol	Td	48 hr	32 hr	48 hr	190 hr	—	—
	Y	5×10^6	1.7 mM	5 mM	0.69 mM	—	—
Pyruvate	Td	24 hr	17 hr	24 hr	40 hr	24 hr	—
	Y	2×10^7	1.3 mM	9 mM	1.2 mM	0.34 mM	trace

glycerol or pyruvate were acetic, propionic and butyric acids. Acetic acid was the most abundant product and butyrate was only observed in the pyruvate cultures. In addition, trace amounts of isobutyric acid were observed at the end of the growth cycle in the lactate and pyruvate cultures. From these observations we conclude that lactate and glycerol are not metabolized via pyruvate but instead by independent pathways. We base this conclusion on three lines of evidence: 1) glycerol and lactate are both more reduced than pyruvate and, therefore, should support greater growth in the presence of an oxidizing agent (i.e. sulfate); 2) butyric acid was produced throughout the growth cycle in the pyruvate culture yet it was not observed in either the lactate or glycerol cultures; and 3) propionic acid was produced throughout the growth cycle in the lactate and glycerol cultures but was not observed in the pyruvate culture until the bulk of the population had ceased growing. We have also observed this more "efficient" growth of pyruvate over lactate in pure culture studies with sulfate reducers. We have further noted that acetate can support the growth of sulfate-reducing bacteria in mixed cultures. More work is needed here because acetate appears to be one of the most abundant end-products of anaerobic metabolism based on both our own work and reports in the literature (Hoering, 1968; Miller et al., 1979).

The data for the non-volatile acids has not been fully analyzed, but end-product analyses on each of the cultures show that no non-volatiles are present at the end of growth. Our experiments permit us to conclude, however, that the bottom waters of Saanich Inlet are definitely carbon limited at least to the extent of the readily metabolizable forms, since the lactate and pyruvate are both readily and completely utilized.

DISCUSSION

Most classical models of sulfur diagenesis are based on the assumption that the rate of sulfate reduction is first order with regard to oxidizable particulate organic carbon (POC). A basic tenet of this assumption is that oxidizable POC, sulfate concentration and the sulfate reduction rates are highest at the the top of the sulfate reduction zone and decrease exponentially with increasing sediment depth (Berner, 1980). However, recently, to explain the observation of concave upwards methane distributions, the anaerobic consumption of methane has been invoked. It has been suggested that this consumption takes place near the bottom of the sulfate reduction zone where sulfate concentrations are low (Barnes and Goldberg, 1976; Reeburgh and Heggie, 1977). If the oxidation of methane was mediated by sulfate-reducing bacteria, one would expect to find a subsurface maximum in the rates of sulfate reduction coincident with the concave methane profile. Devol and Ahmed (1981) have shown in their *in situ* studies of Saanich Inlet sediments utilizing $^{35}SO_4^=$ reduction rate measurements that indeed such subsurface maxima (in the vicinity of 15 cm depth) are clearly observable. The profile they presented obviously cannot be modelled within the framework of current diagenetic models and lead them to conclude that the high rates of sulfate reduction at depth are associated with the anaerobic oxidation of methane, either directly or indirectly through reactions coupled to other substrates.

In this report also we have presented enzymatic evidence that a subsurface peak in both the Cytochrome C reductase activity and the ETS activity is clearly discernible at a depth more or less coincident with the direct sulfate reduction activity measurements of Devol and Ahmed (1981). It would appear that the respiratory activities as measured by the enzymes in our studies are therefore, a good reflection of the biomass or the activity of sulfate reducers in cultures as well as sediments (see also Thauer et al., 1977).

Our results show (Ahmed and King, unpublished results) a generally good correlation between the numbers of sulfate reducers present in the sediments and the sulfate reduction rates as a function of depth. This is to be expected. However, our studies are unique in terms of our analyses of all the other types of microorganisms present in sediments. (We have not yet undertaken a rigorous study of the lower member of the anaerobic ecosystem group, i.e., the methanogenic microorganisms.) Goldhaber and Kaplan (1975) have stated that sulfate reduction requires a complex community of fermentative bacteria capable of degrading polymerized organic matter to simple molecules. Our results indicated that the predominant group of microorganisms present in Saanich sediments belonged to the prolific fermentative family Bacillaceae, as evidenced by their capability to hydrolyze a rather wide variety of complex molecules. The fermentative group of microorganisms derive their energy and cell material via the degradation of relatively complex organic molecules. It is interesting to note that the strains recovered from the sediment surface to 15 cm in the sediment are biochemically more active in the production of acidic end-products than the strains recovered from deeper portions of the core. In addition, more strains recovered above 15 cm produced lactate, the primary substrate for sulfate-reducing bacteria (Sorokin, 1966a, 1966b; Postgate, 1965) in comparison with the strains recovered from below 15 cm, indicating a change in bio-types below this depth. The fermentative organisms provide the sulfate reducers with relatively smaller carbon compounds that they can rather easily oxidize. In return, the sulfate reducers utilize H_2 (which accumulates during the fermentative process) as reducing power enabling the fermentative organisms to continue to grow since a build-up of H_2 can be highly toxic to the fermentative organisms. Thus there appears to be a rather well-concerted symbiotic interaction in operation in marine anoxic environments.

The methanogenic bacteria, although thought to be not as competitive for the transferred hydrogen as the sulfate reducers, also utilize the transferred hydrogen in zones (environments) where sulfate concentration becomes limited and hence the activity of the sulfate reducers is restricted (Abram and Nedwell, 1978a, 1978b; Nedwell, 1982). In addition, the sulfate reducers also release acetate as an end-product of their metabolism (Barnes and Goldberg, 1976). The methanogenic bacteria utilize acetate as a substrate, fermenting it to methane and CO_2 (Baresi et al., 1978). This represents perhaps the final sequence of co-operativity or commensalism in the concerted biological mediation in the overall process of organic matter diagenesis. It is also interesting to point out that recent evidence shows that despite earlier belief to the contrary, the primary sink for the anaerobically generated methane in marine sediments, is sulfate reduction and not aerobic oxidation (Barnes and Goldberg, 1976; Reeburgh, 1976; Reeburgh and Heggie, 1977; Devol and Ahmed, 1981; Zehnder and Brock 1980). In order to successfully maintain the symbiotic relationship the sulfate reducers (and the methanogens) must carry out an efficient interspecies hydrogen transfer to prevent the build-up of the hydrogen from the vicinity of the fermentative bacteria, which may otherwise result in their death. In turn, the sulfate reducers are provided the substrates and the hydrogen which

stimulate their growth. Such ecosystem interrelationship is apparently also maintained in the methanogenic zone (below 15 cm) albeit with a different group of fermentative organisms.

Our laboratory studies with *Desulfovibrio salexigens* isolated from Saanich Inlet clearly point to the utility of such experiments in determining the rates and processes associated with the breakdown of selected organic compounds and to ascertain what parameters are of significant value when determining growth rates. Similarly, the bottom water enrichment experiments can be of significant importance in detecting the fate of the selected organic compounds, and can also provide us with clues to the nature and extent of the intra- and inter-species competition for various substrates. It is expected that such an approach to an integrated study of the various components of the complex anaerobic ecosystem would provide us with a more comprehensive understanding of the nature of organic matter diagenesis in anoxic marine environments. Ecological studies of anaerobic marine environments are of universal importance since such environments or micro-environments are distributed throughout the bottoms of the seas as well as the coastal waters and estuaries. These studies also hold special interest from a point of view of pollution abatement and its management.

Relevance and Significance of such Studies in the Coastal Marine Environments of Pakistan.

The coastal areas of Pakistan, especially around the vicinity of Karachi in particular and the province of Sind in general, is dominated by a variety of woody plants, collectively referred to as mangroves (Saifullah, 1982). The mangrove ecosystem is of significant economic as well as scientific interest to Pakistan for two main reasons: 1) the mangrove habitat forms the spawning grounds for the commercially important shrimp as well as some other fish, and 2) the slow but steady diversion of water flow from the River Indus is strongly influencing the density as well as the size of the mangrove habitats. These concerns have renewed interest in the study and management of the mangrove ecosystem. From a biological point of view, the mangrove plants provide a significant amount of organic matter to the waters in the vicinity. This high organic input can lead to increased biological activity in such waters. This often results in excessive oxygen consumption in the water column and leads to the formation of anoxic sedimentary conditions. Thus in macro- and micro-environments around the mangrove ecosystem, anoxic metabolic processes become of significant importance resulting in the formation of hydrogen sulfide and other end-products of such metabolism. It is, therefore, of significant importance that the physical, chemical as well as biological properties and characteristics of the system are studied in a concerted manner. Such studies should lead to greater understanding of the dynamic changes that are taking place in and around the mangrove ecosystems. In such a way the environmental changes taking place and their effects on a variety of biological species can be carefully assessed. The results of such long-term studies should aid greatly in the formation of more enlightened management policies.

ACKNOWLEDGEMENTS

We wish to thank Drs. J.T. Staley and A.H. Devol for their reviews of the manuscript. This work was supported by research grants OCE 76-23136 AO1 from the National Science Foundation and # N00014-80-C-0252=9 from the Office of Naval Research awarded to S.I.A. This is contribution # 1366 from the School of Oceanography, University of Washington.

REFERENCES

Abram, J.W., and D. B. Nedwell, 1978a. Inhibition of methanogenesis by sulfate reducing bacteria competing for transferred hydrogen. Arch. Microbiol. 117: 89-92.

Abram, J.W., and D.B. Nedwell, 1978b. Hydrogen as a substrate for methanogenesis and sulfate reduction in anaerobic saltmarsh sediment. Arch. Microbiol. 117: 93-97.

Ahmed, S.I., S.L. King and J.R. Clayton, Jr., 1984. Organic matter diagenesis in anoxic sediments of Saanich Inlet: A case history of evolved community interactions. Mar. Chem. (In press).

Anderson, J.J., and A.H. Devol, 1973. Deep water renewal in Saanich Inlet, an intermittently anoxic basin. Estuar. Coastal Mar. Sci. 1: 1-10.

Baresi, L., R.A. Mah, D.M. Ward, and I.R. Kaplan 1978. Methanogenesis from acetate: enrichment studies. Appl. Environ. Microbiol. 36: 186-197.

Barnes, R.O., and E.D. Goldberg, 1976. Methane production and consumption in anoxic marine sediments. Geology. 4: 297-300.

Berner, R.A., 1976. Inclusion of adsorption in the modelling of early diagenesis. Earth Planet. Sci. Lett. 29: 333-340.

Berner, R.A., 1980. Early Diagenesis—A Theoretical Approach. 241 pp. Princeton Univ. Press, Princeton, N.J.

Bramlett, R.N., and H.D. Peck, Jr., 1975. Some physical and kinetic properties of adenylyl sulfate reductase from Desulfovibrio vulgaris. J. Bio. Chem. 250: 2979-2986.

Devol, A.H., and S.I. Ahmed, 1981. Are high rates of sulfate reduction associated with anaerobic oxidation of methane? Nature. 291: 407-408.

Goldhaber, M.B., and I.R. Kaplan, 1975. Controls and consequences of sulfate reduction rates in recent marine sediments. Soil Sci. 119: 42-55.

Hoering, T.C., 1968. Organic acids from the oxidation of recent sediments. Carnegie Inst. Wash. Year Book. 66: 515.

Johnson, V.E., 1981. Isolation, enumeration, classification and identification of bacteria recovered from the sediments of an anoxic fjord, Saanich Inlet, B.C., Canada. M.S. thesis, Department of Oceanography, Univ. of Washington, Seattle, Washington.

Kenner, R.A., and S.I. Ahmed, 1975. Measurements of electron transport in marine phytoplankton. Mar. Biol. 33: 119-127.

King, S.L., Sulfate reduction and organic matter diagenesis in anoxic marine environments. Ph.D. Thesis, University of Washington, School of Oceanography, Seattle, Washington, (in preparation).

Miller, D.C., C.M. Brown, T.H. Pearson, and S.O. Stanley, 1979. Some biologically important low molecular weight organic acids in the sediments of Loch Eil. Mar. Biol. 50: 375-383.

Murray, J.W., V. Grundmanis, and W.M. Smethie, Jr., 1978. Interstitial water chemistry in the sediments of Saanich Inlet. Geochim. Cosmochim. Acta. 42: 1011-1026.

Nedwell, D.B., 1982. The cycling of sulfur in marine and freshwater sediments. In: D.B. Nedwell and C.M. Brown (eds.), Sediment Microbiology. Academic Press, N.Y.

Nissenbaum, A., B.J. Presley, and I.R. Kaplan, 1972. Early diagenesis in a reducing fjord, Saanich Inlet, British Columbia. I. Chemical and isotopic changes in major components of interstitial water. Geochim. Cosmochim. Acta. 36: 1007-1027.

Postgate, J.R., 1965. Recent advances in the study of sulfate-reducing bacteria. Bacteriol. Rev. 29: 425-441.

Reeburgh, W.S., 1976. Methane consumption in Cariaco Trench waters and sediments. Earth Planet. Sci. Let. 28: 337-344.

Reeburgh, W.S., and D.T. Heggie, 1977. Microbial methane consumption reactions and their effect on methane distributions in freshwater and marine sediments. Limnol. Oceanogr. 22: 1-9.

Saifullah, S.M., 1982. Mangrove ecosystem of Pakistan. The Research Conference on Mangroves in the Middle East. Contribution No. 3, Japan Co-operative Center for the Middle East, pp. 69-80. Publication No. 137, Tokyo, Japan.

Sneath, P.H.A., and R.R. Sokal, 1973. Numerical Taxonomy, the Principles and Practice of Numerical Classification W.H. Freeman and Co., San Francisco, California.

Sorokin, Y.I., 1966a. Sources of energy and carbon for biosynthesis in sulfate-reducing bacteria. Mikrobiologiya 35: 643-647.

Sorokin, Y.I., 1966b. Investigations of the structural metabolism of the sulfate-reducing bacteria with ^{14}C. Mikrobiologiya. 35: 806-814.

Thauer, R. K., K. Jungermann, and K. Decker, 1977. Energy conservation in chemotrophic anaerobic bacteria. Bacteriol. Rev. 41: 100-180.

Zehnder, A.J.B., and T.D. Brock, 1980. Anaerobic methane oxidation: occurrence and ecology. Appl. & Environm. Microbiol. 39: 194-204.

INDEX

382

⟨E⟩ Printed by Elite Publishers Limited Karachi, Pakistan